D1207923

Problem Books in Mathematics

Edited by K.A. Bencsáth
P.R. Halmos

Springer
New York
Berlin
Heidelberg
Hong Kong
London
Milan
Paris
Tokyo

Problem Books in Mathematics

Series Editors: K.A. Bencsáth and P.R. Halmos

Pell's Equation
by *Edward J. Barbeau*

Polynomials
by *Edward J. Barbeau*

Problems in Geometry
by *Marcel Berger, Pierre Pansu, Jean-Pic Berry, and Xavier Saint-Raymond*

Problem Book for First Year Calculus
by *George W. Bluman*

Exercises in Probability
by *T. Cacoullos*

Probability Through Problems
by *Marek Capiński and Tomasz Zastawniak*

An Introduction to Hilbert Space and Quantum Logic
by *David W. Cohen*

Unsolved Problems in Geometry
by *Hallard T. Croft, Kenneth J. Falconer, and Richard K. Guy*

Berkeley Problems in Mathematics, (Third Edition)
by *Paulo Ney de Souza and Jorge-Nuno Silva*

Problem-Solving Strategies
by *Arthur Engel*

Problems in Analysis
by *Bernard R. Gelbaum*

Problems in Real and Complex Analysis
by *Bernard R. Gelbaum*

Theorems and Counterexamples in Mathematics
by *Bernard R. Gelbaum and John M.H. Olmsted*

Exercises in Integration
by *Claude George*

(continued after index)

Richard K. Guy

Unsolved Problems in Number Theory

Third Edition

With 18 Figures

Richard K. Guy
Department of Mathematics and Statistics
University of Calgary
2500 University Drive NW
Calgary, Alberta
Canada T2N 1N4
rkg@cpsc.ucalgary.ca

Series Editors:
Katalin A. Bencsáth
Mathematics
School of Science
Manhattan College
Riverdale, NY 10471
USA
katalin.bencsath@manhattan.edu

Paul R. Halmos
Department of Mathematics
Santa Clara University
Santa Clara, CA 95053
USA
phalmos@scuacc.scu.edu

Mathematics Subject Classification (2000): 00A07, 11-01, 11-02

Library of Congress Cataloging-in-Publication Data
    Guy, Richard K.
        Unsolved problems in number theory / Richard Guy.—[3rd ed.].
            p. cm. — (Problem books in mathematics)
        Includes bibliographical references and index.
        ISBN 0-387-20860-7 (acid-free paper)
            1. Number theory.   I. Title.   II. Series.
    QA241.G87 2004
    512.7—dc22   2004045552

ISBN 0-387-20860-7   Printed on acid-free paper.

Printed in the United States of America.   (MVY)

9 8 7 6 5 4 3 2 1   SPIN 10950661

Springer-Verlag is a part of *Springer Science+Business Media*

springeronline.com

Preface to the Third Edition

After 10 years (or should I say 4000 years?) what's new? Too much to accommodate here, even though we've continued to grow exponentially. Sections **A20, C21, D29** and **F32** have been added, and existing sections expanded. The bibliographies are no doubt one of the more useful aspects of the book, but have become so extensive in some places that occasional reluctant pruning has taken place, leaving the reader to access secondary, but at least more accessible, sources.

A useful new feature is the lists, at the ends of about half of the sections, of references to **OEIS**, Neil Sloane's Online Encyclopedia of Integer Sequences. Many thanks to Neil for his suggestion, and for his help with its implementation. As this is a first appearance, many sequences will be missing that ought to be mentioned, and a few that are may be inappropriately placed. As more people make use of this important resource, I hope that they will let me have a steady stream of further suggestions.

I get a great deal of pleasure from the interest that so many people have shown, and consequently the help that they provide. This happens so often that it is impossible to acknowledge all of it here. Many names are mentioned in earlier prefaces. I especially miss Dick Lehmer, Raphael Robinson and Paul Erdős [the various monetary rewards he offered may still be negotiable via Ron Graham].

Renewed thanks to those mentioned earlier and new or renewed thanks to Stefan Bartels, Mike Bennett, David Boyd, Andrew Bremner, Kevin Brown, Kevin Buzzard, Chris Caldwell, Phil Carmody, Henri Cohen, Karl Dilcher, Noam Elkies, Scott Forrest, Dean Hickerson, Dan Hoey, Dave Hough, Florian Luca, Ronald van Luijk, Greg Martin, Jud McCranie, Pieter Moree, Gerry Myerson, Ed Pegg, Richard Pinch, Peter Pleasants, Carl Pomerance, Randall Rathbun, Herman te Riele, John Robertson, Rainer Rosenthal, Renate Scheidler, Rich Schroeppel, Jamie Simpson, Neil Sloane, Jozsef Solymosi, Cam Stewart, Robert Styer, Eric Weisstein, Hugh Williams, David W. Wilson, Robert G. Wilson, Paul Zimmerman, Rita Zuazua: and apologies to those who are omitted.

A special thankyou to Jean-Martin Albert for resuscitating Andy Guy's programme. I am also grateful to the Natural Sciences & Engineering Research Council of Canada for their ongoing support, and to the Department of Mathematics & Statistics of The University of Calgary for having extended their hospitality twenty-one years after retirement. I look forward to the next twenty-one. And thank you to Springer, and to Ina Lindemann and Mark Spencer in particular, for continuing to maintain excellence in both their finished products and their personal relationships.

Calgary 2003-09-16   Richard K. Guy

Preface to the Second Edition

Erdős recalls that Landau, at the International Congress in Cambridge in 1912, gave a talk about primes and mentioned four problems (see **A1**, **A5**, **C1** below) which were unattackable in the present state of science, and says that they still are. On the other hand, since the first edition of this book, some remarkable progress has been made. Fermat's last theorem (modulo some holes that are expected to be filled in), the Mordell conjecture, the infinitude of Carmichael numbers, and a host of other problems have been settled.

The book is perpetually out of date; not always the 1700 years of one statement in **D1** in the first edition, but at least a few months between yesterday's entries and your reading of the first copies off the press. To ease comparison with the first edition, the numbering of the sections is still the same. Problems which have been largely or completely answered are **B47**, **D2**, **D6**, **D8**, **D16**, **D26**, **D27**, **D28**, **E15**, **F15**, **F17** & **F28**. Related open questions have been appended in some cases, but in others they have become exercises, rather than problems.

Two of the author's many idiosyncrasies are mentioned here: the use of the ampersand (&) to denote joint work and remove any possible ambiguity from phrases such as ' ... follows from the work of Gauß and Erdős & Guy'; and the use of the notation

$$¿ \cdots \cdots \cdots ?$$

borrowed from the Hungarians, for a conjectural or hypothetical statement. This could have alleviated some anguish had it been used by the well intentioned but not very well advised author of an introductory calculus text. A student was having difficulty in finding the derivative of a product. Frustrated myself, I asked to see the student's text. He had highlighted a displayed formula stating that the derivative of a product was the product of the derivatives, without noting that the context was 'Why is ... not the right answer?'

The threatened volume on *Unsolved Problems in Geometry* has appeared, and is already due for reprinting or for a second edition.

It will be clear from the text how many have accepted my invitation to use this as a clearing house and how indebted I am to correspondents.

Extensive though it is, the following list is far from complete, but I should at least offer my thanks to Harvey Abbott, Arthur Baragar, Paul Bateman, T. G. Berry, Andrew Bremner, John Brillhart, R. H. Buchholz, Duncan Buell, Joe Buhler, Mitchell Dickerman, Hugh Edgar, Paul Erdős, Steven Finch, Aviezri Fraenkel, David Gale, Sol Golomb, Ron Graham, Sid Graham, Andrew Granville, Heiko Harborth, Roger Heath-Brown, Martin Helm, Gerd Hofmeister, Wilfrid Keller, Arnfried Kemnitz, Jeffrey Lagarias, Jean Lagrange, John Leech, Dick & Emma Lehmer, Hendrik Lenstra, Hugh Montgomery, Peter Montgomery, Shigeru Nakamura, Richard Nowakowski, Andrew Odlyzko, Richard Pinch, Carl Pomerance, Aaron Potler, Herman te Riele, Raphael Robinson, Øystein Rødseth, K. R. S. Sastry, Andrzej Schinzel, Reese Scott, John Selfridge, Ernst Selmer, Jeffrey Shallit, Neil Sloane, Stephane Vandemergel, Benne de Weger, Hugh Williams, Jeff Young and Don Zagier. I particularly miss the impeccable proof-reading, the encyclopedic knowledge of the literature, and the clarity and ingenuity of the mathematics of John Leech.

Thanks also to Andy Guy for setting up the electronic framework which has made both the author's and the publisher's task that much easier. The Natural Sciences and Engineering Research Council of Canada continue to support this and many other of the author's projects.

Calgary 94-01-08   Richard K. Guy

Preface to the First Edition

To many laymen, mathematicians appear to be problem solvers, people who do "hard sums". Even inside the profession we classify ourselves as either theorists or problem solvers. Mathematics is kept alive, much more than by the activities of either class, by the appearance of a succession of unsolved problems, both from within mathematics itself and from the increasing number of disciplines where it is applied. Mathematics often owes more to those who ask questions than to those who answer them. The solution of a problem may stifle interest in the area around it. But "Fermat's Last Theorem", because it is not yet a theorem, has generated a great deal of "good" mathematics, whether goodness is judged by beauty, by depth or by applicability.

To pose good unsolved problems is a difficult art. The balance between triviality and hopeless unsolvability is delicate. There are many simply stated problems which experts tell us are unlikely to be solved in the next generation. But we have seen the Four Color Conjecture settled, even if we don't live long enough to learn the status of the Riemann and Goldbach hypotheses, of twin primes or Mersenne primes, or of odd perfect numbers. On the other hand, "unsolved" problems may not be unsolved at all, or may be much more tractable than was at first thought.

Among the many contributions made by Hungarian mathematician Erdős Pál, not least is the steady flow of well-posed problems. As if these were not incentive enough, he offers rewards for the first solution of many of them, at the same time giving his estimate of their difficulty. He has made many payments, from $1.00 to $1000.00.

One purpose of this book is to provide beginning researchers, and others who are more mature, but isolated from adequate mathematical stimulus, with a supply of easily understood, if not easily solved, problems which they can consider in varying depth, and by making occasional partial progress, gradually acquire the interest, confidence and persistence that are essential to successful research.

But the book has a much wider purpose. It is important for students and teachers of mathematics at all levels to realize that although they are not yet capable of research and may have no hopes or ambitions in that direction, there are plenty of unsolved problems that are well within their comprehension, some of which will be solved in their lifetime. Many amateurs have been attracted to the subject and many successful researchers first gained their confidence by examining problems in euclidean geometry, in number theory, and more recently in combinatorics and graph theory, where it is possible to understand questions and even to formulate them and obtain original results without a deep prior theoretical knowledge.

The idea for the book goes back some twenty years, when I was impressed by the circulation of lists of problems by the late Leo Moser and co-author Hallard Croft, and by the articles of Erdős. Croft agreed to let me help him amplify his collection into a book, and Erdős has repeatedly encouraged and prodded us. After some time, the Number Theory chapter swelled into a volume of its own, part of a series which will contain a volume on Geometry, Convexity and Analysis, written by Hallard T. Croft, and one on Combinatorics, Graphs and Games by the present writer.

References, sometimes extensive bibliographies, are collected at the end of each problem or article surveying a group of problems, to save the reader from turning pages. In order not to lose the advantage of having all references collected in one alphabetical list, we give an Index of Authors, from which particular papers can easily be located provided the author is not too prolific. Entries in this index and in the General Index and Glossary of Symbols are to problem numbers instead of page numbers.

Many people have looked at parts of drafts, corresponded and made helpful comments. Some of these were personal friends who are no longer with us: Harold Davenport, Hans Heilbronn, Louis Mordell, Leo Moser, Theodor Motzkin, Alfred Rényi and Paul Turán. Others are H. L. Abbott, J. W. S. Cassels, J. H. Conway, P. Erdős, Martin Gardner, R. L. Graham, H. Halberstam, D. H. and Emma Lehmer, A. M. Odlyzko, Carl Pomerance, A. Schinzel, J. L. Selfridge, N. J. A. Sloane, E. G. Straus, H. P. F. Swinnerton-Dyer and Hugh Williams. A grant from the National Research Council of Canada has facilitated contact with these and many others. The award of a Killam Resident Fellowship at the University of Calgary was especially helpful during the writing of a final draft. The technical typing was done by Karen McDermid, by Betty Teare and by Louise Guy, who also helped with the proof-reading. The staff of Springer-Verlag in New York has been courteous, competent and helpful.

In spite of all this help, many errors remain, for which I assume reluctant responsibility. In any case, if the book is to serve its purpose it will start becoming out of date from the moment it appears; it has been becoming out of date ever since its writing began. I would be glad to hear from readers. There must be many solutions and references and problems which I don't know about. I hope that people will avail themselves of this clearing house. A few good researchers thrive by rediscovering results for themselves, but many of us are disappointed when we find that our discoveries have been anticipated.

Calgary 81-08-13   Richard K. Guy

Glossary of Symbols

A.P.	arithmetic progression, $a, a+d, \ldots a+kd, \ldots$	A5, A6, E10, E33
$a_1 \equiv a_2 \bmod b$	a_1 congruent to a_2, modulo b; $a_1 - a_2$ divisible by b.	A3, A4, A12, A15, B2, B4, B7, ...
$A(x)$	number of members of a sequence not exceeding x; e.g. number of amicable numbers not exceeding x	B4, E1, E2, E4
c	a positive constant (not always the same!)	A1, A3, A8, A12, B4, B11, ...
d_n	difference between consecutive primes; $p_{n+1} - p_n$	A8, A10, A11
$d(n)$	the number of (positive) divisors of n; $\sigma_0(n)$	B, B2, B8, B12, B18, ...
$d\|n$	d divides n; n is a multiple of d; there is an integer q such that $dq = n$	B, B17, B32, B37, B44, C20, D2, E16
$d \nmid n$	d does not divide n	B, B2, B25, D2, E14, E16, ...
e	base of natural logarithms; $2.718281828459045\ldots$	A8, B22, B39, D12, ...
E_n	Euler numbers; coefficients in series for $\sec x$	B45
$\exp\{..\}$	exponential function	A12, A19, B4, B36, B39, ...
F_n	Fermat numbers; $2^{2^n} + 1$	A3, A12

$f(x) \sim g(x)$	$f(x)/g(x) \to 1$ as $x \to \infty$. $(f, g > 0)$	A1, A3, A8, B33, B41, C1, C17, D7, E2, E30, F26
$f(x) = o(g(x))$	$f(x)/g(x) \to 0$ as $x \to \infty$. $(g > 0)$	A1, A18, A19, B4, C6, C9, C11, C16, C20, D4, D11, E2, E14, F1
$f(x) = O(g(x))$ $f(x) \ll g(x)$	there is a c such that $\|f(x)\| < cg(x)$ for all sufficiently large x $(g(x) > 0)$.	A19, B37, C8, C9, C10, C12, C16, D4, D12, E4, E8, E20, E30, F1, F2, F16 A4, B4, B18, B32, B40, C9, C14, D11, E28, F4
$f(x) = \Omega(g(x))$	there is a $c > 0$ such that there are arbitrarily large x with $\|f(x)\| \geq cg(x)$ $(g(x) > 0)$.	D12, E25
$f(x) \asymp g(x)$ $f(x) = \Theta(g(x))$	there are c_1, c_2 such that $c_1 g(x) \leq f(x) \leq c_2 g(x)$ $(g(x) > 0)$ for all sufficiently large x.	B18 E20
i	square root of -1; $i^2 = -1$	A16
$\ln x$	natural logarithm of x	A1, A2, A3, A5, A8, A12, ...
(m, n)	g.c.d. (greatest common divisor) of m and n; h.c.f. (highest common factor) of m and n	A, B3, B4, B5 B11, D2
$[m, n]$	l.c.m. (least common multiple) of m and n. Also the block of consecutive integers, $m, m+1, \ldots, n$	B35, E2, F14 B24, B26, B32, C12, C16
$m \perp n$	m, n coprime; $(m, n) = 1$; m prime to n.	A, A4, B3, B4, B5, B11, D2
M_n	Mersenne numbers; $2^n - 1$	A3, B11, B38

$n!$	factorial n; $1 \times 2 \times 3 \times \cdots \times n$	A2, B12, B14, B22 B23, B43, ...
$!n$	$0! + 1! + 2! + \ldots + (n-1)!$	B44
$\binom{n}{k}$	n choose k; the binomial coefficient $n!/k!(n-k)!$	B31, B33, C10, D3
$\left(\frac{p}{q}\right)$	Legendre (or Jacobi) symbol	see F5 (A1, A12, F7)
$p^a \| n$	p^a divides n, but p^{a+1} does not divide n	B, B8, B37, F16
p_n	the nth prime, $p_1 = 2$, $p_2 = 3$, $p_3 = 5$, ...	A2, A5, A14, A17 E30
$P(n)$	largest prime factor of n	B30, B46
$\mathbb{Q}$	the field of rational numbers	D2, F7
$r_k(n)$	least number of numbers not exceeding n, which must contain a k-term A.P.	see E10
$s(n)$	sum of aliquot parts (divisors of n other than n) of n; $\sigma(n) - n$	B, B1, B2, B8, B10, ...
$s^k(n)$	kth iterate of $s(n)$	B, B6, B7
$s^*(n)$	sum of unitary aliquot parts of n	B8
$S \bigcup T$	union of sets S and T	E7
$W(k, l)$	van der Waerden number	see E10
$\lfloor x \rfloor$	floor of x; greatest integer not greater than x.	A1, A5, C7, C12, C15, ...
$\lceil x \rceil$	ceiling of x; least integer not less than x.	B24
$\mathbb{Z}$	the integers $\ldots, -2, -1, 0, 1, 2, \ldots$	F14
$\mathbb{Z}_n$	the ring of integers, 0, 1, 2, ..., $n-1$ (modulo n)	E8
γ	Euler's constant; $0.577215664901532\ldots$	A8

ϵ	arbitrarily small positive constant.	A8, A18, A19, B4, B11, ...
ζ_p	p-th root of unity.	D2
$\zeta(s)$	Riemann zeta-function; $\sum_{n=1}^{\infty}(1/n^s)$	D2
π	ratio of circumference of circle to diameter; $3.141592653589793\ldots$	F1, F17
$\pi(x)$	number of primes not exceeding x	A17, E4
$\pi(x;a,b)$	number of primes not exceeding x and congruent to a modulo b	A4
$\prod$	product	A1, A2, A3, A8 A15, ...
$\sigma(n)$	sum of divisors of n; $\sigma_1(n)$	B, B2, B5, B8, B9, ...
$\sigma_k(n)$	sum of kth powers of divisors of n	B, B12, B13, B14
$\sigma^k(n)$	kth iterate of $\sigma(n)$	B9
$\sigma^*(n)$	sum of unitary divisors of n	B8
Σ	sum	A5, A8, A12, B2 B14, ...
$\phi(n)$	Euler's totient function; number of numbers not exceeding n and prime to n	B8, B11, B36, B38, B39, ...
$\phi^k(n)$	kth iterate of $\phi(n)$	B41
ω	complex cube root of 1 $\omega^3 = 1$, $\omega \neq 1$, $\omega^2 + \omega + 1 = 0$	A16
$\omega(n)$	number of distinct prime factors of n	B2, B8, B37
$\Omega(n)$	number of prime factors of n, counting repetitions	B8
$\dot{c}\cdots?$	conjectural or hypothetical statement	A1, A9, B37, C6 E10, E28, F2, F18

Contents

Introduction

Number theory has fascinated both the amateur and the professional for a longer time than any other branch of mathematics, so that much of it is now of considerable technical difficulty. However, there are more unsolved problems than ever before, and though many of these are unlikely to be solved in the next generation, this probably won't deter people from trying. They are so numerous that they have already filled more than one volume: the present book is just a personal sample.

Some good sources of problems in number theory were listed in the Introduction to the first edition, some of which are repeated here, along with more recent references.

Paul Erdős, Problems and results in combinatorial number theory III, *Springer Lecture Notes in Math.*, **626**(1977) 43–72; *MR* **57** #12442.

Paul Erdős, A survey of problems in combinatorial number theory, in *Combinatorial Mathematics, Optimal Designs and their Applications* (Proc. Symp. Colo. State Univ. 1978) *Ann. Discrete Math.*, **6**(1980) 89–115.

P. Erdős, Problems and results in number theory, in Halberstam & Hooley (eds) Recent Progress in Analytic Number Theory, Vol. 1, Academic Press, 1981, 1–13.

Paul Erdős, Problems in number theory, *New Zealand J. Math.*, **26**(1997) 155–160.

Paul Erdős, Some of my new and almost new problems and results in combinatorial number theory, *Number Theory (Eger)*, 1996, de Gruyter, Berlin, 1998, 169–180.

P. Erdős & R. L. Graham, *Old and New Problems and Results in Combinatorial Number Theory*, Monographies de l'Enseignement Math. No. 28, Geneva, 1980. (spelling)

Pál Erdős & András Sárközy, Some solved and unsolved problems in combinatorial number theory, *Math. Slovaca*, **28**(1978) 407–421; *MR* **80i**:10001.

Pál Erdős & András Sárközy, Some solved and unsolved problems in combinatorial number theory II, *Colloq. Math.*, **65** (1993) 201–211; *MR* **99j**:11012.

H. Fast & S. Świerczkowski, *The New Scottish Book*, Wrocław, 1946–1958.

Heini Halberstam, Some unsolved problems in higher arithmetic, in Ronald Duncan & Miranda Weston-Smith (eds.) *The Encyclopaedia of Ignorance*, Pergamon, Oxford & New York, 1977, 191–203.

I. Kátai, Research problems in number theory, *Publ. Math. Debrecen*, **24**(1977) 263–276; *MR* **57** #5940; II, *Ann. Univ. Sci. Budapest. Sect. Comput.*, **16**(1996)

1

223–251; *MR* **98m**:11002.

Victor Klee & Stan Wagon, *Old and New Unsolved Problems in Plane Geometry and Number Theory*, Math. Assoc. of Amer. Dolciani Math. Expositions, **11**(1991).

Proceedings of Number Theory Conference, Univ. of Colorado, Boulder, 1963.

Report of Institute in the Theory of Numbers, Univ. of Colorado, Boulder, 1959.

Joe Roberts, *Lure of the Integers*, Math. Assoc. of America, Spectrum Series, 1992; *MR* **94e**:00004.

András Sárközy, Unsolved problems in number theory, *Period. Math. Hungar.*, **42**(2001) 17–35; *MR* **2002i**:11003.

Daniel Shanks, *Solved and Unsolved Problems in Number Theory*, Chelsea, New York, 2nd ed. 1978; *MR* **80e**:10003.

W. Sierpiński, *A selection of Problems in the Theory of Numbers*, Pergamon, 1964.

Robert D. Silverman, A perspective on computational number theory, in Computers and Mathematics, *Notices Amer. Math. Soc.*, **38**(1991) 562–568.

S. Ulam, *A Collection of Mathematical Problems*, Interscience, New York, 1960.

Throughout this volume, "number" means natural number, i.e.,

$$0, 1, 2, \ldots$$

and c is an absolute positive constant, not necessarily taking the same value at each appearance. We use K. E. Iverson's symbols (popularized by Donald Knuth) "floor"($\lfloor\ \rfloor$) and "ceiling" ($\lceil\ \rceil$)for "the greatest integer not greater than" and "the least integer not less than.' A less familiar symbol may be "$m \perp n$" for "m is prime to n" or "$\gcd(m, n) = 1$."

The book is partitioned, somewhat arbitrarily at times, into six sections:

A. Prime numbers
B. Divisibility
C. Additive number theory
D. Diophantine equations
E. Sequences of integers
F. None of the above.

A. Prime Numbers

We can partition the positive integers into three classes:

the unit 1

the primes 2, 3, 5, 7, 11, 13, 17, 19, 23, 29, 31, 37, ...

the composite numbers 4, 6, 8, 9, 10, 12, 14, 15, 16, ...

A number greater than 1 is **prime** if its only positive divisors are 1 and itself; otherwise it's **composite**. Primes have interested mathematicians at least since Euclid, who showed that there are infinitely many. The largest prime in the Bible is 22273 at *Numbers,* **3** xliii.

The greatest common divisor (gcd) of m and n is denoted by (m, n), e.g., $(36, 66) = 6$, $(14, 15) = 1$, $(1001, 1078) = 77$. If $(m, n) = 1$, we say that m and n are **coprime** and write $m \perp n$; for example $182 \perp 165$.

Denote the n-th prime by p_n, e.g. $p_1 = 2$, $p_2 = 3$, $p_{99} = 523$; and the number of primes not greater than x by $\pi(x)$, e.g., $\pi(2) = 1$, $\pi(3\frac{1}{2}) = 2$, $\pi(1000) = 168$, $\pi(4 \cdot 10^{16}) = 1075292778753150$.

The table on the next page is an extension of that on p. 146 of Conway & Guy, *The Book of Numbers*, and compares $\pi(x)$ with Legendre's Law, $x/\ln x$, Gauß's Guess, $\mathrm{Li}(x) = \int_2^x \frac{t}{\ln t}\, dt$ and Riemann's Refinement,

$$\sum_{k=1}^{\infty} \frac{1}{k} \mu(k) \mathrm{Li}(x^{1/k})$$

where $\mu(x)$ is the **Mobius function@Möbius function**, which is 0 if x contains a repeated factor and $(-1)^\omega$ if x contains ω distinct prime factors

Dusart has shown that

$$\frac{x}{\ln x}\left(1 + \frac{0.992}{\ln x}\right) \leq \pi(x) \leq \frac{x}{\ln x}\left(1 + \frac{1.2762}{\ln x}\right)$$

the first inequality holding for $x \geq 599$ and the second for $x > 1$.

x	$\pi(x)$	$(x/\ln x) - \pi(x)$	$\mathrm{Li}(x) - \pi(x)$	$R(x) - \pi(x)$
10	4	0	2	
10^2	25	-3	5	1
10^3	168	-23	10	0
10^4	1229	-143	17	-2
10^5	9592	-906	38	-5
10^6	78498	-6116	130	29
10^7	664579	-44158	339	88
10^8	5761455	-332774	754	97
10^9	50847534	-2592592	1701	-79
10^{10}	455052511	-20758029	3104	-1828
10^{11}	4118054813	-169923159	11588	-2318
10^{12}	37607912018	-1416705193	38263	-1476
10^{13}	346065536839	-11992858452	108971	-5773
10^{14}	3204941750802	-102838308636	314890	-19200
10^{15}	29844570422669	-891604962452	1052619	73218
10^{16}	279238341033925	-7804289844393	3214632	327052
10^{17}	2623557157654233	-68883734693928	7956589	-598255
10^{18}	24739954287740860	-612483070893536	21949555	-3501366
10^{19}	234057667276344607	-5481624169369961	99877775	23884332
10^{20}	2220819602560918840	-49347193044659702	223744644	-4891826
10^{21}	21127269486018731928	-446579871578168707	597394253	-86432205

Dirichlet's theorem tells us that there are infinitely many primes in any **arithmetic progression**,

$$a, \quad a+b, \quad a+2b, \quad a+3b, \quad \ldots$$

provided $a \perp b$.

A paper of Dickson, not easily accessed, has a misleading title which suggests that he generalizes Dirichlet's theorem to more than one arithmetic progression. His History merely says 'Dickson asked' — a question often asked, before and since, and which would answer numerous other queries such as:

¿ are there infinitely many prime pairs $(p, 2p-1)$?

¿ or infinitely many $(p, 2p+1)$?

(compare **A7**); and has misled some to believe that Schinzel's conjecture

¿ Tout nombre rationnel positif peut être représenté d'une infinite manière sous la forme $(p+1)/(q+1)$ ainsi que sous la forme $(p-1)/(q-1)$ où p et q sont des nombres premiers ?

has been settled, whereas it has not. Matthew Conroy has verified that the first 10^9 integers can be so expressed. For a partial result, see Peter Elliott's paper.

A more sweeping conjecture of Schinzel is his **Hypothesis H**: if $f_1(x)$, $f_2(x)$, ..., $f_k(x)$ are integer-valued polynomials with product $f(x)$, and for every prime p there is an a such that $f(a)$ is not divisible by p, then there are infinitely many n such that $f_1(n)$, $f_2(n)$, ..., $f_(n)$ are all simultaneously prime.

This embodies six of the famous Hardy-Littlewood conjectures (see reference at **A1**) and has been quantized as the **Bateman-Horn conjecture**: if the polynomials are irreducible, with integer coeeficients, and the product of their degrees is h, then the number, $Q(f; N)$ of $n \leq N$ for which the k polynomials simultaneously take prime values is asymptotically

$$Q(f; N) \sim \frac{C}{h} \int_2^N (\ln u)^{-k} \, du$$

where C is the product

$$\prod \left(1 - \frac{1}{p}\right)^{-k} \left(1 - \frac{\omega(p)}{p}\right)$$

taken over all primes p, and $\omega(p)$ is the number of solutions of the congruence $f(x) \equiv 0 \bmod p$ (see reference at **A17**).

Table 7 (**D 27**) can be used as a table of primes < 1000, an entry 1, 3, 5 or 7 indicates a prime in that residue class (see **A4**) modulo 8.

The general problem of determining whether a large number is prime or composite, and in the latter case of determining its factors, has fascinated number theorists down the ages. With the advent of high speed computers, considerable advances have been made, and a special stimulus has recently been provided by the application to cryptanalysis. A recent breakthrough by Agrawal, Kayal & Saxena shows that prime testing can be done in polynomial time, though it may be a while before this becomes of practical value.

Some other references appear after Problem **A3** and in earlier editions of this book.

William Adams & Daniel Shanks, Strong primality tests that are not sufficient, *Math. Comput.*, **39**(1982) 255–300.

Manindra Agrawal, Neeraj Kayal & Nitin Saxena, Primes is in P, *Annals of Math.*, (to appear)

F. Arnault, Rabin-Miller primality test: composite numbers which pass it, *Math. Comput.*, **64**(1995) 355–361.

Stephan Baier, On the Bateman-Horn conjecture, *J. Number Theory*, **96**(2002) 432–448.

Friedrich L. Bauer, Why Legendre made a wrong guess about $\pi(x)$, and how Laguerre's continued fraction for the logarithmic integral improved it, *Math. Intelligencer*, **25**(2003) 7–11.

Folkmar Bornemann, PRIMES is in P: a breakthrough for "Everyman", *Notices Amer. Math. Soc.*, **50**(2003) 545–552; *MR* **2004c**:11237.

Henri Cohen, Computational aspects of number theory, *Mathematics unlimited — 2001 and beyond*, 301–330, Springer, Berlin, 2001; *MR* **2002i**:11126.

Matthew M. Conroy, A sequence related to a conjecture of Schinzel, *J. Integer Seq.*, **4**(2001) no.1 Article 01.1.7; *MR* **2002j**:11017.

Richard E. Crandall & Carl Pomerance, *Prime Numbers: a computational perspective*, Springer, New York, 2001; *MR* **2002a**:11007.

Marc Deléglise & Jo"el Rivat, Computing $\pi(x)$: the Meissel, Lehmer, Lagarias, Miller, Odlyzko method, *Math. Comput.*, **65**(1996) 235–245; *MR* **96d**:11139.

L. E. Dickson, A new extension of Dirichlet's theorem on prime numbers, *Messenger of Mathematics*, **33**(1904) 155–161.

Pierre Dusart, The kth prime is greater than $k(\ln k + \ln\ln k - 1)$ for $k \geq 2$, *Math. Comput.*, **68**(1999) 411–415; *MR* **99d**:11133.

P. D. T. A. Elliott, On primes and powers of a fixed integer, *J. London Math. Soc.*(2), **67**(2003) 365–379; *MR* **2003m**:11151.

Andrew Granville, Unexpected irregularities in the distribution of prime numbers, *Proc. Internat. Congr. Math., Vol. 1, 2 (Zurich, 1994)*, 388–399.

Andrew Granville, Prime possibilities and quantum chaos, *Emissary*, Spring 2002, pp1, 12–18.

Wilfrid Keller, Woher kommen die größten derzeit bekannten Primzahlen? *Mitt. Math. Ges. Hamburg*, **12**(1991) 211–229; *MR* **92j**:11006.

J. C. Lagarias, V. S. Miller & A. M. Odlyzko, Computing $\pi(x)$: the Meissel-Lehmer method, *Math. Comput.*, **44**(1985) 537–560; *MR* **86h**:11111.

J. C. Lagarias & A. M.Odlyzko, Computing $\pi(x)$: an analytic method, *J. Algorithms*, **8**(1987) 173–191; *MR* **88k**:11095.

Arjen K. Lenstra & Mark S. Manasse, Factoring by electronic mail, in Advances in Cryptology—EUROCRYPT'89, *Springer Lect. Notes in Comput. Sci.*, **434**(1990) 355–371; *MR* **91i**:11182.

Hendrik W. Lenstra, Factoring integers with elliptic curves, *Ann. of Math.*(2), **126**(1987) 649–673; *MR* **89g**:11125.

Hendrik W. Lenstra & Carl Pomerance, A rigorous time bound for factoring integers, *J. Amer. Math. Soc.*, **5**(1992) 483–916; *MR* **92m**:11145.

Richard F. Lukes, Cameron D. Patterson & Hugh Cowie Williams, Numerical sieving devices: their history and some applications, *Nieuw Arch. Wisk.*(4), **13**(1995) 113–139; *MR* **99m**:11082.

James K. McKee, Speeding Fermat's factoring method, *Math. Comput.*, **68**(1999) 1729–1737; *MR* **2000b**:11138.

G. L. Miller, Riemann's hypothesis and tests for primality, *J. Comput. System Sci.*, **13**(1976) 300–317; *MR* **58** #470ab.

Peter Lawrence Montgomery, An FFT extension of the elliptic curve method of factorization, PhD dissertation, UCLA, 1992.

Siguna Müller, A probable prime test with very high confidence for $n \equiv 3$ mod 4, *J. Cryptology*, **16**(2003) 117–139.

J. M. Pollard, Theorems on factoring and primality testing, *Proc. Cambridge Philos. Soc.*, **76**(1974) 521–528; *MR* **50** #6992.

J. M. Pollard, A Monte Carlo method for factorization, *BIT*, **15**(1975) 331–334; *MR* **52** #13611.

Carl Pomerance, Recent developments in primality testing, *Math. Intelligencer*, **3**(1980/81) 97–105.

Carl Pomerance, Notes on Primality Testing and Factoring, *MAA Notes* 4(1984) Math. Assoc. of America, Washington DC.

Carl Pomerance (editor), Cryptology and Computational Number Theory, *Proc. Symp. Appl. Math.*, **42** Amer. Math. Soc., Providence, 1990; *MR* **91k**: 11113.

Paulo Ribenboim, *The Book of Prime Number Records*, Springer-Verlag, New York, 1988.

Paulo Ribenboim, *The Little Book of Big Primes*, Springer-Verlag, New York, 1991.

Hans Riesel, *Prime Numbers and Computer Methods for Factorization*, 2nd ed, Birkhäuser, Boston, 1994; *MR* **95h**:11142.

Hans Riesel, Wie schnell kann man Zahlen in Faktoren zerlegen? *Mitt. Math. Ges. Hamburg*, **12**(1991) 253–260.

R. Rivest, A. Shamir & L. Adleman, A method for obtaining digital signatures and public key cryptosystems, *Communications A.C.M.*, Feb. 1978.

Michael Rubinstein & Peter Sarnak, Chebyshev's bias, *Experimental Math.*, **3**(1994) 173–197; *MR* **96d**:11099.

A. Schinzel & W. Sierpiński, Sur certaines hypothèses concernant les nombres premiers, *Acta Arith.*, **4**(1958) 185–208; erratum **5**(1959) 259; *MR* **21** #4936; and see **7**(1961) 1–8; *MR* **24** #A70.

R. Solovay & V. Strassen, A fast Monte-Carlo test for primality, *SIAM J. Comput.*, **6**(1977) 84–85; erratum **7**(1978) 118; *MR* **57** #5885.

Michael Somos & Robert Haas, A linked pair of sequences implies the primes are infinite, *Amer. Math. Monthly*, **110**(2003) 539–540; *MR* **2004c**:11007.

Jonathan Sorenson, Counting the integers cyclotomic methods can factor, *Comput. Sci. Tech. Report*, **919**, Univ. of Wisconsin, Madison, March 1990.

Hugh Cowie Williams, *Édouard Lucas and Primality Testing*, Canad. Math. Soc. Monographs **22**, Wiley, New York, 1998.

H. C. Williams & J. S. Judd, Some algorithms for prime testing using generalized Lehmer functions, *Math. Comput.*, **30**(1976) 867–886.

A1 Prime values of quadratic functions.

Are there infinitely many primes of the form $a^2 + 1$? Probably so, and in fact Hardy and Littlewood (their conjecture E) guessed that the number, $P(n)$, of such primes less than n, was asymptotic to $c\sqrt{n}/\ln n$,

$$\text{¿} \qquad P(n) \sim c\sqrt{n}/\ln n \qquad \text{?}$$

i.e., that the ratio of $P(n)$ to $\sqrt{n}/\ln n$ tends to c as n tends to infinity. The constant c is

$$c = \prod \left\{ 1 - \frac{\left(\frac{-1}{p}\right)}{p-1} \right\} = \prod \left\{ 1 - \frac{(-1)^{(p-1)/2}}{p-1} \right\} \approx 1.3727$$

where $\left(\frac{-1}{p}\right)$ is the Legendre symbol (see **F5**) and the product is taken over all odd primes. They make similar conjectures, differing only in the value of c, for the number of primes represented by more general quadratic expressions. But we don't know of any integer polynomial, of degree greater than one, for which it has been proved that it takes an infinity of prime values. Is there even one prime $a^2 + b$ for each $b > 0$? Sierpiński has shown that for every k there is a b such that there are more than k primes of the form $a^2 + b$.

Fouvry & Iwaniek have shown that there are infinitely many primes $p = l^2 + m^2$ where l is also prime. This provided motivation for Friedlander & Iwaniec to prove that there are infinitely many primes of the form $a^2 + b^4$. Moreover, the frequency of primes follows the expected distribution.

Iwaniec has shown that there are infinitely many n for which $n^2 + 1$ is the product of at most two primes, and his results extend to other irreducible quadratics. With Friedlander he has come close by showing that there are infinitly many primes of shape $n^2 + m^4$.

If $P(n)$ is the largest prime factor of n, Maurice Mignotte has shown that $P(a^2 + 1) \geq 17$ if $a \geq 240$. Note that $239^2 + 1 = 2 \cdot 13^4$ (yet another property of 239). It has been known for 50 years that $P(a^2 + 1) \to \infty$ with a.

Ulam and others noticed that the pattern formed by the prime numbers when the sequence of numbers is written in a "square spiral" seems to favor diagonals which correspond to certain "prime-rich" quadratic polynomials. For example the main diagonal of Figure 1 corresponds to Euler's famous formula $n^2 + n + 41$.

Subject to Hardy & Littlewood's Conjecture F, Jacobson expects that the quadratic $x^2 + x + 3399714628553118047$ will have, asymptotically, a higher density of primes, and he and Hugh Williams expect an even greater density for

$$x^2 + x - 3\,3251810980\,6968781031\,5008525712\,9508857312\,8477514981\,9034998387\,4538507313$$

See also **A17**.

There are some results for expressions (*not* polynomials!) of degree greater than 1, starting with that of Pyateckii-Šapiro, who proved that the number of primes of the form $\lfloor n^c \rfloor$ in the range $1 < n < x$ is $(1 + o(1))x/(1 + c)\ln x$ if $1 \leq c \leq \frac{12}{11}$. This range has been successively extended to $\frac{10}{9}$, $\frac{69}{62}$, $\frac{755}{662}$, $\frac{39}{34}$, $\frac{15}{13}$, $\frac{2817}{2426}$ and $\frac{243}{205}$ by Kolesnik, Graham and Leitmann independently, Heath-Brown, Kolesnik again, Liu & Rivat, Rivat & Sargos and by Rivat & Wu.

421 420 **419** 418 417 416 415 414 413 412 411 410 **409** 408 407 406 405 404 403 402
422 **347** 346 345 344 343 342 341 340 339 338 **337** 336 335 334 333 332 **331** 330 **401**
423 348 **281** 280 279 278 **277** 276 275 274 273 272 **271** 270 **269** 268 267 266 329 400
424 **349** 282 **223** 222 221 220 219 218 217 216 215 214 213 212 **211** 210 265 328 399
425 350 **283** 224 **173** 172 171 170 169 168 **167** 166 165 164 **163** 162 209 264 327 398
426 351 284 225 174 **131** 130 129 128 **127** 126 125 124 123 122 161 208 **263** 326 **397**
427 352 285 226 175 132 **97** 96 95 94 93 92 91 90 121 160 207 262 325 396
428 **353** 286 **227** 176 133 98 **71** 70 69 68 **67** 66 **89** 120 159 206 261 324 395
429 354 287 228 177 134 99 72 **53** 52 51 50 65 88 119 158 205 260 323 394
430 355 288 **229** 178 135 100 **73** 54 **43** 42 49 64 87 118 **157** 204 259 322 393
431 356 289 230 **179** 136 **101** 74 55 44 **41** 48 63 86 117 156 203 258 321 392
432 357 290 231 180 **137** 102 75 56 45 46 **47** 62 85 116 155 202 **257** 320 391
433 358 291 232 **181** 138 **103** 76 57 58 **59** 60 **61** 84 115 154 201 256 319 390
434 **359** 292 **233** 182 **139** 104 77 78 **79** 80 81 82 **83** 114 153 200 255 318 **389**
435 360 **293** 234 183 140 105 106 **107** 108 **109** 110 111 112 **113** 152 **199** 254 **317** 388
436 361 294 235 184 141 142 143 144 145 146 147 148 **149** 150 **151** 198 253 616 387
437 362 295 236 185 186 187 188 189 190 **191** 192 **193** 194 195 196 **197** 252 315 386
438 363 296 237 238 **239** 240 **241** 242 243 244 245 246 247 248 249 250 **251** 314 385
439 364 297 298 299 300 301 302 303 304 305 306 **307** 308 309 310 **311** 312 **313** 384
440 365 366 **367** 368 369 370 371 372 **373** 374 375 376 377 378 **379** 380 381 382 **383**

Figure 1.   Primes (in **bold**) Form Diagonal Patterns.

Alberto Barajas & Rita Zuazua, A sieve for primes of the form n^2+1, preprint, Williams60, Banff, May 2003.

Chen Yong-Gao, Gábor Kun, Gábor Pete, Imre Z. Ruzsa & Ádám Timár, Prime values of reducible polynmials II, *Acta Arith.*, **04**(2002) 117–127.

Cécile Dartyge, Le plus grand facteur de $n^2 + 1$, où n est presque premier, *Acta Arith.*, **76**(1996) 199–226.

Cécile Dartyge, Entiers de la forme $n^2 + 1$ sans grand facteur premier, *Acta Math. Hungar.*, **72**(1996) 1–34.

Cécile Dartyge, Greg Martin & Gérald Tenenbaum, Polynomial values free of large prime factors, *Period. Math. Hungar.*, **43**(2001) 111–119; *MR* **2002c**:11117.

Étienne Fouvry & Henryk Iwaniek, Gaussian primes, *Acta Arith.*, **79**(1997) 249–287; *MR* **98a**:11120.

John Friedlander & Henryk Iwaniec, The polynomial $X^2 + Y^4$ captures its primes. *Ann. of Math.*(2), **148**(1998) 945–1040; *MR* **2000c**:11150a.

Gilbert W. Fung & Hugh Cowie Williams, Quadratic polynomials which have a high density of prime values, *Math. Comput.*, **55**(1990) 345–353; *MR* **90j**:11090.

Martin Gardner, The remarkable lore of prime numbers, *Scientific Amer.*, **210** #3 (Mar. 1964) 120–128.

G. H. Hardy & J. E. Littlewood, Some problems of 'partitio numerorum' III: on the expression of a number as a sum of primes, *Acta Math.*, **44**(1923) 1–70.

D. R. Heath-Brown, The Pyateckii-Šapiro prime number theorem, *J. Number Theory*, **16**(1983) 242–266.

D. R. Heath-Brown, Zero-free regions for Dirichlet L-functions, and the least prime in an arithmetic progression, *Proc. London Math. Soc.*(3) **64**(1992) 265–338.

D. Roger Heath-Brown, Primes represented by x^3+2y^3, *Acta Math.*, **186**(2001) 1–84; *MR* **2002b**:11122.

D. Roger Heath-Brown & B. Z. Moroz, Primes represented by binary cubic forms, *Proc. London Math. Soc.*(3), **84**(2002) 257–288.

Henryk Iwaniec, Almost-primes represented by quadratic polynomials, *Invent. Math.*, **47**(1978) 171–188; *MR* **58** #5553.

M. J. Jacobson, Computational techniques in quadratic fields, M.Sc thesis, Univ. of Manitoba, 1995.

Michael J. Jacobson Jr. & Hugh C. Williams, New quadratic polynomials with high densities of prime values, *Math. Comput.*, **72**(2003) 499–519; *MR* **2003k**:11146.

G. A. Kolesnik, The distribution of primes in sequences of the form $[n^c]$, *Mat. Zametki*(2), **2**(1972) 117–128.

G. A. Kolesnik, Primes of the form $[n^c]$, *Pacific J. Math.*(2), **118**(1985) 437–447.

D. Leitmann, Abschätzung trigonometrischer Summen, *J. reine angew. Math.*, **317**(1980) 209–219.

D. Leitmann, Durchschnitte von Pjateckij-Shapiro-Folgen, *Monatsh. Math.*, **94**(1982) 33–44.

H. Q. Liu & J. Rivat, On the Pyateckii-Šapiro prime number theorem, *Bull. London Math. Soc.*, **24**(1992) 143–147.

Maurice Mignotte, $P(x^2 + 1) \geq 17$ si $x \geq 240$, *C. R. Acad. Sci. Paris Sér. I Math.*, **301**(1985) 661–664; *MR* **87a**:11026.

Carl Pomerance, A note on the least prime in an arithmetic progression, *J. Number Theory*, **12**(1980) 218–223.

I. I. Pyateckii-Šapiro, On the distribution of primes in sequences of the form $[f(n)]$ (Russian), *Mat. Sbornik N.S.*, **33**(1953) 559–566; *MR* **15**, 507.

Joël Rivat & Patrick Sargos, Nombres premiers de la forme $\lfloor n^c \rfloor$, *Canad. J. Math.*, **53**(2001) 414–433: *MR* **2002a**:11107.

J. Rivat & J. Wu, Prime numbers of the form $[n^c]$, *Glasg. Math. J.*, **43**(2001) 237–254; *MR* **2002k**:11154.

Daniel Shanks, On the conjecture of Hardy and Littlewood concerning the number of primes of the form $n^2 + a$, *Math. Comput.*, **14**(1960) 321–332.

W. Sierpiński, Les binômes $x^2 + n$ et les nombres premiers, *Bull. Soc. Roy. Sci. Liège*, **33**(1964) 259–260.

E. R. Sirota, Distribution of primes of the form $p = [n^c] = [t^d]$ in arithmetic progressions (Russian), *Zap. Nauchn. Semin. Leningrad Otdel. Mat. Inst. Steklova*, **121**(1983) 94–102; *Zbl.* **524**.10038.

Panayiotis G. Tsangaris, A sieve for all primes of the form $x^2 + (x + 1)^2$, *Acta Acad. Paedagog. Agriensis Sect. Mat. (N.S.)* bf25(1998) 39–53 (1999); *MR* **2001c**:11044.

OEIS: A000101, A000847, A002110, A002386, A005867, A008407, A008996, A020497, A022008, A030296, A040976, A048298, A049296, A051160-A051168, A058188, A058320, A059861-A059865.

A2 Primes connected with factorials.

Are there infinitely many primes of the form $n! \pm 1$ or of the form $p\# \pm 1$, where $p\#$ is the product, primorial p, of the primes $2 \cdot 3 \cdot 5 \cdots p$ up to p.

Discoveries since the second edition by Harvey Dubner and others have brought the lists to:

$n! + 1$ is prime for $n = 1, 2, 3, 11, 27, 37, 41, 73, 77, 116, 154, 320, 340,$ 399, 427, 872, 1477, 6380.

$n! - 1$ is prime for $n = 3, 4, 6, 7, 12, 14, 30, 32, 33, 38, 94, 166, 324,$ 379, 469, 546, 974, 1963, 3507, 3610, 6917, 21480.

$p\# + 1$ is prime for $p = 2, 3, 5, 7, 11, 31, 379, 1019, 1021, 2657, 3229,$ 4547, 4787, 11549, 13649, 18523, 23801, 24029, 42209, 145823, 366439, 392113.

$p\# - 1$ is prime for $p = 3, 5, 11, 13, 41, 89, 317, 337, 991, 1873, 2053,$ 2377, 4093, 4297, 4583, 6569, 13033, 15877.

Let q be the least prime greater than $p\#$. Then Reo F. Fortune conjectured that $q - p\#$ is prime (or 1) for all primes p. It is clear that it can only be divisible by primes greater than p, and Selfridge observes that the truth of the conjecture would follow from Schinzel's formulation of Cramer's conjecture, that for $x > 7.1374035$ there is always a prime between x and $x + (\ln x)^2$. Stan Wagon has calculated the first 100 fortunate primes:

```
 3   5   7  13  23  17  19  23  37    61  67  61  71  47 107  59   61 109  89 103
79 151 197 101 103 233 223 127 223   191 163 229 643 239 157 167  439 239 199 191
199 383 233 751 313 773 607 313 383   293 443 331 283 277 271 401  307 331 379 491
331 311 397 331 353 419 421 883 547  1381 457 457 373 421 409 1061 523 499 619 727
457 509 439 911 461 823 613 617 1021  523 941 653 601 877 607 631  733 757 877 641
```

under the assumption that the very large probable primes involved are genuine primes. The answers to the questions are probably "yes", but it does not seem conceivable that such conjectures will come within reach either of computers or of analytical tools in the foreseeable future. Schinzels conjecture@Schinzel's conjecture has been attributed to Cramér, but Cramér conjectured (see reference at **A8**)

$$\text{¿} \qquad \limsup_{n \to \infty} \frac{p_{n+1} - p_n}{(\ln p_n)^2} = 1 \qquad ?$$

Schinzel notes that this doesn't imply the existence of a prime between x and $x + (\ln x)^2$, even for sufficiently large x.

More hopeful, but still difficult, is the following conjecture of Erdős and Stewart: are $1!+1 = 2$, $2!+1 = 3$, $3!+1 = 7$, $4!+1 = 5^2$, $5!+1 = 11^2$ the only cases where $n! + 1 = p_k^a p_{k+1}^b$ and $p_{k-1} \le n < p_k$? [Note that $(a, b) = (1, 0)$, $(1, 0)$, $(0, 1)$, $(2, 0)$ and $(0, 2)$ in these five cases.] Flammenkamp & Luca announced on 98-12-16 that they had proved the conjecture.

Erdős also asks if there are infinitely many primes p for which $p - k!$ is composite for each k such that $1 \le k! < p$; for example, $p = 101$ and $p = 211$. He suggests that it may be easier to show that there are infinitely many integers n $(l! < n \le (l+1)!)$ all of whose prime factors are greater than l, and for which all the numbers $n - k!$ $(1 \le k \le l)$ are composite.

One of the few necessary and sufficient theorems about prime numbers is **Wilson's theorem**: A necessary and sufficient condition that $m > 1$ should be prime is that m divides $(m-1)! + 1$. A curiosity is that p^2 divides $(p-1)! + 1$ for $p = 5$, 13 and 563. Crandall, Dilcher & Pomerance (reference at **A3**) show that there are no other such **Wilson primes** below $5 \cdot 10^8$.

Javier Soria asked for what p is $(1 + (p-1)!)/p$ also prime? E.g., 5, 7, 11, 29, 773, ... Mike Oakes found just two more such numbers, 1321 and 2621, up to 30941, which yield probable primes. Are there infinitely many? Several, including Noam Elkies, Dean Hickerson, Carl Pmerance and Bjorn Poonen gave a heuristic argument suggesting that there are.

Subbarao followed Erdős and defined a **Pillai prime** as a prime p for which there is an n such that $n! + 1 \equiv 0 \pmod{p}$, but $p \not\equiv 1 \pmod{n}$. The Pillai primes < 100 are 23, 29, 59, 61, 67, 71, 79 and 83. G. E. Hardy & Subbarao prove that there are infinitely many Pillai primes, and infinitely many associated values of n, which they modestly call **EHS numbers**; the first few are 8, 9, 13, 14, 15, 16, 17, 18, 19, 22. Do the Pillai primes have an asymptotic density? Experimental results suggest that they do, and that it may be between 0.5 and 0.6. On the other hand there seems no reason why it should not be 1. Is the density of EHS numbers $\frac{1}{2}$? If $g(p)$ is the number of $n < p$ for which $n! + 1 \equiv 0 \pmod{p}$, is $\limsup g(p) = \infty$? Perhaps $g(p) \to \infty$ for almost all primes p. The density, call it e_k, of primes p for which $g(p) = k$ probably exists for each k and $\sum_{k=1}^{\infty} = 1$.

Erdős asked if there were many p for which $n! \pmod{p}$ has $p-2$ nonzero values. Perhaps $p = 5$ is the only one. Theorem 114 of Hardy & Wright implies that such p are $\equiv 1 \pmod{4}$. Erdős also asked, if $A(x)$ is the number of composite $u < x$ for which $n! + 1 \equiv 0 \pmod{u}$, is $A(x) = o(x^\epsilon)$? Examples of such u are 25, 121, 721.

Hardy & Subbarao believe that, if $f(p)$ is the least integer for which $f(p)! + 1 \equiv 0 \pmod{p}$, then there are infinitely many p for which $f(p) = p - 1$, but that the number of such $p \leq x$ is $o(x/\ln x)$. Erdős believed that $f(p)/p \to 0$ for almost all p.

Takashi Agoh, Karl Dilcher & Ladislav Skula, Wilson quotients for composite moduli, *Math. Comput.*, **67**(1998) 843–861; *MR* **98h**:11003.

I. O. Angell & H. J. Godwin, Some factorizations of $10^n \pm 1$, *Math. Comput.*, **28**(1974) 307–308.

Chris K. Caldwell & Yves Gallot, On the primality of $n! \pm 1$ and $2 \times 3 \times 5 \times \cdots \times p \pm 1$, *Math. Comput.*, **71**(2002) 441–448; *MR* **2002g**:11011.

Alan Borning, Some results for $k! \pm 1$ and $2 \cdot 3 \cdot 5 \cdots p \pm 1$, *Math. Comput.*, **26**(1972) 567–570.

J. P. Buhler, R. E. Crandall & M. A. Penk, Primes of the form $n! \pm 1$ and $2 \cdot 3 \cdot 5 \cdots p \pm 1$, *Math. Comput.*, **38**(1982) 639–643; corrigendum, Wilfrid Keller, **40**(1983) 727; *MR* **83c**:10006, **85b**:11119.

Chris K. Caldwell, On the primality of $n! \pm 1$ and $2 \cdot 3 \cdot 5 \cdots p \pm 1$, *Math.*

Comput., **64**(1995) 889–890; *MR* **95g**:11003.

Chris K. Caldwell & Harvey Dubner, Primorial, factorial and multifactorial primes, *Math. Spectrum*, **26**(1993-94) 1–7.

Chris Caldwell & Yves Gallot, On the primality of $n! \pm 1$ and $2 \times 3 \times 5 \times \cdots \times p \pm 1$, *Math. Comput.*, **71**(2002) 441–448.

Harvey Dubner, Factorial and primorial primes, *J. Recreational Math.*, **19** (1987) 197–203.

Martin Gardner, Mathematical Games, *Sci. Amer.*, **243**#6(Dec. 1980) 18–28.

Solomon W. Golomb, The evidence for Fortune's conjecture, *Math. Mag.*, **54**(1981) 209–210.

G. E. Hardy & M. V. Subbarao, A modified problem of Pillai and some related questions, *Amer. Math. Monthly*, **109**(2002) 554–559.

S. Kravitz & D. E. Penney, An extension of Trigg's table, *Math. Mag.*, **48**(1975) 92–96.

Le Mao-Hua & Chen Rong-Ji, A conjecture of Erdős and Stewart concerning factorials, *J. Math. (Wuhan)* **23**(2003) 341–344.

S. S. Pillai, Question 1490, *J. Indian Math. Soc.*, **18**(1930) 230.

Mark Templer, On the primality of $k! + 1$ and $2 * 3 * 5 * \cdots * p + 1$, *Math. Comput.*, **34**(1980) 303–304.

Miodrag Živković, The number of primes $\sum_{i=1}^{n}(-1)^{n-i}i!$ is finite, *Math. Comput.*, **68**(1999) 403–409; *MR* **99c**:11163.

OEIS: A002110, A005235, A006862, A035345-035346, A037155, A046066, A055211, A057016, A057019–057020, A057034, A058020, A058024, A67362–067365.

A3 Mersenne primes. Repunits. Fermat numbers. Primes of shape $k \cdot 2^n + 1$.

Primes of special form have been of perennial interest, especially the **Mersenne primes** $2^p - 1$. Here p is necessarily prime, but that is *not* a sufficient condition! $2^{11} - 1 = 2047 = 23 \cdot 89$. They are connected with perfect numbers (see **B1**).

The powerful Lucas Lehmer test@Lucas-Lehmer test, in conjunction with successive generations of computers, and more sophisticated techniques in using them, continues to add to the list of primes p for which $2^p - 1$ is also prime:

2, 3, 5, 7, 13, 17, 19, 31, 61, 89, 107, 127, 521, 607, 1279, 2203, 2281, 3217, 4253, 4423, 9689, 9941, 11213, 19937, 21701, 23209, 44497, 86243, 110503, 132049, 216091, 756839, 859433, 1257787, 1398269, 2976221, 3021377, 6972593, 13466917, ...

The last five entries were found by GIMPS,the Great Internet Mersenne Prime Search. A 40th member, $2^{20996011} - 1$, was discovered by Michael Shafer on 2003-11-17.

The number of Mersenne primes is undoubtedly infinite, but proof is again hopelessly beyond reach. Suppose $M(x)$ is the number of primes $p \leq x$ for which $2^p - 1$ is prime. Find a convincing heuristic argument for the size of $M(x)$. Gillies gave one suggesting that $M(x) \sim c \ln x$. H. W. Lenstra, Pomerance and Wagstaff all believe this and in fact suggest that

$$¿ \quad M(x) \sim e^{\gamma} \log x \quad ?$$

where the log is to base 2.

The largest known prime is usually a Mersenne prime, but in early 1992 the record was $391581 \cdot 2^{216193} - 1$, discovered by J. Brown, L.C. Noll, B. Parady, G. Smith, J. Smith & S. Zarantonello.

D. H. Lehmer puts $S_1 = 4$, $S_{k+1} = S_k^2 - 2$, supposes that $2^p - 1$ is a Mersenne prime, notes that $S_{p-2} \equiv 2^{(p+1)/2}$ or $-2^{(p+1)/2} \bmod 2^p - 1$ and asks: which? Selfridge observes that the problem was asked in 1954 by Raphael Robinson. In a 94-06-22 email, Franz Lemmermeier provides much evidence, later further strengthened by Robert Harley, that if $2^p - 1$ is written in the form $4a^2 + 27b^2$, then the sign is that of the Jacobi symbol $\left(\frac{2}{a}\right)$ (see **F5**). Sun Zhi-Hong had made a similar conjecture on 88-09-19.

Selfridge conjectures that if n is a prime of the form $2^k \pm 1$ or $2^{2k} \pm 3$, then $2^n - 1$ and $(2^n + 1)/3$ are either both prime or both composite. Moreover if both are prime, then n is of one of those forms. Is this an example of the Law of Small Numbers? Dickson, on p. 28 of Vol. 1 of his *History*, says:

> In a letter to Tannery (*l'Intermédiaire des math.*, **2**(1895) 317) Lucas stated that Mersenne (1644, 1647) implied that a necessary and sufficient condition that $2^p - 1$ be a prime is that p be a prime of one of the forms $2^{2n} + 1$, $2^{2n} \pm 3$, $2^{2n+1} - 1$. Tannery expressed his belief that the theorem was empirical and due to Frenicle, rather than to Fermat.

If p is a prime, is $2^p - 1$ always **squarefree** (does it never contain a repeated factor)? This seems to be another unanswerable question. It is safe to conjecture that the answer is "No!" This *could* be settled by computer if you were lucky. As D. H. Lehmer has said about various factoring methods, "Happiness is just around the corner". Selfridge puts the computational difficulties in perspective by proposing the problem: find fifty more numbers like 1093 and 3511. [Fermat's theorem tells us that if p is prime, then p divides $2^p - 2$. The primes 1093 and 3511 are the only two less than 1.25×10^{15} (Josh Knauer & Jörg Richstein, May 2003) for which p^2 divides $2^p - 2$.] It is not known if there are infinitely many **Wieferich primes**, p, for which p^2 divides $2^p - 2$. It is not even known if there are infinitely many p for which p^2 *does not* divide $2^p - 2$ — although Silverman deduced this from the very powerful "ABC conjecture" (see **B19**) and Ribenboim has extended this to other second order recurring sequences. Karl Dilcher reports that $3^{1093} \equiv 3 \pmod{1093^2 + 1093 + 1}$.

The so-called **repunits**, $(10^p - 1)/9$, are prime for $p = 2$, 19, 23, 317, 1031 and probably 49081. Repunits other than 1 are known never to be squares and Rotkiewicz has shown that they are not cubes. When are they squarefree? The primes 3, 487 and 56598313 are the only ones less than 2^{32} for which p^2 divides $10^p - 10$. Peter Montgomery lists cases where p^2 divides $a^{p-1} - 1$ for $a < 100$ and $p < 2^{32}$.

Repunits are the simplest example of **palindromes**, which read the same backwards as forwards.

Dubner & Broadhurst found the 15601-digit palidromic prime

$$\underbrace{18080108081808010808\ldots10808180808}_{\text{1808010808 repeated 1560 times}}1$$

Which reads upside-down as well as back to front, yielding a 4-group of symmetries. Broadhurst has also used repunits to create other curiosities. For example, with $R_n = (10^n - 1)/9$, he announced on 2003-06-23 that

$$[7532 \cdot 10^{14} \cdot R_{2160}/R_4 + 75753575332572] \cdot R_{26088}/R_{2174} + 1$$

is a palindromic prime with each its 26088 digits prime, and on 2003-08-05 he beat his own record with the 30913-digit specimen

$$34 \cdot R_{30912} - 1 - c \cdot 10^{1266} \cdot R_{30912}/R_{2576}$$

where $c = 40044000444004004400000000440040044400044004$.

Charles Nicol and John Selfridge ask if the sequence of concatenations of the natural numbers in base 10,

$$1, 12, 123, 1234, 12345, 123456, 1234567, 12345678, 123456789,$$

contains infinitely many primes. Robert Baillie has found that there are no such primes out to $n = 1000$. Nicol & Mike Filaseta find that the first prime among the reverse concatenations is

$$82818079787776\cdots1110987654321$$

The questions can also be asked for other scales of notation; for example $123456101112 13_7 = 1318 70666077_{10}$ is prime.

It is hard to put an upper bound on the number of questions that one may ask about the digits of primes. De Koninck asks for a proof that for $k \geq 2$, k not a multiple of 3, there is always a prime whose (decimal) digits sum to k, and that for $k \geq 4$ there are infinitely many such. If $\rho(k)$ is the smallest prime with digital sum k, is $\rho(k) \equiv 9 \pmod{10}$ for $k > 25$? Is it $\equiv 99 \pmod{100}$ for $k > 38$? And $\equiv 999 \pmod{1000}$ for $k > 59$?

David Wilson asks if there are infinitely many **deletable primes**, i.e., primes from which a decimal digit may be deleted, leaving either the empty set or another deletable prime. If so, what proportion of primes are deletable? Here are the first sixty deletable primes:

$$2 \quad 3 \quad 5 \quad 7 \quad 13 \quad 17 \quad 23 \quad 29 \quad 31 \quad 37 \quad 43 \quad 47 \quad 53 \quad 59 \quad 67$$
$$71 \quad 73 \quad 79 \quad 83 \quad 97 \quad 103 \, 107 \, 113 \, 127 \, 131 \, 137 \, 139 \, 157 \, 163 \, 167$$
$$173 \, 179 \, 193 \, 197 \, 223 \, 229 \, 233 \, 239 \, 263 \, 269 \, 271 \, 283 \, 293 \, 307 \, 311$$
$$313 \, 317 \, 331 \, 337 \, 347 \, 353 \, 359 \, 367 \, 373 \, 379 \, 383 \, 397 \, 431 \, 433 \, 439$$

Richard McIntosh notes that the largest known prime of shape $n = (2^{4p} + 1)/17$ has $p = 317$ and that n is composite for all primes p with $317 < p < 10000$. Are there any more such primes?

Wagstaff observes that the only primes < 180 for which $(p^p - 1)/(p-1)$ is prime are $p = 2, 3, 19$ and 31; for $(p^p + 1)/(p + 1)$ they are $p = 3, 5, 17$ and 157. In our second edition it was misstated that $(7^7 - 1)/(7 - 1)$ is prime.

The **Fermat numbers**, $F_n = 2^{2^n} + 1$ are also of continuing interest; they are prime for $0 \le n \le 4$ and composite for $5 \le n \le 32$ and for many larger values of n. Hardy & Wright give a heuristic argument which suggests that only a finite number of them are prime.

It has been conjectured that the Fermat numbers are squarefree. It was verified by Gostin & McLaughlin that 82 of the 85 then known factors of the 71 known composite Fermat numbers were not repeated. At least 250 prime factors of 214 different Fermat numbers are now known. Lenstra, Lenstra, Manasse & Pollard have completely factored the ninth, and R. P. Brent the eleventh, Fermat number. The author has bet John Conway that another Fermat number will be completely factored before 2016-09-30. If not, he pays up on his 100th birthday, so, to echo Sam Wagstaff, keep those factors coming!

$F_5 = 641 \cdot 6700417$
$F_6 = 274177 \cdot 67280421310721$
$F_7 = 59649589127497217 \cdot 5704689200685129054721$
$F_8 = 1238926361552897 \cdot 93461639715357977769163558199606896584051237541638188580280321$
$F_9 = 2424833 \cdot 7455602825647884208337395736200454918783366342659 \cdot p_{99}$
$F_{10} = 45592577 \cdot 6487031809 \cdot 4659775785220018543264560743076778192897 \cdot p_{252}$
$F_{11} = 319489 \cdot 974849 \cdot 167988556341760475137 \cdot 3560841906445833920513 \cdot p_{564}$
$F_{12} = 114689 \cdot 26017793 \cdot 63766529 \cdot 190274191361 \cdot 1256132134125569 \cdot c_{1187}$
$F_{13} = 2710954639361 \cdot 2663848877152141313 \cdot$
$\qquad 3603109844542291969 \cdot 319546020820551643220672513 \cdot c_{2391}$
$F_{14} = c_{4933}$
$F_{15} = 1214251009 \cdot 2327042503868417 \cdot 168768817029516972383024127016961 \cdot c_{9808}$
$F_{16} = 825753601 \cdot c_{19720} \quad F_{17} = 31065037602817 \cdot c_{39444}$
$F_{18} = 13631489 \cdot c_{78906} \quad F_{19} = 70525124609 \cdot 646730219521 \cdot c_{157804}$

where p_n, c_n respectively denote n-digit prime and composite numbers.

Because of their special interest as potential factors of Fermat numbers, and because proofs of their primality are comparatively easy, numbers of the form $k \cdot 2^n + 1$ have received special attention, at least for small values of k. For example, large primes were found by Harvey Dubner and Wilfrid Keller, the record on 84-09-05 for a non-Mersenne prime being $(k, n) = (5, 23473)$ by Keller. Another of his discoveries, $(k, n) = (289, 18502)$ is amusing in that it may be written as $(18496, 18496)$, a Cullen prime (**B20**) and as $(17 \cdot 2^{9251})^2 + 1$, a prime of shape $a^2 + 1$ (**A1**).

As we mentioned, the record has since been beaten with a prime of shape $k \cdot 2^n - 1$ with $(k, n) = (391581, 216193)$. See also **B21**.

On 2003-02-22 John Cosgrave discovered that $3 \cdot 2^{2145353} + 1$ divides $F_{2145351}$. This was the largest known prime that is not a Mersenne prime, but on 2003-10-12 he found that $3 \cdot 2^{2478785} + 1$ divides $F_{2478782}$.

Hugh Williams has found, for $r = 3$, 5, 7 and 11, all values of $n \leq 500$ for which $(r - 1)r^n - 1$ is prime:

$r = 3 \quad n = 1, 2, 3, 7, 8, 12, 20, 23, 27, 35, 56, 62, 68, 131, 222, 384, 387$

$r = 5 \quad n = 1, 3, 9, 13, 15, 25, 39, 69, 165, 171, 209, 339$

$r = 7 \quad n = 1, 2, 7, 18, 55, 69, 87, 119, 141, 189, 249, 354$

$r = 11 \quad n = 1, 3, 37, 119, 255, 355, 371, 497$

Pi Xin-Ming found the generalized Fermat prime $1632^{1024} + 1$.

The only known primes of shape $n^n + (n{+}1)^n$ are 3, 13 and 881; since n must be a power of 2, the analogy with Fermat primes suggests that there are no more; there are no others with $n < 2^{15}$. Cino Hilliard reports that $(n{+}1)^n - n^{n-1}$ is prime for $n = 2$, 5, 6, 9, 24; and $(n{+}1)^n + n^{n-1}$ for $n = 2$, 3, 8, 9, 15, 16, 219.

We are very unlikely to know for sure that the **Fibonacci sequence**

$$1, 1, \mathbf{2}, \mathbf{3}, \mathbf{5}, 8, \mathbf{13}, 21, 34, 55, \mathbf{89}, 144, \mathbf{233}, 377, 610, 987, \mathbf{1597}, \ldots,$$

where $u_1 = u_2 = 1$ and $u_{r+1} = u_r + u_{r-1}$, contains infinitely many primes. Similarly for the related **Lucas sequence**

$$1, \mathbf{3}, 4, \mathbf{7}, \mathbf{11}, 18, \mathbf{29}, \mathbf{47}, 76, 123, \mathbf{199}, 322, \mathbf{521}, 843, 1364, \ldots,$$

in which L_r is prime for $r = 0$, 2, 4, 5, 7, 8, 11, 13, 16, 17, 19, 31, 37, 41, 47, 53, 61, 71, 79, 113, 313, 353, 503, 613, 617, 863, 1097, 1361, 4787, 4793, 5851, 7741, 8467, 10691, 12251, 13963, 14449, 19469, 35449, 36779, 44507, 51169, 56003, 81671, 89849, 94823, 140057, 148091, 159521, 183089, 193201, 202667 (the last ten being only 'probable primes'). It is believed that most Lucas Lehmer sequence@Lucas-Lehmer sequences defined by second-order recurrence relations with $u_1 \perp u_2$ also contain an infinity of primes. However, Graham has shown that the sequence with

$$u_0 = 1786\,772701\,928802\,632268\,715130\,455793$$

$$u_1 = 1059\,683225\,053915\,111058\,165141\,686995$$

contains no primes at all! Knuth notes that Graham's numbers should have been given as

$$u_0 = 331\,635635\,998274\,737472\,200656\,430763$$

$$u_1 = 1510\,028911\,088401\,971189\,590305\,498785$$

and gives the smaller example

$$u_1 = 49463\,435743\,205655, \quad u_2 = 62638\,280004\,239857$$

John Nicol gives an even smaller example $u_0 = 8983542533631 = 3 \cdot 223 \cdot 13428314699$, $u_1 = 248272649660939 = 17 \cdot 155081 \cdot 94171907$. An almost as small pair, $u_1 = 3794765361567513$, $u_2 = 20615674205555510$, had been earlier found by Wilf (letter, *Math. Mag.*, **63**(1990) 284).

Raphael Robinson considers the Lucas sequence (sometimes called the **Pell sequence**) $u_0 = 0$, $u_1 = 1$, $u_{r+1} = 2u_r + u_{r-1}$ and defines the **primitive part**, L_r, by

$$u_r = \prod_{d|r} L_d$$

He notes that $L_7 = 13^2$ and $L_{30} = 31^2$ and asks if there is any larger r for which L_r is a square.

The following table shows the ranks, r, of the Fibonacci numbers u_r which are prime. $f = e^\gamma \log_\tau r$, where τ is the golden ratio. Row M shows values of M for which $2^M - 1$ is prime. $g = e^\gamma \log_2 M$

n	1	2	3	4	5	6	7	8	9	10	11	12	13	14	15	16	17	18
r	3	4	5	7	11	13	17	23	29	43	47	83	131	137	359	431	433	449
f	4.1	5.1	6.0	7.2	8.9	9.5	10.5	11.6	12.5	13.9	14.3	16.4	18.0	18.2	21.8	22.5	22.5	22.6
M	2	3	5	7	13	17	19	31	61	89	107	127	521	607	1279	2203	2281	3217
g	1.8	2.8	4.1	5.0	6.6	7.3	7.6	8.8	10.6	11.5	12.0	12.4	16.1	16.5	18.4	19.8	19,9	20.8

n	19	20	21	22	23	24	25	26	27	28	29	30
r	509	569	571	2971	4723	5387	9311	9677	14431	25561	30757	35999
f	23.1	23.5	23.5	29.6	31.3	31.8	33.8	34.0	35.4	37.6	38.2	38.8
M	4253	4423	9689	9941	11213	19937	21701	23209	44497	86243	110503	132049
g	21.5	21.6	23.6	23.7	24.0	25.4	25.7	25.8	27.5	29.2	29.8	30.3

n	31	32	33	34	35	36	37	38	39	40
r	37511	50833	81839	104911	130021	148091	201107			
f	39.0	40.1	41.9	42.8	43.6	44.1	45.2			
M	216091	756839	859433	1257787	1398269	2976221	3021377	6972593	13466917	20996011
g	31.6	34.8	35.1	36.1	36.4	38.3	38.3	40.5	42.2	43.2

Takashi Agoh, On Fermat and Wilson quotients, *Exposition. Math.*, **14**(1996) 145–170; *MR* **97e**:11008.

Takashi Agoh, Karl Dilcher & Ladislav Skula, Fermat quotients for composite moduli, *J. Number Theory*, **66**(1997) 29–50; *MR* **98h**:11002.

R. C. Archibald, *Scripta Math.*, **3**(1935) 117.

A. O. L. Atkin & N. W. Rickert, Some factors of Fermat numbers, *Abstracts Amer. Math. Soc.*, **1**(1980) 211.

P. T. Bateman, J. L. Selfridge & S. S. Wagstaff, The new Mersenne conjecture, *Amer. Math. Monthly*, **96**(1989) 125–128; *MR* **90c**:11009.

Anders Björn & Hans Riesel, Factors of generalized Fermat numbers, *Math. Comput.*, **67**(1998) 441–446; *MR* **98e**:11008.

Wieb Bosma, Explicit primality criteria for $h \cdot 2^k \pm 1$, *Math. Comput.*, **61**(1993) 97–109.

R. P. Brent, Factorization of the eleventh Fermat number, *Abstracts Amer. Math. Soc.*, **10**(1989) 89T-11-73.

R. P. Brent, R. E. Crandall, K. Dilcher & C. van Halewyn, Three new factors of Fermat numbers, *Math. Comput.*, **69**(2000) 1297–1304; *MR* **2000j**:11194.

R. P. Brent & J. M. Pollard, Factorization of the eighth Fermat number, *Math. Comput.*, **36**(1981) 627–630; *MR* **83h**:10014.

John Brillhart & G. D. Johnson, On the factors of certain Mersenne numbers, I, II *Math. Comput.*, **14**(1960) 365–369, **18**(1964) 87–92; *MR* **23**#A832, **28**#2992.

John Brillhart, D. H. Lehmer, J. L. Selfridge, Bryant Tuckerman & Samuel S. Wagstaff, Factorizations of $b^n \pm 1$, $b = 2, 3, 5, 6, 7, 10, 11, 12$ up to high powers, *Contemp. Math.*, **22**. Amer. Math. Soc., Providence RI, 1983, 1988; *MR* **84k**:10005, **90d**:11009.

John Brillhart, Peter L. Montgomery & Robert D. Silverman, Tables of Fibonacci and Lucas factorizations, *Math. Comput.*, **50**(1988) 251–260 & S1–S15.

John Brillhart, J. Tonascia & P. Weinberger, On the Fermat quotient, in *Computers in Number Theory*, Academic Press, 1971, 213–222.

Chris K. Caldwell & Yves Gallot, On the primality of $n! \pm 1$ and $2 \times 3 \times 5 \times \cdots \times p \pm 1$, *Math. Comput.*, **71**(2002) 441–448 (electronic).

W. N. Colquitt & L. Welsh, A new Mersenne prime, *Math. Comput.*, **56**(1991) 867–870; *MR* **91h**:11006.

Curtis Cooper & Manjit Parihar, On primes in the Fibonacci and Lucas sequences, *J. Inst. Math. Comput. Sci.*, **15**(2002) 115–121.

R. Crandall, K. Dilcher & C. Pomerance, A search for Wieferich and Wilson primes, *Math. Comput.*, **66**(1997) 433–449; *MR* **97c**:11004.

Richard E. Crandall, J. Doenias, C. Norrie & Jeff Young, The twenty-second Fermat Number is composite, *Math. Comput.*, **64**(1995) 863–868; *MR* **95f**:11104.

Richard E. Crandall, Ernst W. Mayer & Jason S. Papadopoulos, The twenty-fourth Fermat number is composite, *Math. Comput.*, **72**(2003) 1555–1572; *MR* **2004c**:11236.

Karl Dilcher, Fermat numbers, Wieferich and Wilson primes: computations and generalizations, *Public-key cryptography and computational number theory* (*Warsaw* 2000) 29–48, de Gruyter, Berlin, 2001; *MR* **2002j**:11004.

Harvey Dubner, Generalized Fermat numbers, *J. Recreational Math.*, **18** (1985–86) 279–280.

Harvey Dubner, Generalized repunit primes, *Math. Comput.*, **61** (1993) 927–930.

Harvey Dubner, Repunit R_{49081} is a probable prime, *Math. Comput.*, **71**(2002) 833–835; *MR* **2002k**:11224.

Harvey Dubner & Yves Gallot, Distribution of generalized Fermat prime numbers, *Math. Comput.*, **71**(2002) 825–832; *MR* **2002j**:11156.

Harvey Dubner & Wilfrid Keller, Factors of generalized Fermat numbers, *Math. Comput.*, **64**(1995) 397–405; *MR* **95c**11010.

Harvey Dubner & Wilfrid Keller, New Fibonacci and Lucas primes, *Math. Comput.*, **68**(1999) 417–427, S1–S12; *MR* **99c**11008.

John R. Ehrman, The number of prime divisors of certain Mersenne numbers, *Math. Comput.*, **21**(1967) 700–704; *MR* **36**#6368.

Peter Fletcher, William Lindgren & Carl Pomerance, Symmetric and asymmetric primes, *J. Number Theory*, **58**(1996) 89–99; *MR* **97c**:11007.

Donald B. Gillies, Three new Mersenne primes and a statistical theory, *Math. Comput.*, **18**(1964) 93–97; *MR* **28**#2990.

Gary B. Gostin, New factors of Fermat numbers, *Math. Comput.*, **64**(1995) 393–395; *MR* **95c**:11151.

Gary B. Gostin & Philip B. McLaughlin, Six new factors of Fermat numbers, *Math. Comput.*, **38**(1982) 645–649; *MR* **83c**:10003.

R. L. Graham, A Fibonacci-like sequence of composite numbers, *Math. Mag.*, **37**(1964) 322–324; *Zbl* **125**, 21.

Richard K. Guy, The strong law of small numbers, *Amer. Math. Monthly* **95**(1988) 697–712; *MR* **90c**:11002 (see also *Math. Mag.*, **63**(1990) 3–20; *MR* **91a**:11001.

Kazuyuki Hatada, Mod 1 distribution of Fermat and Fibonacci quotients and values of zeta functions at $2 - p$, *Comment. Math. Univ. St. Paul.* **36**(1987), no. 1, 41–51; *MR* **88i**:11085.

G. L. Honaker & Chris K. Caldwell, Palindromic prime pyramids, *J. Recreational Math.*, **30**(1999-2000) 169–176.

A. Grytczuk, M. Wójtowicz & Florian Luca, Another note on the greatest prime factors of Fermat numbers, *Southeast Asian Bull. Math.*, **25**(2001) 111–115; *MR* **2002g**:11008.

Wilfrid Keller, Factors of Fermat numbers and large primes of the form $k.2^n + 1$, *Math. Comput.*, **41**(1983) 661–673; *MR* **85b**:11117.

Wilfrid Keller, New factors of Fermat numbers, *Abstracts Amer. Math. Soc.*, **5**(1984) 391

Wilfrid Keller, The 17th prime of the form $5.2^n + 1$, *Abstracts Amer. Math. Soc.*, **6**(1985) 121.

Christoph Kirfel & Øystein J. Rødseth, On the primality of $wh \cdot 3^h + 1$, Selected papers in honor of Helge Tverberg, *Discrete Math.*, **241**(2001) 395–406.

Joshua Knauer & Jörg Richstein, The continuing search for Weiferich primes, *Math. Comput.*, (2004)

Donald E. Knuth, A Fibonacci-like sequence of composite numbers, *Math. Mag.*, **63**(1990) 21–25; *MR* **91e**:11020.

M. Kraitchik, *Sphinx*, 1931, 31.

Michal Křížek & Jan Chleboun, Is any composite Fermat number divisible by the factor $5h2^n + 1$, *Tatra Mt. Math. Publ.*, **11**(1997) 17–21; *MR* **98j**:11003.

Michal Křížek & Lawrence Somer, A necessary and sufficient condition for the primality of Fermat numbers, *Math. Bohem.*, **126**(2001) 541–549; *MR* **2004a**:11004.

D. H. Lehmer, *Sphinx*, 1931, 32, 164.

D. H. Lehmer, On Fermat's quotient, base two, *Math. Comput.*, **36**(1981) 289–290; *MR* **82e**:10004.

A. K. Lenstra, H. W. Lenstra, M. S. Manasse & J. M. Pollard, The factorization of the ninth Fermat number, *Math. Comput.*, **61**(1993) 319–349; *MR* **93k**:11116.

J. C. P. Miller & D. J. Wheeler, Large prime numbers, *Nature*, **168**(1951) 838.

Satya Mohit & M. Ram Murty, Wieferich primes and Hall's conjecture, *C.R. Math. Acad. Sci. Soc. Roy. Can.*, **20**(1998) 29–32; *MR* **98m**:11004.

Peter L. Montgomery, New solutions of $a^{p-1} \equiv 1 \bmod p^2$, *Math. Comput.*, **61**(1993) 361–363; *MR* **94d**:11003.

Pieter Moree & Peter Stevenhagen, Prime divisors of the Lagarias sequence, *J. Théor. Nombres Bordeaux*, **13**(2001) 241–251; *MR* **2002c**:11016.

Thorkil Naur, New integer factorizations, *Math. Comput.*, **41**(1983) 687–695; *MR* **85c**:11123.

Rudolf Ondrejka, Titanic primes with consecutive like digits, *J. Recreational Math.*, **17**(1984-85) 268–274.

Pi Xin-Ming, Generalized Fermat primes for $b \leq 2000$, $m \leq 10$, *J. Math. (Wuhan)* **22**(2002) 91–93; *MR* **2003a**:11008.

Paulo Ribenboim, On square factors of terms of binary recurring sequences and the *ABC* conjecture, *Publ. Math. Debrecen*, **59**(2001) 459–469; *MR* **2002i**:11033.

Herman te Riele, Walter Lioen & Dik Winter, Factorization beyond the googol with MPQS on a single computer, *CWI Quarterly*, **4**(1991) 69–72.

Hans Riesel & Anders Bjørn, Generalized Fermat numbers, in *Mathematics of Computation 1943–1993*, Proc. Symp. Appl. Math., **48**(1994) 583–587; *MR* **95j**:11006.

Raphael M. Robinson, Mersenne and Fermat numbers, *Proc. Amer. Math. Soc.*, **5**(1954) 842–846; *MR* **16**, 335.

A. Rotkiewicz, Note on the diophantine equation $1 + x + x^2 + \ldots + x^n = y^m$, *Elem. Math.*, **42**(1987) 76.

Daniel Shanks & Sidney Kravitz, On the distribution of Mersenne divisors, *Math. Comput.*, **21**(1967) 97–100; *MR* **36**:#3717.

Hiromi Suyama, Searching for prime factors of Fermat numbers with a microcomputer (Japanese) *bit*, **13**(1981) 240-245; *MR* **82c**:10012.

Hiromi Suyama, Some new factors for numbers of the form $2^n \pm 1$, 82T-10-230 *Abstracts Amer. Math. Soc.*, **3**(1982) 257; IV, **5**(1984) 471.

Hiromi Suyama, The cofactor of F_{15} is composite, 84T-10-299 *Abstracts Amer. Math. Soc.*, **5**(1984) 271.

Paul M. Voutier, Primitive divisors of Lucas and Lehmer sequences, *Math. Comput.*, **64**(1995) 869–888; II *J. Théorie des Nombres de Bordeaux*, **8**(1996) 251–274; III *Math. Proc. Cambridge Philos. Soc.*, **123**(1998) 407–419; *MR* **95f**:11022, **98h**:11037, **99b**:11027.

Samuel S. Wagstaff, Divisors of Mersenne numbers, *Math. Comput.*, **40**(1983) 385–397; *MR* **84j**:10052.

Samuel S. Wagstaff, The period of the Bell exponential integers modulo a prime, in *Math. Comput. 1943–1993* (Vancouver, 1993), Proc. Sympos. Appl. Math., **48**(1994) 595–598; *MR* **95m**:11030.

H. C. Williams, The primality of certain integers of the form $2Ar^n - 1$, *Acta Arith.*, **39**(1981) 7–17; *MR* **84h**:10012.

H. C. Williams, How was F_6 factored? *Math. Comput.*, **61**(1993) 463–474.

Samuel Yates, Titanic primes, *J. Recreational Math.*, **16**(1983-84) 265–267.

Samuel Yates, Sinkers of the Titanics, *J. Recreational Math.*, **17**(1984-85) 268–274.

Samuel Yates, Tracking Titanics, in *The Lighter Side of Mathematics*, Proc. Strens Mem. Conf., Calgary 1986, Math. Assoc. of America, Washington DC, *Spectrum* series, 1993, 349–356.

Jeff Young, Large primes and Fermat factors, *Math. Comput.*, **67**(1998) 1735–1738; *MR* **99a**:11010.

Jeff Young & Duncan Buell, The twentieth Fermat number is composite, *Math. Comput.*, **50**(1988) 261–263.

OEIS: A000215, A001220, A001771, A002235, A003307, A005541, A007540, A007908, A008555, A019434, A046865-046867, A014491, A049093-049096, A050922, A051179, A056725, A056826, A079906-079907, A080603, A080608.

A4 The prime number race.

A number a is said to be **congruent** to c, **modulo** a positive number b, written $a \equiv c \bmod b$, if b is a divisor of $a - c$. S. Chowla conjectured that if $a \perp b$, then there are infinitely many pairs of consecutive primes such that $p_n \equiv p_{n+1} \equiv a \bmod b$. The case $b = 4$, $a = 1$ follows from a theorem of Littlewood. Bounds between which such consecutive primes occur have been given in this case, and for $b = 4$, $a = 3$ by Knapowski & Turán. Turán observed that it would be of interest (in connexion with the Riemann hypothesis, for example) to discover long sequences of consecutive primes $\equiv 1 \bmod 4$. Den Haan found the nine primes

11593, 11597, 11617, 11621, 11633, 11657, 11677, 11681, 11689.

Four sequences of 10 such primes end at 373777, 495461, 509521 and 612217 and a sequence of 11 ends at 766373. Stephane Vandemergel has discovered no fewer than 16 consecutive primes of shape $4k + 1$; They are $207622000 + 273$, 297, 301, 313, 321, 381, 409, 417, 421, 489, 501, 517, 537, 549, 553, 561.

Thirteen consecutive primes congruent to 3 mod 4 are $241000 + 603$, 639, 643, 651, 663, 667, 679, 687, 691, 711, 727, 739 and 771.

If $p(b, a)$ is the least prime in the arithmetic progression $a + nb$, with $a \perp b$, then Linnik showed that there is a constant L, now called **Linnik's constant** such that $p(b, a) \ll b^L$. Pan Cheng-Tung, Chen Jing-Run, Matti Jutila, Chen Jing-Run, Matti Jutila, S. Graham, Chen Jing-Run, Chen Jing-Run & Liu Jian-Min, and Wang Wei₃ have successively improved the best known value of L to 5448, 777, 550, 168, 80, 36, 17, 13.5, and 8, and Heath-Brown has recently established the remarkable result $L \leq 5.5$.

Elliott & Halberstam have shown that

$$p(b, a) < \phi(b)(\ln b)^{1+\delta}$$

almost always.

In the other direction it is known (see the papers of Prachar, Schinzel and Pomerance) that, given a, there are infinitely many values of b for which

$$p(b, a) > \frac{cb \ln b \ln \ln b \ln \ln \ln \ln b}{(\ln \ln \ln b)^2}$$

where c is an absolute constant.

Turán was particularly interested in the **prime number race**. Let $\pi(n; a, b)$ be the number of primes $p \leq n$, $p \equiv a \bmod b$. Is it true that for every a and b with $a \perp b$, there are infinitely many values of n for which

$$\pi(n; a, b) > \pi(n; a_1, b)$$

for every $a_1 \not\equiv a \bmod b$? Knapowski & Turán settled special cases, but the general problem is wide open.

Chebyshev noted that $\pi(n; 1, 3) < \pi(n; 2, 3)$ and $\pi(n; 1, 4) \leq \pi(n; 3, 4)$ for small values of n. Leech, and independently Shanks & Wrench, discovered that the latter inequality is reversed for $n = 26861$ and Bays & Hudson that the former is reversed for two sets, each of more than 150 million integers, between $n = 608981813029$ and $n = 610968213796$.

Bays & Hudson found that $\pi(n; 1, 4) > \pi(n; 3, 4)$ for $n = 1488478427089$.

Bach & Sorenson have shown, under the assumption of the generalized Riemann hypothesis, that if m, q are integers with $m \perp q$, then then there is a prime $p \equiv m \bmod q$ satisfying $p < 1.7(q \ln q)^2$.

Carlos B. Rivera F. has found 19 consecutive primes of shape $4k + 1$: $297779117 + 4c$ for $c = 0, 11, 14, 15, 18, 20, 21, 24, 33, 45, 48, 53, 56, 69, 71,$ 80, 90, 98, 99, and 20 consecutive primes of shape $4k - 1$: $727334879 + 4c$ for $c = 0, 2, 3, 5, 15, 21, 26, 36, 38, 50, 51, 57, 62, 63, 71, 83, 87, 117, 120,$ 125.

Eric Bach & Jonathan Sorenson, Explicit bounds for primes in residue classes, in *Math. Comput. 1943-1993* (Vancouver, 1993), Proc. Sympos. Appl. Math., (1994) –.

Carter Bays & Richard H. Hudson, The appearance of tens of billions of integers x with $\pi_{24,13}(x) < \pi_{24,1}(x)$ in the vicinity of 10^{12}, *J. reine angew. Math.*, **299/300**(1978) 234–237; *MR* **57** #12418.

Carter Bays & Richard H. Hudson, Details of the first region of integers x with $\pi_{3,2}(x) < \pi_{3,1}(x)$, *Math. Comput.*, **32**(1978) 571–576; *MR* **57** #16175.

Carter Bays & Richard H. Hudson, Numerical and graphical description of all axis crossing regions for the moduli 4 and 8 which occur before 10^{12}, *Internat. J. Math. Sci.*, **2**(1979) 111–119; *MR* **80h**:10003.

Carter Bays, Kevin Ford, Richard H. Hudson & Michael Rubinstein, Zeros of Dirichlet L-functions near the real axis and Chebyshev's bias, *J. Number Theory*, **87**(2001) 54–76; *MR* **2001m**:11148.

P. L. Chebyshev, Lettre de M. le Professeur Tschébychev à M. Fuss sur un nouveaux Théorème relatif aux nombres premiers contenus dans les formes $4n+1$ et $4n + 3$, *Bull. Classe Phys. Acad. Imp. Sci. St. Petersburg*, **11**(1853) 208.

Chen Jing-Run, On the least prime in an arithmetical progression, *Sci. Sinica*, **14**(1965) 1868–1871; *MR* **32** #5611.

Chen Jing-Run, On the least prime in an arithmetical progression and two theorems concerning the zeros of Dirichlet's L-functions, *Sci. Sinica*, **20**(1977) 529–562; *MR* **57** #16227.

Chen Jing-Run, On the least prime in an arithmetical progression and theorems concerning the zeros of Dirichlet's L-functions II, *Sci. Sinica*, **22**(1979) 859–889; *MR* **80k**:10042.

Chen Jing-Run & Liu Jian-Min, On the least prime in an arithmetic progression III, IV, *Sci. China Ser. A*, **32**(1989) 654–673, 792–807; *MR* **91h**:11090ab.

Andrey Feuerverger & Greg Martin, Biases in the Shanks-Rényi prime number race, *Experiment. Math.*, **9**(2000) 535–570; **2002d**:11111.

K. Ford & R. H. Hudson, Sign changes in $\pi_{q,a}(x) - \pi_{q,b}(x)$, *Acta Arith.*, **100**(2001) 297–314; *MR* **2003a**:11121.

Kevin Ford & Sergeĭ Konyagin, The prime number race and zeros of L-functions off the critical line, *Duke Math. J.*, **113**(2002) 313–330; *MR* **2003i**:11135.

S. Graham, On Linnik's constant, *Acta Arith.*, **39**(1981) 163–179; *MR* **83d**:10050.

Andrew Granville & Carl Pomerance, On the least prime in certain arithmetic progressions, *J. London Math. Soc.*(2), **41**(1990) 193–200; *MR* **91i**:11119.

D. R. Heath-Brown, Siegel zeros and the least prime in an arithmetic progression, *Quart. J. Math. Oxford Ser.*(2), **41**(1990) 405–418; *MR* **91m**:11073.

D. R. Heath-Brown, Zero-free regions for Dirichlet L-functions and the least prime in an arithmetic progression, *Proc. London Math. Soc.*(3), **64**(1992) 265–338; *MR* **93a**:11075.

Richard H. Hudson, A common combinatorial principle underlies Riemann's formula, the Chebyshev phenomenon, and other subtle effects in comparative prime number theory I, *J. reine angew. Math.*, **313**(1980) 133–150.

Richard H. Hudson, Averaging effects on irregularities in the distribution of primes in arithmetic progressions, *Math. Comput.*, **44**(1985) 561–571; *MR* **86h**:11064.

Richard H. Hudson & Alfred Brauer, On the exact number of primes in the arithmetic progressions $4n \pm 1$ and $6n \pm 1$, *J. reine angew. Math.*, **291**(1977) 23–29; *MR* **56** #283.

Matti Jutila, A new estimate for Linnik's constant, *Ann. Acad. Sci. Fenn. Ser. A I No.* 471 (1970), 8pp.; *MR* **42** #5939.

Matti Jutila, On Linnik's constant, *Math. Scand.*, **41**(1977) 45–62; *MR* **57** #16230.

Jerzy Kaczorowski, A contribution to the Shanks-Rényi race problem, *Quart. J. Math. Oxford Ser.* (2) **44**(1993) 451–458; *MR* **94m**:11105.

Jerzy Kaczorowski, On the Shanks–Rényi race problem mod 5, *J. Number Theory* **50**(1995) 106–118; *MR* **94m**:11105.

Jerzy Kaczorowski, On the distribution of primes (mod 4), *Analysis*, **15**(1995) 159–171; *MR* **96h**:11095.

Jerzy Kaczorowski, On the Shanks–Rényi race problem, *Acta Arith.*, **74**(1996) 31–46; *MR* **96k**:11113.

S. Knapowski & P. Turán, Über einige Fragen der vergleichenden Primzahltheorie, *Number Theory and Analysis*, Plenum Press, New York, 1969, 157–171.

S. Knapowski & P. Turán, On prime numbers $\equiv$ 1 resp. 3 mod 4, *Number Theory and Algebra*, Academic Press, New York, 1977, 157–165; *MR* **57** #5926.

John Leech, Note on the distribution of prime numbers, *J. London Math. Soc.*, **32**(1957) 56–58.

U. V. Linnik, On the least prime in an arithmetic progression I. The basic theorem. II. The Deuring-Heilbronn phenomenon. *Rec. Math. [Mat. Sbornik] N.S.*, **15(57)**(1944) 139–178, 347–368; *MR* **6**, 260bc.

J. E. Littlewood, Distribution des nombres premiers, *C.R. Acad. Sci. Paris*, **158**(1914) 1869–1872.

Greg Martin, Asymmetries in the Shanks-Rényi prime number race, *Number theory for the millenium, II (Urbana IL, 2000)* 403–415, AKPeters, Natick MA,2002; *MR* **2004b**:11134.

Pieter Moree, Chebyshev's bias for composite numbers, *Math. Comput.*, **73** (2004) 425–449.

Pan Cheng-Tung, On the least prime in an arithmetic progression, *Sci. Record* (*N.S.*) **1**(1957) 311–313; *MR* **21** #4140.

Carl Pomerance, A note on the least prime in an arithmetic progression, *J. Number Theory*, **12**(1980) 218–223; *MR* **81m**:10081.

K. Prachar, Über die kleinste Primzahl einer arithmetischen Reihe, *J. reine angew. Math.*, **206**(1961) 3–4; *MR* **23** #A2399; and see Andrzej Schinzel, Remark on the paper of K. Prachar, **210**(1962) 121–122; *MR* **27** #118.

Imre Z. Ruzsa, Consecutive primes modulo 4, *Indag. Math. (N.S.)*, **12**(2001) 489–503.

Daniel Shanks, Quadratic residues and the distribution of primes, *Math. Tables Aids Comput.*, **13**(1959) 272–284.

Wang Wei$_3$, On the least prime in an arithmetic progression, *Acta Math. Sinica*(**N.S.**), **7**(1991) 279–289; *MR* **93c**:11073.

OEIS: A035498, A035518, A035525.

A5 Arithmetic progressions of primes.

How long can an arithmetic progression be which consists only of primes? Table 1 shows progressions of n primes, a, $a+d$, ..., $a+(n-1)d$, discovered by James Fry, V.A. Golubev, Andrew Moran, Paul Pritchard, S.C. Root, W.N. Seredinskii, S. Weintraub, Jeff Young and Anthony Thyssen (see the first edition for earlier, smaller discoveries). Of course, the common difference must have every prime $p \leq n$ as a divisor (unless $n = a$). It is conjectured that n can be as large as you like. This would follow if it were possible to improve Szemerédi's theorem (see **E10**).

STOP PRESS (04-04-09): Ben Green & Terence Tao have made just such an improvement, and it seems virtually certain that they have proved that you can indeed have arbitrarily long arithmetic progressions of primes.

More generally, Erdős conjectures that if $\{a_i\}$ is any infinite sequence of integers for which $\sum 1/a_i$ is divergent, then the sequence contains arbitrarily long arithmetic progressions. He offered $3000.00 for a proof or disproof of this conjecture.

In 1993 Paul Pritchard coordinated an effort which discovered 22 primes in arithmetic progression, starting with $a = 11410337850553$ and having common difference $d = 4609098694200$,i.e., $a + 21d = 108201410428753$.

Sierpiński defines $g(x)$ to be the maximum number of terms in a progression of primes not greater than x. The least x, $l(x)$, for which $g(x)$ takes the values

$$g(x) = 0 \quad 1 \quad 2 \quad 3 \quad 4 \quad 5 \quad 6 \quad 7 \quad 8 \quad 9 \quad 10 \quad \ldots$$

is

$$l(x) = 1 \quad 2 \quad 3 \quad 7 \quad 23 \quad 29 \quad 157 \quad 1307 \quad 1669 \quad 1879 \quad 2089 \quad \ldots$$

Günter Löh has searched for arithmetic progressions of primes with first term q and length q. Examples are $(q, d) = (7, 150), (11, 1536160080)$ and $(13, 9918821194590)$. On 2001-11-07 Phil Carmody announced his find of $(17, 341976204789992332560)$.

Table 1. Long Arithmetic Progressions of Primes.

n	d	a	$a + (n-1)d$	source
12	30030	23143	353473	G, 1958
13	510510	766439	6892559	S, 1965
14	2462460	46883579	78895559	
16	9699690	53297929	198793279	
16	223092870	2236133941	5582526991	R, 1969
17	87297210	3430751869	4827507229	W, 1977
18	717777060	4808316343	17010526363	P
19	4180566390	8297644387	83547839407	P
19	13608665070	244290205469	489246176729	F, Mar 1987
20	2007835830	803467381001	841616261771	F, Mar 1987
20	7643355720	1140997291211	1286221049891	F, Mar 1987
20	18846497670	214861583621	572945039351	Y&F, 87-09-01
20	1140004565700	1845449006227	23505535754527	M&P, Nov 1990
20	19855265430	24845147147111	25222397190281	M&P, Nov 1990
21	1419763024680	142072321123	28537332814723	M&P, 90-11-30
22	4609098694200	11410337850553	108201410428753	MPT, Mar 1993

Pomerance produces the "prime number graph" by plotting the points (n, p_n) and shows that for every k we can find k primes whose points are collinear.

Grosswald has shown that there are long arithmetic progressions consisting only of **almost primes**, in the following sense. There are infinitely many arithmetic progressions of k terms, each term being the product of at most r primes, where

$$r \leq \lfloor k \ln k + 0.892k + 1 \rfloor.$$

He has also shown that the Hardy-Littlewood estimate is of the right order of magnitude for 3-term arithmetic progressions of primes.

Jim Fougeron & Hans Rosenthal have found a 7-term AP of 1286-digit primes, $(260708115 + 31232070k) \cdot 3001\# + 1$ for $k = 0,1,2,3,4,5,6$, where $3001\#$ means the product of all the primes from 2 to 3001.

Harvey Dubner and others found the longest known arithmetic progression of palindromic primes:

$$74295029087X0X0X78092059247$$

for $X = 0, 1, 2, \ldots, 9$ [See *Coll. Math. J.*, **30**(1999) 292].

See also **A9**.

P. D. T. A. Elliott & H. Halberstam, The least prime in an arithmetic progression, in *Studies in Pure Mathematics* (*Presented to Richard Rado*), Academic Press, London, 1971, 59–61; *MR* **42** #7609.

Christian Elsholtz, Triples of primes in arithmetic progressions, *Q. J. Math.*, **53**(2002) 393–395.

P. Erdős & P. Turán, On certain sequences of integers, *J. London Math. Soc.*, **11**(1936) 261–264.

John B. Friedlander & Henryk Iwaniec, Exceptional characters and prime numbers in arithmetic progressions, *Int. Math. Res. Not.*, **2003** no. 37, 2033–2050.

J. Gerver, The sum of the reciprocals of a set of integers with no arithmetic progression of k terms, *Proc. Amer. Math. Soc.*, **62**(1977) 211–214.

Joseph L. Gerver & L. Thomas Ramsey, Sets of integers with no long arithmetic progressions generated by the greedy algorithm, *Math. Comput.*, **33**(1979) 1353–1359.

V. A. Golubev, Faktorisation der Zahlen der Form $x^3 \pm 4x^2 + 3x \pm 1$, *Anz. Oesterreich. Akad. Wiss. Math.-Naturwiss. Kl.*, **1969** 184–191 (see also 191–194; 297–301; **1970**, 106–112; **1972**, 19–20, 178–179).

Ben Green & Terence Tao, The primes contain arbitrarily long arithmetic progression, April 2004 preprint.

Emil Grosswald, Long arithmetic progressions that consist only of primes and almost primes, *Notices Amer. Math. Soc.*, **26**(1979) A451.

Emil Grosswald, Arithmetic progressions of arbitrary length and consisting only of primes and almost primes, *J. reine angew. Math.*, **317**(1980) 200–208.

Emil Grosswald, Arithmetic progressions that consist only of primes, *J. Number Theory*, **14**(1982) 9–31.

Emil Grosswald & Peter Hagis, Arithmetic progressions consisting only of primes, *Math. Comput.*, **33**(1979) 1343–1352; *MR* **80k**:10054.

H. Halberstam, D. R. Heath-Brown & H.-E. Richert, On almost-primes in short intervals, in *Recent Progress in Analytic Number Theory*, Vol. 1, Academic Press, 1981, 69–101; *MR* **83a**:10075.

D. R. Heath-Brown, Almost-primes in arithmetic progressions and short intervals, *Math. Proc. Cambridge Philos. Soc.*, **83**(1978) 357–375; *MR* **58** #10789.

D. R. Heath-Brown, Three primes and an almost-prime in arithmetic progression, *J. London Math. Soc.*, (2) **23**(1981) 396–414.

D. R. Heath-Brown, Siegel zeros and the least prime in an arithmetic progression, *Quart. J. Math. Oxford Ser.*(2), **41**(1990) 405–418; *MR* **91m**:11073.

D. R. Heath-Brown, Zero-free regions for Dirichlet L-functions, and the least prime in an arithmetic progression, *Proc. London Math. Soc.*(3), **64**(1992) 265–338; *MR* **93a**:11075.

Edgar Karst, 12–16 primes in arithmetical progression, *J. Recreational Math.*, **2**(1969) 214–215.

Edgar Karst, Lists of ten or more primes in arithmetical progression, *Scripta Math.*, **28**(1970) 313–317.

Edgar Karst & S. C. Root, Teilfolgen von Primzahlen in arithmetischer Progression, *Anz. Oesterreich. Akad. Wiss. Math.-Naturwiss. Kl.*, **1972**, 19–20 (see also 178–179).

A. Kumchev, The difference between consecutive primes in an arithmetic progression, *Q. J. Math.*, **53**(2002) 479–501.

U. V. Linnik, On the least prime in an arithmetic progression, I. The basic theorem, *Rec. Math.* [*Mat. Sbornik*] *N.S.*, **15(57)**(1944) 139–178; II. The Deuring-Heilbronn phenomenon, 347–368; *MR* **6**, 260bc.

Carl Pomerance, The prime number graph, *Math. Comput.*, **33**(1979) 399–408; *MR* **80d**:10013.

Paul A. Pritchard, Andrew Moran & Anthony Thyssen, Twenty-two primes in arithmetic progression, *Math. Comput.*, **64**(1995) 1337–1339; *MR* **95j**:11003.

Olivier Ramaré & Robert Rumely, Primes in arithmetic progressions, *Math. Comput.*, **65**(1996) 397–425; *MR* **97a**:11144.

W. Sierpiński, Remarque sur les progressions arithmétiques, *Colloq. Math.*, **3**(1955) 44–49.

Sol Weintraub, Primes in arithmetic progression, *BIT* **17**(1977) 239–243.

K. Zarankiewicz, Problem 117, *Colloq. Math.*, **3**(1955) 46, 73.

OEIS: A005115.

A6 Consecutive primes in A.P.

It has even been conjectured that there are arbitrarily long arithmetic progressions of *consecutive* primes, such as

$$251, 257, 263, 269 \quad \text{and} \quad 1741, 1747, 1753, 1759.$$

Jones, Lal & Blundon discovered the sequence $10^{10} + 24493 + 30k$ ($0 \le k \le 4$) of five consecutive primes, and Lander & Parkin, soon after, found six such primes, $121174811 + 30k$ ($0 \le k \le 5$). They also established that $9843019 + 30k$ ($0 \le k \le 4$) is the least progression of five terms, that there are 25 others less than $3 \cdot 10^8$, but no others of length six.

On 95-09-15 Harvey Dubner emailed that he & Harry Nelson had recently found 7 consecutive primes in arithmetic progression. The common difference is 210, and the first prime is $x + Nm + 1$ where m is the product of the first 48 primes,

3670097 3182733191 6465034565 5501367323 3980031295 5331782619 4624570399 8807331115 7667212930

x is a solution of 48 modular equations,

1189306 1343242550 4731600916 6253605398 9417322887 0159415462 9760140568 0908210746 0202605690

and success occurred with $N = 2968677222$. The first prime is the 97-digit number

1089533431247059310875780378922957732908036492993138195385213105561742150447308967213141717486151

In Nov 1997 Harvey Dubner, Tony Forbes, Nik Lygeros, Michel Mizony & Paul Zimmermann announced that they had found 8 consecutive primes in AP. On 98-01-15, Manfred Toplic informed Harvey Dubner, Tony Forbes, Paul Zimmermann, Nik Lygeros & Michel Mizony that their project, which involved about 100 people using about 200 computers and took about two months, was a success, and that he had just found 9 consecutive primes in A.P. On 98-03-02, Paul Zimmermann said "we've just found 10 consecutive primes in arithmetic progression, the first one being (93 digits)

100996972469714247637786655587969840329509324689190041803603417758904341703348882159067229719

with common difference 210. It was the same team as that for 9 primes in January, with about 100 helpers all around the world."

It is not known if there are infinitely many sets of three *consecutive* primes in arithmetic progression, but S. Chowla has demonstrated this without the restriction to consecutive primes. Schinzel writes that Chowla was anticipated by van der Corput.

Harry Nelson has collected the $100.00 prize that Martin Gardner offered to the first discoverer of a 3 × 3 magic square whose nine entries are consecutive primes. These are *not* in arithmetic progression, of course. The central prime is 1480028171 and the others are this ±12, ±18, ±30 and ±42. He found more than 20 other such squares.

On 2002-08-25 David Wilson sent a 3 × 3 magic square with all prime entries and minimal row sum 177.

$$\begin{array}{ccc} 17 & 89 & 71 \\ 113 & 59 & 5 \\ 47 & 29 & 101 \end{array}$$

For more on magic squares see **D15**.

S. Chowla, There exists an infinity of 3-combinations of primes in A.P., *Proc. Lahore Philos. Soc.*, **6** no. 2(1944) 15–16; *MR* **7**, 243.

J. G. van der Corput, Über Summen von Primzahlen und Primzahlquadraten, *Math. Annalen*, **116**(1939) 1–50; *Zbl.* **19**, 196.

H. Dubner, T. Forbes, N. Lygeros, M. Mizony, H. Nelson & P. Zimmermann, Ten consecutive primes in arithmetic progression. *Math. Comput.* **71**(2002) 1323–1328; *MR* **2003d**:11137.

H. Dubner & H. Nelson, Seven consecutive primes in arithmetic progression, *Math. Comput.*, **66**(1997) 1743–1749; *MR* **98a**:11122.

P. Erdős & A. Rényi, Some problems and results on consecutive primes, *Simon Stevin*, **27**(1950) 115–125; *MR* **11**, 644.

M. F. Jones, M. Lal & W. J. Blundon, Statistics on certain large primes, *Math. Comput.*, **21**(1967) 103–107; *MR* **36** #3707.

L. J. Lander & T. R. Parkin, Consecutive primes in arithmetic progression, *Math. Comput.*, **21**(1967) 489.

H. L. Nelson, There is a better sequence, *J. Recreational Math.*, **8**(1975) 39–43.

S. Weintraub, Consectutive primes in arithmetic progression, *J. Recreational Math.*, **25**(1993) 169–171.

OEIS: A031217.

A7 Cunningham chains.

Primes p such that $2p + 1$ is also prime are known as **Sophie Germain primes**. It is believed, but not known, that there are infinitely many. Dubner has found such primes, of form $3003c \cdot 10^b - 1$ for $(c, b) = (7014, 2110)$, (581436,2581), (15655515,2999), (5199545,3529), (488964,4003) and (1803301,4526). Indlekofer & Járai (ref. at **A8**) found a Sophie Germain prime with 5847 decimal digits, and, on 2002-08-22 Michael Angel, Dirk Augustin & Paul Jobling gave the 24432-digit specimen $1213822389 \cdot 2^{81131} - 1$.

A common method for proving that p is a prime involves the factorization of $p - 1$. If $p - 1 = 2q$, where q is another prime, the size of the problem has only been reduced by a factor of 2, so it's interesting to observe **Cunningham chains** of primes with each member one more than twice the previous one. D. H. Lehmer found just three such chains of 7 primes with least member $< 10^7$:

  1122659, 2245319, 4490639, 8981279, 17962559, 35925119, 71850239

  2164229, 4328459, 8656919, 17313839, 34627679, 69255359, 138510719

  2329469, 4658939, 9317879, 18635759, 37271519, 74543039, 149086079

and two others with least members 10257809 and 10309889. The factorization of $p + 1$ can also be used to prove that p is prime. Lehmer found seven chains of length 7 based on $p + 1 = 2q$. The first three had least members 16651, 67651 and 165901, but the second of these must be discarded, since the fifth member is $1082401 = 601 = 1801$ (curiously, this is a divisor of $2^{25} - 1$).

Lalout & Meeus found chains of length 8 of each kind, starting with 19099919 and 15514861, and these are the smallest of this length. Günter Löh has found many new chains: the least of length 9 start with 85864769 and 857095381; of length 10 with 26089808579 and 205528443121; of length 11 with 665043081119 and 1389122693971; of length 12 with 554688278429 and 216857744866621; and a chain of length 13 of the second kind starts with 758083947856951. A count of all chains of the first kind starting below 10^{11} and of length 6, 7, 8, 9, 10 gave the respective frequencies 19991, 2359, 257, 21, 2.

Warut Roonguthai found the Cunningham chain of length 3: $p = 651358155 \times 2^{3291} - 1$, 2p+1, 4p+3.

Tony Forbes (see reference at **A9**) has found several chains of length 15 and one of length 16, beginning with $p = 3203000719597029781$.

Phil Carmody & Paul Jobling announced on 2002-03-01 that $810433818265726529160 \cdot 2^k - 1$ is prime for $0 \leq k \leq 15$.

In 1969 Daniel Shanks wrote to D. H. & Emma Lehmer about **quadratic chains**. Teske & Williams defined a **Shanks chain** of primes to be one satisfying $p_{i+1} = ap_i^2 - b$ for $1 \leq i < k$ and a, b are integers. They showed that, for all but 56 pairs of a, b values with $ab \leq 1000$, any corresponding Shanks chain must have bounded length. Shanks's original suggestion of $p_{i+1} = 4p_i^2 - 17$ yields a 4-chain if $p_1 = 3$ and a 5-chain if $p_1 = 303593$, but it 'can be seen (mod 59) that no such chain has length 17. It seems certain that such chains cannot be of arbitrary length. Are there any of length 7?

Takashi Agoh, On Sophie Germain primes, *Tatra Mt. Math. Publ.*, **20**(2000) 65–73; *MR* **2002d**:11008.

Harvey Dubner, Large Sophie Germain primes, *Math. Comput.*, **65**(1996) 393–396; *MR* **96d**:11008.

Claude Lalout & Jean Meeus, Nearly-doubled primes, *J. Recreational Math.*, **13** (1980/81) 30–35.

D. H. Lehmer, Tests for primality by the converse of Fermat's theorem, *Bull. Amer. Math. Soc.*, **33**(1927) 327–340.

D. H. Lehmer, On certain chains of primes, *Proc. London Math. Soc.*, **14A** (Littlewood 80 volume, 1965) 183–186.

Günter Löh, Long chains of nearly doubled primes, *Math. Comput.*, **53**(1989) 751–759; *MR* **90e**:11015.

Edlyn Teske & Hugh C. Williams, A note on Shanks's chains of primes, *Algorithmic number theory (Leiden, 2000)*, *Springer Lecture Notes in Comput. Sci.* **1838**(2000) 563–580; *MR* **2002k**:11228.

A8 Gaps between primes. Twin primes.

There are many problems concerning the gaps between consecutive primes. Write $d_n = p_{n+1} - p_n$ so that $d_1 = 1$ and all other d_n are even. How large and how small can d_n be? Rankin has shown that

$$d_n > \frac{c \ln n \ln \ln n \ln \ln \ln \ln n}{(\ln \ln \ln n)^2}$$

for infinitely many n and Erdős offers \$5,000 for a proof or disproof that the constant c can be taken arbitrarily large. Rankin's best value is $c = e^\gamma$ where γ is Euler's constant: Maier & Pomerance have improved this by a factor $k \approx 1.31256$, the root of the equation $4/k - e^{-4/k} = 3$, and Pintz has made a further improvement to $c = 2e^\gamma > 3.562$.

It may be worth observing, since the multiplicity of $\ln s$ occasionally confuses, that, infinitely often, the gap d_n exceeds the average gap, $\ln n$, by an arbitrarily large factor, M, say. [Put $n = e^{e^{M^2}}$.] It's hard to believe that there are infinitely many gaps $> 10^{100} \ln n$.

On 2004-01-07 Hans Rosenthal reported that a gap of 675034 between two probabilistic primes can be found at
http://www.trnicely.net/gaps/g675034.html.

R.C. Baker & Harman prove that $p_{n+1} - p_n \ll p_n^{0.535}$, and, with Pintz, improve the exponent to 0.525, improvements over Heath-Brown's 0.55.

In March 2003, it was thought that Dan Goldston and Cem Yildirim had proved that $p_{n+1} - p_n$ is infinitely often as small as $(\ln p_n)^{4/5}$, but it was subsequently discovered that an error term was large enough to invalidate the result. Can their methods yield an improvement on our very limited knowledge of this problem?

A very famous conjecture is the Twin Prime Conjecture, that $d_n = 2$ infinitely often. If $n > 6$, are there always twin primes between n and $2n$? Conjecture B of Hardy and Littlewood (cf. **A1**) is that $P_k(n)$, the number of pairs of primes less than n and differing by an even number k, is given asymptotically by

$$P_k(n) \sim \frac{2cn}{(\ln n)^2} \prod \left(\frac{p-1}{p-2} \right)$$

where the product is taken over all odd prime divisors of k (and so is empty and taken to be 1 when k is a power of 2) and $c = \prod(1 - 1/(p-1)^2)$ taken over all odd primes, so that $2c \approx 1.32032$. If $\pi_{1,2}(n)$ is the number of primes p such that $p + 2$ has at most two prime factors, then Fouvry & Grupp have shown that

$$\pi_{1,2}(n) \geq 0.71 \times \frac{2cn}{(\ln n)^2}$$

and 0.71 has been improved to 1.015 by Liu and then to 1.05 by Wu.

On p. 11 of the July 2002 London Math. Society's newsletter, A. A. Mullin asks us to prove that there are infinitely many perfect squares which exceed an odd bicomposite (product of two distinct primes) by 1.

The large twin primes $9 \cdot 2^{211} \pm 1$ were discovered by the Lehmers and independently by Riesel. Crandall & Penk found twin primes with 64, 136, 154, 203 and 303 digits, Williams found $156 \cdot 5^{202} \pm 1$, Baillie $297 \cdot 2^{546} \pm 1$, Atkin & Rickert

$$694503810 \cdot 2^{2304} \pm 1 \quad \text{and} \quad 1159142985 \cdot 2^{2304} \pm 1$$

and in 1989 Brown, Noll, Parady, Smith, Smith & Zarantonello found

$$663777 \cdot 2^{7650} \pm 1, \quad 571305 \cdot 2^{7701} \pm 1, \quad 1706595 \cdot 2^{11235} \pm 1.$$

On 93:08:16 Harvey Dubner announced a new record with

$$2^{4025} \cdot 3 \cdot 5^{4020} \cdot 7 \cdot 11 \cdot 13 \cdot 79 \cdot 223 \pm 1,$$

numbers with 4030 decimal digits. Indlekofer & Járai found $697053813 \cdot 2^{16352} \pm 1$ in November 1994, and a year later found the 11713-digit pair

$242206083 \cdot 2^{38880} \pm 1$. Meanwhile, on 95-07-25, Tony Forbes had found $6797727 \times 2^{15328} \pm 1$.

Jack Brennan noted that $k \cdot 2^n \pm 1$ were primes for $k = 3 \cdot 5 \cdot 7 \cdot 11 \cdot 13 \cdot 13487$ and the twelve values $n = 2, 12, 17, 28, 31, 33, 42, 55, 62, 86, 89, 91$. Phil Carmody found that $k = 3^2 \cdot 5^3 \cdot 7 \cdot 11 \cdot 13^2 \cdot 23 \cdot 503 \cdot 4129$ gave 14 such twins with $n \le 470$ and that $k = 3 \cdot 5 \cdot 7 \cdot 11 \cdot 13 \cdot 23 \cdot 37 \cdot 97 \cdot 460891$ gave 15 twins, with $n = 1, 13, 15, 17, 18, 22, 29, 35, 39, 43, 60, 189, 385, 705, 979$. His request for a set of 16 has not so far been answered.

Richard Brent counted 224376048 primes p less than 10^{11} for which $p+2$ is also prime; about 9% more than predicted by Conjecture B. Nicely, who, in course of his work cunningly detected the notorious Pentium chip flaw, counted 135780321665 twin prime pairs $< 10^{14}$, the sum of whose reciprocals is

$$B(10^{14}) = 1.8202449681302705288947178386195338283464 9\ldots.$$

Brun showed that the sum of the reciprocals of twin primes is convergent, to sum B, say. Nicely estimates

$$B = 1.9021605778 \pm 2.1 \times 10^{-9}$$

and, based on a subsequent calculation to 2.5×10^{14},

$$B = 1.9021605803 \pm 1.3 \times 10^{-9}$$

Pascal Sebah's August 2002 calculation of $\pi_2(10^{16}) = 10304195697298$ yields $B = 1.902160583104\ldots$.

Bombieri & Davenport have shown that

$$\liminf \frac{d_n}{\ln n} < \frac{2 + \sqrt{3}}{8} \approx 0.46650$$

(no doubt the real answer is zero; of course the truth of the Twin Prime Conjecture would imply this); G.Z. Pilt'yaĭ has improved the constant on the right to $(2\sqrt{2} - 1)/4 \approx 0.45711$; Uchiyama to $(9 - \sqrt{3})/16 \approx 0.454256$; Huxley to $(4\sin\theta + 3\theta)/(16\sin\theta) \approx 0.44254$, where $\theta + \sin\theta = \pi/4$, and later to 0.4394; and Helmut Maier to 0.248.

Huxley has also shown that

$$d_n < p_n^{7/12 + \epsilon},$$

Heath-Brown & Iwaniec have improved the exponent to 11/20; Mozzochi to 0.548; and Lou & Yao to 6/11. Cramér proved, using the Riemann hypothesis, that

$$\sum_{n < x} d_n^2 < cx(\ln x)^4.$$

Erdős conjectures that the right-hand side should be $cx(\ln x)^2$, but thinks that there is no hope of a proof. The Riemann hypothesis implies that $d_n < p_n^{1/2+\epsilon}$.

A. S. Peck shows that the sum $\sum d_n$ of those d_n which are $> n$, taken over the interval $x \le p_n \le 2x$ is $\ll x^{25/36+\epsilon}$, improving on Heath-Brown's previous best exponent of $3/4$.

Jie Wu shows that if $\theta > 0.973$, then the number of primes p in the interval $[x, x + x^\theta]$ such that $p + 2$ has at most two prime factors is at least $cx^\theta/(\ln x)^2$. Salerno & Vitolo have a similar result.

Sándor has shown that $\liminf_{n\to\infty} \sqrt[4]{p_n}(\sqrt{p_{n+1}} - \sqrt{p_n}) = 0$ and that $\liminf nd_{n-1}(d_n/p_n)^2 = 0$.

Erdős & Nathanson show that

$$\sum_{n=2}^{\infty} \frac{1}{n(\ln \ln n)^c(p_{n+1} - p_n)}$$

converges for $c > 2$ and conjecture that it diverges for $c = 2$.

Dorin Andrica conjectures that, for all natural n,

$$¿ \qquad \sqrt{p_{n+1}} - \sqrt{p_n} < 1 \qquad ?$$

Dan Grecu has verified this for $p_n < 10^6$. In *Amer. Math. Monthly*, **83**(1976) 61, it is given as a difficult unsolved problem that

$$¿ \qquad \lim_{n\to\infty} (\sqrt{p_{n+1}} - \sqrt{p_n}) = 0 \qquad ?$$

If true, this implies Andrica's conjecture for large enough n, which, is comparable with that of Cramér, mentioned in **A2**, and with the following one of Shanks, who has given a heuristic argument which supports the conjecture that if $p(g)$ is the first prime that follows a gap of g between consecutive primes, then $\ln p(g) \sim \sqrt{g}$. Record gaps between consecutive primes have been observed by Lehmer, Lander & Parkin, Brent, Weintraub, Young & Potler and others. Table 2 illustrates Shanks's conjecture. The last several entries are taken from T. R. Nicely's web page
http://www.trnicely.net/gaps .

On 2001-12-13 Harvey Dubner found a gap of 119738 between two 3396-digit primes; on 2003-02-15 Marcel Martin announced that Jose Luis Gomez Pardo had proved the primality of a 5878-digit number which is the upper bound of a gap of size 233822; and on 2004-01-15 Andersen & Rosenthal announced a prime gap of 1001548 between two probabilistic primes with 43429 digits.

Chen Jing-Run showed that, for x large enough, there is always a number with at most two prime factors in the interval $[x - x^\alpha, x]$ for any value of $\alpha \ge 0.477$. Halberstam, Heath-Brown & Richert (see reference at **A5**) showed that in such an interval with $\alpha = 0.455$ there are at least $x^\alpha/121 \ln x$

numbers with at most two prime factors, and Iwaniec & Laborde reduced the exponent to $\alpha = 0.45$. Further improvements to the value of α are 0.4436 (Fouvry), 0.44 (Wu), 0.4386 (Li) and 0.436 (Liu).

Victor Meally used the phrase **prime deserts**. He notes that below 373 the commonest gap is 2; below 467 there are 24 gaps of each of 2, 4 and 6; below 563 the commonest gap is 6, as it is between 10^{14} and $10^{14} + 10^8$ and probably also from 2 to 10^{14}. He asks: when does 30 take over as the commonest gap?

Conway & Odlyzko call d a **champion** for x if it occurs most frequently as the difference between consective primes $\leq x$. There may be more than one champion for the same x: $C(135) = 4$, $C(100) = \{2, 4\}$. They suggest that the only champions are 4 and the prime factorials 2, 6, 30, 210, 2310, ... Do champions $\to \infty$? Does each prime p divide all champions for $x \geq x_0(p)$?

Table 2. Earliest large gaps between consecutive primes.

g	$p(g)$	$(\ln p)^2$	$g/(\ln p)^2$
456	25056082543	573.33	0.7953
464	42652618807	599.09	0.7745
468	127976335139	654.09	0.7155
474	182226896713	672.29	0.7051
486	241160624629	686.90	0.7075
490	297501076289	697.95	0.7021
500	303371455741	698.98	0.7153
514	304599509051	699.19	0.7351
516	416608696337	715.85	0.7208
532	461690510543	721.36	0.7375
534	614487454057	736.80	0.7247
540	738832928467	746.84	0.7230
582	1346294311331	779.99	0.7462
588	1408695494197	782.53	0.7514
602	1968188557063	801.35	0.7512
652	2614941711251	817.52	0.7975
674	7177162612387	876.27	0.7692
716	13829048560417	915.53	0.7821
766	19581334193189	936.70	0.8178
778	42842283926129	985.24	0.7897
804	90874329412297	1033.01	0.7783
806	171231342421327	1074.14	0.7504
906	218209405437449	1090.09	0.8311
916	1189459969826399	1204.94	0.7602
924	1686994940956727	1229.32	0.7516
1132	1693182318747503	1229.58	0.9206
1184	43841547845542243	1468.37	0.8063
1198	55350776431904441	1486.29	0.8060

A **cluster prime** p is one such that every even number less than $p - 2$ is the difference of two primes both $\leq p$. The first 23 odd primes, 3, 5, 7, ..., 89 are all cluster primes. The smallest non-cluster prime is 97 (88 is not the difference of two primes ≤ 97). Blecksmith & Selfridge ask if there are infinitely many cluster primes.

Maier & Stewart have shown that there are long intervals which contain fewer primes than the average for intervals of such length.

A. O. L. Atkin & N. W. Rickert, On a larger pair of twin primes, Abstract 79T-A132, *Notices Amer. Math. Soc.*, **26**(1979) A-373.

R. C. Baker & Glyn Harman, The difference between consecutive primes, *Proc. London Math. Soc.*(3) **72**(1996) 261–280; *MR* **96k**:11111.

R. C. Baker, Glyn Harman & J. Pintz, The difference between consecutive primes, II, *Proc. London Math. Soc.*(3) **83**(2001) 522–562; *MR* **2002f**:11125.

Danilo Bazzanella, Primes in almost all short intervals, *Boll. Un. Mat. Ital.*(7), **9**(1995) 233–249; *MR***99c**:11089.

Richard Blecksmith, Paul Erdős & J. L. Selfridge, Cluster primes, *Amer. Math. Monthly*, **106**(1999) 43–48; *MR* **2000a**:11126.

E. Bombieri & H. Davenport, Small differences between prime numbers, *Proc. Roy. Soc. Ser. A*, **293**(1966)1–18; *MR* **33** #7314.

Richard P. Brent, The first occurrence of large gaps between successive primes, *Math. Comput.*, **27** (1973) 959–963; *MR* **48** #8360; (and see *Math. Comput.*, **35** (1980) 1435–1436.

Richard P. Brent, Irregularities in the distribution of primes and twin primes, *Math. Comput.*, **29**(1975) 43–56.

V. Brun, Le crible d'Eratosthène et le théorème de Goldbach, *C. R. Acad. Sci. Paris*, **168**(1919) 544–546.

J. H. Cadwell, Large intervals between consecutive primes, *Math. Comput.*, **25**(1971) 909–913.

Cai Ying-Chun & Lu Ming-Gao, On the upper bound for $\pi_2(x)$, *Acta Arith.*, **110**(2003) 275–298.

Chen Jing-Run, On the distribution of almost primes in an interval II, *Sci. Sinica*, **22**(1979) 253–275; *Zbl.*, 408.10030.

Chen Jing-Run & Wang Tian-Ze, *Acta Math. Sinica*, **32**(1989) 712–718; *MR* **91e**:11108.

Chen Jing-Run & Wang Tian-Ze, On distribution of primes in an arithmetical progression, *Sci. China Ser. A*, **33**(1990) 397–408; *MR* **91k**:11078.

H. Cramér, On the order of magnitude of the difference between consecutive prime numbers, *Acta Arith.*, **2**(1937) 23–46.

Pamela A. Cutter, Finding prime pairs with particular gaps, *Math. Comput.*, **70**(2001) 1737–1744 (electronic); *MR* **2002c**:11174.

Tamás Dénes, Estimation of the number of twin primes by application of the complementary prime sieve, *Pure Math. Appl.*, **13**(2002) 325–331; *MR* **2003m**: 11153.

Christian Elsholtz, On cluster primes, *Acta Arith.*, **109**(2003) 281–284; *MR* **2004c**:11169.

Paul Erdős & Melvyn B. Nathanson, On the sums of the reciprocals of the differences between consecutive primes, *Number Theory* (*New York, 1991–1995*, Springer, New York, 1996, 97–101; *MR* **97h**:11094.

Anthony D. Forbes, A large pair of twin primes, *Math. Comput.*, **66**(1997)451–455; *MR* **99c**:11111.

Étienne Fouvry, in *Analytic Number Theory* (*Tokyo 1988*) 65–85, Lecture Notes in Math., **1434**, Springer, Berlin, 1990; *MR* **91g**:11104.

É. Fouvry & F. Grupp, On the switching principle in sieve theory, *J. reine angew. Math.*, **370**(1986) 101–126; *MR* **87j**:11092.

J. B. Friedlander & J. C. Lagarias, On the distribution in short intervals of integers having no large prime factor, *J. Number Theory*, **25**(1987) 249–273.

J. W. L. Glaisher, On long successions of composite numbers, *Messenger of Math.*, **7**(1877) 102-106, 171-176.

D. A. Goldston, On Bombieri and Davenport's theorem concerning small gaps between primes, *Mathematika*, **39**(1992) 10–17; *MR* **93h**:11102.

Glyn Harman, Short intervals containing numbers without large prime factors, *Math. Proc. Cambridge Philos. Soc.*, **109**(1991) 1–5; *MR* **91h**:11093.

Jan Kristian Haugland, Large prime-free intervals by elementary methods, *Normat*, **39**(1991) 76–77.

D. R. Heath-Brown, The differences between consecutive primes III, *J. London Math. Soc.*(2) **20**(1979) 177–178; *MR* **81f**:10055.

Y. K. Huen, The twin prime problem revisited, *Internat. J. Math. Ed. Sci. Tech.*, **28**(1997) 825–834.

Martin Huxley, An application of the Fouvry-Iwaniec theorem, *Acta Arith.*, **43**(1984) 441–443.

Karl-Heinz Indlekofer & Antal Járai, Largest known twin primes, *Math. Comput.*, **65**(1996) 427–428; *MR* **96d**:11009.

Karl-Heinz Indlekofer & Antal Járai, Largest known twin primes and Sophie Germain primes, *Math. Comput.*, **68**(1999) 1317–1324; *MR* **99k**:11013.

H. Iwaniec & M. Laborde, P_2 in short intervals, *Ann. Inst. Fourier(Grenoble)*, **31**(1981) 37–56; *MR* **83e**:10061.

Jia Chao-Hua, Three primes theorem in a short interval VI (Chinese), *Acta Math. Sinica*, **34**(1991) 832–850; *MR* **93h**:11104.

Jia Chao-Hua, Difference between consecutive primes, *Sci. China Ser. A*, **38**(1995) 1163–1186; *MR* **99m**:11081.

Jia Chao-Hua, Almost all short intervals containing prime numbers, *Acta Arith.*, **76**(1996) 21–84; *MR* **97e**:11110.

Li Hong-Ze, Almost primes in short intervals, *Sci. China Ser. A*, **37**(1994) 1428–1441; *MR* **96e**:11116.

Li Hong-Ze, Primes in short intervals, *Math. Proc. Cambridge Philos. Soc.*, **122**(1997) 193–205; *MR* **98e**:11103.

Liu Hong-Quan, On the prime twins problem, *Sci. China Ser. A*, **33**(1990) 281–298; *MR* **91i**:11125.

Liu Hong-Quan, Almost primes in short intervals, *J. Number Theory*, **57**(1996) 303–302; *MR* **97c**:11092.

Lou Shi-Tuo & Qi Yao, Upper bounds for primes in intervals (Chinese), *Chinese Ann. Math. Ser. A*, **10**(1989) 255–262; *MR* **91d**:11112.

Helmut Maier, Small differences between prime numbers, *Michigan Math. J.*, **35**(1988) 323–344.

Helmut Maier, Primes in short intervals, *Michigan Math. J.*, **32**(1985) 221–225.

Helmut Maier & Carl Pomerance, Unusually large gaps between consecutive primes, *Théorie des nombres*, (Quebec, PQ, 1987), de Gruyter, 1989, 625–632; *MR* **91a**:11045: and see *Trans. Amer. Math. Soc.*, **322**(1990) 201–237; *MR* **91b**:11093.

H. Maier & Cam Stewart, On intervals with few prime numbers, (2003 preprint)

Hiroshi Mikawa, On prime twins in arithmetic progressions, *Tsukuba J. Math.*, **16**(1992) 377–387; *MR* **94e**:11101.

C. J. Mozzochi, On the difference between consecutive primes, *J. Number Theory* **24** (1986) 181–187.

Thomas R. Nicely, Enumeration to 10^{14} of the twin primes and Brun's constant, *Virginia J. Sci.*, **46**(1995) 195–204; reviewed by Richard P. Brent, *Math. Comput.*, **66**(1997) 924–925; *MR* **97e**:11014.

Thomas R. Nicely, New maximal prime gaps and first occurrences, *Math. Comput.*, **68**(1999) 1311–1315; *MR* **99i**:11004.

Thomas R. Nicely, A new error analysis for Brun's constant, *Virginia J. Sci.*, **52**(2001) 45–55; *MR* **2003d**:11184.

Bertil Nyman & Thomas R. Nicely, New prime gaps between 10^{15} and 5×10^{16}, *J. Integer Seq.*, **6**(2003) Article 03.3.1, 6 pp. (electronic).

Bodo K. Parady, Joel F. Smith & Sergio E. Zarantonello, Largest known twin primes, *Math. Comput.*, **55**(1990) 381–382; *MR* **90j**:11013.

A. S. Peck, Differences between consecutive primes, *Proc. London Math. Soc.*,(3) **76**(1998) 33–69; *MR* **98i**:11071.

G. Z. Pil'tyaĭ, The magnitude of the difference between consecutive primes (Russian), *Studies in number theory*, **4**, Izdat. Saratov. Univ., Saratov, 1972, 73–79; *MR* **52** #13680.

János Pintz, Very large gaps between consecutive primes, *J. Number Theory*, **63**(1997) 286–301; *MR* **98c**:11092.

Olivier Ramaré & Yannick Saouter, Short effective intervals containing primes, *J. Number Theory*, **98**(2003) 10–33; *MR* **2004a**:11095.

S. Salerno & A. Vitolo, $p + P_2$ in short intervals, *Note Mat.*, **13**(1993) 309–328; *MR* **96c**:11109.

József Sándor, Despre şirusi, serü şi aplicatü în teoria numerelor prime (On sequences, series and applications in prime number theory. Romanian), *Gaz. Mat. Perf. Met.*, **6**(1985) 38–48.

Daniel Shanks, On maximal gaps between successive primes, *Math. Comput.*, **18**(1964) 646–651; *MR* **29** #4745.

Daniel Shanks & John W. Wrench, Brun's constant, *Math. Comput.*, **28**(1974) 293–299; corrigenda, ibid. **28**(1974) 1183; *MR* **50** #4510.

S. Uchiyama, On the difference between consecutive prime numbers, *Acta Arith.*, **27** (1975) 153–157.

Nigel Watt, Short intervals almost all containing primes, *Acta Arith.*, **72**(1995) 131–167; *MR* **96f**:11115.

Wu Jie[1], Sur la suite des nombres premiers jumeaux, *Acta Arith.*, **55**(1990) 365–394; *MR* **91j**:11074.

Wu Jie[1], Thórmes généralisés de Bombieri-Vinogradov dans les petits intervalles, *Quart. J. Math. Oxford Ser.*(2), **44**(1993) 109–128; *MR* **93m**:11090.

Jeff Young & Aaron Potler, First occurrence prime gaps, *Math. Comput.*, **52**(1989) 221–224.

Yu Gang[1], The differences between consecutive primes, *Bull. London Math. Soc.*, **28**(1996) 242–248; *MR* **97a**:11142.

Alessandro Zaccagnini, A note on large gaps between consecutive primes in arithmetic progressions, *J. Number Theory*, **42**(1992) 100–102.

OEIS: A000101, A000847, A002110, A002386, A005867, A008407, A008996, A020497, A022008, A030296, A040976, A048298, A049296, A051160–A051168, A058188, A058320, A059861–A059865.

A9 Patterns of primes.

A conjecture more general than Chowla's (see **A4**) is that there are infinitely many sets of consecutive primes of any given pattern, provided that there are no congruence relations which rule them out. It seems likely, for example, that there are infinitely many triples of primes $\{6k - 1, 6k + 1, 6k + 5\}$ and $\{6k + 1, 6k + 5, 6k + 7\}$. This would be even harder to settle than the Twin Prime Conjecture, but its plausibility is of interest, since Hensley & Richards have shown that it is incompatible with the well-known conjecture (also due to Hardy & Littlewood)

$$\text{¿¿¿} \qquad \pi(x + y) \le \pi(x) + \pi(y) \qquad ???$$

for all integers x, $y \ge 2$. We've put more queries than usual round this, since it is very likely to be false. Indeed, there's some hope of finding values of x and y which contradict it. However there's an alternative conjecture,

$$\text{¿} \qquad \pi(x + y) \le \pi(x) + 2\pi(y/2) \qquad ?$$

that the Hensley-Richards method doesn't comment on.

Montgomery & Vaughan showed that

$$\pi(x + y) - \pi(x) \le 2y/\ln y$$

and Iwaniec observed that for each θ, $0 < \theta < 1$, there is an $\eta(\theta) > \theta$ such that

$$\pi(x + x^\theta) - \pi(x) < (2 + \epsilon)x^\theta/(\eta(\theta)\ln x)$$

for sufficiently large x and he found that $\eta(\theta) = \frac{5}{3}\theta - \frac{2}{9}$ for $\theta > \frac{1}{3}$, and that $\eta(\theta) = (1 + \theta)/2$ for $\theta > \frac{1}{2}$. Lou & Yao improve this in part by showing that $\eta(\theta) = (100\theta - 45)/11$ for $\frac{6}{11} < \theta \le \frac{11}{20}$.

On 2002-08-20 David Broadhurst announced that the number $N = (kn(n + 1) + 210)(n - 1)/35$ with $k = 61504372896$ and $n = 5119$ times the product of all primes not exceeding 3163, yields the three 4019-digit primes $N + 5$, $N + 7$, $N + 11$.

C. W. Trigg reported that in 1978 M. A. Penk found four primes p, $p + 2$, $p + 6$ and $p + 8$ where

$$p = 8023591500031216055575513808675195603443569 71.$$

H. F. Smith noted that the pattern 11, 13, 17, 19, 23, 29, 31, 37 is repeated at least three times, starting with the primes 15760091, 25658841 and 93625991. In none of these cases is the number corresponding to 41 a prime, although $n - 11$, $n - 13$, $\ldots n - 41$ are all primes for $n = 88830$ and 855750.

Leech gave as an unsolved problem to find 33 consecutive numbers greater than 11 which include 10 primes. In 1961 Herschel Smith found

20 such sets and also 5 examples of 37 consecutive numbers containing 11 primes. Smith writes that Selfridge noted some errors in his 1957 paper. Sten Säfholm found primes

$$\{n + 11, \ldots, n + 43\} \quad \text{for} \quad n = 33081664140$$

and rediscovered Smith's first three examples, that each of

$$\{n - 11, \ldots, n - 43\} \quad \text{is prime for} \quad n = 9853497780,$$

for $n = 21956291910$ and for $n = 22741837860$. Leech wondered why the latter sets seem to occur more readily than the former. My guess is that this is just a more complicated version of the prime number race (see **A4**) and that with much more high-powered telescopes we'd see the balance being redressed (infinitely often). Dimitrios Betsis & Sten Säfholm have found many more patterns, culminating in $\{n + 11, \ldots, n + 61\}$ for $n = 21817283854511250$ and $\{n - 11, \ldots, n - 61\}$ for $n = 79287805466244270$. Two large prime 15-plets were found over Sep-Oct 2003 by Jens Kruse Andersen. Add the 15 primes from 11 to 67 to either of the numbers 125103001259595590131218845037 or 110091624923387985733407523482.

Warut Roonguthai searched for the smallest n-digit prime $p = 10^{n-1} + k$ such that $p + 2$, $p + 6$ and $p + 8$ are also prime, i.e., $\{p, p + 2, p + 6, p + 8\}$ is the smallest n-digit prime quadruplet:

n	100	200	300	400	500
k	349781731	21156403891	140159459341	34993836001	883750143961

In none of these cases is $p + 12$ or $p - 4$ a prime.

Erdős asks, for each k, what is the smallest l for which p_k, p_{k+1}, $\ldots$, p_{k+l-1} is the only set of l consecutive primes with this pattern. E.g., the pattern 3, 5, 7 cannot occur again. The pattern 5, 7, 11, 13, 17 repeats at 101, 103, 107, 109, 113 and no doubt occurs infinitely often, But considerations mod 5 show that the pattern 5, 7, 11, 13, 17, 19 does not occur again. $(p_k, l) = (2, 2)$, $(3, 3)$, $(5, 6)$, $\ldots$.

See also **A5**.

Antal Balog, The prime k-tuplets conjecture on the average, *Analytic Number Theory (Allerton Park, IL, 1989)*, 47–75, *Progr. Math.*, **85**, Birkhäuser, Boston, 1990.

Pierre Dusart, Sur la conjecture $\pi(x+y) \leq \pi(x)+\pi(y)$, *Acta Arith.*, **102**(2002) 295–308.

Paul Erdős & Ian Richards, Density functions for prime and relatively prime numbers, *Monatsh. Math.*, **83**(1977) 99–112; *Zbl.* **355**.10034.

Anthony D. Forbes, Prime clusters and Cunningham chains, *Mathy. Comput.*, **68**(1999) 1739–1747; *MR* **99m**:11007.

Anthony D. Forbes, Large prime triplets, *Math. Spectrum*, **29**(1996-97) 65.

Anthony D. Forbes, Prime k-tuplets — 15, *M500*, **156**(July 1997) 14–15.

John B. Friedlander & Andrew Granville, Limitations to the equi-distribution of primes I, IV, III, *Ann. of Math.*(2), **129**(1989) 363–382; *MR* **90e**:11125; *Proc. Roy. Soc. London Ser. A*, **435**(1991) 197–204; *MR* **93g**:11098; *Compositio Math.*, **81**(1992) 19–32.

John B. Friedlander, Andrew Granville, Adolf Hildebrandt & Helmut Maier, Oscillation theorems for primes in arithmetic progressions and for sifting functions, *J. Amer. Math. Soc.*, **4**(1991) 25–86; *MR* **92a**:11103.

R. Garunkštis, On some inequalities concerning $\pi(x)$, *Experiment. Math.*, **11**(2002) 297–301; *MR* **2003k**:11143.

D. R. Heath-Brown, Almost-prime k-tuples, *Mathematika*, **44**(1997) 245–266.

Douglas Hensley & Ian Richards, On the incompatibility of two conjectures concerning primes, *Proc. Symp. Pure Math.*, (Analytic Number Theory, St. Louis, 1972) **24** 123–127.

H. Iwaniec, On the Brun-Titchmarsh theorem, *J. Math. Soc. Japan*, **34**(1982) 95–123; *MR* **83a**:10082.

John Leech, Groups of primes having maximum density, *Math. Tables Aids Comput.*, **12**(1958) 144–145; *MR* **20** #5163.

H. L. Montgomery & R. C. Vaughan, The large sieve, *Mathematika*, **20**(1973) 119–134; *MR* **51** #10260.

Laurenţiu Panaitopol, On the inequality $p_a p_b > p_{ab}$, *Bull. Math. Soc. Sci. Math. Roumanie* (*N.S.*, **41**(**89**)(1998) 135–139.

Laurenţiu Panaitopol, An inequality related to the Hardy-Littlewood conjecture, *An. Univ. Bucureşti Mat.*, **49**(2000) 163–166; *MR* **2003c**:11114.

Laurenţiu Panaitopol, Checking the Hardy-Littlewood conjecture in special cases, *Rev. Roumaine Math. Pures Appl.*, **46**(2001) 465–470; *MR* **2003e**:11098.

Ian Richards, On the incompatibility of two conjectures concerning primes; a discussion of the use of computers in attacking a theoretical problem, *Bull. Amer. Math. Soc.*, **80**(1974) 419–438.

Herschel F. Smith, On a generalization of the prime pair problem, *Math. Tables Aids Comput.*, **11**(1957) 249–254; *MR* **20** #833.

Warut Roonguthai, Large prime quadruplets, *M500* **153**(Dec.1996) 4–5.

Charles W. Trigg, A large prime quadruplet, *J. Recreational Math.*, **14** (1981/82) 167.

Sheng-Gang Xie, The prime 4-tuplet problem (Chinese. English summary), *Sichuan Daxue Xuebao*, **26**(1989) 168–171; *MR* **91f**:11066.

OEIS: A008407, A020497-020498, A023193, A066081.

A10 Gilbreath's conjecture.

Define d_n^k by $d_n^1 = d_n$ and $d_n^{k+1} = |d_{n+1}^k - d_n^k|$, that is, the successive absolute differences of the sequence of primes (Figure 2). N. L. Gilbreath conjectured (and Hugh Williams notes that Proth, long before, claimed to have proved) that $d_1^k = 1$ for all k. This was verified for $k < 63419$ by Killgrove & Ralston. Odlyzko has checked it for primes up to $\pi(10^{13}) \approx 3 \cdot 10^{11}$; he only needed to examine the first 635 differences.

Hallard Croft and others have suggested that it has nothing to do with primes as such, but will be true for any sequence consisting of 2 and odd numbers, which doesn't increase too fast, or have too large gaps. Odlyzko discusses this.

```
1   2   3   4    5   6   7   8   9  10  11  12  13  14  15  16  17  18  19  20  21  22  23  24
2   3   5   7   11  13  17  19  23  29  31  37  41  43  47  53  59  61  67  71  73  79  83  89
  1   2   2   4    2   4   2   4   6   2   6   4   2   4   6   6   2   6   4   2   6   4   6
    1   0   2    2   2   2   2   2   4   4   2   2   2   2   0   4   4   2   2   4   2   2
      1   2    0   0   0   0   0   2   0   2   0   0   0   2   4   0   2   0   2   2   0
        1    2   0   0   0   0   2   2   2   2   0   0   2   2   4   2   2   2   0   2
          1    2   0   0   0   2   0   0   0   2   0   2   0   2   2   0   0   2   2
            1    2   0   0   2   2   0   0   2   2   2   2   2   0   2   0   2   0
              1    2   0   2   0   2   0   2   0   0   0   0   2   2   2   2
                1    2   2   2   2   2   2   2   0   0   0   2   0   0   0   0
                  1    0   0   0   0   0   0   2   0   0   2   2   0   0   0
```

Figure 2. Successive Absolute Differences of the Sequence of Primes.

R. B. Killgrove & K. E. Ralston, On a conjecture concerning the primes, *Math. Tables Aids Comput.*, **13**(1959) 121–122; *MR* **21** #4943.

Andrew M. Odlyzko, Iterated absolute values of differences of consecutive primes, *Math. Comput.*, **61**(1993) 373–380; *MR* **93k**:11119.

F. Proth, Sur la série des nombres premiers, *Nouv. Corresp. Math.*, **4**(1878) 236–240.

OEIS: A036261-036262, A036277, A054977.

A11 Increasing and decreasing gaps.

Since the proportion of primes gradually decreases, albeit somewhat erratically, $d_m < d_{m+1}$ infinitely often, and Erdős & Turán have shown that the same is true for $d_n > d_{n+1}$. They also shown that the values of n for which $d_n > d_{n+1}$ have positive lower density, but it is not known if there are infinitely many decreasing or increasing sets of *three* consecutive values of d_n. If there were not, then there is an n_0 so that for every i and $n > n_0$ we have $d_{n+2i} > d_{n+2i+1}$ and $d_{n+2i+1} < d_{n+2i+2}$. Erdős offers $100.00 for a proof that such an n_0 does not exist. He and Turán could not even prove that for $k > k_0$, $(-1)^r(d_{k+r+1} - d_{k+r})$ can't always have the same sign.

P. Erdős, On the difference of consecutive primes, *Bull. Amer. Math. Soc.*, **54**(1948) 885–889; *MR* **10**, 235.

P. Erdős & P. Turán, On some new questions on the distribution of prime numbers, *Bull. Amer. Math. Soc.*, **54**(1948) 371–378; *MR* **9**, 498.

Laurenţiu Panaitopol, Erdős-Turán problem for prime numbers in arithmetic progression, *Math. Rep, (Bucur.)*, **1**(**51**)(1999) 595–599; *MR* **2002d**:11112.

A12 Pseudoprimes. Euler pseudoprimes. Strong pseudoprimes.

Pomerance, Selfridge & Wagstaff call an odd composite n for which $a^{n-1} \equiv 1 \bmod n$ a **pseudoprime to base** a (psp(a)). This usage is introduced to avoid the clumsy "composite pseudoprime" which appears throughout the literature. Odd composite n which are psp(a) for every a prime to n are **Carmichael numbers** (see **A13**). An odd composite n is an **Euler pseudoprime to base** a (epsp(a)) if $a \perp n$ and $a^{(n-1)/2} \equiv \left(\frac{a}{n}\right) \bmod n$, where $\left(\frac{a}{n}\right)$ is the Jacobi symbol (see **F5**). Finally, an odd composite n with $n-1 = d \cdot 2^s$, d odd, is a **strong pseudoprime to base** a (spsp(a)) if $a^d \equiv 1 \bmod n$ (otherwise $a^{d \cdot 2^r} \equiv -1 \bmod n$ for some r, $0 \le r < s$). These definitions are illustrated by a Venn diagram (Figure 3) which displays the smallest member of each set.

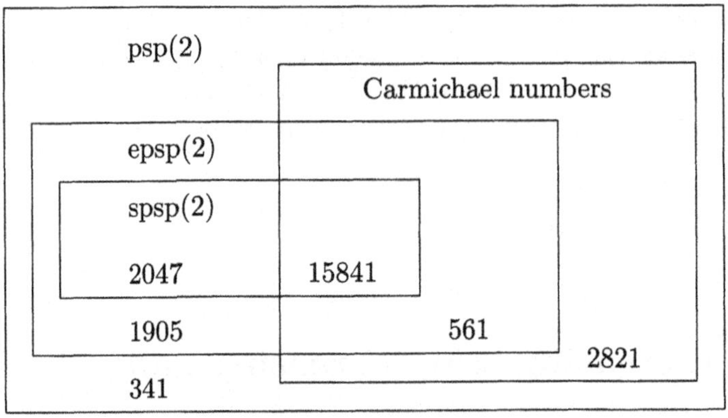

Figure 3. Relationships of Sets of psps with Least Element in Each Set.

The following values of $P_2(x)$, $E_2(x)$, $S_2(x)$ and $C(x)$ — the numbers of psp(2), epsp(2), spsp(2) and Carmichael numbers less than x, respectively — were given by Pomerance, Selfridge & Wagstaff and extended to 10^{13} by Pinch, who makes some random observations: the only twin pseudoprimes up to 10^{13} are 4369, 4371: there are 54 non-squarefree pseudoprimes up to 10^{13} — all multiples of $1194649 = 1093^2$ or $12327121 = 3511^2$; and asks are there any other square psp?

x	10^4	10^5	10^6	10^7	10^8	10^9	10^{10}	10^{11}	10^{12}	10^{13}
$P_2(x)$	22	78	245	750	2057	5597	14884	38975	101629	264239
$E_2(x)$	12	36	114	375	1071	2939	7706	20417	53332	139597
$S_2(x)$	5	16	46	162	488	1282	3291	8607	22412	58897
$C(x)$	7	16	43	105	255	646	1547	3605	8241	19279

The least spsp's whose bases are the first k primes, $0 < k \le 7$ are 2047, 1373653, 25326001, 3215031751, 2152302898747, 3474749660383,

341550071728321, where this last is also an spsp19. Zhang & Tang conjecture that $3825123056546413051 = 149491 \cdot 747451 \cdot 34233211$ is the candidate for $k = 9$, 10 and 11.

Lehmer and Erdős showed that, for sufficiently large x,

$$c_1 \ln x < P_2(x) < x \exp\{-c_2(\ln x \ln \ln x)^{1/2}\}$$

and Pomerance improved these bounds to

$$\exp\{(\ln x)^{5/14}\} < P_2(x) < x \exp\{(-\ln x \ln \ln \ln x)/2 \ln \ln x\}$$

and has a heuristic argument that the true estimate is the upper bound with the 2 omitted. The exponent 5/14 has been improved to 85/207 by Pomerance, using a result of Friedlander.

There are also examples of even numbers such that $2^n \equiv 2 \bmod n$. Lehmer found $161038 = 2 \cdot 73 \cdot 1103$ and Beeger showed that there are infinitely many. The next few are 215326, 2568226, 3020626, 7866046, 9115426, 49699666, 143742226, ...

If F_n is the Fermat number $2^{2^n} + 1$, Cipolla showed that $F_{n_1} F_{n_2} \cdots F_{n_k}$ is psp(2) if $k > 1$ and $n_1 < n_2 < \ldots < n_k < 2^{n_1}$.

If $P_n^{(a)}$ is the n-th psp(a), Szymiczek has shown that $\sum 1/P_n^{(2)}$ is convergent, while Mąkowski has shown that $\sum 1/\ln P_n^{(a)}$ is divergent. Rotkiewicz has a booklet on pseudoprimes which contains 58 problems and 20 conjectures.

For example, Problem #22 asks if there is a pseudoprime of form $2^N - 2$. Wayne McDaniel answers this affirmatively with $N = 465794$. Rotkiewicz has shown that the congruence $2^{n-2} \equiv 1 \bmod n$ has infinitely many composite solutions n. Shen Mok-Kong found five such less than a million, each of which ended in 7. McDaniel and Zhang Ming-Zhi have given the examples $73 \cdot 48544121$ and $524287 \cdot 13264529$ which each show that 3 is also a possible final digit.

Selfridge, Wagstaff & Pomerance offer $500.00 + $100.00 + $20.00 for a composite $n \equiv 3$ or $7 \bmod 10$ which divides both $2^n - 2$ and the Fibonacci number u_{n+1} (see **A3**) or $20.00 + $100.00 + $500.00 for a proof that there is no such n.

Shen Mok-Kong has shown that there are infinitely many k such that $2^{n-k} \equiv 1 \bmod n$ has infinitely many composite solutions n, and Kiss & Phong have shown that this is so for all $k \geq 2$ and for all $a \geq 2$ in place of 2.

The Lucas sequence $V_n = V_n(b, c)$ associated with the pair (b, c) is given by $V_0 = 2$, $V_1 = b$, $V_{n+2} = bV_{n+1} - cV_n$. For every prime p and integer m, $V_p(m, -1) \equiv m \pmod{p}$. An odd composite integer n with $V_n(m, -1) \equiv m \pmod{n}$ is a **Fibonacci pseudoprime** of the m-th kind. A **strong Fibonacci pseudoprime** is a Fibonacci pseudoprime of the m-th kind for all m. To be a strong Fibonacci pseudoprime, n must be squarefree, and $p-1$ must divide $n-1$ for each p that divides n, so that n must be a Carmichael number [**A13**].

Somer shows that if an even Fibonacci pseudoprime exists, it is $> 28 \cdot 10^{12}$.

Richard Pinch has found the smallest strong Dickson pseudoprime with parameter $c = -1$ to be $17 \cdot 31 \cdot 41 \cdot 43 \cdot 89 \cdot 97 \cdot 167 \cdot 331$.

Interesting is the Perrin sequence,

$$a_0 = 3, 0, 2, 3, 2, 5, 5, 7, 10, 12, 17, 22, 29, 39, 51, 68, 90, 119, 158, 209, \ldots$$

generated by the cubic recurrence $a_{n+3} = a_n + a_{n+1}$, which has the property that p divides a_p whenever p is prime. This gives rise to the concept of **Perrin pseudoprime**; the first two such are $271441 = 521^2$ and $904631 = 7 \cdot 13 \cdot 9941$. Adams & Shanks give a more restrictive definition (not satisfied by 271441 or 904631) and partitioned them into classes S, I and Q. Kurtz, Shanks & Williams listed the 55 S-Perrin pseudoprimes $< 10^9$ and Arno showed that there were none of type I or $Q < 10^{14}$. If $L > 0$ and $M \neq 0$ are coprime and α, β are the roots of $z^2 - \sqrt{L}z + M$, define $D = L - 4M$ and the **Lehmer sequences** $P_n = (\alpha^n - \beta^n)/(\alpha^e - \beta^e)$ where $e = 1$ or 2 according as n is odd or even, and $V_n = (\alpha^n + \beta^n)/(\alpha + \beta)^{2-e}$. For each n define d and s by $n - \left(\frac{DL}{n}\right) = d \cdot 2^s$ with d odd. An odd composite number n is a **strong Lehmer pseudoprime** for the bases α and β if $n \perp DL$ and either n divides P_d or n divides $V_{d \cdot 2^r}$ for some r, $0 \leq r < s$. Rotkiewicz proves that if α/β is not a root of unity, then every A.P. $ax + b$ with $a \perp b$ contains infinitely many odd strong Lehmer pseudoprimes for the bases α and β.

Stan Wagon notices, in Daniel Bleichenbacher's list of pseudoprimes to 10^{17}, the example $n = 134670080641 = 211873 \cdot 635617$, which is a k-strong pseudoprime for $k = 2$, 3 and 5, but not for $k = 6$. So, the first 'strong pseudoprime witness' to the composite nature of n is 6, which is not a prime.

Shyam Sunder Gupta is interested in palindromic pseudoprimes, base 2. These include 101101, 129921, 1837381, of which the first is a Carmichael number. In March & October 2002 he adds 127665878878566721 and $1037998220228997301 = 41 \cdot 101 \cdot 211 \cdot 251 \cdot 757 \cdot 1321 \cdot 4733$

William Adams & Daniel Shanks, Strong primality tests that are not sufficient, *Math. Comput.*, **39**(1982) 255–300.

Richard André-Jeannin, On the existence of even Fibonacci pseudoprimes with parameters P and Q, *Fibonacci Quart.*, **34**(1996) 75–78; *MR* **96m**:11013.

F. Arnault, The Rabin-Monier theorem for Lucas pseudoprimes, *Math. Comput.*, **66**(1997) 869–881; *MR* **97f**:11009.

Steven Arno, A note on Perrin pseudoprimes, *Math. Comput.*, **56**(1991) 371–376. 371–376; *MR* **91k**:11011.

N. G. W. H. Beeger, On even numbers m dividing $2^m - 2$, *Amer. Math. Monthly*, **58**(1951) 553–555; *MR* **13**, 320.

Jerzy Browkin, Some new kinds of pseudoprimes, *Math. Comput.*, **73**(2004) 1031–1037.

Paul S. Bruckman, On the infinitude of Lucas pseudoprimes, *Fibonacci Quart.*, **32** (1994) 153–154; *MR* **95c**:11011.

Paul S. Bruckman, Lucas pseudoprimes are odd, *Fibonacci Quart.*, **32** (1994) 155–157; *MR* **95c**:11012.

Paul S. Bruckman, On a conjecture of Di Porto and Filipponi, *Fibonacci Quart.*, **32**(1994) 158–159; *MR* **95c**:11013.

Paul S. Bruckman, Some interesting subsequences of the Fibonacci and Lucas pseudoprimes, *Fibonacci Quart.*, **34**(1996) 332–341; *MR* **97e**:11022.

Bui Minh-Phong, On the distribution of Lucas and Lehmer pseudoprimes. Festscrift for the 50th birthday of Karl-Heinz Indlekofer, *Ann. Univ. Sci. Budapest. Sect. Comput.*, **14**(1994) 145–163; *MR* **96d**:11106.

Walter Carlip, Eliot Jacobson & Lawrence Somer, Pseudoprimes, perfect numbers and a problem of Lehmer, *Fibonacci Quart.*, **36**(1998) 361–371; *MR* **99g**:11013.

R. D. Carmichael, On composite numbers P which satisfy the Fermat congruence $a^{P-1} \equiv 1 \bmod P$, *Amer. Math. Monthly*, **19**(1912) 22–27.

M. Cipolla, Sui numeri composti P che verificiano la congruenza di Fermat, $a^{P-1} \equiv 1 \pmod{P}$, *Annali di Matematica*, **9**(1904) 139–160.

John H. Conway, Richard K. Guy, William A. Schneeberger & N. J. A. Sloane, The primary pretenders, *Acta. Arith.*, **78**(1997) 307–313; *MR* (**97k**:11004.

Adina Di Porto, Nonexistence of even Fibonacci pseudoprimes of the 1st kind, *Fibonacci Quart.*, **31**(1993) 173–177; *MR*94d:11007.

L. A. G. Dresel, On pseudoprimes related to generalized Lucas sequences, *Fibonacci Quart.*, **35**(1997) 35–42; *MR* **98b**:11010.

P. Erdős, On the converse of Fermat's theorem, *Amer. Math. Monthly*, **56** (1949) 623–624; *MR* **11**, 331.

P. Erdős, On almost primes, *Amer. Math. Monthly*, **57**(1950) 404–407; *MR* **12**, 80.

P. Erdős, P. Kiss & A. Sárközy, A lower bound for the counting function of Lucas pseudoprimes, *Math. Comput.*, **51**(1988) 259–279; *MR* **89e**:11011.

P. Erdős & C. Pomerance, The number of false witnesses for a composite number, *Math. Comput.*, **46**(1986) 259–279; *MR* **87i**:11183.

J. B. Friedlander, Shifted primes without large prime factors, in *Number Theory and Applications*, (Proc. NATO Adv. Study Inst., Banff, 1988), Kluwer, Dordrecht, 1989, 393–401;

Horst W. Gerlach, On liars and witnesses in the strong pseudoprimality test, *New Zealand J. Math.*, **24**(1995) 5–15.

Daniel M. Gordon & Carl Pomerance, The distribution of Lucas and elliptic pseudoprimes, *Math. Comput.*, **57**(1991) 825–838; *MR* **92h**:11081; corrigendum **60**(1993) 877; *MR* **93h**:11108.

S. Gurak, Pseudoprimes for higher-order linear recurrence sequences, *Math. Comput.*, **55**(1990) 783–813.

S. Gurak, On higher-order pseudoprimes of Lehmer type, *Number Theory* (*Halifax NS*, 1994) 215–227, *CMS Conf. Proc.*, **15**, Amer. Math. Soc., 1995; *MR* **96f**:11167.

Hideji Ito, On elliptic pseudoprimes, *Mem. College Ed. Akita Univ. Natur. Sci.*, **46**(1994) 1–7; *MR* **95e**:11009.

Gerhard Jaeschke, On strong pseudoprimes to several bases, *Math. Comput.*, **61**(1993) 915–926; *MR* **94d**:11004.

I. Joó, On generalized Lucas pseudoprimes, *Acta Math. Hungar.*, **55**(1990) 315–322.

I. Joó & Phong Bui-Minh, On super Lehmer pseudoprimes, *Studia Sci. Math. Hungar.*, **25**(1990) 121–124; *MR* **92d**:11109.

Kim Su-Hee & Carl Pomerance, The probability that a random probable prime is composite, *Math. Comput.*, **53**(1989) 721–741; *MR* **90e**:11190.

Péter Kiss & Phong Bui-Minh, On a problem of A. Rotkiewicz, *Math. Comput.*, **48**(1987) 751–755; *MR* **88d**:11004.

G. Kowol, On strong Dickson pseupoprimes, *Appl. Algebra Engrg. Comm. Comput.*, **3**(1992) 129–138; *MR* **96g**:11005.

D. H. Lehmer, On the converse of Fermat's theorem, *Amer. Math. Monthly*, **43**(1936) 347–354; II **56**(1949) 300–309; *MR* **10**, 681.

Rudolf Lidl, Winfried B. Müller & Alan Oswald, Some remarks on strong Fibonacci pseudoprimes, *Appl. Algebra Engrg. Comm. Comput.*, **1**(1990) 59–65; *MR* **95k**:11015.

Andrzej Mąkowski, On a problem of Rotkiewicz on pseudoprime numbers, *Elem. Math.*, **29**(1974) 13.

A. Mąkowski & A. Rotkiewicz, On pseudoprime numbers of special form, *Colloq. Math.*, **20**(1969) 269–271; *MR* **39** #5458.

Wayne L. McDaniel, Some pseudoprimes and related numbers having special forms, *Math. Comput.*, **53**(1989) 407–409; *MR* **89m**:11006.

F. Morain, Pseudoprimes: a survey of recent results, *Eurocode '92, Springer Lecture Notes*, **339**(1993) 207–215; *MR* **95c**:11172.

Siguna Müller, On strong Lucas pseudoprimes, *Contributions to general algebra*, **10** (*Klagenfurt*, 1997) 237–249, Heyn, Klagenfurt, 1998; *MR* **99k**:11008.

Siguna Müller, A note on strong Dickson pseudoprimes, *Appl. Algebra Engrg. Comm. Comput.*, **9**(1998) 247–264; *MR* **99j**:11008.

Winfried B. Müller & Alan Oswald, Generalized Fibonacci pseudoprimes and probablr primes, *Applications of Fibonacci numbers, Vol.5* (*St. Andrews*, 1992), 459–464, Kluwer Acad. Publ.. Dordrecht, 1993; *MR* **95f**:11105.

R. Perrin, Item 1484, *L'Intermédiaire des mathématicians*, **6**(1899) 76–77; see also E. Malo, *ibid.*, **7**(1900) 312 & E. B. Escott, *ibid.*, **8**(1901) 63–64.

Richard G. E. Pinch, The pseudoprimes up to 10^{13}, *Algorithmic Number Theory* (*Leiden*, 2000), *Springer Lecture Notes in Comput. Sci.*, **1838**(2000) 459–473; *MR* **2002g**:11177.

Carl Pomerance, A new lower bound for the pseudoprime counting function, *Illinois J. Math.*, **26**(1982) 4–9; *MR* **83h**:10012.

Carl Pomerance, On the distribution of pseudoprimes, *Math. Comput.*, **37** (1981) 587–593; *MR* **83k**:10009.

Carl Pomerance, Two methods in elementary analytic number theory, in *Number Theory and Applications*, (Proc. NATO Adv. Study Inst., Banff, 1988), Kluwer, Dordrecht, 1989, 135–161;

Carl Pomerance, John L. Selfridge & Samuel S. Wagstaff, The pseudoprimes to $25 \cdot 10^9$, *Math. Comput.*, **35**(1980) 1003–1026; *MR* **82g**:10030.

A. Rotkiewicz, *Pseudoprime Numbers and their Generalizations*, Student Association of the Faculty of Sciences, Univ. of Novi Sad, 1972; *MR* **48** #8373; *Zbl.* 324.10007.

A. Rotkiewicz, Sur les diviseurs composés des nombres $a^n - b^n$, *Bull. Soc. Roy. Sci. Liège*, **32**(1963) 191–195; *MR* **26** #3645.

A. Rotkiewicz, Sur les nombres pseudopremiers de la forme $ax + b$, *Comptes Rendus Acad. Sci. Paris*, **257**(1963) 2601–2604; *MR* **29** #61.

A. Rotkiewicz, Sur les formules donnant des nombres pseudopremiers, *Colloq. Math.*, **12**(1964) 69–72; *MR* **29** #3416.

A. Rotkiewicz, Sur les nombres pseudopremiers de la forme $nk + 1$, *Elem. Math.*, **21**(1966) 32–33; *MR* **33** #112.

A. Rotkiewicz, On Euler-Lehmer pseudoprimes and strong Lehmer pseudoprimes with parameters L, Q in arithmetic progressions, *Math. Comput.*, **39**(1982) 239–247; *MR* **83k**:10004.

A. Rotkiewicz, On the congruence $2^{n-2} \equiv 1(\mathrm{mod}\ n)$, *Math. Comput.*, **43** (1984) 271–272; *MR* **85e**:11005.

A. Rotkiewicz, On strong Lehmer pseudoprimes in the case of negative discriminant in arithmetic progressions, *Acta Arith.*, **68**(1994) 145–151; *MR* **96h**:11008.

A. Rotkiewicz, Arithmetical progressions formed by k different Lehmer pseudoprimes, *Rend. Circ. Mat. Palermo* (2), **43**(1994) 391–402; *MR* **96f**:11026.

A. Rotkiewicz, On Lucas pseudoprimes of the form $ax^2 + bxy + cy^2$, *Applications of Fibonacci numbers, Vol.* 6 (*Pullman WA* 1994), 409–421, Kluwer Acad. Publ., Dordrecht 1996; *MR***9??**:110??.

A. Rotkiewicz, On Lucas pseudoprimes of the form $ax^2 + bxy + cy^2$, *Applications of Fibonacci numbers, Vol.* 6 (*Pullman WA,* 1994) 409–421, Kluwer Acad. Publ. Dordrecht, 1996; *MR* **97d**:11008.

A. Rotkiewicz, On the theorem of Wójcik, *Glasgow Math. J.*, **38**(1996) 157–162; *MR* **97i**:11010.

A. Rotkiewicz, On Lucas pseudoprimes of the form $ax^2 + bxy + cy^2$ in arithmetic progressions $AX + B$ with a prescribed value of the Jacobi symbol, *Acta Math. Inform. Univ. Ostraviensis*, **10**(2002) 103–109; *MR* **2003i**:11004.

A. Rotkiewicz & K. Ziemak, On even pseudoprimes, *Fibonacci Quart.*, **33** (1995) 123–125; *MR* **96c**:11005.

Shen Mok-Kong, On the congruence $2^{n-k} \equiv 1(\mathrm{mod}\ n)$, *Math. Comput.*, **46**(1986) 715–716; *MR* **87e**:11005.

Lawrence Somer, On even Fibonacci pseudoprimes, *Applications of Fibonacci numbers*, 4(1990), Kluwer, 1991, 277–288; *MR* **94b**:11014.

George Szekeres, Higher order pseudoprimes in primality testing, *Combinatorics, Paul Erdős is eighty, Vol.*2 (*Keszthely,* 1993) 451–458, *Bolyai Soc. Math. Stud.*, **2** János Bolyai Math. Soc., Budapest, 1996.

K. Szymiczek, On prime numbers p, q and r such that pq, pr and qr are pseudoprimes, *Colloq. Math.*, **13**(1964–65) 259–263; *MR* **31** #4757.

K. Szymiczek, On pseudoprime numbers which are products of distinct primes, *Amer. Math. Monthly*, **74**(1967) 35–37; *MR* **34** #5746.

S. S. Wagstaff, Pseudoprimes and a generalization of Artin's conjecture, *Acta Arith.*, **41**(1982) 151–161; *MR* **83m**:10004.

Masataka Yorinaga, Search for absolute pseudoprime numbers (Japanese), *Sûgaku*, **31** (1979) 374–376; *MR* **82c**:10008.

Zhang Zhen-Xiang, Finding strong pseudoprimes to several bases, *Math. Comput.*, **70**(2001) 863–872; *MR* **2001g**:11009.

Zhang Zhen-Xiang & Min Tang, Finding strong pseudoprimes to several bases, II, *Math. Comput.*, **72**(2003) 2085–2097; *MR* **2004c**:11008.

OEIS: A001262, A001567, A005935-005939, A006935, A006945, A006970, A047713.

A13 Carmichael numbers.

The Carmichael numbers ($\text{psp}(a)$ for all a prime to n, n composite) must be the product of at least three odd prime factors. As long ago as 1899 Korselt had given a necessary and sufficient condition for n to be a Carmichael number; namely that n be squarefree and such that $(p-1)|(n-1)$ for each p that divides n. The smallest example is $561 = 3 \cdot 11 \cdot 17$. More generally, if $p = 6k + 1$, $q = 12k + 1$ and $r = 18k + 1$ are each prime, then pqr is a Carmichael number. It seems certain that there are infinitely many such triples of primes, but beyond our means to prove it. Alford, Granville & Pomerance have shown (by a different method!) that there are infinitely many Carmichael numbers, in fact, for sufficiently large x, more than x^β of them less than x, where

$$\beta = \frac{5}{12} \left(1 - \frac{1}{2\sqrt{e}}\right) > 0.290306 > \frac{2}{7}$$

Erdős had conjectured that $(\ln C(x))/\ln x$ tends to 1 as x tends to infinity and he improved a result of Knödel to show that

$$C(x) < x \exp\{-c \ln x \ln \ln \ln x / \ln \ln x\}.$$

Then Pomerance, Selfridge & Wagstaff (see **A12**) proved this with $c = 1 - \epsilon$ and give a heuristic argument supporting the conjecture that the reverse inequality holds with $c = 2 + \epsilon$.

They found 2163 Carmichael numbers $< 25 \cdot 10^9$ and Jaeschke finds 6075 more between that bound and 10^{12}; seven of these have eight prime factors. Richard Pinch has corrected these counts to

	8241	19279	44706	105212	246683	such numbers
$<$	10^{12}	10^{13}	10^{14}	10^{15}	10^{16}	

A fair-sized specimen is

$$2013745337604001 = 17 \cdot 37 \cdot 41 \cdot 131 \cdot 251 \cdot 571 \cdot 4159.$$

J. R. Hill found the large Carmichael number pqr where $p = 5 \cdot 10^{19} + 371$, $q = 2p - 1$ and $r = 1 + (p - 1)(q + 2)/433$. Wagstaff produced a 321-digit example, and Woods & Huenemann one of 432 digits. Dubner has continued to beat this and his own records, a 3710-digit specimen being $N = PQR$ where $P = 6M + 1$, $Q = 12M + 1$ and $R = 1 + (PQ - 1)/X$ are primes given by $M = (TC - 1)^A/4$, T the product of the odd primes up to 47, $C = 141847$, $A = 41$ and $X = 123165$. But Günter Löh & Wolfgang Niebuhr have developed new algorithms which completely eclipse these by producing a Carmichael number with no fewer than 1101518 prime factors, a number of 16142049 decimal digits!

David Broadhurst has found a 60351-digit Carmichael number with the minimum number, three, of prime factors, eclipsing Carmody's earlier 30052-digit record. Harvey Dubner found a 3003-digit 4-component specimen, eclipsing the previous 1093 digits.

Alford, Granville & Pomerance proved that there are infinitely many Carmichael numbers n with the stronger requirement that $p|n$ implies that $(p^2 - 1)|(n - 1)$, but didn't know of any examples. Sid Graham found 18 such numbers, the smallest being

$$589398328999039533470037072001 = 29 \cdot 31 \cdot 37 \cdot 43 \cdot 53 \cdot 67 \cdot 79 \cdot 89 \cdot 97 \cdot 151 \cdot 181 \cdot 191 \cdot 419 \cdot 881 \cdot 883$$

Richard Pinch had already found the smallest of all:

$$443372888629441 = 17 \cdot 31 \cdot 41 \cdot 43 \cdot 89 \cdot 97 \cdot 167 \cdot 331$$

Graham found 17 other numbers that satisfy the slightly weaker condition $\frac{p^2-1}{2} \mid (n-1)$.

Granville conjectures that if $C_k(x)$ is the number of Carmichael numbers with k prime factors, then $C_k(x) \ll x^{1/k+o(1)}$ as $x \to \infty$. Balasubramanian & Nataraj show that $C_3(x) \ll x^{5/14+o(1)}$.

Marko shows that for each $k \geq 3$ Schinzel's conjecture (see **A2**) implies that there are infinitely many Carmichael numbers with k distinct prime factors.

Pinch lists one more Carmichael number with 6 prime factors and 2 more with 7 prime factors than does Jaesche, and discusses other discrepancies with earlier tables. He asks if there is an n with $p \mid n$ implies both $p - 1 \mid n - 1$ (so Carmichael) and $p + 1 \mid n + 1$. There is none up to 10^{16}. And is it true that $C_k(x)$, the number of Carmichael numbers up to x with exactly k prime factors is 'essentially' $x^{1/k}$, i.e., is bounded above and below by this times powers of $\ln x$:

$$¿ \quad \ln C_k(x) = (1/k)\ln x + O(\ln \ln x) \quad ?$$

Zhang Ming-Zhi found a Carmichael number with over a thousand prime factors and Löh & Niebuhr raised this to more than a million. Alford

& Grantham show that there are Carmichael numbers with k factors for each k, $3 \le k \le 19565220$.

W. R. Alford & Jon Grantham, There are Carmichael numbers with many prime factors, (2003 preprint)

W. Red Alford, Andrew Granville & Carl Pomerance, On the difficulty of finding reliable witnesses, in *Algorithmic Number Theory (Ithaca, NY, 1994)*, 1–16, *Lecture Notes in Comput. Sci.*, **877**, Springer, Berlin, 1994; *MR* **96d**:11136.

W. Red Alford, Andrew Granville & Carl Pomerance, There are infinitely many Carmichael numbers, *Ann. of Math.*(2), **139**(1994) 703–722; *MR* **95k**:11114.

François Arnault, Constructing Carmichael numbers which are strong pseudo-primes to several bases, *J. Symbolic Comput.*, **20**(1995) 151–161; *MR* **96k**:11153.

Robert Baillie & Samuel S. Wagstaff, Lucas pseudoprimes, *Math. Comput.*, **35**(1980) 1391–1417.

Ramachandran Balasubramanian & S. V. Nagaraj, Density of Carmichael numbers with three prime factors, *Math. Comput.*, **66**(1997) 1705–1708; *MR* **98d**:11110.

N. G. W. H. Beeger, On composite numbers n for which $a_{n-1} \equiv 1 \bmod n$ for every a prime to n, *Scripta Math.*, **16**(1950) 133–135.

R. D. Carmichael, Note on a new number theory function, *Bull. Amer. Math. Soc.*, **16**(1909–10) 232–238.

Harvey Dubner, A new method for producing large Carmichael numbers, *Math. Comput.*, **53**(1989) 411–414; *MR* **89m**:11013.

Harvey Dubner, Carmichael numbers and Egyptian fractions, *Math. Japon.*, **43**(1996) 411–419; *MR* **97d**:11011.

Harvey Dubner, Carmichael numbers of the form $(6m+1)(12m+1)(18m+1)$, *J. Integer Seq.*, **5**(2002) A02.2.1, 8pp. (electronic); *MR* **2003g**:11144.

H. J. A. Duparc, On Carmichael numbers, *Simon Stevin*, **29**(1952) 21–24; *MR* **14**, 21f.

P. Erdős, On pseudoprimes and Carmichael numbers, *Publ. Math. Debrecen*, **4**(1956) 201–206; *MR* **18**, 18.

Andrew Granville, Prime testing and Carmichael numbers, *Notices Amer. Math. Soc.*, **39**(1992) 696–700.

Andrew Granville & Carl Pomerance, Two contradictory conjectures concerning Carmichael numbers, *Math. Comput.*, **71**(2002) 883–908; *MR* **2003d**:11148.

D. Guillaume, Table de nombres de Carmichael inférieurs à 10^{12}, preprint, May 1991

D. Guillaume & F. Morain, Building Carmichael numbers with a large number of prime factors and generalization to other numbers, preprint June, 1992.

Jay Roderick Hill, Large Carmichael numbers with three prime factors, Abstract 79T-A136, *Notices Amer. Math. Soc.*, **26**(1979) A-374.

Gerhard Jaeschke, The Carmichael numbers to 10^{12}, *Math. Comput.*, **55** (1990) 383–389; *MR* **90m**:11018.

I. Joó & Phong Bui-Minh, On super Lehmer pseudoprimes, *Studia Sci. Math. Hungar.*, **25**(1990) 121–124.

W. Keller, The Carmichael numbers to 10^{13}, *Abstracts Amer. Math. Soc.*, **9**(1988) 328–329.

W. Knödel, Eine obere Schranke für die Anzahl der Carmichaelschen Zahlen kleiner als x, *Arch. Math.*, **4**(1953) 282–284; *MR* **15**, 289 (and see *Math. Nachr.*, **9**(1953) 343–350).

A. Korselt, Problème chinois, *L'intermédiaire des math.*, **6**(1899) 142–143.

D. H. Lehmer, Strong Carmichael numbers, *J. Austral. Math. Soc. Ser. A*, **21**(1976) 508–510.

G. Löh, Carmichael numbers with a large number of prime factors, *Abstracts Amer. Math. Soc.*, **9**(1988) 329; II (with W. Niebuhr) **10**(1989) 305.

Günter Löh & Wolfgang Niebuhr, A new algorithm for constructing large Carmichael numbers, *Math. Comput.*, **65**(1996) 823–836; *MR* **96g**:11123.

František Marko, A note on pseudoprimes with respect to abelian linear recurring sequence, *Math. Slovaca*, **46**(1996) 173–176; *MR* **97k**:11009.

Rex Matthews, Strong pseudoprimes and generalized Carmichael numbers, *Finite fields: theory, applications and algorithms* (*Las Vegas NV* 1993), 227–223, *Contemp. Math.*, **168** Amer. Math. Soc., 1994; *MR* **95g**:11125.

Siguna M. S. Müller, Carmichael numbers and Lucas tests, *Finite fields: theory, applications and algorithms* (*Waterloo ON*, 1997) 193–202, *Contemp. Math.*, **225**, Amer. Math. Soc., 1999; *MR* **99m**:11008.

R. G. E. Pinch, The Carmichael numbers up to 10^{15}, *Math. Comput.*, **61** (1993) 381–391; *MR* **93m**:11137.

A. J. van der Poorten & A. Rotkiewicz, On strong pseudoprimes in arithmetic progressions, *J. Austral. Math. Soc. Ser. A*, **29**(1980) 316–321.

S. S. Wagstaff, Large Carmichael numbers, *Math. J. Okayama Univ.*, **22** (1980) 33–41; *MR* **82c**:10007.

H. C. Williams, On numbers analogous to Carmichael numbers, *Canad. Math. Bull.*, **20**(1977) 133–143.

Dale Woods & Joel Huenemann, Larger Carmichael numbers, *Comput. Math. Appl.*, **8**(1982) 215–216; *MR* **83f**:10017.

Masataka Yorinaga, Numerical computation of Carmichael numbers, I, II, *Math. J. Okayama Univ.*, **20**(1978) 151–163, **21**(1979) 183–205; *MR* **80d**:10026, **80j**:10002.

Masataka Yorinaga, Carmichael numbers with many prime factors, *Math. J. Okayama Univ.*, **22**(1980) 169–184; *MR* **81m**:10018.

Zhang Ming-Zhi, A method for finding large Carmichael numbers, *Sichuan Daxue Xuebao*, **29**(1992) 472–479; *MR* **93m**:11009.

Zhang Zhen-Xiang, A one-parameter quadratic-base version of the Baillie-PSW probable prime test, *Math. Comput.*, **71**(2002) 1699–1734.

Zhu Wen-Yu, Carmichael numbers that are strong pseudoprimes to some bases, *Sichuan Daxue Xuebao*, **34**(1997) 269–275.

OEIS: A001567, A002997, A033502, A046025.

A14 "Good" primes and the prime number graph.

Erdős and Straus called the prime p_n **good** if $p_n^2 > p_{n-i}p_{n+i}$ for all i, $1 \leq i \leq n - 1$; for example, 5, 11, 17 and 29. Pomerance used the "prime number graph" (see **A5**) to show that there are infinitely many good primes. He asks the following questions. Is it true that the set of n for which p_n is good has density 0? Are there infinitely many n with $p_n p_{n+1} > p_{n-i}p_{n+1+i}$ for all i, $1 \leq i \leq n - 1$? Are there infinitely many n with $p_n + p_{n+1} > p_{n-i} + p_{n+1+i}$ for all i, $1 \leq i \leq n - 1$? Does the set of n for which $2p_n < p_{n-i} + p_{n+i}$ for all i, $1 \leq i \leq n - 1$ have density 0? (Pomerance proved that there were infinitely many such n.) Is $\limsup\{\min_{0<i<n}(p_{n-i} + p_{n+i}) - 2p_n\} = \infty$?

Erdős has given another definition of good prime: call p good if every even number $2r \leq p - 3$ can be written as the difference of two primes not greater than p. The first bad prime is 97. Selfridge & Blecksmith have tables of good primes up to 10^{37}.

OEIS: A028388, A046868-046869.

A15 Congruent products of consecutive numbers.

Erdős, in a letter dated 79-10-31, observes that $3 \cdot 4 \equiv 5 \cdot 6 \cdot 7 \equiv 1 \bmod 11$ and asks for the least prime p such that there are integers a, k_1, k_2, k_3 and

$$\prod_{i=1}^{k_1}(a + i) \equiv \prod_{i=1}^{k_2}(a + k_1 + i) \equiv \prod_{i=1}^{k_3}(a + k_1 + k_2 + i) \equiv 1 \bmod p.$$

He suggests that such primes p exist for any number of such congruent products.

Mąkowski sends examples corresponding to rows $n = 5$ and 6 in the table below [compare **F11**] and says that tables of indices can be used to find others. W. Narkiewicz also sends these examples, together with those in rows $n = 7$, 8 and 9 below. Landon Noll & Chuck Simmons generalize the problem slightly by asking for solutions of

$$q_1! \equiv q_2! \equiv \ldots \equiv q_n! \bmod p$$

and they give the least prime p for which there is a solution with n terms.

n	p	q_1	q_2	q_3	q_4	q_5	q_6	q_7	q_8	q_9	q_{10}	q_{11}
1	2	0										
2	2	0	1									
3	5	0	1	3								
4	17	0	1	5	11							
5	17	0	1	5	11	15						
6	23	0	1	4	8	11	21					
7	71	8	10	20	52	62	64	71				
8	599	29	51	123	184	251	290	501	540			
9	599	29	51	123	184	251	290	501	540	556		
10	3011	0	1	611	723	749	805	2205	2261	2287	2399	
11	3011	0	1	611	723	749	805	2205	2261	2287	2399	3009

Andrzej Mąkowski, On a number-theoretic problem of Erdős, *Elem. Math.*, **38** (1983) 101–102.

A16 Gaussian and Eisenstein-Jacobi primes.

Prime numbers can be defined in fields other than the rational field. In the complex number field they are called **Gaussian primes**. Many problems on ordinary primes can be reformulated for Gaussian primes.

Gaussian integers $a + bi$, where a, b are integers and $i^2 = -1$, behave like ordinary integers in the sense that there is **unique factorization** (apart from order, **units** ($\pm 1, \pm i$) and **associates**; the associates of 7, for example, are 7, -7, $7i$ and $-7i$).

Primes of shape $4k - 1$, i.e., 3, 7, 11, 19, 23, ..., are still primes in the ring of Gaussian integers, but the other ordinary primes can be factored into Gaussian primes:

$$2 = (1 + i)(1 - i), \qquad 5 = (2 + i)(2 - i) = -(2i - 1)(21 + 1), \text{ etc.}$$

$$13 = (2 + 3i)(2 - 3i), \qquad 17 = (4 + i)(4 - i), \qquad 29 = (5 + 2i)(5 - 2i), \ldots .$$

The Gaussian primes $\pm 1 \pm i$, $\pm 1 \pm 2i$, $\pm 2 \pm i$, ± 3, $\pm 3i$, $\pm 2 \pm 3i$, $\pm 3 \pm 2i$, $\pm 4 \pm i$, $\pm 1 \pm 4i$, $\pm 5 \pm 2i$, $\pm 2 \pm 5i$, ... make a pleasing pattern (Figure 4) when drawn on an Argand diagram, which has been used for tiling floors and weaving tablecloths.

Motzkin and Gordon asked if one can "walk" from the origin to infinity using the Gaussian primes as "stepping stones" and taking steps of bounded length. Presumably not. Jordan & Rabung have shown that steps of length at least 4 are necessary.

Gethner, Wagon & Wick have produced a moat of width $\sqrt{26}$.

Haugland argues that a corresponding walk on the Eisenstein-Jacobi primes would require unbounded steps.

The **Eisenstein-Jacobi integers** $a + b\omega$, where a, b are integers and ω is a complex cube root of unity, $\omega^2 + \omega + 1 = 0$, also enjoy unique factorization. The primes again form a pattern (Figure 5), this time with hexagonal symmetry, because there are six units, ± 1, $\pm\omega$, $\pm\omega^2$. The prime 2 and those of shape $6k-1$ (5, 11, 17, 23, 29, 41, ...,) are still Eisenstein-Jacobi primes, but 3 and those of shape $6k+1$ can be factored:

$$3 = (1-\omega)(1-\omega^2), \qquad 7 = (2-\omega)(2-\omega^2), \qquad 13 = (3-\omega)(3-\omega^2),$$

$$19 = (3-2\omega)(3-2\omega^2), \quad 31 = (5-\omega)(5-\omega^2), \quad 37 = (4-3\omega)(4-3\omega^2), \; \ldots .$$

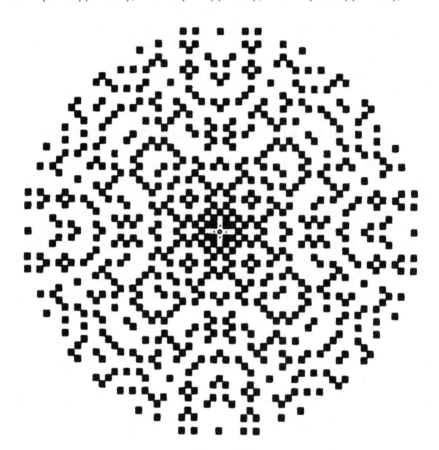

Figure 4. The Gaussian Primes with Norm Less Than 1000.

John Leech asks for long arithmetic progressions of Gaussian primes and also of Eisenstein-Jacobi primes. He finds nine in Figure 4 and twelve in Figure 5. He later found the arithmetic progression

$$-8-13i, \; -3-8i, \; 2-3i, \; 7+2i, \; \ldots, \; 37+32i$$

of ten Gaussian primes, the last three of which are outside Figure 4.

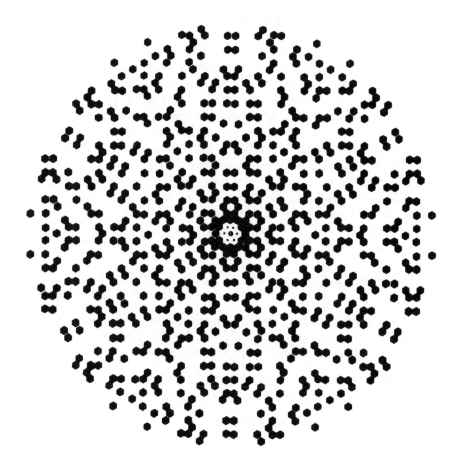

Figure 5. The Eisenstein-Jacobi Primes.

Jan Kristian Haugland, A walk on complex primes (Norwegian), *Normat*, **43**(1995) 168–170, 192; *MR* **96h**:11105.

Guediri Hedy, Sur le répartition des nombres de Gauss dans une progression arithmétique, *Math. Montisnigri*, **4**(1995) 19–26.

Ellen Gethner, Stan Wagon & Brian Wick, A stroll through the Gaussian primes, *Amer. Math. Monthly*, **105**(1998) 327–337; *MR* **99b**:11006.

J. H. Jordan & J. R. Rabung, A conjecture of Paul Erdős concerning Gaussian primes, *Math. Comput.*, **24**(1970) 221–223.

John Renze, Stan Wagon & Brian Wick, The Gaussian zoo, *Experiment. Math.*, **10**(2001) 161–173; *MR* **2002g**:11110.

OEIS: A004016, A027206, A035019, A055025-055029, A055664-055668, A064152.

A17 Formulas for primes.

Perhaps the philosopher's stone of number theory is a formula for p_n, or for $\pi(x)$, or for a necessary and sufficient condition for primality. Wilsons theorem@Wilson's theorem, that $(p-1)! \equiv -1 \pmod{p}$, seems to be unique (is Vantieghem's result, that $p > 2$ is prime if and only if $\prod_{d=1}^{p-1}(2^d - 1) \equiv p \bmod 2^p - 1$, equivalent to it?); but even that is useless for computation. C. P. Willans and C. P. Wormell used it to give formulas which use only elementary functions, but which are too clumsy to print here. The Mann-Shanks algorithm is another curiosity, of even less practical value. Matiyasevich and other logicians have used Wilson's theorem and their solution of Hilbert's tenth problem to produce polynomials the positive part of whose range is exactly the set of primes.

Three theorems of Boris Stechkin may be worth recording. They are based on the function

$$S(n) = \# \left\{ m \; : \; 2 \le m \le n, \; (m-1) \left| \left\lfloor \frac{n(m-1)}{m} \right\rfloor \right. \right\}$$

(1) $n-1$ is prime just if $S(n) = d(n)$, the number of divisors of n,

(2) $n \pm 1$ are twin primes just if $S(n) + S(n+1) = 2d(n)$.

(3) $p < q$ odd primes implies $S(q) - S(q-1) + S(q-2) - \ldots - S(p+1) = 0$.

The numerous papers on this topic vary widely in their sophistication and in their aim. It seems desirable to distinguish between

1. A formula for $\pi(x)$ as a function of x.

2. A formula for p_n as a function of n.

3. A necessary and sufficient condition for n to be prime.

4. A function that is prime for each member of its domain.

5. A function (the positive part of) whose range consists only of primes, or consists of all of the primes.

6. A function whose range contains a high density of primes.

7. A formula for the largest prime divisor of n.

8. A formula for the prime factors of n.

9. A formula for the smallest prime greater than n.

10. A formula for p_{n+1} in terms of p_1, p_2, ..., p_n.

11. An algorithm for generating the primes.   And so on ....

Examples of each can be found in the references. We have already mentioned (in **A1**) Euler's famous formula $n^2 + n + 41$. In some sense this is best possible, but quadratic expressions with positive discriminant can yield even longer sequences of prime values (though some of them may be negative). Gilbert Fung gives $47n^2 - 1701n + 10181$, $0 \le n \le 42$, $\Delta = 979373$ and Russell Ruby $36n^2 - 810n + 2753$, $0 \le n \le 44$, $\Delta = 2^2 3^2 7213$.

The first 1000 values of Euler's formula include 581 primes. Edgar Karst beats this with 598 values of $2n^2 - 199$ and in a 91-01-01 letter, Stephen Williams announces 602 prime values of $2n^2 - 1000n - 2609$. The corresponding numbers among the first 10000 values are 4148, 4373 and 4151. However, what is significant is not the actual density over the first so many values, which clearly has to tend to zero in all cases, but the **asymptotic** density, which, if we believe Hardy & Littlewood (see **A1**), is always $c\sqrt{n}/\ln n$, and the best that can be done is to make the value of c as large as possible. Shanks has calculated $c = 3.3197732$ for Euler's formula and $c = 3.6319998$ for a polynomial $x^2 + x + 27941$ found by Beeger. Fung & Williams (see reference at **A1** and the references they give) have achieved $c = 5.0870883$ with the formula $x^2 + x + 132874279528931$.

If Δ is the discriminant of the quadratic, then the Legendre symbol (see **F5**) $\left(\frac{\Delta}{p}\right)$ takes the value 1 for very few of the small primes, p.

Sierpiński observes that it follows from Fermat's theorem that if n is prime, then n divides

$$1^{n-1} + 2^{n-1} + \ldots + (n-1)^{n-1} + 1.$$

Is the converse true? Giuga conjectured so, and verified it for $n \le 10^{1000}$ and Bedocchi verified it to $n \le 10^{1700}$. Giuga observed that a counterexample would be a Carmichael number (**A12**, **A13**), that $p|n$ would imply that $(p-1)|(n-1)$ and that

$$\sum_{p|n} \frac{1}{p} - \frac{1}{n}$$

must be an integer, so that n contains at least eight distinct prime factors. An equivalent conjecture is that

$$nB_{n-1} \equiv -1 \bmod n$$

where the **Bernoulli numbers** B_k are the coefficients in the expansion of $x/(e^x - 1) = \sum_{k \ge 0} B_k x^k / k!$ (compare **D2**).

For a good survey paper, see Bateman & Diamond.

The Borweins & Girgensohn list ten open problems. The first is a statement of Giuga's conjecture in the form: show that no integer exists which is both a Giuga number and a Carmichael number, or, more generally, that no Giuga sequence can be a Carmichael sequence. The last is the Bernoulli number conjecture, which they attribute to Takashi Agoh. Here, a **Giuga number** is a composite number n all of whose prime divisors, p, satisfy $p|(\frac{n}{p} - 1)$. A **Giuga sequence** is a finite increasing sequence $\{n_1, \ldots, n_m\}$ such that

$$\sum_{i=1}^{m} \frac{1}{n_i} - \prod_{i=1}^{m} \frac{1}{n_i} \in \mathbb{N}.$$

A **Carmichael sequence** is the corresponding generalization of Carmichael numbers (**A13**). The smallest Giuga number contains at least 12055 decimal digits. Borwein & Wong increase this to 13800 digits.

Agoh shows that if n is a Giuga number, then $a^{n-p} \equiv 1 \bmod p^3$ for each p which divides n and each a not divisible by p. Also, if the **Fermat quotient** is $q_p(a) = (a^{p-1} - 1)/p$ and the **Giuga quotient** is $G_p(a, n) = (a^{n-p} - 1)/p^3$, then

$$(p - 1)G_p(a, n) \equiv \frac{n - p}{p^2} q_p(a) \bmod p$$

for such n, p and a.

A century-and-a-half-old conjecture is that there is always a prime between x^2 and $(x + 1)^2$. See Dickson's *History*, I, pp.435–436, and, for references to Sylvester and Schur, Erdős's famous 1934 paper; also his 1955 paper. This is implied by the conjecture that there is a prime in the interval $(x, x + x^{1/2}]$. An easier problem is to let $P(x)$ denote the largest prime factor of $\prod_{x < n \le x + x^{1/2}} n$ and show that $P(x) > x^c$ for as large a value of $c \le 1$ as possible. The values $c = \frac{15}{26}$, $\frac{5}{8}$, 0.66, $\frac{1}{12}\sqrt{69} = 0.6922185\ldots$, 0.7, 0.71, 0.723, and 0.732 have been obtained by Ramachandra, Ramachandra again (and see Jutila), S. W. Graham, Jia Chao-Hua, R. C. Baker, Jia again, Jia and Liu Hong-Quan independendently, and by Baker & Harman.

Gerry Myerson asks us to prove that there is a positive constant c such that

$$\#\{n \le x : [(4/3)^n] \text{ is composite}\} > cx$$

and says that there are heuristic arguments supporting much sharper estimates for the numbers of primes and composites in an initial segment of the sequence $[(4/3)^n]$, but the statement above would already be far better than any known result. He also asks us to prove that for every non-zero real α there is a positive integer n (and hence infinitely many) such that $[10^n \alpha]$ is composite. Equivalently, show that there is no infinite sequence of primes, each obtained from the previous by tacking a single (decimal) digit on at the end. He says that the analogous result has been proved for bases 2 through 6.

Ulrich Abel & Hartmut Siebert, Sequences with large numbers of prime values, *Amer. Math. Monthly* **100**(1993) 167–169; *MR* **94b**:11092.

William W. Adams, Eric Liverance & Daniel Shanks, Infinitely many necessary and sufficient conditions for primality, *Bull. Inst. Combin. Appl.*, **3**(1991) 69–76; *MR* **90e**:11011.

Takashi Agoh, On Giuga's conjecture, *Manuscripta Math.*, **87**(1995) 501–510; *MR* **96f**:11005.

Takashi Agoh, Karl Dilcher & Ladislav Skula, Wilson quotients for composite moduli, *Math. Comput.*, **67**(1998) 843–861.

Jesús Almansa & Leonardo Prieto, New formulas for the nth prime, (spanish) *Lect. Math.*, **15**(1994) 227–231.

A. R. Ansari, On prime representing function, *Ganita*, **2**(1951) 81–82; *MR* **15**, 11.

R. C. Baker, The greatest prime factor of the integers in an interval, *Acta Arith.*, **47**(1986) 193–231; *MR* **88a**:11086.

R. C. Baker & Glyn Harman, Numbers with a large prime factor, *Acta Arith.*, **73**(1995) 119–145; *MR* **97a**:11138.

Thøger Bang, A function representing all prime numbers, *Norsk Mat. Tidsskr.*, **34**(1952) 117–118; *MR* **14**, 621.

V. I. Baranov & B. S. Stechkin, *Extremal Combinatorial Problems and their Applications*, Kluwer, 1993, Problem 2.22.

Paul T. Bateman & Harold G. Diamond, A hundred years of prime numbers, *Amer. Math. Monthly*, **103**(1996) 729–741.

Paul T. Bateman & Roger A. Horn, A heuristic asymptotic formula concerning the distribution of prime numbers, *Math. Comput.*, **16**(1962) 363–367; *MR* **26** #6139.

Christoph Baxa, Über Gandhis Primzahlformel, *Elem. Math.*, **47**(1992) 82–84; *MR* **93h**: 11007.

E. Bedocchi, Nota ad una congettura sui numeri primi, *Riv. Mat. Univ. Parma*(4), **11**(1985) 229–236.

David Borwein, Jonathan M. Borwein, Peter B. Borwein & Roland Girgensohn, Giuga's conjecture on primality, *Amer. Math. Monthly*, **103**(1996) 40–50; *MR* **97b**:11004.

Jonathan M. Borwein & Erick Bryce Wong, A survey of results relating to Giuga's conjecture on primality, *Advances in Mathematical Sciences; CRM's 25 years (Montreal, P.Q., 1994)* 13–27, *CRM Proc. Lecture Notes*, **11**, Amer. Math. Soc., Providence RI, 1997; *MR* **98i**:11005.

Nigel Boston & Marshall L. Greenwood, Quadratics representing primes, *Amer. Math. Monthly*, **102**(1995) 595–599; *MR* **96g**:11154.

J. Braun, Das Fortschreitungsgesetz der Primzahlen durch eine transcendente Gleichung exakt dargestellt, *Wiss. Beilage Jahresber. Fr. W. Gymn. Trier*, 1899, 96 pp.

Rodrigo de Castro Korgi, Myths and realities about formulas for calculating primes (Spanish), *Lect. Math.*, **14**(1993) 77–101; *MR* **95c**:11009.

R. Creighton Buck, Prime representing functions, *Amer. Math. Monthly*, **53**(1946) 265.

John H. Conway, Problem 2.4, *Math. Intelligencer*, **3**(1980) 45.

L. E. Dickson, *History of the Theory of Numbers*, Carnegie Institute, Washington, 1919, 1920, 1923; reprinted Stechert, New York, 1934; Chelsea, New York, 1952, 1966, Vol. I, Chap. XVIII.

Underwood Dudley, History of a formula for primes, *Amer. Math. Monthly*, **76**(1969) 23–28; *MR* **38** #4270.

Underwood Dudley, Formulas for primes, *Math. Mag.*, **56**(1983) 17–22.

D. D. Elliott, A prime generating function, *Two-Year Coll. Math. J.*, **14**(1983) 57.

David Ellis, Some consequences of Wilson's theorem, *Univ. Nac. Tucumán Rev. Ser. A*, **12**(1959) 27–29; *MR* **21** #7179.

P. Erdős, A theorem of Sylvester and Schur, *J. London Math. Soc.*, **9**(1934) 282–288; *Zbl.* **10**, 103.

P. Erdős, On consecutive integers, *Nieuw Arch. Wisk.* (3), **3**(1955) 124–128; *MR* **17**, 461f.

[For the Fung & Williams reference, see **A1**.]

Reijo Ernvall, A formula for the least prime greater than a given integer, *Elem. Math.*, **30**(1975) 13–14; *MR* **54** #12616.

Robin Forman, Sequences with many primes, *Amer. Math. Monthly* **99**(1992) 548–557; *MR* **93e**:11104.

J. M. Gandhi, Formulae for the n-th prime, *Proc. Washington State Univ. Conf. Number Theory*, Pullman, 1971, 96–106; *MR* **48** #218.

Betty Garrison, Polynomials with large numbers of prime values, *Amer. Math. Monthly*, **97**(1990) 316–317; *MR* **91i**:11124.

James Gawlik, A function which distinguishes the prime numbers, *Math. Today (Southend-on-Sea)* **33**(1997) 21.

Giuseppe Giuga, Sopra alcune proprietà caratteristiche dei numeri primi, *Period. Math.* (4), **23**(1943) 12–27; *MR* **8**, 11.

Giuseppe Giuga, Su una presumibile proprietà caratteristica dei numeri primi, *Ist. Lombardo Sci. Lett. Rend. Cl. Sci. Mat. Nat.*(3), **14(83)**(1950) 511–528; *MR* **13**, 725.

P. Goetgheluck, On cubic polynomials giving many primes, *Elem. Math.*, **44**(1989) 70–73; *MR* **90j**:11014.

Solomon W. Golomb, A direct interpretation of Gandhi's formula, *Amer. Math. Monthly*, **81**(1974) 752–754; *MR* **50** #7003.

Solomon W. Golomb, Formulas for the next prime, *Pacific J. Math.*, **63**(1976) 401–404; *MR* **53** #13094.

R. L. Goodstein, Formulae for primes, *Math. Gaz.*, **51**(1967) 35–36.

H. W. Gould, A new primality criterion of Mann and Shanks, *Fibonacci Quart.*, **10**(1972) 355–364, 372; *MR* **47** #119.

S. W. Graham, The greatest prime factor of the integers in an interval, *J. London Math. Soc.* (2), **24**(1981) 427–440; *MR* **82k**:10054.

Andrew Granville, Unexpected irregularities in the distribution of prime numbers, *Proc. Internat. Congr. Math.*, Vol. 1, 2 (*Zurich*, 1994) 388–399, Birkhäuser Basel, 1995; *MR* **97d**:11139.

Andrew Granville & Sun Zhi-Wei, Values of Bernoulli polynomials, *Pacific J. Math.*, **172**(1996) 117–137; *MR* **98b**:11018.

Silvio Guiasu, Is there any regularity in the distribution of prime numbers at the beginning of the sequence of positive integers? *Math. Mag.*, **68**(1995) 110–121.

Richard K. Guy, Conway's prime producing machine, *Math. Mag.*, **56**(1983) 26–33; *MR* **84j**:10008.

G. H. Hardy, A formula for the prime factors of any number, *Messenger of Math.*, **35**(1906) 145–146.

V. C. Harris, A test for primality, *Nordisk Mat. Tidskr.*, **17**(1969) 82; *MR* **40** #4197.

E. Härtter, Über die Verallgemeinerung eines Satzes von Sierpiński, *Elem. Math.*, **16** (1961) 123–127; *MR* **24** #A1869.

Olga Higgins, Another long string of primes, *J. Recreational Math.*, **14** (1981/82) 185.

C. Isenkrahe, Ueber eine Lösung der Aufgabe, jede Primzahl als Function der vorhergehenden Primzahlen durch einen geschlossenen Ausdruck darzustellen, *Math. Ann.*, **53**(1900) 42–44.

Jia Chao-Hua, The greatest prime factor of the integers in a short interval, *Acta Math. Sinica*, **29**(1986) 815–825; *MR* **88f**:11083; II, **32**(1989) 188–199; *MR* **90m**:11135; III, *ibid.* (*N.S.*), **9**(1993) 321–336; *MR* **95c**:11107.

James P. Jones, Formula for the n-th prime number, *Canad. Math. Bull.*, **18**(1975) 433–434; *MR* **57** #9641.

James P. Jones, Daihachiro Sato, Hideo Wada & Douglas Wiens, Diophantine representation of the set of prime numbers, *Amer. Math. Monthly*, **83**(1976) 449–464; *MR* **54** #2615.

James P. Jones & Yuri V. Matiyasevich, Proof of recursive unsolvability of Hilbert's tenth problem, *Amer. Math. Monthly*, **98**(1991) 689–709; *MR* **92i**:03050.

Matti Jutila, On numbers with a large prime factor, *J. Indian Math. Soc.* (*N.S.*), **37**(1973) 43–53(1974); *MR* **50** #12936.

Steven Kahan, On the smallest prime greater than a given positive integer, *Math. Mag.*, **47**(1974) 91–93; *MR* **48** #10964.

Sushil Kumar Karmakar, An algorithm for generating prime numbers, *Math. Ed.* (*Siwan*), **28**(1994) 175.

E. Karst, The congruence $2^{p-1} \equiv 1 \bmod p^2$ and quadratic forms with a high density of primes, *Elem. Math.*, **22**(1967) 85–88.

John Knopfmacher, Recursive formulae for prime numbers, *Arch. Math. (Basel)*, **33** (1979/80) 144–149; *MR* **81j**:10008.

Masaki Kobayashi, Prime producing quadratic polynomials and class-number one problem for real quadratic fields, *Proc. Japan Acad. Ser. A Math. Sci.*, **66**(1990) 119–121; *MR* **91i**:11140.

L. Kuipers, Prime-representing functions, *Nederl. Akad. Wetensch. Proc.*, **53**(1950) 309–310 = *Indagationes Math.*, **12**(1950) 57–58; *MR* **11**, 644.

J. C. Lagarias, V. S. Miller & A. M. Odlyzko, Computing $\pi(x)$: the Meissel-Lehmer method, *Math. Comput.*, **44**(1985) 537–560.

Klaus Langmann, Eine Formel für die Anzahl der Primzahlen, *Arch. Math. (Basel)*, **25**(1974) 40; *MR* **49** #4951.

A.-M. Legendre, *Essai sur la Théorie des Nombres*, Duprat, Paris, 1798, 1808.

D. H. Lehmer, On the function $x^2 + x + A$, *Sphinx*, **6**(1936) 212–214; **7**(1937) 40.

Liu Hong-Kuan, The greatest prime factor of the integers in an interval, *Acta Arith.*, **65**(1993) 301–328; *MR* **95d**:11117.

S. Louboutin, R. A. Mollin & H. C. Williams, Class numbers of real quadratic fields, continued fractions, reduced ideals, prime-producing polynomials and quadratic residue covers, *Canad. J. Math.*, **44**(1992) 1–19.

H. B. Mann & Daniel Shanks, A necessary and sufficient condition for primality and its source, *J. Combin. Theory Ser. A*, **13**(1972) 131–134; *MR* **46** #5225.

J.-P. Massias & G. Robin, Effective bounds for some functions involving prime numbers, Preprint, Laboratoire de Théorie des Nombres et Algorithmique, 123 rue A. Thomas, 87060 Limoges Cedex, France.

Yuri V. Matiyasevich, Primes are enumerated by a polynomial in 10 variables, *Zap. Naučn. Sem. Leningrad. Otdel. Mat. Inst. Steklov*,**68**(1977) 62–82, 144–145; *MR* **58** #21534; English translation: *J. Soviet Math.*, **15**(1981) 33–44.

W. H. Mills, A prime-representing function, *Bull. Amer. Math. Soc.*, **53**(1947) 604; *MR* **8**, 567.

Richard A. Mollin, Prime valued polynomials and class numbers of quadratic fields, *Internat. J. Math. Math. Sci.*, **13**(1990) 1–11; *MR* **91c**:11060.

Richard A. Mollin, Prime-producing quadratics, *Amer. Math. Monthly*, **104**(1997) 529–544; *MR* **98h**:11113.

Richard A. Mollin & Hugh Cowie Williams, Quadratic nonresidues and prime-producing polynomials, *Canad. Math. Bull.*, **32**(1989) 474–478; *MR* **91a**:11009. [see also *Number Theory*, de Gruyter, 1989, 654–663 and *Nagoya Math. J.*, **112**(1988) 143–151.]

Leo Moser, A prime-representing function, *Math. Mag.*, **23**(1950) 163–164.

Joe L. Mott & Kermit Rose, Prime-producing cubic polynomials, *Ideal theoretic methods in commutative algebra (Columbia MO 1999)* 281–317, *Lecture Notes in Pure and Appl. Math.*, **220** Dekker, New York, 2001; *MR* **2002c**:11140.

Gerry Myerson, Problems **96:16** and **96:17**, Western Number Theory Problems, 96-12-16 & 19.

K. S. Namboodiripad, A note on formulae for the n-th prime, *Monatsh. Math.*, **75**(1971) 256–262; *MR* **46** #126.

T. B. M. Neill & M. Singer, The formula for the Nth prime, *Math. Gaz.*, **49**(1965) 303.

Ivan Niven, Functions which represent prime numbers, *Proc. Amer. Math. Soc.*, **2**(1951) 753–755; *MR* **13**, 321a.

O. Ore, On the selection of subsequences, *Proc. Amer. Math. Soc.*, **3**(1952) 706–712; *MR* **14**, 256.

Joaquin Ortega Costa, The explicit formula for the prime number function $\pi(x)$, *Revista Mat. Hisp.-Amer.*(4), **10**(1950) 72–76; *MR* **12**, 392b.

Laurenţiu Panaitopol, On the inequality $p_a p_b > p_{ab}$, *Bull. Math. Soc. Sci. Math. Roumanie (N.S.)*, **41(89)**(1998) 135–139; *MR* **2003c**:11115.

Makis Papadimitriou, A recursion formula for the sequence of odd primes, *Amer. Math. Monthly*, **82**(1975) 289; *MR* **52** #246.

Carlos Raitzin, The exact count of the prime numbers that do not exceed a given upper bound (Spanish), *Rev. Ingr.*, **1**(1979) 37–43; *MR* **82e**:10074.

K. Ramachandra, A note on numbers with a large prime factor, *J. London Math. Soc.* (2), **1**(1969) 303–306; *MR* **40** #118; II *J, Indian Math. Soc. (N.S.)*, **34**(1970) 39–48(1971); *MR* **45** #8616.

Stephen Regimbal, An explicit formula for the k-th prime number, *Math. Mag.*, **48**(1975) 230–232; *MR* **51** #12676.

Paulo Ribenboim, Selling primes, *Math. Mag.*, **68**(1995) 175–182.

Paulo Ribenboim, Are there functions that generate prime numbers? *College Math. J.*, **28**(1997) 352–359; *MR* **98h**:11019.

B. Riemann, Über die Anzahl der Primzahlen unter einer gegebenen Grösse, *Monatsber. Königl. Preuss. Akad. Wiss. Berlin* 1859, 671–680; also in *Gesammelte Mathematische Werke*, Teubner, Leipzig, 1892, pp. 145–155.

J. B. Rosser & L. Schoenfeld, Approximate formulas for some functions of prime numbers, *Illinois J. Math.*, **6**(1962) 64–94.

Michael Rubinstein, A formula and a proof of the infinitude of the primes, *Amer. Math. Monthly* **100**(1993) 388–392.

W. Sierpiński, *Elementary Number Theory*, (ed. A. Schinzel) PWN, Warszawa, 1987, p. 218.

W. Sierpiński, Sur une formule donnant tous les nombres premiers, *C.R. Acad. Sci. Paris*, **235**(1952) 1078–1079; *MR* **14**, 355.

W. Sierpiński, Les binômes $x^2 + n$ et les nombres premiers, *Bull. Soc. Royale Sciences Liège*, **33**(1964) 259–260.

B. R. Srinivasan, Formulae for the n-th prime, *J. Indian Math. Soc. (N.S.)*, **25**(1961) 33–39; *MR* **26** #1289.

B. R. Srinivasan, An arithmetical function and an associated formula for the n-th prime. I, *Norske Vid. Selsk. Forh. (Trondheim)*, **35**(1962) 68–71; *MR* **27** #101.

Garry J. Tee, Simple analytic expressions for primes, and for prime pairs, *New Zealand Math. Mag.*, **9**(1972) 32–44; *MR* **45** #8601.

E. Teuffel, Eine Rekursionsformel für Primzahlen, *Jber. Deutsch. Math. Verein.*, **57**(1954) 34–36; *MR* **15**, 685.

John Thompson, A method for finding primes, *Amer. Math. Monthly*, **60** (1953) 175; *MR* **14**, 621.

P. G. Tsangaris & James P. Jones, An old theorem on the GCD and its application to primes, *Fibonacci Quart.*, **30**(1992) 194–198; *MR* **93e**:11004.

Charles Vanden Eynden, A proof of Gandhi's formula for the n-th prime, *Amer. Math. Monthly*, **79**(1972) 625; *MR* **46** #3425.

E. Vantieghem, On a congruence only holding for primes, *Indag. Math.(N.S.)*, **2**(1991) 253–255; *MR* **92e**:11005.

C. P. Willans, On formulae for the Nth prime number, *Math. Gaz.*, **48**(1964) 413–415.

Erick Wong, Computations on normal families of primes, M.Sc. thesis, Simon Fraser Univ., August 1997.

C. P. Wormell, Formulae for primes, *Math. Gaz.*, **51**(1967) 36–38.

E. M. Wright, A prime representing function, *Amer. Math. Monthly*, **58**(1951) 616–618; *MR* **13**, 321b.

E. M. Wright, A class of representing functions, *J. London Math. Soc.*, **29** (1954) 63–71; *MR* **15**, 288d.

Don Zagier, *Die ersten 50 Millionen Primzahlen*, Beihefte zu Elemente der Mathematik **15**, Birkhäuser, Basel 1977.

OEIS: A055004, A055023, A055030-055032.

A18 The Erdős-Selfridge classification of primes.

Erdős & Selfridge classify the primes as follows: p is in class 1 if the only prime divisors of $p+1$ are 2 or 3; and p is in class r if every prime factor of $p+1$ is in some class $\leq r-1$, with equality for at least one prime factor. For example:

class 1:   2 3 5 7 11 17 23 31 47 53 71 107 127 191 431 647 863 971
           1151 2591 4373 6143 6911 8191 8747 13121 15551 23327 27647 62207 ...
class 2:   13 19 29 41 43 59 61 67 79 83 89 97 101 109 131 137 139 149 167 179 197
           199 211 223 229 239 241 251 263 269 271 281 283 293 307 317 319 359 367 373
           377 383 419 439 449 461 467 499 503 509 557 563 577 587 593 599 619 641 643
           659 709 719 743 751 761 769 809 827 839 881 919 929 953 967 979 991 1019 ...
class 3:   37 103 113 151 157 163 173 181 193 227 233 257 277 311 331 337 347 353
           379 389 397 401 409 421 457 463 467 487 491 521 523 541 547 571 601 607 613
           631 653 683 701 727 733 773 787 811 821 829 853 857 859 877 883 911 937 947
           983 997 1009 1013 1031 ...
class 4:   73 313 443 617 661 673 677 691 739 757 823 887 907 941 977 1093 ...
class 5:   1021 1321 1381 1459 1877 2467 2503 2657 2707 3253 3313 3547 ...
class 6:   2917 4933 5413 7507 8167 8753 10567 10627 11047 11261 11677 ...
class 7:   15013 16333 22093 24841 43321 49003 52517 54721 62533 63761 ...
class 8:   49681 ...

It's easy to prove that the number of primes in class r, not exceeding n, is $o(n^\epsilon$ for every $\epsilon > 0$ and all r. Prove that there are infinitely many primes in each class. If $p_1^{(r)}$ denotes the least prime in class r, so that $p_1^{(1)} = 2$, $p_1^{(2)} = 13$, $p_1^{(3)} = 37$, $p_1^{(4)} = 73$ and $p_1^{(5)} = 1021$, then Erdős thought that $(p_1^{(r)})^{1/r} \to \infty$, while Selfridge thought it quite likely to be bounded.

A similar classification arises if $p+1$ is replaced by $p-1$:

class 1:   2 3 5 7 13 17 19 37 73 97 109 163 193 433 487 577 769 1153 ...
class 2:   11 29 31 41 43 53 61 71 79 101 103 113 127 131 137 149 151 157 181 191
           197 211 223 229 239 241 251 257 271 281 293 307 313 337 379 389 401 409 421
           439 443 449 459 491 521 541 547 571 593 601 613 631 641 647 653 673 677 701
           751 757 761 773 811 877 883 911 919 937 953 971 1009 1021 ...
class 3:   23 59 67 83 89 107 173 199 227 233 263 311 317 331 349 353 367 373 383
           397 419 431 463 479 503 509 523 563 569 587 607 617 619 661 683 727 733 739
           743 787 809 821 823 853 859 881 887 907 929 947 977 983 991 1031 1033 ...
class 4:   47 139 167 179 269 277 347 461 467 499 599 643 691 709 797 827 829 839
           857 863 967 997 1013 1019 ...
class 5:   283 359 557 659 941 ...
class 6:   719 1319 ...
class 7:   1439 ...

for which similar answers are to be expected. Are corresponding classes equally dense? There is a connexion with Cunningham chains (**A7**).

P. Erdős, Problems in number theory and combinatorics, *Congr. Numer.* *XVIII*, Proc. 6th Conf. Numer. Math., Manitoba, 1976, 35–58 (esp. p. 53); *MR* **80e**:10005.

OEIS: A005105-005113, A048135-048136.

A19 Values of n making $n - 2^k$ prime. Odd numbers not of the form $\pm p^a \pm 2^b$.

Erdős conjectures that 4, 7, 15, 21, 45, 75 and 105 are the only values of n for which $n - 2^k$ is prime for all k such that $2 \leq 2^k < n$. Mientka & Weitzenkamp have verified this for $n < 2^{44}$ and Uchiyama & Yorinaga have extended this to 2^{77}. Vaughan has proved that there are not too many such numbers, less than $x \exp\{-(\ln x)^c\}$ of them less than x, but he was unable to show that there were less than $x^{1-\epsilon}$.

Erdős also conjectures that for infinitely many n, all the integers $n - 2^k$, $1 \leq 2^k < n$ are squarefree (see also **F13**).

If we denote by $A(x)$ the number of $n \leq x$ for which all $n - 2^k$ are prime, $2 \leq 2^k < n$, then Hooley showed that the extended Riemann hypothesis implies that $A(x) = O(x^c)$ with an explicit $c < 1$, and Narkiewicz improved this to $c < \frac{1}{2}$.

Cohen & Selfridge ask for the least positive odd number *not* of the form $\pm p^a \pm 2^b$, where p is prime, $a \geq 0$, $b \geq 1$ and any choice of signs may be made. They observe that the number is greater than 2^{18}, but at most

$$6120\ 6699060672\ 7677809211\ 5601756625\ 4819576161\text{-}$$
$$6319229817\ 3436854933\ 4512406741\ 7420946855\ 8999326569.$$

Sun Zhi-Wei proved that if

$$x \equiv 47867742232066880047611079 \bmod 66483034025018711639862527490$$

then it is not of the form $\pm p^a \pm q^b$ with p, q prime.

Crocker proved that there are infinitely many odd integers *not* of the form $2^k + 2^l + p$, where p is prime. Erdős suggests that there may be cx of them less than x, but can $> x^\epsilon$ be proved? Can we show that covering congruences (**F13**) do not help here? I.e., does $p + 2^u + 2^v$ (or $p + 2^u + 2^v + 2^w$) meet every arithmetic progression? More generally, Erdős asks if, for each r, there are infinitely many odd integers not the sum of a prime and r or fewer powers of 2. Is their density positive? Do they contain an infinite arithmetic progression? In the opposite direction, Gallagher has proved that for every $\epsilon > 0$ there is a sufficiently large r so that the lower density of sums of primes with r powers of 2 is greater than $1 - \epsilon$.

Erdős also asks if there is an odd integer *not* of the form $2^k + s$ where s is squarefree. Jud McCranie searched for such numbers but found none below $1.4 \cdot 10^9$.

Let $f(n)$ be the number of representations of n as a sum $2^k + p$, and let $\{a_i\}$ be the sequence of values of n for which $f(n) > 0$. Does the density of $\{a_i\}$ exist? Erdős showed that $f(n) > c \ln \ln n$ infinitely often, but could not decide if $f(n) = o(\ln n)$. He conjectures that $\limsup(a_{i+1} - a_i) = \infty$. This would follow if there are covering systems with arbitrarily large least moduli.

Vassilev-Missana showed that there are infinitely many primes p such that $p + 2^k$ is composite for all positive k and infinitely many such that $|p - 2^k|$ is so. Is there a prime p such that both are true? This is answered positively by Sun Zhi-Wei. Brüdern & Perelli show, assuming the generalized Riemann hypothesis, that the number of integers not exceeding x that are not representable as a sum of a prime and a k-th power is $\ll x^{1-1/25k+\epsilon}$.

Carl Pomerance notes that for $n = 210$, $n - p$ is prime for all p, $n/2 < p < n$, and asked if there is any other such n. With help from Deshouillers, Granville & Narkiewicz he later answered this negatively.

Hardy & Littlewood conjectured in 1923 that every sufficiently large integer is either a square or the sum of a prime and a square. Davenport & Heilbronn proved this in 1937 for almost all positive integers. If $E(X)$ is the number of not so expressible numbers less than X, then Wang showed that $E(X) \ll X^{0.99}$, and Li improved the exponent to 0.982.

Jörg Brüdern & Alberto Perelli, The addition of primes and power, *Canad. J. Math.*, **48**(1996) 512–526; *MR* **97h**:11118.

R. Brünner, A. Perelli & J. Pintz, The exceptional set for the sum of a prime and a square, *Acta Math. Hungar.*, **53**(1989) 347–365; *MR* **91b**:11104.

Chen Yong-Gao, On integers of the forms $k - 2^n$ and $k2^n + 1$, *J. Number Theory* **89**(2001) 121–125; *MR* **2002b**:11020.

Fred Cohen & J. L. Selfridge, Not every number is the sum or difference of two prime powers, *Math. Comput.*, **29**(1975) 79–81; *MR* **51** #12758.

R. Crocker, On the sum of a prime and of two powers of two, *Pacific J. Math.*, **36**(1971) 103–107; *MR* **43** #3200.

P. Erdős, On integers of the form $2^r + p$ and some related problems, *Summa Brasil. Math.*, **2**(1947-51) 113–123; *MR* **13**, 437.

Patrick X. Gallagher, Primes and powers of 2, *Inventiones Math.*, **29**(1975) 125–142; *MR* **52** #315.

C. Hooley, *Applications of Sieve Methods*, Academic Press, 1974, Chap. VIII.

Koichi Kawada, On the representation of numbers as the sum of a prime and a k th power. Interdisciplinary studies on number theory (Japanese) *Sūrikaiseki-kenkyūsho Kōkyūroku No.* **837**(1993) 92–100; *MR* **95k**:11129.

Li Hong-Ze, The exceptional set for the sum of a prime and a square, *Acta Math. Hungar.*, **99**(2003) 123–141; *MR* **2004c**:11184.

Donald E. G. Malm, A graph of primes, *Math. Mag.*, **66**(1993) 317–320; *MR* **94k**:11132.

Walter E. Mientka & Roger C. Weitzenkamp, On f-plentiful numbers, *J. Combin. Theory*, **7**(1969) 374–377; *MR* **42** #3015.

W. Narkiewicz, On a conjecture of Erdős, *Colloq. Math.*, **37**(1977) 313–315; *MR* **58** #21971.

A. Perelli & A. Zaccagnini, On the sum of a prime and a k th power, *Izv. Ross. Akad. Nauk Ser. Mat.*, **59**(1995) 185–200; *MR* **96f**:11134.

A. de Polignac, Recherches nouvelles sur les nombres premiers, *C. R. Acad. Sci. Paris*, **29**(1849) 397–401, 738–739.

Sun Zhi-Wei, On integers not of the form $\pm p^a \pm q^b$, *Proc. Amer. Math. Soc.*, **128**(2000) 997–1002; *MR* **2000i**:11157.

Sun Zhi-Wei & Le Mao-Hua, Integers not of the form $c(2^a + 2^b) + p^\alpha$, *Acta Arith.*, **99**(2001) 183–190; *MR* **2002e**:11043.

Sun Zhi-Wei & Yang Si-Man, A note on integers of the form $2^n + cp$, *Proc. Edinb. Math. Soc.*(2) **45**(2002) 155–160; *MR* **2002j**:11117.

Saburô Uchiyama & Masataka Yorinaga, Notes on a conjecture of P. Erdős, I, II, *Math. J. Okayama Univ.*, **19**(1977) 129–140; **20**(1978) 41–49; *MR* **56** #11929; **58** #570.

Mladen V. Vassilev-Missana, Note on "extraordinary primes", *Notes Number Theory Discrete Math.*, **1**(1995) 111–113; *MR* **97g**:11004.

R. C. Vaughan, Some applications of Montgomery's sieve, *J. Number Theory*, **5**(1973) 64–79; *MR* **49** #7222.

V. H. Vu, High order complementary bases of primes, *Integers*, **2**(2002) A12; *MR* **2003k**:11017.

Wang Tian-Ze, On the exceptional set for the equation $n = p + k^2$, *Acta Math. Sinica (N.S.)*, **11**(1995) 156–167; *MR* **97i**:11104.

A. Zaccagnini, Additive problems with prime numbers, *Rend. Sem. Mat. Univ. Politec. Torino*, **53**(1995) 471–486; *MR* **98c**:11109

A20 Symmetric and asymmetric primes.

Given a pair of odd primes p, q, let the number of lattice points (m, n) in the rectangle $0 < m < p/2$, $0 < n, q/2$ and below the diagonal through $(0,0)$ be denoted by $S(q, p)$. Then the number above the diagonal is $S(p, q)$, $(-1)^{S(q,p)} = \left(\frac{q}{p}\right)$, the Legendre symbol (see **F5**) and the identity $S(p, q) + S(q, p) = \frac{1}{4}(p-1)(q-1)$ is related to the law of quadratic reciprocity. Fletcher, Lindgren & Pomerance call a pair (p, q) such that $S(p, q) = S(q, p)$ **symmetric**, and show that a pair is symmetric just if $|p - q| = (p-1, q-1)$. They call a member of a symmetric pair a **symmetric prime** and show that the number of them less than x is at most $x/(\ln x)^{1.027}$. They conjecture that this number is $x/(\ln x)^{\sigma+o(1)}$, where $\sigma = 2 - (1 + \ln \ln 2)/\ln 2 \approx 1.08607$.

Peter Fletcher, William Lindgren & Carl Pomerance, Symmetric and asymmetric primes, *J. Number Theory*, **58**(1996) 89–99; *MR* **97c**:11007.

B. Divisibility

We will denote by $d(n)$ the number of positive divisors of n, by $\sigma(n)$ the sum of those divisors, and by $\sigma_k(n)$ the sum of their kth powers, so that $\sigma_0(n) = d(n)$ and $\sigma_1(n) = \sigma(n)$. We use $s(n)$ for the sum of the **aliquot parts** of n, i.e., the positive divisors of n other than n itself, so that $s(n) = \sigma(n) - n$. The number of distinct prime factors of n will be denoted by $\omega(n)$ and the total number, counting repetitions, by $\Omega(n)$.

Iteration of various arithmetic functions will be denoted, for example, by $s^k(n)$, which is defined by $s^0(n) = n$ and $s^{k+1}(n) = s(s^k(n))$ for $k \geq 0$.

We use the notation $d \mid n$ to mean that d divides n, and $e \nmid n$ to mean that e does not divide n. The notation $p^k \| n$ is used to imply that $p^k \mid n$ but $p^{k+1} \nmid n$. By $[m, n]$ we will mean the consecutive integers $m, m+1, \ldots, n$.

B1 Perfect numbers.

A **perfect number** is such that $n = s(n)$. Euclid knew that $2^{p-1}(2^p - 1)$ was perfect if $2^p - 1$ is prime. For example, 6, 28, 496, ... ; see the list of Mersenne primes in **A3**. Euler showed that these were the only even perfect numbers.

The existence or otherwise of odd perfect numbers is one of the more notorious unsolved problems of number theory. Euler showed that they have shape $p^\alpha m^2$ where p is prime and $p \equiv \alpha \equiv 1 \pmod 4$. Touchard showed that they were of shape $12m + 1$ or $36m + 9$.

The lower bound for an odd perfect number has now been pushed to 10^{300} by Brent, Cohen & te Riele. Brandstein has shown that the largest prime factor is > 500000 and Hagis that the second largest is > 1000. Cohen has shown that it contains a component (prime power divisor) $> 10^{20}$, and Sayers that there are at least 29 prime factors (not necessarily distinct). Iannucci & Sorli improve this to $\Omega(n) \geq 37$.

Pomerance has shown that an odd perfect number with at most k distinct factors is less than

$$(4k)^{(4k)^{2^{k^2}}}$$

but Heath-Brown has much improved this by showing that if n is an odd number with $\sigma(n) = an$, then $n < (4d)^{4^k}$, where d is the denominator of a and k is the number of distinct prime factors of n. In particular, if n_k is an odd perfect number with k distinct prime factors, then $n < 4^{4^k}$. Roger Cook further improves this to $n_k < C^{4^k}$ where $C = 3^{512/511} \approx 3.006$ and, for $k = 8$, to $n_8 < D^{4^8}$ with $D = 2^{16/15} \approx 2.094$ if $195 \mid n_8$ or $D = 195^{1/7} \approx 2.123$ otherwise. A recent claim by Nielsen is that $n_k < 2^{4^k}$.

Starni has shown that if $n = p^\alpha M^2$ is an odd perfect number with p prime, $p \perp M$, $p \equiv \alpha \equiv 1 \bmod 4$, $\alpha + 2$ prime, $(\alpha + 2) \perp (p - 1)$, then $M^2 \equiv 0 \bmod \alpha + 2$.

Iannucci has shown that the second largest prime divisor of an odd perfect number exceeds 10^4, that the third largest exceeds 10^2, and (with Sorli) that there are at least 37 not necessarily distinct prime factors.

John Leech asks for examples of spoof odd perfect numbers, like Descartes's

$$3^2 7^2 11^2 13^2 22021$$

which is perfect if you pretend that 22021 is prime.

Greg Martin offers the following 'proof' that 4680 is perfect. Write 4680 as $2^3 \cdot 3^2 \cdot (-5) \cdot (-13)$. Then $\sigma(4680) =$
$(1 + 2 + 2^2 + 2^3)(1 + 3 + 3^2)(1 + (-5))(1 + (-13)) = 9360 = 2 \cdot 4680$.
He asks: if you allow $\sigma(-p^n) = \sum_{j=0}^{n}(-p)^j$, are there others? Dennis Eichhorn and Peter Montgomery found $-84 = 2^2(3)(-7)$ and $-120 = 2^3(3)(-5)$ and noted that $\sigma((-2)^5(3)(7)) = (-2)^5(3)(7)$. Martin also defines $\tilde{\sigma}(p^r) = p^r - p^{r-1} + p^{r-2} - \cdots + (-1)^r$, and asks about $\tilde{\sigma}$-(k)-perfect numbers. For $k = 2$ there are 2, 12, 40, 252, 880, 10880, 75852. For $k = 3$ there are at least 40, including 30240 and $2^{10}3^45^411 \cdot 13^2 \cdot 31 \cdot 61 \cdot 157 \cdot 521 \cdot 683$.

Are there $\tilde{\sigma}$-k-perfect numbers with $k \geq 4$?

Are there infinitely many $\tilde{\sigma}$-k-perfect numbers?

Are there any odd $\tilde{\sigma}$-3-perfect numbers? Such a number must be a square.

For many earlier references, see the first edition of this book.

Jennifer T. Betcher & John H. Jaroma, An extension of the results of Servais and Cramer on odd perfect and odd multiply perfect numbers, *Amer. Math. Monthly*, **110**(2003) 49–52; *MR* **2003k**:11006.

Michael S. Brandstein, New lower bound for a factor of an odd perfect number, #82T-10-240, *Abstracts Amer. Math. Soc.*, **3**(1982) 257.

Richard P. Brent & Graeme L. Cohen, A new lower bound for odd perfect numbers, *Math. Comput.*, **53**(1989) 431–437.

R. P. Brent, G. L. Cohen & H. J. J. te Riele, Improved techniques for lower bounds for odd perfect numbers, *Math. Comput.*, **57**(1991) 857–868; *MR* **92c**:11004.

E. Chein, An odd perfect number has at least 8 prime factors, PhD thesis, Penn. State Univ., 1979.

Chen Yi-Ze & Chen Xiao-Song, A condition for an odd perfect number to have at least 6 prime factors $\equiv 1 \bmod 3$, *Hunan Jiaoyu Xueyuan Xuebao* (*Ziran Kexue*), **12**(1994) 1–6; *MR* **96e**:11007.

Graeme L. Cohen, On the largest component of an odd perfect number, *J. Austral. Math. Soc. Ser. A*, **42**(1987) 280–286.

Roger Cook, Factors of odd perfect numbers, *Number Theory* (*Halifax NS*, 1994) 123–131, *CMS Conf. Proc.*, **15** Amer. Math. Soc., 1995; *MR* **96f**:11009.

Roger Cook, Bounds for odd perfect numbers, *Number Theory* (*Ottawa ON*, 1996) 67–71, *CRM Proc. Lecture Notes*, **19** Amer. Math. Soc., 1999; *MR* **2000d**:11010.

Simon Davis, A rationality condition for the existence of odd perfect numbers, *Int. J. Math. Math. Sci.*, **2003** 1261–1293; *MR* **2004b**:11007.

John A. Ewell, On necessary conditions for the existence of odd perfect numbers, *Rocky Mountain J. Math.*, **29**(1999) 165–175.

Aleksander Grytczuk & Marek Wójtowicz, There are no small odd perfect numbers, *C.R. Acad. Sci. Paris Sér. I Math.*, **328**(1999) 1101–1105.

P. Hagis, Sketch of a proof that an odd perfect number relatively prime to 3 has at least eleven prime factors, *Math. Comput.*, **40**(1983) 399–404.

P. Hagis, On the second largest prime divisor of an odd perfect number, *Lecture Notes in Math.*, **899**, Springer-Verlag, New York, 1971, pp. 254–263.

Peter Hagis & Graeme L. Cohen, Every odd perfect number has a prime factor which exceeds 10^6, *Math. Comput*, **67**(1998) 1323–1330; *MR* **98k**:11002.

Judy A. Holdener, A theorem of Touchard on odd perfect numbers, *Amer. Math. Monthly*, **109**(2002) 661–663; *MR* **2003d**:11012.

Douglas E. Iannucci, The second largest prime divisor of an odd perfect number exceeds ten thousand, *Math. Comput.*, **68**(1999) 1749–1760; *MR* **2000i**:11200.

Douglas E. Iannucci, The third largest prime divisor of an odd perfect number exceeds one hundred, *Math. Comput.*, **69**(2000) 867–879; *MR* **2000i**:11201.

D. E. Iannucci & R. M. Sorli, On the total number of prime factors of an odd perfect number, *Math. Comput.*, **72**(2003) 2077–2084; *MR* **2004b**:11008.

Abd El-Hamid M. Ibrahim, On the search for perfect numbers, *Bull. Fac. Sci. Alexandria Univ.*, **37**(1997) 93–95; *J. Inst. Math. Comput. Sci. Comput. Sci. Ser.*, **10**(1999) 55–57.

Paul M. Jenkins, Odd perfect numbers have a prime factor exceeding 10^7, *Math. Comput.*, **72**(2003) 1549–1554; *MR* **2004a**:11002.

Masao Kishore, Odd perfect numbers not divisible by 3 are divisible by at least ten distinct primes, *Math. Comput.*, **31**(1977) 274–279; *MR* **55** #2727.

Masao Kishore, Odd perfect numbers not divisible by 3. II, *Math. Comput.*, **40**(1983) 405–411; *MR* **84d**:10009.

Masao Kishore, On odd perfect, quasiperfect, and odd almost perfect numbers, *Math. Comput.*, **36**(1981) 583–586; *MR* **82h**:10006.

Pace P. Nielsen, An upper bound for odd perfect numbers, *Elect. J. Combin. Number Theory*, **3**(2003) A14. http://www.integers-ejcnt.org/vol3.html

J. Touchard, On prime numbers and perfect numbers, *Scripta Math.*, **19**(1953) 35–39; *MR* **14**, 1063b.

M. D. Sayers, An improved lower bound for the total number of prime factors of an odd perfect number, M.App.Sc. Thesis, NSW Inst. Tech., 1986.

Paulo Starni, On the Euler's factor of an odd perfect number, *J. Number Theory*, **37**(1991) 366–369; *MR* **92a**:11010.

Paulo Starni, Odd perfect numbers: a divisor related to the Euler's factor, *J. Number Theory* 44(1993) 58–59; *MR* **94c**:11003.

OEIS: A000396, A058007.

B2 Almost perfect, quasi-perfect, pseudoperfect, harmonic, weird, multiperfect and hyperperfect numbers.

Perhaps because they were frustrated by their failure to disprove the existence of odd perfect numbers, numerous authors have defined a number of closely related concepts and produced a raft of problems, many of which seem no more tractable than the original.

For a perfect number, $\sigma(n) = 2n$. If $\sigma(n) < 2n$, n is called **deficient**. A problem in *Abacus* was to prove sum of divisors that every number $n > 3$ is the sum of two deficient numbers, or to find a number that was not. If $\sigma(n) > 2n$, then n is called **abundant**. If $\sigma(n) = 2n - 1$, n has been called **almost perfect**. Powers of 2 are almost perfect; it is not known if any other numbers are. If $\sigma(n) = 2n + 1$, n has been called **quasi-perfect**. Quasi-perfect numbers must be odd squares, but no one knows if there are any. Masao Kishore shows that $n > 10^{30}$ and that $\omega(n) \geq 6$. Hagis & Cohen have improved these results to $n > 10^{35}$ and $\omega(n) \geq 7$. Cattaneo originally claimed to have proved that $3 \nmid n$, but Sierpiński and others have observed that his proof is fallacious. Kravitz, in a letter, makes a more general conjecture, that there are no numbers whose **abundance**, $\sigma(n) - 2n$, is an odd square. In this connexion Graeme Cohen writes that it is interesting that

$$\sigma(2^2 3^2 5^2) = 3(2^2 3^2 5^2) + 11^2$$

and that if $\sigma(n) = 2n + k^2$ with $n \perp k$, then $\omega(n) \geq 4$ and $n > 10^{20}$. He has also shown that if $k < 10^{10}$ then $\omega(n) \geq 6$, and that if $k < 44366047$ then n is primitive abundant (see below). Later, relaxing the condition $n \perp k$, he finds the solution

$$n = 2 \cdot 3^2 \cdot 238897^2, \quad k = 3^2 \cdot 23 \cdot 1999$$

and five solutions $n = 2^2 \cdot 7^2 \cdot p^2$, with

$p =$	53	277	541	153941	358276277
$k =$	$7 \cdot 29$	$5 \cdot 7 \cdot 23$	$5 \cdot 7 \cdot 43$	$5 \cdot 7 \cdot 103 \cdot 113$	$5 \cdot 7 \cdot 227 \cdot 229 \cdot 521$

He verifies that the first of these last five is the smallest integer with odd square abundance. Sidney Kravitz has since sent two more solutions,

$$n = 2^3 \cdot 3^2 \cdot 1657^2, \quad k = 3 \cdot 11 \cdot 359,$$

$$n = 2^4 \cdot 31^2 \cdot 7992220179128893^2, \quad k = 44498798693247589.$$

In the latter, 31 divides k. Erdős asks for a characterization of the large numbers for which $|\sigma(n) - 2n| < C$ for some constant C. For example, $n = 2^m$: for other infinite families, see Mąkowski's two papers.

Wall, Crews & Johnson showed that the density of abundant numbers lies between 0.2441 and 0.2909. In an 83-08-17 letter Wall claimed to have narrowed these bounds to 0.24750 and 0.24893. Erdős asks if the density is irrational.

Paul Zimmerman reports that Marc Deléglise has improved the bounds for the density of abundant numbers to 0.2477 ± 0.0003.

http://www.mathsoft.com/asolve/constant/abund/abund.html

Sándor has shown that, for sufficiently large n, there is a deficient number between n and $n + (\ln n)^2$.

Sierpiński called a number **pseudoperfect** if it was the sum of *some* of its divisors; e.g., $20 = 1 + 4 + 5 + 10$. Erdős has shown that their density exists and says that presumably there are integers n which are not pseudoperfect, but for which $n = ab$ with a abundant and b having many prime factors: can b in fact have many factors $< a$?

For $n \geq 3$ Abbott lets $l = l(n)$ be the least integer for which there are n integers $1 \leq a_1 < a_2 < \ldots < a_n = l$ such that $a_i | s = \sum a_i$ for each i (so that s is pseudoperfect). He can show that $l(n) > n^{c_1 \ln \ln n}$ for some $c_1 > 0$ and all $n \geq 3$ and that $l(n) < n^{c_2 \ln \ln n}$ for some $c_2 > 0$ and infinitely many n.

Call a number **primitive abundant** if it is abundant, but all its proper divisors are deficient, and **primitive pseudoperfect** if it is pseudoperfect, but none of its proper divisors are. If the harmonic mean, $H(n) = nd(n)/\sigma(n)$, of all the positive divisors of n is an integer, Pomerance called n a **harmonic number**. A. & E. Zachariou called these "Ore numbers" and they called primitive pseudoperfect numbers "irreducible semiperfect". They note that every multiple of a pseudoperfect number is pseudoperfect and that the pseudoperfect numbers and the harmonic numbers both include the perfect numbers as a proper subset. The last result is due to Ore. All numbers $2^m p$ with $m \geq 1$ and p a prime between 2^m and 2^{m+1} are primitive pseudoperfect, but there are such numbers not of this form, e.g., 770. There are infinitely many primitive pseudoperfect numbers that are not harmonic numbers. The smallest odd primitive pseudoperfect number is 945. Erdős showed that the number of odd primitive pseudoperfect numbers is infinite. He also showed that, for sufficiently large n, the number of primitive abundant numbers less than n was bounded above and below by functions of the form $n \exp(-c\sqrt{\ln n \ln \ln n})$. Ivić showed that the constants could be taken to be $12^{-\frac{1}{2}} - \epsilon$ and $6^{\frac{1}{2}} + \epsilon$ and Avidon improved these to $1 - \epsilon$ and $2^{\frac{1}{2}} + \epsilon$.

García extended the list of harmonic numbers to include all 45 which are $< 10^7$, and he found more than 200 larger ones. The least one, apart from 1 and the perfect numbers, is 140. Are any of them squares, apart from 1? Are there infinitely many of them? If so, find upper and lower bounds on the number of them that are $< x$. Kanold has shown that their density is zero, and Pomerance that a harmonic number of the form $p^a q^b$ (p and q primes) is an even perfect number. If $n = p^a q^b r^c$ is harmonic, is it even?

Which values does the harmonic mean take? The specific questions that we asked earlier have been answered. Ore's own conjecture, that every harmonic number > 1 is even, implies that there are no odd perfect numbers!

Cohen listed all 52 harmonic numbers of the form $2^a m$, where m is odd and squarefree and $1 \leq a \leq 11$; 45 of them have $a = 8$. He also shows that there are just 13 harmonic numbers n with $H(n) = nd(n)/\sigma(n) \leq 13$:

$$H(n) = 1 \quad 2 \quad 3 \quad 5 \quad 6 \quad 5 \quad 8 \quad 9 \quad 11 \quad 10 \quad 7 \quad 13 \quad 13$$
$$n = \quad 1 \quad 6 \quad 28 \quad 140 \quad 270 \quad 496 \quad 672 \quad 1638 \quad 2970 \quad 6200 \quad 8128 \quad 105664 \quad 2^{12}8191$$

Cohen & Sorli defined a harmonic seed to be a harmonic number which has no smaller proper unitary divisor which is harmonic. It is not known if there are infinitely many harmonic seeds, not even that there are infinitely many harmonic numbers.

Goto & Shibata found all harmonic numbers with mean up to 300. They ask several questions. Is there a nontrivial odd harmonic number? Are there infinitely many harmonic seeds with just three distinct prime factors? If not, find them all. Does every harmonic number have a unique harmonic seed? Is there a powerful (see B16) harmonic number? Is there a deficient harmonic number?

A positive integer n is an **arithmetic number** if the arithmetic mean, $A(n) = \sigma(n)/d(n)$, of its positive divisors is an integer. Goto & Shibata show that a harmonic number is also arithmetic just if $H(n)$, the harmonic mean of the divisors, divides n. Examples are 140, 270, 672. They note that in all examples with $H(n) = 3p < 300$, n is arithmetic, and ask if this is true for all prime values of p.

Bateman, Erdős, Pomerance & Straus show that the set of arithmetic numbers has density 1, that the set for which $\sigma(n)/d(n)^2$ is an integer has density $\frac{1}{2}$, and that the number of rationals $r \leq x$ of the form $\sigma(n)/d(n)$ is $o(x)$. They ask for an asymptotic formula for

$$\frac{1}{x} \sum 1$$

where the sum is taken over those $n \leq x$ for which $d(n)$ does *not* divide $\sigma(n)$. They also note that the integers n for which $d(n)$ divides $s(n) = \sigma(n) - n$, have zero density, because for almost all n, $d(n)$ and $\sigma(n)$ are divisible by a high power of 2, while n is divisible only by a low power of 2.

David Wilson, in a 98-08-27 email, conjectured that $\sigma(n) \neq kn+5$. Dan Hoey lists the following values of n for which $\sigma(n) = kn + r$ and suggests that there are only finitely many when r is odd.

r	Values of n for which $\sigma(n) \equiv r \bmod n$
2	20, 104, 464, 650, 1952, 130304, 522752
3	4, 18
4	9, 12, 70, 88, 1888, 4030, 5830, 32128, 521728, 1848964
6	25, 180, 8925
7	8, 196
8	10, 49, 56, 368, 836, 11096, 17816, 45356, 77744, 91388,
	128768, 254012, 388076, 2087936, 2291936
9	15

Benkoski has called a number **weird** if it is abundant but not pseudo-perfect. For example, 70 is not the sum of any subset of

$$1 + 2 + 5 + 7 + 10 + 14 + 35 = 74$$

There are 24 primitive weird numbers less than a million: 70, 836, 4030, 5830, 7192, ... Nonprimitive weird numbers include $70p$ with p prime and $p > \sigma(70) = 144$; $836p$ with $p = 421$, 487, 491, or p prime and ≥ 557; also $7192 \cdot 31$. Some large weird numbers were found by Kravitz, and Benkoski & Erdős showed that their density is positive. Here the open questions are: are there infinitely many primitive abundant numbers which are weird? Is every odd abundant number pseudoperfect (i.e., not weird)? Can $\sigma(n)/n$ be arbitrarily large for weird n? Benkoski & Erdős conjecture "no" in answer to the last question and Erdős offers \$10 and \$25 respectively for solutions to the last two questions.

He also asks if there are extra-weird numbers n for which $\sigma(n) > 3n$, but n is not the sum of distinct divisors of n in two ways without repetitions. For example, 180 does not qualify, because although $\sigma(180) = 546$, 180=30+60+90 and is the sum of all its other divisors except 6.

One definition of a **practical** number, m, is for every n, $1 \leq n \leq \sigma(m)$ to be expressible as the sum of distinct divisors of m. E.g., 6 is practical, since 4=1+3, 5=2+3, 7=1+6, 8=2+6, 9=3+6, 10=1+3+6, 11=2+3+6, 12=1+2+3+6. Erdős showed in 1950 that the practical numbers have zero asymptotic density. It is known that if $P(x)$ is the number of practical numbers less than x, then

$$x \exp(-\alpha(\ln \ln x)^2) \ll P(x) \ll x/(\ln x)^\beta$$

for some positive constants α and β. The lower bound is due to Margenstein and the upper bound to Hausman & Shapiro. Melfi has shown that every even number is the sum of two practical numbers and that there are infinitely many practical numbers m such that $m \pm 2$ are also practical.

Numbers have been called **multiply perfect**, **multiperfect** or **k-fold perfect** if $\sigma(n) = kn$ with k an integer. For example, ordinary perfect numbers are 2-fold perfect and 120 is 3-fold perfect. Dickson's *History* records a long interest in such numbers. Lehmer has remarked that if n is odd, then n is perfect just if $2n$ is triperfect.

Selfridge and others have observed that there are just six known 3-perfect numbers and they come from $2^h - 1$ for $h = 4$, 6, 9, 10, 14, 15. For example, the third one is illustrated by

$$\sigma(2^8 \cdot 7 \cdot 73 \cdot 37 \cdot 19 \cdot 5) = (2^9 - 1)(2^3)(37 \cdot 2)(19 \cdot 2)(5 \cdot 2^2)(2 \cdot 3).$$

It appears that there may be a similar explanation for the 36 known 4-perfect numbers, the last of which was published by Poulet as long ago as 1929.

For many years the largest known value of k was 8, for which Alan L. Brown gave three examples and Franqui & García two others.

In late 1992 and early 1993, half a dozen examples with $k = 9$ had already been found by Fred Helenius. The smallest is
$2^{114} \cdot 3^{35} \cdot 5^{17} \cdot 7^{12} \cdot 11^4 \cdot 13^5 \cdot 17^3 \cdot 19^8 \cdot 23^2 \cdot 29^2 \cdot 31^2 \cdot 37^4 \cdot 41 \cdot 43 \cdot 47^2 \cdot 53 \cdot$
$\cdot 61^2 \cdot 67 \cdot 71 \cdot 73 \cdot 79^2 \cdot 83^2 \cdot 89^2 \cdot 97 \cdot 103 \cdot 109 \cdot 127 \cdot 131^2 \cdot 151 \cdot 157 \cdot 167 \cdot 179^2 \cdot$
$\cdot 197 \cdot 211 \cdot 227 \cdot 331 \cdot 347 \cdot 367 \cdot 379 \cdot 443 \cdot 523 \cdot 599 \cdot 709 \cdot 757 \cdot 829 \cdot 1151 \cdot 1699 \cdot$
$\cdot 1789 \cdot 2003 \cdot 2179 \cdot 2999 \cdot 3221 \cdot 4271 \cdot 4357 \cdot 4603 \cdot 5167 \cdot 8011 \cdot 8647 \cdot 8713 \cdot$
$\cdot 14951 \cdot 17293 \cdot 21467 \cdot 29989 \cdot 110563 \cdot 178481 \cdot 530713 \cdot 672827 \cdot 4036961 \cdot$
$\cdot 218834597 \cdot 16148168401 \cdot 151871210317 \cdot 2646507710984041$

On 97-05-13 Ron Sorli found a 10-fold perfect number, and George Woltman found an 11-fold one on 2001-03-13.

In 1992 we knew of 700 k-perfect numbers with $k \geq 3$. In January, 1993, this number leapt to about 1150 from the discoveries of Fred Helenius which included 114 7-perfect, 327 8-perfect and two 9-perfect numbers. He continued to find dozens of new ones each month, so that it is even less possible to keep this section of the book up-to-date than it is elsewhere; in March 1993 the total neared 1300; a postscript of a 93-09-08 letter from Schroeppel gave 1526; by the time he mailed it next day it was 1605. At the end of 2002, there were 8! known multiperfect numbers. These include 1 (for $k = 1$) and 39 Mersenne primes ($k = 2$). The numbers of others are

$k =$	3	4	5	6	7	8	9	10	11	Total
	6	36	65	245	516	1134	2074	923	1	5000

Can k be as large as we wish? Erdős conjectured that $k = o(\ln \ln n)$. It has even been suggested that there may be only finitely many k-perfect numbers for each $k \geq 3$. The first five numbers in the above table may well be complete.

If n is an odd triperfect number, then McDaniel, Cohen, Kishore, Bugulov, Kishore, Cohen & Hagis, Reidlinger, and Kishore have respectively shown that $\omega(n) \geq 9$, 9, 10, 11, 11, 11, 12, and 12; while Beck

& Najar, Alexander, and Cohen & Hagis have respectively shown that
$n > 10^{50}, 10^{60}, 10^{70}$. Cohen & Hagis have shown that the largest prime
factor of n is at least 100129 and that the second largest is at least 1009.

Shigeru Nakamura writes that Bugulov showed, in 1966, that odd k-
perfect numbers contain at least ω distinct prime factors, where $(k, \omega) =$
(3,11), (4,21), (5,54) [incorrectly stated in *MR* **37** #5139 & rNT A32-96].
Nakamura claims to prove that for an even k-perfect number,

$$\omega > \max\{k^3/81 + \tfrac{5}{3},\ k^5/2500 + 2.9,\ k^{10}/(14 \cdot 10^8) + 2.9999\}$$

and for an odd k-perfect number,

$$\omega > \max\{k^5/60 + \tfrac{47}{12},\ k^5/50 - 20.8,\ 737k^{10}/10^9 + 11.5\}.$$

These improve the results of Cohen & Hendy and of Reidlinger; he also
gives the improvements $(k, \omega) = (4, 23), (5,56), (6,142), (7,373)$ to those of
Bugulov.

Minoli & Bear say that n is k-**hyperperfect** if $n = 1 + k \sum d_i$, where
the summation is taken over all proper divisors, $1 < d_i < n$, so that
$k\sigma(n) = (k+1)n+k-1$. For example, 21, 2133 and 19521 are 2-hyperperfect
and 325 is 3-hyperperfect. They conjecture that there are k-hyperperfect
numbers for every k.

Cohen & te Riele call numbers (m, k)-**perfect** if $\sigma^m(n) = kn$; e.g.,
perfect numbers are $(1, 2)$-perfect, multiperfect numbers are $(1, k)$-perfect;
$(2, 2)$-perfect numbers have been called superperfect and $(2, k)$-perfect numbers multiply superperfect. They tabulate all (m, k)-perfect numbers n for
$(m, n) = (2, < 10^9), (3, < 2 \cdot 10^8), (4, < 10^8)$ and prove that the equation $\sigma^2(2n) = 2\sigma^2(n)$ has infinitely many solutions. They ask: for any
fixed m, are there infinitely many (m, k)-perfect numbers? and: is every n
(m, k)-perfect for some m? For $n \in [1, 400]$ they list the least such m.

Eswarathasan & Levine define $a(n)/b(n) = \sum_i = 1^n 1/i$ with $a(n) \perp$
$b(n)$ and $a(0) = 0$, and for p prime consider the sets $J(p) = \{n \geq 0 : p \mid$
$a(n)\}$ and $I(p) = \{n \geq 0 : p$ does not divide $b(n)\}$. Then $J(2) = \{0\}$,
$J(3) = \{0, 2, 7, 22\}$ and, for $p \geq 5$, $J(p) \supseteq 0, p - 1, p^2 - p, p^2 - 1$; they give
necessary and sufficient conditions for equality and show that the primes
less than 200 which satisfy these conditions are 5, 13, 17, 23, 41, 67, 73,
79, 107, 113, 139, 149, 157, 179, 191, 193. [$I(5) = \{1, 2, 3, 4, \ldots\}$] was
the subject of Putnam problem 1997 B3.] It is conjectured that there are
infinitely many such primes. In contrast

$$J(7) = \{0, 6, 42, 48, 295, 299, 337, 341, 2096, 2390, 14675, 16735, 102728\},$$

but it is also conjectured that $J(p)$ is always finite.

Ron Graham asks if $s(n) = \lfloor n/2 \rfloor$ implies that n is 2 or a power of 3.
Luo Shi-Le & Le Mao-Hua have given a partial answer.

Erdős lets $f(n)$ be the smallest integer for which $n = \sum_{i=1}^{k} d_i$ for some k, where $1 = d_1 < d_2 < \dots d_l = f(n)$ is the increasing sequence of divisors of $f(n)$. Is $f(n) = o(n)$? Or is this true only for almost all n, with $\limsup f(n)/n = \infty$?

n	1	2	3	4	5	6	7	8	9	10	11	12	13	14	15	16	17	18	19	20	21	22	23	24	25	26	27	28
$f(n)$	1	-	2	3	-	5	4	7	15	12	21	6	9	13	8	12	30	10	42	19	18	20	57	14	36	46	30	12

Erdős defined n_k to be the smallest integer for which if you partition the proper divisors of n_k into k classes, n_k will always be the sum of distinct divisors from the same class. Clearly $n_1 = 6$, but he was not even able to prove the existence of n_2.

H. Abbott, C. E. Aull, Ezra Brown & D. Suryanarayana, Quasiperfect numbers, *Acta Arith.*, **22**(1973) 439–447; *MR* **47** #4915; corrections, **29**(1976) 427–428.

Leon Alaoglu & Paul Erdős, On highly composite and similar numbers, *Trans. Amer. Math. Soc.*, **56**(1944) 448–469; *MR* **6**, 117b.

L. B. Alexander, Odd triperfect numbers are bounded below by 10^{60}, M.A. thesis, East Carolina University, 1984.

M. M. Artuhov, On the problem of odd h-fold perfect numbers, *Acta Arith.*, **23**(1973) 249–255.

Michael R. Avidon, On the distribution of primitive abundant numbers, *Acta Arith.*, **77**(1996) 195–205; *MR* **97g**:11100.

Paul T. Bateman, Paul Erdős, Carl Pomerance & E.G. Straus, The arithmetic mean of the divisors of an integer, in *Analytic Number Theory (Philadelphia, 1980)* 197–220, *Lecture Notes in Math.*, **899**, Springer, Berlin - New York, 1981; *MR* **84b**:10066.

Walter E. Beck & Rudolph M. Najar, A lower bound for odd triperfects, *Math. Comput.*, **38**(1982) 249–251.

S. J. Benkoski, Problem E2308, *Amer. Math. Monthly*, **79**(1972) 774.

S. J. Benkoski & P. Erdős, On weird and pseudoperfect numbers, *Math. Comput.*, **28**(1974) 617–623; *MR* **50** #228; corrigendum, S. Kravitz, **29**(1975) 673.

Alan L. Brown, Multiperfect numbers, *Scripta Math.*, **20**(1954) 103–106; *MR* **16**, 12.

E. A. Bugulov, On the question of the existence of odd multiperfect numbers (Russian), *Kabardino–Balkarsk. Gos. Univ. Ucen. Zap.*, **30**(1966) 9–19.

William Butske, Lynda M. Jaje & Daniel R. Mayernik, On the equation $\sum_{p|N}(1/p) + (1/N) = 1$, pseudoperfect numbers, and perfectly weighted graphs, *Math. Comput.*, **69**(2000) 407–420; *MR* **2000i**:11053.

David Callan, Solution to Problem 6616, *Amer. Math. Monthly*, **99**(1992) 783–789.

R. D. Carmichael & T. E. Mason, Note on multiply perfect numbers, including a table of 204 new ones and the 47 others previously published, *Proc. Indiana Acad. Sci.*, **1911** 257–270.

Paolo Cattaneo, Sui numeri quasiperfetti, *Boll. Un. Mat. Ital.*(3), **6**(1951) 59–62; *Zbl.* **42**, 268.

Cheng Lin-Feng, A result on multiply perfect number, *J. Southeast Univ. (English Ed.)*, **18**(2002) 265–269.

Graeme L. Cohen, On odd perfect numbers II, multiperfect numbers and quasiperfect numbers, *J. Austral. Math. Soc. Ser. A*, **29**(1980) 369–384; *MR* **81m**:10009.

Graeme L. Cohen, The non-existence of quasiperfect numbers of certain forms, *Fibonacci Quart.*, **20**(1982) 81–84.

Graeme L. Cohen, On primitive abundant numbers, *J. Austral. Math. Soc. Ser. A*, **34**(1983) 123–137.

Graeme L. Cohen, Primitive α-abundant numbers, *Math. Comput.*, **43**(1984) 263–270.

Graeme L. Cohen, Stephen Gretton and his multiperfect numbers, Internal Report No. 28, School of Math. Sciences, Univ. of Technology, Sydney, Australia, Oct 1991.

G. L. Cohen, Numbers whose positive divisors have small integral harmonic mean, *Math. Comput.*, **66**(1997) 883–891; *MR* **97f**:11007.

G. L. Cohen & Deng Moujie, On a generalisation of Ore's harmonic numbers, *Nieuw Arch. Wisk.*(4), **16**(1998) 161–172; *MR* **2000k**:11008.

G. L. Cohen & H. J. J. te Riele, Iterating the sum-of-divisors function, *Experiment Math.*, **5**(1996) 91–100; errata, **6**(1997) 177; *MR* **97m**:11007.

Graeme L. Cohen & Ronald M. Sorli, Harmonic seeds, *Fibonacci Quart.*, **36**(1998) 386–390; errata, **39**(2001) 4; *MR* **99j**:11002.

G. L. Cohen & P. Hagis, Results concerning odd multiperfect numbers, *Bull. Malaysian Math. Soc.*, **8**(1985) 23–26.

G. L. Cohen & M. D. Hendy, On odd multiperfect numbers, *Math. Chronicle*, **9**(1980) 120–136; **10**(1981) 57–61.

Philip L. Crews, Donald B. Johnson & Charles R. Wall, Density bounds for the sum of divisors function, *Math. Comput.*, **26**(1972) 773–777; *MR* **48** #6042; Errata **31**(1977) 616; *MR* **55** #286.

J. T. Cross, A note on almost perfect numbers, *Math. Mag.*, **47**(1974) 230–231.

Jean-Marie De Koninck & Aleksandar Ivić, On a sum of divisors problem, *Publ. Inst. Math. (Beograd)(N.S.)* **64(78)**(1998) 9–20; *MR* **99m**:11103.

Jean-Marie De Koninck & Imre Kátai, On the frequency of k-deficient numbers, *Publ. Math. Debrecen*, **61**(2002) 595–602; *MR* **2004a**:11098.

Marc Deléglise, Encadrement de la densité des nombres abondants, (submitted).

P. Erdős, On the density of the abundant numbers, *J. London Math. Soc.*, **9**(1934) 278–282.

P. Erdős, Problems in number theory and combinatorics, *Congressus Numerantium XVIII, Proc. 6th Conf. Numerical Math. Manitoba*, 1976, 35–58 (esp. pp. 53–54); *MR* **80e**:10005.

A. Eswarathasan & E. Levine, p-integral harmonic sums, *Discrete Math.*, **91**(1991), 249–259; *MR* **93b**:11039.

Benito Franqui & Mariano García, Some new multiply perfect numbers, *Amer. Math. Monthly*, **60**(1953) 459–462; *MR* **15**, 101.

Benito Franqui & Mariano García, 57 new multiply perfect numbers, *Scripta Math.*, **20**(1954) 169–171 (1955); *MR* **16**, 447.

Mariano García, A generalization of multiply perfect numbers, *Scripta Math.*, **19**(1953) 209–210; *MR* **15**, 199.

Mariano García, On numbers with integral harmonic mean, *Amer. Math. Monthly*, **61** (1954) 89–96; *MR* **15**, 506d, 1140.

T. Goto & S. Shibata, All numbers whose positive divisors have integral harmonic mean up to 300, *Math. Comput.*, **73**(2004) 475–491.

Peter Hagis, The third largest prime factor of an odd multiperfect number exceeds 100, *Bull. Malaysian Math. Soc.*, **9**(1986) 43–49.

Peter Hagis, A new proof that every odd triperfect number has at least twelve prime factors, *A tribute to Emil Grosswald: number theory and related analysis*, 445–450 *Contemp. Math.*, **143** Amer. Math. Soc., 1993. 43–49; *MR* **93e**:11003.

Peter Hagis & Graeme L. Cohen, Some results concerning quasiperfect numbers, *J. Austral. Math. Soc. Ser. A*, **33**(1982) 275–286.

B. E. Hardy & M. V. Subbarao, On hyperperfect numbers, Proc. 13th Manitoba Conf. Numer. Math. Comput., *Congressus Numerantium*, **42**(1984) 183–198; *MR* **86c**:11006.

Miriam Hausman & Harold N. Shapiro, On practical numbers, *Comm. Pure Appl. Math.*, **37**(1984) 705–713; *MR* **86a**:11036.

B. Hornfeck & E. Wirsing, Über die Häufigkeit vollkommener Zahlen, *Math. Ann.*, **133**(1957) 431–438; *MR* **19**, 837; see also **137**(1959) 316–318; *MR* **21** #3389.

Aleksandar Ivić, The distribution of primitive abundant numbers, *Studia Sci. Math. Hungar.*, **20**(1985) 183–187; *MR* **88h**:11065.

R. P. Jerrard & Nicholas Temperley, Almost perfect numbers, *Math. Mag.*, **46**(1973) 84–87.

H.-J. Kanold, Über mehrfach vollkommene Zahlen, *J. reine angew. Math.*, **194**(1955) 218–220; II **197**(1957) 82–96; *MR* **17**, 238; **18**, 873.

H.-J. Kanold, Über das harmonische Mittel der Teiler einer natürlichen Zahl, *Math. Ann.*, **133**(1957) 371–374; *MR* **19** 635f.

H.-J. Kanold, Einige Bemerkungen über vollkommene und mehrfach vollkommene Zahlen, *Abh. Braunschweig. Wiss. Ges.*, **42**(1990/91) 49–55; *MR* **93c**: 11002.

David G. Kendall, The scale of perfection, *J. Appl. Probability*, **19A**(1982) 125–138; *MR* **83d**:10007.

Masao Kishore, Odd triperfect numbers, *Math. Comput.*, **42**(1984) 231–233; *MR* **85d**:11009.

Masao Kishore, Odd triperfect numbers are divisible by eleven distinct prime factors, *Math. Comput.* **44**(1985) 261–263; *MR* **86k**:11007.

Masao Kishore, Odd triperfect numbers are divisible by twelve distinct prime factors, *J. Autral. Math. Soc. Ser. A*, **42**(1987) 173–182; *MR* **88c**:11009.

Masao Kishore, Odd integers N with 5 distinct prime factors for which $2 - 10^{-12} < \sigma(N)/N < 2 + 10^{-12}$, *Math. Comput.*, **32**(1978) 303–309.

M. S. Klamkin, Problem E1445*, *Amer. Math. Monthly*, **67**(1960) 1028; see also **82**(1975) 73.

Sidney Kravitz, A search for large weird numbers, *J. Recreational Math.*, **9**(1976-77) 82–85.

Richard Laatsch, Measuring the abundancy of integers, *Math. Mag.*, **59** (1986) 84–92.

G. Lord, Even perfect and superperfect numbers, *Elem. Math.*, **30**(1975) 87–88; *MR* **51** #10213.

Luo Shi-Le & Le Mao-Hua, On Graham's problem concerning the sum of the aliquot parts (Chinese), *J. Jishou Univ. Nat. Sci. Edu.*, **20**(1999) 8–11; *MR* **2000i**:11010; and see **23**(2002) 1–3, 65; *MR* **2003g**:11002.

A. Mąkowski, Remarques sur les fonctions $\theta(n)$, $\phi(n)$ et $\sigma(n)$, *Mathesis*, **69**(1960) 302–303.

A. Mąkowski, Some equations involving the sum of divisors, *Elem. Math.*, **34**(1979) 82; *MR* **81b**:10004.

Wayne L. McDaniel, On odd multiply perfect numbers, *Boll. Un. Mat. Ital.* (4), **3**(1970) 185–190; *MR* **41** #6764.

Guiseppe Melfi, On 5-tuples of twin practical numbers, *Boll. Unione Mat. Ital. Sez. B Artic. Ric. Mat.*(8), **2**(1999) 723–734; *MR* **2000j**:11138.

W. H. Mills, On a conjecture of Ore, *Proc. Number Theory Conf.*, Boulder CO, 1972, 142–146.

D. Minoli, Issues in non-linear hyperperfect numbers, *Math. Comput.*, **34** (1980) 639–645; *MR* **82c**:10005.

Daniel Minoli & Robert Bear, Hyperperfect numbers, *Pi Mu Epsilon J.*, **6**#3(1974-75) 153–157.

Shigeru Nakamura, On k-perfect numbers (Japanese), *J. Tokyo Univ. Merc. Marine(Nat. Sci.)*, **33**(1982) 43–50.

Shigeru Nakamura, On some properties of $\sigma(n)$, *J. Tokyo Univ. Merc. Marine(Nat. Sci.)*, **35**(1984) 85–93.

John C. M. Nash, Hyperperfect numbers. *Period. Math. Hungar.*, **45**(2002) 121–122.

Oystein Ore, On the averages of the divisors of a number, *Amer. Math. Monthly*, **55**(1948) 615–619; *MR* **10** 284a.

Seppo Pajunen, On primitive weird numbers, *A collection of manuscripts related to the Fibonacci sequence*, 18th anniv. vol., Fibonacci Assoc., 162–166.

Carl Pomerance, On a problem of Ore: Harmonic numbers (unpublished typescript); see Abstract *709-A5, *Notices Amer. Math. Soc.*, **20**(1973) A-648.

Carl Pomerance, On multiply perfect numbers with a special property, *Pacific J. Math.*, **57**(1975) 511–517.

Carl Pomerance, On the congruences $\sigma(n) \equiv a \bmod n$ and $n \equiv a \bmod \phi(n)$, *Acta Arith.*, **26**(1975) 265–272.

Paul Poulet, *La Chasse aux Nombres*, Fascicule I, Bruxelles, 1929, 9–27.

Herwig Reidlinger, Über ungerade mehrfach vollkommene Zahlen [On odd multiperfect numbers], *Österreich. Akad. Wiss. Math.-Natur. Kl. Sitzungsber. II*, **192**(1983) 237–266; *MR* **86d**:11018.

Herman J. J. te Riele, Hyperperfect numbers with three different prime factors, *Math. Comput.*, **36**(1981) 297–298.

Neville Robbins, A class of solutions of the equation $\sigma(n) = 2n + t$, *Fibonacci Quart.*, **18**(1980) 137–147 (misprints in solutions for $t = 31, 84, 86$).

Richard F. Ryan, A simpler dense proof regarding the abundancy index, *Math. Mag.*, **76**(2003) 299–301.

Richard F. Ryan, Results concerning uniqueness for $\sigma(x)/x = \sigma(p^n q^m)/p^n q^m$ and related topics, *Int. Math. J.*, **2**(2002) 497–514; *MR* **2002k**:11007.

József Sándor, On a method of Galambos and Kátai concerning the frequency of deficient numbers, *Publ. Math. Debrecen*, **39**(1991) 155–157; *MR* **92j**:11111.

M. Satyanarayana, Bounds of $\sigma(N)$, *Math. Student*, **28**(1960) 79–81.

H. N. Shapiro, Note on a theorem of Dickson, *Bull. Amer. Math. Soc.*, **55**(1949) 450–452.

H. N. Shapiro, On primitive abundant numbers, *Comm. Pure Appl. Math.*, **21**(1968) 111–118.

W. Sierpiński, Sur les nombres pseudoparfaits, *Mat. Vesnik*, **2**(17)(1965) 212–213; *MR* **33** #7296.

W. Sierpiński, *Elementary Theory of Numbers* (ed. A. Schinzel), PWN–Polish Scientific Publishers, Warszawa, 1987, pp. 184–186.

R. Steuerwald, Ein Satz über natürlich Zahlen N mit $\sigma(N) = 3N$, *Arch. Math.*, **5**(1954) 449–451; *MR* **16** 113h.

D. Suryanarayana, Quasi-perfect numbers II, *Bull. Calcutta Math. Soc.*, **69** (1977) 421–426; *MR* **80m**:10003.

Charles R. Wall, The density of abundant numbers, Abstract 73T–A184, *Notices Amer. Math. Soc.*, **20**(1973) A-472.

Charles R. Wall, A Fibonacci-like sequence of abundant numbers, *Fibonacci Quart.*, **22**(1984) 349; *MR* **86d**:11018.

Charles R. Wall, Phillip L. Crews & Donald B. Johnson, Density bounds for the sum of divisors function, *Math. Comput.*, **26**(1972) 773–777.

Paul A. Weiner, The abundancy ratio, a measure of perfection, *Math. Mag.*, **73**(2000) 307–310.

Motoji Yoshitake, Abundant numbers, sum of whose divisors is equal to an integer times the number itself (Japanese), *Sūgaku Seminar*, **18**(1979) no. 3, 50–55.

Andreas & Eleni Zachariou, Perfect, semi-perfect and Ore numbers, *Bull. Soc. Math. Grèce*(N.S.), **13**(1972) 12–22; *MR* **50** #12905.

OEIS: A000396, A001599-001600, A002975, A005100-005101, A005231, A005820, A005835, A006036-006038, A007539, A007592-007593, A019279, A027687, A033630, A034897-034898, A038536, A045768, A046060-046061, A050413-050415, A071395.

B3 Unitary perfect numbers.

If d divides n and $d \perp n/d$, call d a **unitary divisor** of n. A number n which is the sum of its unitary divisors, apart from n itself, is a **unitary perfect number**. There are no odd unitary perfect numbers, and Subbarao conjectures that there are only a finite number of even ones. He, Carlitz & Erdős each offer \$10.00 for settling this question and Subbarao offers 10¢ for each new example. If $n = 2^a m$, where m is odd and has r distinct prime factors, then Subbarao and others have shown that, apart from $2 \cdot 3$, $2^2 \cdot 3 \cdot 5$, $2 \cdot 3^2 \cdot 5$ and $2^6 \cdot 3 \cdot 5 \cdot 7 \cdot 13$, there are no unitary perfect numbers with $a \leq 10$, or with $r \leq 6$. S. W. Graham has shown that the first and third are the only unitary perfect numbers of shape $2^a m$ with m odd and squarefree, and Jennifer DeBoer that the second is the only one of shape $2^a 3^2 m$ with $m \perp 6$ and squarefree.

Wall has found the unitary perfect number

$$2^{18} \cdot 3 \cdot 5^4 \cdot 7 \cdot 11 \cdot 13 \cdot 19 \cdot 37 \cdot 79 \cdot 109 \cdot 157 \cdot 313$$

and shown that it is the fifth such. He can prove that any other unitary perfect number has an odd component greater than 2^{15}. Frey has shown that if $N = 2^m p_1^{a_1} \ldots p_r^{a_r}$ is unitary perfect with $N \perp 3$, then $m > 144$, $r > 144$ and $N > 10^{440}$.

Peter Hagis investigates **unitary multiperfect numbers**: there are no odd ones. Write $\sigma^*(n)$ for the sum of the unitary divisors of n. If $\sigma^*(n) = kn$ and n contains t distinct odd prime factors, then $k = 4$ or 6 implies $n > 10^{110}$, $t \geq 51$ and $2^{49}|n$; $k \geq 8$ implies $n > 10^{663}$ and $t \geq 247$; while k odd and $k \geq 5$ imply $n > 10^{461}$, $t \geq 166$ and $2^{166}|n$.

Sitaramaiah & Subbarao call a number **unitary superperfect** if it satisfies the equation $\sigma^*(\sigma^{ast}(n)) = 2n$. They find 22 such numbers below 10^8.

Cohen calls a divisor d of an integer n a 1-**ary divisor** of n if $d \perp n/d$, and he calls d a k-**ary divisor** of n (for $k > 1$), and writes $d|_k n$, if the greatest common $(k-1)$-ary divisor of d and n/d is 1 (written $(d, n/d)_{k-1} = 1$). In this notation $d|n$ and $d \parallel n$ are written $d|_0 n$ and $d|_1 n$. He also calls p^x an **infinitary divisor** of $p^y (y > 0)$ if $p^x|_{y-1}p^y$. This gives rise to infinitary analogs of earlier concepts. Write $\sigma_\infty(n)$ for the sum of the infinitary divisors of n. He found 14 infinitary perfect numbers, i.e., with $\sigma_\infty(n) = kn$ and $k = 2$; 13 numbers with $k = 3$; 7 with $k = 4$; and two with $k = 5$. There are no odd ones, and he conjectures that there are no infinitary multiperfect numbers not divisible by 3.

Note that Suryanarayana (who also uses the term 'k-ary divisor') and Alladi give *different* generalizations of unitary divisors.

K. Alladi, On arithmetic functions and divisors of higher order, *J. Austral. Math. Soc. Ser. A*, **23**(1977) 9–27.

Eckford Cohen, Arithmetical functions associated with the unitary divisors of an integer, *Math. Z.*, **74**(1960) 66–80; *MR* **22** #3707.

Eckford Cohen, The number of unitary divisors of an integer, *Amer. Math. Monthly*, **67**(1960) 879–880; *MR* **23** # A124.

Graeme L. Cohen, On an integer's infinitary divisors, *Math. Comput.*, **54** (1990) 395–411.

Graeme Cohen & Peter Hagis, Arithmetic functions associated with the infinitary divisors of an integer, *Internat. J. Math. Math. Sci.*, (to appear).

J. L. DeBoer, On the non-existence of unitary perfect numbers of certain type, *Pi Mu Epsilon J.* (submitted).

H. A. M. Frey, Über unitär perfekte Zahlen, *Elem. Math.*, **33**(1978) 95–96; *MR* **81a**:10007.

S. W. Graham, Unitary perfect numbers with squarefree odd part, *Fibonacci Quart.*, **27**(1989) 317–322; *MR* **90i**:11003.

Peter Hagis, Lower bounds for unitary multiperfect numbers, *Fibonacci Quart.*, **22**(1984) 140–143; *MR* **85j**:11010.

Peter Hagis, Odd nonunitary perfect numbers, *Fibonacci Quart.*, **28** (1990) 11–15; *MR* **90k**:11006.

Peter Hagis & Graeme Cohen, Infinitary harmonic numbers, *Bull. Austral. Math. Soc.*, **41**(1990) 151–158; *MR* **91d**:11001.

József Sándor, On Euler's arithmetical function, *Proc. Alg. Conf. Braşov 1988*, 121–125.

V. Sitaramaiah & M. V. Subbarao, On the equation $\sigma^*(\sigma^{ast}(n)) = 2n$, *Utilitas Math.*, **53**(1998) 101–124; *MR* **99a**:11009.

V. Sitaramaiah & M. V. Subbarao, On unitary multiperfect numbers, *Nieuw Arch. Wisk.*(4), **16**(1998) 57–61; *MR* **99h**:11008.

V. Siva Rama Prasad & D. Ram Reddy, On unitary abundant numbers, *Math. Student*, **52**(1984) 141–144 (1990) *MR* **91m**:11002.

V. Siva Rama Prasad & D. Ram Reddy, On primitive unitary abundant numbers, *Indian J. Pure Appl. Math.*, **21**(1990) 40–44; *MR* **91f**:11004.

M. V. Subbarao, Are there an infinity of unitary perfect numbers? *Amer. Math. Monthly*, **77**(1970) 389–390.

M. V. Subbarao & D. Suryanarayana, Sums of the divisor and unitary divisor functions, *J. reine angew. Math.*, **302**(1978) 1–15; *MR* **80d**:10069.

M. V. Subbarao & L. J. Warren, Unitary perfect numbers, *Canad. Math. Bull.*, **9**(1966) 147–153; *MR* **33** #3994.

M. V. Subbarao, T. J. Cook, R. S. Newberry & J. M. Weber, On unitary perfect numbers, *Delta*, **3**#1(Spring 1972) 22–26.

D. Suryanarayana, The number of k-ary divisors of an integer, *Monatsh. Math.*, **72**(1968) 445–450.

Charles R. Wall, The fifth unitary perfect number, *Canad. Math. Bull.*, **18**(1975) 115–122. See also *Notices Amer. Math. Soc.*, **16**(1969) 825.

Charles R. Wall, Unitary harmonic numbers, *Fibonacci Quart.*, **21**(1983) 18–25.

Charles R. Wall, On the largest odd component of a unitary perfect number, *Fibonacci Quart.*, **25**(1987) 312–316; *MR* **88m**:11005.

OEIS: A002827, A034460, A034448.

B4 Amicable numbers.

Unequal numbers m, n are called **amicable** if each is the sum of the aliquot parts of the other, i.e., $\sigma(m) = \sigma(n) = m + n$. Several thousand (in 2003, 'million') such pairs are known. The smaller member, 220, of the smallest pair, occurs in *Genesis*, xxxii, 14, and intrigued the Greeks and Arabs and many others since. For their history see the articles of Lee & Madachy. The *Genesis* reference, from the King James Bible, is achieved by amalgamating 200 females and 20 males. Aviezri Fraenkel writes that in his Pentateuch, they occur at xxxii, 15, and gives the more convincing occurrences of 220 in *Ezra* viii, 20 and in *1 Chronicles* xv, 6; and of 284 in *Nehemiah* xi, 18. He notes that the three places are amicably related: all are connected to the tribe of Levi, whose name derives from the wish of Levi's mother to be amicably related to his father (*Genesis* xxix, 34).

It is not known if there are infinitely many, but it is believed that there are. In fact Erdős conjectured that the number, $A(x)$, of such pairs with $m < n < x$ is at least $x^{1-\epsilon}$. He improved a result of Kanold to show that $A(x) = o(x)$ and his method can be used to obtain $A(x) \leq cx/\ln\ln\ln x$. Pomerance obtained the further improvement

$$A(x) \leq x\exp\{-c(\ln\ln\ln x \ln\ln\ln\ln x)^{1/2}\}.$$

Erdős conjectured that $A(x) = o(x/(\ln x)^k)$ for every k whereupon Pomerance proved the stronger result

$$A(x) \leq x\exp\{-(\ln x)^{1/3}\}.$$

This implies that the sum of the reciprocals of the amicable numbers is finite, a fact not earlier known. He also notes that his proof can be modified to give the slightly stronger result

$$A(x) \ll x\exp\{-c(\ln x \ln\ln x)^{1/3}\}.$$

Herman te Riele has found all 1427 amicable pairs whose lesser members are less than 10^{10}. He remarks that the quantity $A(x)(\ln x)^3/x^{1/2}$ "remains very close to 174.6", but I suspect that a much more powerful telescope would require the exponent $1/2$ to be increased much nearer to 1. Moews & Moews have continued the complete search to beyond $2 \cdot 10^{11}$.

Some large amicable pairs, with 32, 40, 81 and 152 decimal digits, discovered by te Riele, are mentioned by Kaplansky under "Mathematics" in the 1975 *Encyclopedia Britannica Yearbook*. The largest previously known had 25 decimal digits. More recently te Riele has constructed, from a "mother" list of 92 known amicable pairs, more than 2000 new pairs of sizes up to 38 decimal digits, and five pairs with from 239 to 282 digits. The largest amicable pair known in mid-1993 had 1041 decimal digits; it was found in July 1988 by Holger Wiethaus, a student at Dortmund. On 97-10-04 Mariano Garcia found a pair each of whose members has 4829 digits, and soon raised this to 5577. These exceeded the pair with 3766 digits found by Frank Zweers the previous August. On 2003-06-06 Paul Jobling announced the 8684-digit pair $(abp^{145}q_1, aqp^{145}q_2)$ where $a = 3^3 \cdot 5^2 \cdot 19 \cdot 31 \cdot 757 \cdot 3329$, $b = 1511 \cdot 72350721629 \cdot 2077116867246979$, $p = 454299173560665115110417717001$, $q = 2328969556742338358853240430207843780057919$, $q_1 = 23289695567468813505888470813589479577450920 \cdot p^{145} - 1$ and $q_2 = 2272247272342510733026822333800 \cdot p^{145} - 1$.

Elvin J. Lee has given half a dozen rules for amicable pairs of type $(2^npq, 2^nrs)$ where p, q, r, s are primes of appropriate shape. E.g.,

$$p = 3 \cdot 2^{n-1} - 1, \quad q = 35 \cdot 2^{n+1} - 29, \quad r = 7 \cdot 2^{n-1} - 1, \quad s = 15 \cdot 2^{n+1} - 13,$$

but the simultaneous discovery of four such primes is a rare event.

Borho, Hoffman & te Riele have made considerable advances, both with proliferation of generalized Thabit rules, and with actual computation. Of the 1427 amicable pairs mentioned above, all but 17 have $m+n \equiv 0 \bmod 9$. The smallest exception is Poulet's pair

$$2^4 \cdot 331 \cdot \left\{ \begin{array}{l} 19 \cdot 6619 \\ 199 \cdot 661 \end{array} \right.$$

with $m + n \equiv 5 \bmod 9$: te Riele gives the first examples

$$2^4 \cdot \left\{ \begin{array}{l} 19^2 \cdot 103 \cdot 1627 \\ 3847 \cdot 16763 \end{array} \right. \quad \text{and} \quad 2^2 \cdot 19 \cdot \left\{ \begin{array}{l} 13^2 \cdot 37 \cdot 43 \cdot 139 \\ 41 \cdot 151 \cdot 6709 \end{array} \right.$$

with m, n even, $m + n \equiv 3 \bmod 9$.

It is not known if an amicable pair exists with m and n of opposite parity, or with $m \perp n$. Bratley & McKay conjectured that both members of all odd amicable pairs are divisible by 3, but Battiato & Borho produced 15 counterexamples with from 36 to 73 decimal digits. In an 87-05-15 letter te Riele announced a 33-digit specimen (misquoted in UPINT2)

$$5 \cdot 7^2 \cdot 11^2 \cdot 13 \cdot 17 \cdot 19^3 \cdot 23 \cdot 37 \cdot 181 \left\{ \begin{array}{l} 101 \cdot 8693 \cdot 19479382229 \\ 365147 \cdot 47307071129 \end{array} \right.$$

Is this the smallest such pair? Is there an odd amicable pair with one member, but not both, divisible by 3? No pair has been discovered with different smallest prime factors.

Yasutoshi Kohmoto found pairs with each member coprime to 30, for example

$7^2 \cdot 11 \cdot 13^2 \cdot 17^4 \cdot 19^3 \cdot 23 \cdot 29^2 \cdot 31 \cdot 37 \cdot 43 \cdot 59 \cdot 61 \cdot 67 \cdot 83 \cdot 97 \cdot 139 \cdot 173 \times$
$181 \cdot 331 \cdot 349 \cdot 577 \times 661 \cdot 1321 \cdot 4349 \cdot 11093 \cdot 41519 \cdot 43973 \cdot 44371 \cdot 88741 \times$
$15223567 \cdot 91341401 \cdot 264271333 \cdot 1281651920873 \cdot 47031498888355607 \times$
$\quad\quad 41277542598611381429 \cdot 4870750026636143008621 \times$

either $\quad\quad 179 \cdot 15256639719248840486146283019032958 21727019$

or $\quad\quad 274619514946479128750633094342593247910863599$

An old conjecture of Charles Wall is that odd amicable pairs must be incongruent modulo 4.

On p. 169 of *Mathematical Magic Show*, Vintage Books, 1978, Martin Gardner makes a conjecture about the digital roots of amicable numbers. Lee confirms this in part by showing that if $(2^n pqr, 2^n stu)$ is an amicable pair whose sum is not divisible by 9, then each number is congruent to 7, modulo 9.

Unitary amicable numbers, that is, pairs (m, n) such that the sum of the unitary divisors of each is equal to their sum, $m + n$, have been studied by Peter Hagis and by Mariano García. If an amicable pair is squarefree, then it is also a unitary amicable pair. Hagis found 76 such pairs and 32 others which are not square free. The smallest is (114,126). He asks if there are infinitely many such pairs. In particular, are there infinitely many "twin" pairs, such as (197340,286500) and (241110,242730) whose sums are both 483840, a very smooth number, as is the sum, $2^{11} \cdot 3^5 \cdot 5^4 \cdot 7 \cdot 11$ of the other twin pairs that he lists. He conjectures that no relatively prime pair exists, and asks for an example of an odd non-squarefree pair (which would not, therefore, be amicable).

Yasutoshi Kohmoto, on 2004-01-30, announced the 317-digit pair
$2^4 \cdot 3 \cdot 5^4 \cdot 7^5 \cdot 11 \cdot 13 \cdot 19^2 \cdot 23 \cdot 29 \cdot 31 \cdot 97 \cdot 101 \cdot 127 \cdot 137 \cdot 151 \cdot 181 \cdot 191 \cdot 227 \times$
$251 \cdot 313^2 \cdot 1523 \cdot 17569 \cdot 18119 \cdot 22193 \cdot 42767 \cdot 133157 \cdot 1594471 \cdot 3592427 \times$
$12755767 \cdot 16563721580414291 \cdot 369213334428491989954037$
$11076400032854759699862110099 \cdot 50932682932260257055099576060765094 3$
$\times$ either $7567093741752835889818512302299461638848622 51 \times$
1064593171879941377731001687180268527474016787661602696170176449715407158002429 87
or $53 \cdot 140131365588015479441083561153693732079099 21 \times$
1064593171879941377731001687180268542468072905579258996366117493160636597043310 51

Cohen & te Riele call (a,b) a ϕ-**amicable pair** with multiplier k if $\phi(a) = \phi(b) = (a+b)/k$ for some integer $k \geq 1$, where ϕ is Euler's totient function (see **B36**). They computed all such pairs with larger member $\leq 10^9$ and found 812 pairs whose gcd is squarefree.

J. Alanen, O. Ore & J. G. Stemple, Systematic computations on amicable numbers, *Math. Comput.*, **21**(1967) 242–245; MR **36** #5058.

M. M. Artuhov, On some problems in the theory of amicable numbers (Russian), *Acta Arith.*, **27**(1975) 281–291.

S. Battiato, *Über die Produktion von 37803 neuen befreundeten Zahlenpaaren mit der Brütermethode*, Master's thesis, Wuppertal, June 1988.

Stefan Battiato & Walter Borho, Breeding amicable numbers in abundance II, *Math. Comput.*, **70**(2001) 1329–1333; MR **2002b**:11011.

S. Battiato & W. Borho, Are there odd amicable numbers not divisible by three? *Math. Comput.*, **50**(1988) 633–636; MR **89c**:11015.

W. Borho, On Thabit ibn Kurrah's formula for amicable numbers, *Math. Comput.*, **26**(1972) 571–578.

W. Borho, Befreundete Zahlen mit gegebener Primteileranzahl, *Math. Ann.*, **209**(1974) 183–193.

W. Borho, Eine Schranke für befreundete Zahlen mit gegebener Teileranzahl, *Math. Nachr.*, **63**(1974) 297–301.

W. Borho, Some large primes and amicable numbers, *Math. Comput.*, **36** (1981) 303–304.

W. Borho & H. Hoffmann, Breeding amicable numbers in abundance, *Math. Comput.*, **46**(1986) 281–293.

P. Bratley & J. McKay, More amicable numbers, *Math. Comput.*, **22**(1968) 677–678; MR **37** #1299.

P. Bratley, F. Lunnon & J. McKay, Amicable numbers and their distribution, *Math. Comput.*, **24**(1970) 431–432.

B. H. Brown, A new pair of amicable numbers, *Amer. Math. Monthly* **46** (1939) 345.

Graeme L. Cohen, Stephen F. Gretton & Peter Hagis, Multiamicable numbers, *Math. Comput.*, **64**(1995) 1743–1753; *MR* **95m**:11012.

Graeme L. Cohen & H. J. J. te Riele, On ϕ-amicable pairs, *Math. Comput.*, **67**(1998) 399–411; *MR* **98d**:11009.

Patrick Costello, Four new amicable pairs, *Notices Amer. Math. Soc.*, **21** (1974) A-483.

Patrick Costello, Amicable pairs of Euler's first form, *Notices Amer. Math. Soc.*, **22**(1975) A-440.

Patrick Costello, Amicable pairs of the form $(i, 1)$, *Math. Comput.*, **56**(1991) 859–865; *MR* **91k**:11009.

Patrick Costello, New amicable pairs of type (2,2) and type (3,2), *Math. Comput.*, **72**(2003) 489–497; *MR* **2003i**:11006.

P. Erdős, On amicable numbers, *Publ. Math. Debrecen*, **4**(1955) 108–111; *MR* **16**, 998.

P. Erdős & G. J. Rieger, Ein Nachtrag über befreundete Zahlen, *J. reine angew. Math.*, **273**(1975) 220.

E. B. Escott, Amicable numbers, *Scripta Math.*, **12**(1946) 61–72; *MR* **8**, 135.

L. Euler, De numeris amicabilibus, *Opera Omnia*, Ser.1, Vol.2, Teubner, Leipzig & Berlin, 1915, 63–162.

M. García, New amicable pairs, *Scripta Math.*, **23**(1957) 167–171; *MR* **20** #5158.

Mariano García, New unitary amicable couples, *J. Recreational Math.*, **17** (1984-5) 32–35.

Mariano García, Favorable conditions for amicability, *Hostos Community Coll. Math. J.*, New York, Spring 1989, 20–25.

Mariano García, K-fold isotopic amicable numbers, *J. Recreational Math.*, **19**(1987) 12–14.

Mariano Garcia, New amicable pairs of Euler's first form with greatest common factor a prime times a power of 2, *Nieuw Arch. Wisk.*(4), **17**(1999) 25–27; *MR* **2000d**:11158.

Mariano Garcia, A million new amicable pairs, *J. Integer seq.*, **4**(2001) no.2 Article 01.2.6 3pp.(electronic).

Mariano Garcia, The first known type (7,1) amicable pair, *Math. Comput.*, **72**(2003) 939–940; *MR* **2003j**:11007.

A. A. Gioia & A. M. Vaidya, Amicable numbers with opposite parity, *Amer. Math. Monthly*, **74**(1967) 969–973; correction **75**(1968) 386; *MR* **36** #3711, **37** #1306.

Peter Hagis, On relatively prime odd amicable numbers, *Math. Comput.*, **23**(1969) 539–543; *MR* **40** #85.

Peter Hagis, Lower bounds for relatively prime amicable numbers of opposite parity, *Math. Comput.*, **24**(1970) 963–968.

Peter Hagis, Relatively prime amicable numbers of opposite parity, *Math. Mag.*, **43**(1970) 14–20.

Peter Hagis, Unitary amicable numbers, *Math. Comput.*, **25**(1971) 915–918.

H.-J. Kanold, Über die Dichten der Mengen der vollkommenen und der befreundeten Zahlen, *Math. Z.*, **61**(1954) 180–185; *MR* **16**, 337.

H.-J. Kanold, Über befreundete Zahlen I, *Math. Nachr.*, **9**(1953) 243–248; II *ibid.*, **10** (1953) 99–111; *MR* **15**, 506.

H.-J. Kanold, Über befreundete Zahlen III, *J. reine angew. Math.*, **234**(1969) 207–215; *MR* **39** #122.

E. J. Lee, Amicable numbers and the bilinear diophantine equation, *Math. Comput.*, **22**(1968) 181–187; *MR* **37** #142.

E. J. Lee, On divisibility by nine of the sums of even amicable pairs, *Math. Comput.*, **23**(1969) 545–548; *MR* **40** #1328.

E. J. Lee & J. S. Madachy, The history and discovery of amicable numbers, part 1, *J. Recreational Math.*, **5**(1972) 77–93; part 2, 153–173; part 3, 231–249.

O. Ore, *Number Theory and its History*, McGraw-Hill, New York, 1948, p. 89.

Carl Pomerance, On the distribution of amicable numbers, *J. reine angew. Math.*, **293/294**(1977) 217–222; II **325**(1981) 183–188; *MR* **56** #5402, **82m**: 10012.

P. Poulet, 43 new couples of amicable numbers, *Scripta Math.*, **14**(1948) 77.

H. J. J. te Riele, Four large amicable pairs, *Math. Comput.*, **28**(1974) 309–312.

H. J. J. te Riele, On generating new amicable pairs from given amicable pairs, *Math. Comput.*, **42**(1984) 219–223.

Herman J. J. te Riele, New very large amicable pairs, in *Number Theory* Noordwijkerhout 1983, *Springer Lecture Notes in Math.*, **1068**(1984) 210–215.

H. J. J. te Riele, Computation of all the amicable pairs below 10^{10}, *Math. Comput.*, **47**(1986) 361–368 & S9–S40.

H. J. J. te Riele, A new method for finding amicable pairs, in *Mathematics of Computation 1943–1993* (Vancouver, 1993), *Proc. Sympos. Appl. Math.* **48**(1994) 577–581; *MR* **95m**:11013.

H. J. J. te Riele, W. Borho, S. Battiato, H. Hoffmann & E.J. Lee, *Table of Amicable Pairs between 10^{10} and 10^{52}*, Centrum voor Wiskunde en Informatica, Note NM-N8603, Stichting Math. Centrum, Amsterdam, 1986.

Charles R. Wall, *Selected Topics in Number Theory*, Univ. of South Carolina Press, Columbia SC, 1974, P.68.

Dale Woods, Construction of amicable pairs, #789-10-21, *Abstracts Amer. Math. Soc.*, **3**(1982) 223.

S. Y. Yan, 68 new large amicable pairs, *Comput. Math. Appl.*, **28**(1994) 71–74; *MR* **96a**:11008; errata, **32**(1996) 123–127; *MR* **97h**:11005.

S. Y. Yan & T. H. Jackson, A new large amicable pair, *Comput. Math. Appl.*, **27**(1994) 1–3; *MR* **94m**:11012.

B5 Quasi-amicable or betrothed numbers.

García has called a pair of numbers (m, n), $m < n$, **quasi-amicable** if

$$\sigma(m) = \sigma(n) = m + n + 1.$$

For example, (48,75), (140,195), (1575,1648), (1050,1925) and (2024,2295). Rufus Isaacs, noting that each of m and n is the sum of the *proper* divisors of the other (i.e., omitting 1 as well as the number itself) has much more appropriately named them **betrothed numbers**.

Mąkowski gave examples of betrothed numbers and also of **amicable triples**

$$\sigma(a) = \sigma(b) = \sigma(c) = a + b + c,$$

e.g., $2^2 3^2 5 \cdot 11$, $2^5 3^2 7$, $2^2 3^2 71$. Similarly, in a 92-07-20 letter, Yasutoshi Kohmoto calls the set $\{a, b, c, d\}$ **quadri-amicable** if

$$\sigma(a) = \sigma(b) = \sigma(c) = \sigma(d) = a + b + c + d.$$

As examples which are not multiples of 3 he gives

$a = x \cdot 173 \cdot 1933058921 \cdot 149 \cdot 103540742849$ $b = x \cdot 173 \cdot 1933058921 \cdot 15531111427499$
$c = 336352252427 \cdot 149 \cdot 103540742849$ $d = 336352252427 \cdot 15531111427499$

where x is the product of

$5^9 \cdot 7^2 \cdot 11^4 \cdot 17^2 \cdot 19 \cdot 29^2 \cdot 67 \cdot 71^2 \cdot 109 \cdot 131 \cdot 139 \cdot 179 \cdot 307 \cdot 431 \cdot 521 \cdot 653 \cdot 1019 \cdot 1279 \cdot 2557 \cdot 3221 \cdot 5113 \cdot 6949$

with a perfect number $2^{p-1} M_p$, $M_p = 2^p - 1$ being a Mersenne prime (see **A3**) with $p > 3$.

Hagis & Lord have found all 46 pairs of betrothed numbers with $m < 10^7$. All of them are of opposite parity. No pairs are known with m, n having the same parity. If there are such, then $m > 10^{10}$. If $m \perp n$, then mn contains at least four distinct prime factors, and if mn is odd, then mn contains at least 21 distinct prime factors.

Beck & Najar call such pairs *reduced* amicable pairs, and call numbers m, n such that

$$\sigma(m) = \sigma(n) = m + n - 1$$

augmented amicable pairs. They found 11 augmented amicable pairs. They found no reduced or augmented *unitary* amicable or sociable numbers (see **B8**) with $n < 10^5$.

Walter E. Beck & Rudolph M. Najar, More reduced amicable pairs, *Fibonacci Quart.*, **15**(1977) 331–332; *Zbl.* **389**.10004.

Walter E. Beck & Rudolph M. Najar, Fixed points of certain arithmetic functions, *Fibonacci Quart.*, **15**(1977) 337–342; *Zbl.* **389**.10005.

Walter E. Beck & Rudolph M. Najar, Reduced and augmented amicable pairs to 10^8, *Fibonacci Quart.*, **31**(1993) 295–298; *MR* **94g**:11005.

Peter Hagis & Graham Lord, Quasi-amicable numbers, *Math. Comput.*, **31** (1977) 608–611; *MR* **55** #7902; *Zbl.* **355**.10010.

M. Lal & A. Forbes, A note on Chowla's function, *Math. Comput.*, **25**(1971) 923–925; *MR* **45** #6737; *Zbl.* **245**.10004.

Andrzej Mąkowski, On some equations involving functions $\phi(n)$ and $\sigma(n)$, *Amer. Math. Monthly*, **67**(1960) 668–670; correction **68**(1961) 650; *MR* **24** #A76.

OEIS: A003502-003503, A005276.

B6 Aliquot sequences.

Since some numbers are abundant and some deficient, it is natural to ask what happens when you iterate the function $s(n) = \sigma(n) - n$ and produce an **aliquot sequence**, $\{s^k(n)\}$, $k = 0, 1, 2, \ldots$. Catalan and Dickson conjectured that all such sequences were bounded, but we now have heuristic arguments and experimental evidence that some sequences, perhaps almost all of those with n even, go to infinity. The smallest n for which there was ever doubt was 138, but D. H. Lehmer eventually showed that after reaching a maximum

$$s^{117}(138) = 179931\,895322 = 2 \cdot 61 \cdot 929 \cdot 1587569$$

the sequence terminated at $s^{177}(138) = 1$. The next value for which there continues to be real doubt is 276. A good deal of computation by Lehmer, subsequently assisted by Godwin, Selfridge, Wunderlich and others, pushed the calculation as far as $s^{469}(276)$, which was quoted in the first edition. Thomas Struppeck factored this term and computed two more iterates. Andy Guy wrote a PARI program which started from scratch and overnight verified all the earlier calculations and reached $s^{487}(276)$.

The first few sequences whose fate was unknown are the "Lehmer six" starting from 276, 552, 564, 660, 840 and 966. Our program found that the 840 sequence hit the prime $s^{746}(840) = 601$ and established a new record

$$s^{287}(840) = 3\,463982\,260143\,725017\,429794\,136098\,072146\,586526\,240388$$

$$= 2^2 \cdot 64970467217 \cdot 6237379309797547 \cdot 2136965558478112990003$$

for the maximum of a terminating sequence. This has since been beaten by Mitchell Dickerman who found that the 1248 sequence has length 1075 after reaching a maximum $s^{583}(1248) =$

$$1231\,636691\,923602\,991963\,829388\,638861\,714770\,651073\,275257\,065104 = 2^4 p$$

of 58 digits, and by Paul Zimmerman who found that the 446580 sequence terminates at step 4736 with the prime 601.

Godwin investigated the fourteen main sequences starting between 1000 and 2000 whose outcome was unknown and discovered that the sequence 1848 terminated. We have found that those for 2580, 2850, 4488, 4830, 6792, 7752, 8862 and 9540 also terminate.

Wieb Bosma showed that the 3556 sequence teminated. Benito & Verona have shown that the 4170, 7080 and 8262 sequences each terminate, the first of which had a (then) record maximum of 84 decimal digits: $s^{289}(4170) =$

$$3295610803424772127472036928633662138338387031588588223270640321920936903214888891836=$$
$$2^2 \cdot 41 \cdot 97 \cdot 20374357 \cdot 1559593537 \cdot 6519660739549760813421075978322876520913951561749905234983331163$$

Wolfgang Creyaufmüller has made extensive calculations, the results of which may be seen at

`http://home.t-online.de/home/wolfgang.creyaufmueller/aliquot`.

Comparatively few have terminated, and many open-ended sequences have appeared. His limits of computation generally exceed 10^{80}. The current numbers of open-ended sequences in the interval $\big((k-1)10^5, k \cdot 10^5\big)$ is

$k =$	1	2	3	4	5	6	7	8	9	10
	922	975	938	877	917	971	958	971	982	985

which support the Guy-Selfridge conjecture as opposed to that of Catalan-Dickson. Creyaufmüller's graphs are convincing even though 564 almost bit the dust on one occasion. In March 2004, his page gave statistics for the Lehmer five as:

Sequence	276	552	564	660	966
Length so far	1356	854	3098	578	579
# of digits	124	130	127	127	121

Much of the later calculation has been done by Paul Zimmermann. Creyaufmüller ceased calculating the 389508 sequence when it reached 101 digits at step 7135.

H. W. Lenstra has proved that it is possible to construct arbitrarily long monotonic increasing aliquot sequences. See the quadruple paper cited under **B41**. The last of the following references has a bibliography of 60 items concerning the iteration of number-theoretic functions.

Jack Alanen, Empirical study of aliquot series, *Math. Rep.*, **133** Stichting Math. Centrum Amsterdam, 1972; see *Math. Comput.*, **28**(1974) 878–880.

Manuel Benito, Wolfgang Creyaufmüller, Juan L. Varona & Paul Zimmermann, Aliquot sequence 3630 ends after reaching 100 digits, *Experiment Math.*, **11**(2002) 201–206; *MR* **2003j**:11150.

Manuel Benito & Juan L. Varona, Advances in aliquot sequences, *Math. Comput.*, **68**(1999) 389–393; *MR* **99c**:11162.

E. Catalan, Propositions et questions diverses, *Bull. Soc. Math. France*, **16** (1887–88) 128–129.

John Stanley Devitt, Aliquot Sequences, MSc thesis, The Univ. of Calgary, 1976; see *Math. Comput.*, **32**(1978) 942–943.

J. S. Devitt, R. K. Guy & J. L. Selfridge, Third report on aliquot sequences, *Congr. Numer.* XVIII, Proc. 6th Manitoba Conf. Numer. Math., 1976, 177–204; *MR* **80d**:10001.

L. E. Dickson, Theorems and tables on the sum of the divisors of a number, *Quart. J. Math.*, **44**(1913) 264–296.

Paul Erdős, On asymptotic properties of aliquot sequences, *Math. Comput.*, **30**(1976) 641–645.

Andrew W. P. Guy & Richard K. Guy, A record aliquot sequence, in *Mathematics of Computation 1943–1993* (Vancouver, 1993), *Proc. Sympos. Appl. Math.*, (1994) Amer. Math. Soc., Providence RI, 1984.

Richard K. Guy, Aliquot sequences, in *Number Theory and Algebra*, Academic Press, 1977, 111–118; *MR* **57** #223; *Zbl.* **367**.10007.

Richard K. Guy & J. L. Selfridge, Interim report on aliquot sequences, *Congr. Numer.* V, Proc. Conf. Numer. Math., Winnipeg, 1971, 557–580; *MR* **49** #194; *Zbl.* **266**.10006.

Richard K. Guy & J. L. Selfridge, Combined report on aliquot sequences, The Univ. of Calgary Math. Res. Rep. **225**(May, 1974).

Richard K. Guy & J. L. Selfridge, What drives an aliquot sequence? *Math. Comput.*, **29**(1975) 101–107; *MR* **52** #5542; *Zbl.* **296**.10007. Corrigendum, *ibid.*, **34**(1980) 319–321; *MR* **81f**:10008; *Zbl.* **423**.10005.

Richard K. Guy & M. R. Williams, Aliquot sequences near 10^{12}, *Congr. Numer.* XII, Proc. 4th Manitoba Conf. Numer. Math., 1974, 387–406; *MR* **52** #242; *Zbl.* **359**.10007.

Richard K. Guy, D. H. Lehmer, J. L. Selfridge & M. C. Wunderlich, Second report on aliquot sequences, *Congr. Numer.* IX, Proc. 3rd Manitoba Conf. Numer. Math., 1973, 357–368; *MR* **50** #4455; *Zbl.* **325**.10007.

H. W. Lenstra, Problem 6064, *Amer. Math. Monthly*, **82**(1975) 1016; solution **84** (1977) 580.

G. Aaron Paxson, Aliquot sequences (preliminary report), *Amer. Math. Monthly*, **63**(1956) 614. See also *Math. Comput.*, **26** (1972) 807–809.

P. Poulet, La chasse aux nombres, Fascicule I, Bruxelles, 1929.

P. Poulet, Nouvelles suites arithmétiques, *Sphinx*, Deuxième Année (1932) 53–54.

H. J. J. te Riele, A note on the Catalan-Dickson conjecture, *Math. Comput.*, **27**(1973) 189–192; *MR* **48** #3869; *Zbl.* **255**.10008.

H. J. J. te Riele, Iteration of number theoretic functions, Report NN 30/83, Math. Centrum, Amsterdam, 1983.

OEIS: A008885-008892, A014360-014365.

B7 Aliquot cycles. Sociable numbers.

Aliquot cycles or **sociable numbers.** Poulet discovered two cycles of numbers, showing that $s^k(n)$ can have the periods 5 and 28, in addition to 1 and 2. For $k \equiv 0, 1, 2, 3, 4 \bmod 5$, $s^k(12496)$ takes the values

$$12496 = 2^4 \cdot 11 \cdot 71, \quad 14288 = 2^4 \cdot 19 \cdot 47, \quad 15472 = 2^4 \cdot 967,$$

$$14536 = 2^3 \cdot 23 \cdot 79, \quad 14264 = 2^3 \cdot 1783.$$

For $k \equiv 0, 1, \ldots, 27 \bmod 28$, $s^k(14316)$ takes the values

14316	19116	31704	47616	83328	177792	295488
629072	589786	294896	358336	418904	366556	274924
275444	243760	376736	381028	285778	152990	122410
97946	48976	45946	22976	22744	19916	17716

After a gap of over 50 years, and the advent of high-speed computing, Henri Cohen discovered nine cycles of period 4, and Borho, David and Root also discovered some. Recently Moews & Moews have made an exhaustive search for such cycles with greatest member less than 10^{10}. There are twenty-four: their smallest members are

1264460	7169104	46722700	330003580	2387776550	4424606020
2115324	18048976	81128632	498215416	2717495235	4823923384
2784580	18656380	174277820	1236402232	2879697304	5373457070
4938136	28158165	209524210	1799281330	3705771825	8653956136

Moews & Moews give five larger 4-cycles, and, in a 90-09-01 letter, another whose least member is:

$$2^6 \cdot 79 \cdot 1913 \cdot 226691 \cdot 207722852483$$

They also found an 8-cycle:

1095447416	1259477224	1156962296	1330251784
1221976136	1127671864	1245926216	1213138984

Ren Yuanhua had already found three of the 4-cycles and Achim Flammenkamp had also found many of them, as well as a second 8-cycle:

1276254780	2299401444	3071310364	2303482780
2629903076	2209210588	2223459332	1697298124

and a 9-cycle:

805984760	1268997640	1803863720	2308845400	3059220620
3367978564	2525983930	2301481286	1611969514	

Moews & Moews have continued their exhaustive search to uncover all cycles, of any length, whose member preceding the largest member is less than $3.6 \cdot 10^{10}$. There are three more 4-cycles, with least members

$$15837081520, \qquad 17616303220, \qquad 21669628904,$$

and a 6-cycle, all of whose members are odd:

$21548919483 = 3^5 \cdot 7^2 \cdot 13 \cdot 17 \cdot 19 \cdot 431,$ $23625285957 = 3^5 \cdot 7^2 \cdot 13 \cdot 19 \cdot 29 \cdot 277,$
$24825443643 = 3^2 \cdot 7^2 \cdot 11 \cdot 13 \cdot 19 \cdot 20719,$ $26762383557 = 3^4 \cdot 7^2 \cdot 13 \cdot 19 \cdot 27299,$
$25958284443 = 3^2 \cdot 7^2 \cdot 13 \cdot 19 \cdot 167 \cdot 1427,$ $23816997477 = 3^2 \cdot 7^2 \cdot 13 \cdot 19 \cdot 218651.$

Their continued search to 1.03×10^{11} produced four more 4-cycles. A recent count is

length	4	5	6	8	9	28	Total
number	110	1	2	2	1	1	127

50 four-cycles having been discovered by Blankenagel, Borho & vom Stein. See J. O. M. Pedersen's pages at
http://amicable.adsl.dk/aliquot/sociable.txt

It has been conjectured that there are no 3-cycles. On the other hand it has been conjectured that for each k there are infinitely many k-cycles.

Karsten Blankenagel, Walter Borho & Axel vom Stein, New amicable four-cycles, *Math. Comput.*, **72**(2003) 2071–2076; *MR* **2004c**:11006.

Walter Borho, Über die Fixpunkte der *k*-fach iterierten Teilersummenfunktion, *Mitt. Math. Gesellsch. Hamburg*, **9**(1969) 34–48; *MR* **40** #7189.

Achim Flammenkamp, New sociable numbers, *Math. Comput.*, **56**(1991) 871–873.

David Moews & Paul C. Moews, A search for aliquot cycles below 10^{10}, *Math. Comput.*, **57**(1991) 849–855; *MR* **92e**:11151.

David Moews & Paul C. Moews, A search for aliquot cycles and amicable pairs, *Math. Comput.*, **61**(1993) 935–938.

OEIS: A003416.

B8 Unitary aliquot sequences.

The ideas of aliquot sequence and aliquot cycle can be adapted to the case where only the *unitary* divisors are summed, leading to **unitary aliquot sequences** and . We use $\sigma^*(n)$ and $s^*(n)$ for the analogs of $\sigma(n)$ and $s(n)$ when just the unitary divisors are summed (compare **B3**).

Are there unbounded unitary aliquot sequences? Here the balance is more delicate than in the ordinary aliquot sequence case. The only sequences which deserve serious consideration are those involving odd multiples of 6, which is a unitary perfect number as well as an ordinary one. Now the sequences tend to increase if $3\|n$, but decrease when a higher power of 3 is present, and it is a moot point as to which situation will dominate. Once a term of a sequence is $6m$, with m odd, then $\sigma^*(6m)$ is an even multiple of 6, making $s^*(6m)$ an odd multiple of 6 again, except in the extremely rare case that m is 4 raised to an odd power.

te Riele pursued all unitary aliquot sequences for $n < 10^5$. The only one which did not terminate or become periodic was 89610. Later calculations showed that this reached a maximum,

$$645\,856907\,610421\,353834 = 2 \cdot 3^2 \cdot 13 \cdot 19 \cdot 73 \cdot 653 \cdot 3047409443791$$

at its 568th term, and terminated at its 1129th.

One can hardly expect typical behavior until the expected number of prime factors is large. Since this number is $\ln\ln n$, such sequences are well beyond computer range. Of 80 sequences examined near 10^{12}, all have terminated or become periodic. One sequence exceeded 10^{23}.

Unitary amicable pairs and unitary sociable numbers may occur rather more frequently than their ordinary counterparts. Lal, Tiller & Summers found cycles of periods 1, 2, 3, 4, 5, 6, 14, 25, 39 and 65. Examples of unitary amicable pairs are (56430,64530) and (1080150,1291050), while (30,42,54) is a 3-cycle and

$$(1482, 1878, 1890, 2142, 2178)$$

is a 5-cycle.

Cohen (see **B3** for definitions and a reference) finds 62 infinitary amicable pairs with smaller member less than a million, eight infinitary aliquot cycles of order 4 and three of order 6. The only other such cycle of order less than 17 and least member less than a million is of order 11:

448800, 696864, 1124448, 1651584, 3636096, 6608784,

5729136, 3736464, 2187696, 1572432, 895152.

A type of aliquot sequence which can be unbounded has been suggested by David Penney & Carl Pomerance and is based on Dedekind's function: see **B41**. It was in fact the subject of Chapter 7 of te Riele's thesis.

Erdős, looking for a number-theoretic function whose iterates might be bounded, suggested defining $w(n) = n \sum 1/p_i^{\alpha_i}$ where $n = \prod p_i^{\alpha_i}$, and $w^k(n) = w(w^{k-1}(n))$. Note that $w(n) \perp n$. Can it be proved that $w^k(n)$, $k = 1, 2, \ldots$, is bounded? Is $|\{w(n) : 1 \leq n \leq x\}| = o(x)$?

Erdős & Selfridge called n a **barrier** for a number-theoretic function $f(m)$ if, for all $m < n$, $m + f(m) \leq n$. Euler's ϕ-function (see **B36**) and $\sigma(m)$ increase too fast to have barriers, but does $\omega(m)$ have infinitely many barriers? The numbers 2, 3, 4, 5, 6, 8, 9, 10, 12, 14, 17, 18, 20, 24, 26, 28, 30, ..., are barriers for $\omega(m)$. Does $\Omega(m)$ have infinitely many barriers? Selfridge observes that 99840 is the largest barrier for $\Omega(m)$ that is $< 10^5$. Mąkowski observes that $n = 1$ is a barrier for every function, and that 2 is a barrier for every function $f(n)$ with $f(1) = 1$; in particular for $d(m)$, the number of divisors of m. The inequality

$$\max\{d(n-1) + n - 1, d(n-2) + n - 2\} \geq n + 2$$

holds for $n \geq 7$, but not for $n = 6$. But $d(n-1) + n - 1 \geq n + 1$ for $n \geq 3$, so $d(m)$ has no barriers ≥ 3. Does

$$\max_{m<n}(m + d(m)) = n + 2$$

have infinitely many solutions? It is very doubtful. The first few are $n = 5$, 8, 10, 12, 24; Jud McCranie found no others below 10^{10}.

Paul Erdős, A mélange of simply posed conjectures with frustratingly elusive solutions, *Math. Mag.*, **52**(1979) 67–70.

P. Erdős, Problems and results in number theory and graph theory, *Congressus Numerantium* **27**, Proc. 9th Manitoba Conf. Numerical Math. Comput., 1979, 3–21.

Richard K. Guy & Marvin C. Wunderlich, Computing unitary aliquot sequences – a preliminary report, *Congressus Numerantium* **27**, Proc. 9th Manitoba Conf. Numerical Math. Comput., 1979, 257–270.

P. Hagis, Unitary amicable numbers, *Math. Comput.*, **25**(1971) 915–918; *MR* **45** #8599.

Peter Hagis, Unitary hyperperfect numbers, *Math. Comput.*, **36**(1981) 299–301.

M. Lal, G. Tiller & T. Summers, Unitary sociable numbers, *Congressus Numerantium* **7**, Proc. 2nd Manitoba Conf. Numerical Math., 1972, 211–216: *MR* **50** #4471.

Rudolph M. Najar, The unitary amicable pairs up to 10^8, *Internat. J. Math. Math. Sci.*, **18**(1995) 405–410; *MR* **96c**:11011.

H. J. J. te Riele, *Unitary Aliquot Sequences*, MR139/72, Mathematisch Centrum, Amsterdam, 1972; reviewed *Math. Comput.*, **32**(1978) 944–945; *Zbl.* 251. 10008.

H. J. J. te Riele, *Further Results on Unitary Aliquot Sequences*, NW12/73, Mathematisch Centrum, Amsterdam, 1973; reviewed *Math. Comput.*, **32**(1978) 945.

H. J. J. te Riele, *A Theoretical and Computational Study of Generalized Aliquot Sequences*, MCT72, Mathematisch Centrum, Amsterdam, 1976; reviewed *Math. Comput.*, **32**(1978) 945–946; *MR* **58** #27716.

C. R. Wall, Topics related to the sum of unitary divisors of an integer, PhD thesis, Univ. of Tennessee, 1970.

OEIS: A005236, A068597, A087281.

B9 Superperfect numbers.

Suryanarayana defines **superperfect numbers** n by $\sigma^2(n) = 2n$, i.e., $\sigma(\sigma(n)) = 2n$. He and Kanold show that the even ones are just the numbers 2^{p-1} where $2^p - 1$ is a Mersenne prime. Are there any odd superperfect numbers? If so, Kanold shows that they are perfect squares, and Dandepat and others that n or $\sigma(n)$ is divisible by at least three distinct primes.

More generally, Bode defines m-**superperfect numbers** as numbers n for which $\sigma^m(n) = 2n$, and shows that for $m \geq 3$ there are no even m-superperfect numbers. He also shows that for $m = 2$ there is no superperfect number $< 10^{10}$. Hunsucker & Pomerance have raised this bound to 7×10^{24} and have shown that if n is an odd super perfect number, then $n\sigma(n)$ has at least 5 distinct prime factors, and that the number of distinct prime factors in n, together with the number of distinct prime factors in $\sigma(n)$ is at least 7. These results are announced in the paper with Dandapat.

If $\sigma^2(n) = 2n+1$, it would be consistent with earlier terminology to call n quasi-superperfect. The Mersenne primes are such. Are there others? Are there "almost superperfect numbers" for which $\sigma^2(n) = 2n - 1$?

Erdős asks if $(\sigma^k(n))^{1/k}$ has a limit as $k \to \infty$. He conjectures that it is infinite for each $n > 1$.

Schinzel asks if $\liminf \sigma^k(n)/n < \infty$ for each k, as $n \to \infty$, and observes that it follows for $k = 2$ from a deep theorem of Rényi. Mąkowski & Schinzel give an elementary proof for $k = 2$ that the limit is 1. Helmut Maier has used sieve methods to prove the result for $k = 3$.

Sitaramaiah & Subbarao call a number **unitary superperfect** if it satisfies the equation $\sigma^*(\sigma^*(n)) = 2n$. They note that the equation $\sigma^*(\sigma^*(n)) = 2n + 1$ has no solutions and that $\sigma^*(\sigma^*(n)) = 2n - 1$ has only $n = 1$ and 3. They list the unitary superperfect numbers less than 10^8: 2, 9, 165, 238, 1640, 4320, 10250, 10824, 13500, 23760, 58500, 66912, 425880, 520128, 873180, 931392, 1899744, 2129400, 2253888, 3276000, 4580064, 4668300, and they conjecture that there are infinitely many unitary superperfect numbers, of which only a finite number are odd. They also list the solutions $n = 10$, 30, 288, 660, 720, 2146560 of $\sigma^*(\sigma^*(n)) = kn$ for $k = 3$ and the solution $n = 18$ for $k = 4$.

Dieter Bode, Über eine Verallgemeinerung der volkommenen Zahlen, Dissertation, Braunschweig, 1971.

G. G. Dandapat, J. L. Hunsucker & C. Pomerance, Some new results on odd perfect numbers, *Pacific J. Math.*, **57**(1975) 359–364; **52** #5554.

P. Erdős, Some remarks on the iterates of the ϕ and σ functions, *Colloq. Math.*, **17**(1967) 195–202.

J. L. Hunsucker & C. Pomerance, There are no odd super perfect numbers less than $7 \cdot 10^{24}$, *Indian J. Math.*, **17**(1975) 107–120; *MR* **82b**:10010.

H.-J. Kanold, Über "Super perfect numbers," *Elem. Math.*, **24**(1969) 61–62; *MR* **39** #5463.

Graham Lord, Even perfect and superperfect numbers, *Elem. Math.*, **30** (1975) 87–88.

Helmut Maier, On the third iterates of the ϕ- and σ-functions, *Colloq. Math.*, **49**(1984) 123–130.

Andrzej Mąkowski, On two conjectures of Schinzel, *Elem. Math.*, **31**(1976) 140–141.

A. Schinzel, Ungelöste Probleme Nr. 30, *Elem. Math.*, **14**(1959) 60–61.

V. Sitaramaiah & M. V. Subbarao, On the equation $\sigma^*(\sigma^*(n)) = 2n$, *Utilitas Math.*, **53**(1998) 101–124; *MR* **99a**:11009.

D. Suryanarayana, Super perfect numbers, *Elem. Math.*, **24**(1969) 16–17; *MR* **39** #5706.

D. Suryanarayana, There is no superperfect number of the form $p^{2\alpha}$, *Elem. Math.*, **28**(1973) 148–150; *MR* **48** #8374.

B10 Untouchable numbers.

Erdős has proved that there are infinitely many n such that $s(x) = n$ has no solution. Alanen calls such n **untouchable**. In fact Erdős showed that the untouchable numbers have positive lower density. Here are the untouchable numbers less than 1000:

```
  2    5   52   88   96  120  124  146  162  178  188  206  210  216  238  246
248  262  268  276  288  290  292  304  306  322  324  326  336  342  372  406
408  426  430  448  472  474  498  516  518  520  530  540  552  556  562  576
584  612  624  626  628  658  668  670  714  718  726  732  738  748  750  756
766  768  782  784  792  802  804  818  836  848  852  872  892  894  896  898
902  916  926  936  964  966  976  982  996
```

In view of the plausibility of the Goldbach conjecture (**C1**), it seems likely that 5 is the only odd untouchable number since if $2n + 1 = p + q + 1$ with p and q prime, then $s(pq) = 2n + 1$. Can this be proved independently? Are there arbitrarily long sequences of consecutive even numbers which are untouchable? How large can the gaps between untouchable numbers be?

Wouter Meeussen, in a 98-08-27 email, calls an integer $m < n$ **unrelated** to n if it is neither a divisor of n nor relatively prime to it, and defines the function $W(n) = n - d(n) - \phi(n) + 1$, the number of integers unrelated to n. For $n < 14$ the only examples are: 4 unrelated to 6; 6 unrelated to 8 and 9; 4, 6, 8 unrelated to 10; and 8, 9, 10 unrelated to 12, so that $W(6) = W(8) = W(9) = 1$ and $W(10) = W(12) = 3$. Meeussen asks if any of the numbers

2, 13, 67, 93, 123, 133, 141, 173, 187, 193, 205, 217, 229, 245, 253, 257, 283, 285, 293, 303, 317, 319, 325, 333, 341, 389, 393, 397, 405, 415, 427, 445, 453, 467, 473, 483, 491, 493, 509, 525, 527, 533, 537, 549, 557, 571, 573, 581, 587, 589, 595, 609, 621, 635, 643, 653, 655, 667, 669, 673, 679, 685, 701, 709, 723, 765, 777, 779, 789, 797, 811, 813, 833, 843, 845, 869, 877, 893, 899, 901, 907, 915, 921, 941, 957, 973, 997, ...

can occur as a value of $W(n)$.

Felice Russo found the following **unitary untouchable numbers**: 2, 3, 4, 5, 7, 374, 702, 758, 998, i.e., numbers n for which $s^*(x) = n$ has no solution, $s^*(x) = \sigma^*(x) - x$ being the sum of the unitary divisors of x other than x itself. They are the only ones < 1000. David Wilson found 862 unitary untouchable numbers $\leq 10^5$. What is a good estimate for their number $\leq x$?

P. Erdős, Über die Zahlen der Form $\sigma(n) - n$ und $n - \phi(n)$, *Elem. Math.*, **28**(1973) 83–86; *MR* **49** #2502.

Paul Erdős, Some unconventional problems in number theory, *Astérisque*, **61**(1979) 73–82; *MR* **81h**:10001.

OEIS: A005114, A063948.

B11 Solutions of $m\sigma(m) = n\sigma(n)$.

Leo Moser observed that while $n\phi(n)$ determines n uniquely, $n\sigma(n)$ does not. [$\phi(n)$ is Euler's totient function; see **B36**.] For example, $m\sigma(m) = n\sigma(n)$ for $m = 12$, $n = 14$. The multiplicativity of $\sigma(n)$ now ensures an infinity of solutions, $m = 12q$, $n = 14q$, where $q \perp 42$. So Moser asked if there is an infinity of *primitive* solutions, in the sense that (m^*, n^*) is *not* a solution for any $m^* = m/d$, $n^* = n/d$, $d > 1$. The example we've given is the least of the set $m = 2^{p-1}(2^q - 1)$, $n = 2^{q-1}(2^p - 1)$, where $2^p - 1$, $2^q - 1$ are distinct Mersenne primes, so that only a finite number of such solutions is known. Another set of solutions is $m = 2^7 \cdot 3^2 \cdot 5^2 \cdot (2^p - 1)$, $n = 2^{p-1} \cdot 5^3 \cdot 17 \cdot 31$, where $2^p - 1$ is a Mersenne prime other than 3 or 31; also $p = 5$ gives a primitive solution on deletion of the common factor 31. There are other solutions, such as $m = 2^4 \cdot 3 \cdot 5^3 \cdot 7$, $n = 2^{11} \cdot 5^2$ and $m = 2^9 \cdot 5$, $n = 2^3 \cdot 11 \cdot 31$. An example with $m \perp n$ is $m = 2^5 \cdot 5$, $n = 3^3 \cdot 7$. If $m\sigma(m) = n\sigma(n)$, is m/n bounded?

Erdős observed that if n is squarefree, then integers of the form $n\sigma(n)$ are distinct. He also proved that the number of solutions of $m\sigma(m) = n\sigma(n)$ with $m < n < x$ is $cx + o(x)$. In answer to the question, are there three distinct numbers l, m, n such that $l\sigma(l) = m\sigma(m) = n\sigma(n)$, Mąkowski observes that for distinct Mersenne primes M_{p_i}, $1 \leq i \leq s$, we have $n_i\sigma(n_i)$ is constant for $n_i = A/M_{p_i}$, where $A = \prod_{j=1}^{s} M_{p_j}$. Is there an infinity of primitive solutions of the equation $\sigma(a)/a = \sigma(b)/b$? Without restricting the solutions to being primitive, Erdős showed that their number with $a < b < x$ is at least $cx + o(x)$; with the restriction $a \perp b$ no solution is known at all.

Erdős believes that the number of solutions of $x\sigma(x) = n$ is less than $n^{\epsilon/\ln\ln n}$ for every $\epsilon > 0$, and says that the number may be less than $(\ln n)^c$.

Jean-Marie De Koninck asks if $n = 1782$ is the only non-trivial solution of $\sigma(n) = (\text{rad } n)^2$, where $\text{rad } n$, the radical of n, is its greatest squarefree divisor:

$$\sigma(1782) = \sigma(2 \cdot 3^4 \cdot 11) = (2 + 1) \cdot \frac{3^5 - 1}{3 - 1} \cdot (11 + 1) = (2 \cdot 3 \cdot 11)^2$$

At the conclusion of the article by Huard, Ou, Spearman & Williams on convolution sums of divisor functions is mentioned the possibility of finding further identities and of connexions with Ramanujan's τ-function.

Nicolae Ciprian Bonciocat, Congruences for the convolution of divisor sum function, *Bull. Greek Math. Soc.*, **46**(2002) 161–170; MR **2003e**:11111.

P. Erdős, Remarks on number theory II: some problems on the σ function, *Acta Arith.*, **5**(1959) 171–177; MR **21** #6348.

James G. Huard, Zhiming M. Ou, Blair K. Spearman & Kenneth S. Williams, Elementary evaluation of certain convolution sums involving divisor functions, *Number Theory for the Millenium II (Urbana IL*, 2000) 229–274, AKPeters, Natick MA¡ 2002.

B12 Analogs with $d(n)$, $\sigma_k(n)$.

Analogous questions may be asked with $\sigma_k(n)$ in place of $\sigma(n)$, where $\sigma_k(n)$ is the sum of the k-th powers of the divisors of n. For example, are there distinct numbers m and n such that $m\sigma_2(m) = n\sigma_2(n)$? For $k = 0$ we have

$md(m) = nd(n)$ for $(m, n) = (18, 27)$, $(24, 32)$, $(56, 64)$ and $(192, 224)$. The last pair can be supplemented by 168 to give three distinct numbers such that $ld(l) = md(m) = nd(n)$. There are primitive solutions (m, n) of shape

$$m = 2^{qt-1}p, \qquad n = 2^{pt \cdot 2^{tu}-1}q$$

where p and $q = u + p \cdot 2^{tu}$ are primes, but it does not immediately follow that these are infinitely numerous. Many other solutions can be constructed; for example $(2^{70}, 2^{63} \cdot 71)$, $(3^{19}, 3^{17} \cdot 5)$ and $(5^{51}, 5^{49} \cdot 13)$.

Bencze proves the inequalities

$$\frac{n^k + 1}{2} \geq \frac{\sigma_k(n)}{\sigma_{k-l}(n)} \geq \sqrt{n^l}$$

for $0 \leq l \leq k$ and gives no fewer than 60 applications.

Mihály Bencze, A contest problem and its application (Hungarian), *Mat. Lapok Ifjúsági Folyóirat (Románia)*, **91**(1986) 179–186.

OEIS: A000005, A033950, A036762-036763, A039819, A051278-051280.

B13 Solutions of $\sigma(n) = \sigma(n+1)$.

Sierpiński has asked if $\sigma(n) = \sigma(n+1)$ infinitely often. Hunsucker, Nebb & Stearns extended the tabulations of Mąkowski and of Mientka & Vogt and have found just 113 solutions

14, 206, 957, 1334, 1364, 1634, 2685, 2974, 4364, ...

less than 10^7. They also obtain statistics concerning the equation $\sigma(n) = \sigma(n+l)$, of which Mientka & Vogt had asked: for what l (if any) is there an infinity of solutions? They found many solutions if l is a factorial, but only two solutions for $l = 15$ and $l = 69$. They also ask whether, for each l and m, there is an n such that $\sigma(n) + m = \sigma(n+l)$.

Jud McCranie found 832 solutions of $\sigma(n) = \sigma(n+1)$ for $n < 4.25 \times 10^9$ and 2189 solutions of $\sigma(n) = \sigma(n+2)$ in the same range. An example is

$$4236745811 = 64399 \times 65789$$
$$4236745813 = 64499 \times 65687$$

for which $\sigma(n) = \sigma(n+2) = 4236876000 = 2^5 \cdot 3^2 \cdot 5^3 \cdot 7 \cdot 11 \cdot 23 \cdot 43$. He found no solutions for $\sigma(n) = \sigma(n+1) = \sigma(n+2)$ in that range.

Hunsucker, Nebb & Stearns conjectured that if $\sigma(n) = \sigma(n+1)$ and neither n nor $n+1$ is squarefree, then $n \equiv 0$ or $-1 \bmod 4$ but Haukkanen gave the counterexamples $n = 52586505 = 3^2 \cdot 5 \cdot 71 \cdot 109 \cdot 151$, $n+1 = 2 \cdot 7^2 \cdot 43 \cdot 12479$ and $n = 164233250 = 2 \cdot 5^3 \cdot 353 \cdot 1861$, $n+1 = 3^5 \cdot 7^2 \cdot 13 \cdot 1061$.

He also observed that for no $n \leq 2 \cdot 10^8$ is $\sigma(n) = \sigma(n+1) = \sigma(n+2)$.

If n and $n+2$ are twin primes, then $\sigma(n+2) = \sigma(n) + 2$. Mąkowski found the composite solutions $n = 434, 8575, 8825$ and Haukkanen showed that these were the only ones $\leq 2 \cdot 10^8$.

One can ask corresponding questions for $\sigma_k(n)$, the sum of the k-th powers of the divisors of n [For $k = 0$, see **B15**.] The only solution of $\sigma_2(n) = \sigma_2(n+1)$ is $n = 6$, since $\sigma_2(2n) > \sigma_2(2n+1)$ for $n > 7$ and $\sigma_2(2n) > 5n^2 > (\pi^2/8)(2n-1)^2 > \sigma_2(2n-1)$. Note that $\sigma_2(24) = \sigma_2(26)$; Erdős doubts that $\sigma_2(n) = \sigma_2(n+2)$ has infinitely many solutions, and thinks that $\sigma_3(n) = \sigma_3(n+2)$ has no solutions at all. De Koninck shows that $\sigma_2(n) = \sigma_2(n+l)$ has only finitely many solutions for l odd, whereas Schinzel's Hypothesis H (see **A**) implies that there are infinitely many solutions for l even.

J.-M. De Koninck, On the solutions of $\sigma_2(n) = \sigma_2(n+l)$, *Ann. Univ. Sci. Budapest. Sect. Comput.*, **21**(2002) 127–133; *MR* **2003h**:11007.

Richard K. Guy & Daniel Shanks, A constructed solution of $\sigma(n) = \sigma(n+1)$, *Fibonacci Quart.*, **12**(1974) 299; *MR* **50** #219.

Pentti Haukkanen, Some computational results concerning the divisor functions $d(n)$ and $\sigma(n)$, *Math. Student*, **62**(1993) 166-168; *MR* **90j**:11006.

John L. Hunsucker, Jack Nebb & Robert E. Stearns, Computational results concerning some equations involving $\sigma(n)$, *Math. Student*, **41**(1973) 285–289.

W. E. Mientka & R. L. Vogt, Computational results relating to problems concerning $\sigma(n)$, *Mat. Vesnik*, **7**(1970) 35–36.

OEIS: A002961.

B14 Some irrational series.

Is $\sum_{n=1}^{\infty}(\sigma_k(n)/n!)$ irrational? It is for $k = 1$ and 2.

Erdős established the irrationality of the series

$$\sum_{n=1}^{\infty} \frac{1}{2^n - 1} = \sum_{n=1}^{\infty} \frac{d(n)}{2^n}$$

and Peter Borwein showed that

$$\sum_{n=1}^{\infty} \frac{1}{q^n + r} \quad \text{and} \quad \sum_{n=1}^{\infty} \frac{(-1)^n}{q^n + r}$$

are irrational if q is an integer other than 0, ± 1, and r is a rational other than 0 or $-q^n$.

Peter B. Borwein, On the irrationality of $\sum 1/(q^n + r)$, *J. Number Theory*, **37**(1991) 253–259.

Peter B. Borwein, On the irrationality of certain series, *Math. Proc. Cambridge Philos. Soc.*, **112**(1992) 141–146; *MR* **93g**:11074.

P. Erdős, On arithmetical properties of Lambert series, *J. Indian Math. Soc.*(*N.S.*) **12**(1948) 63–66.

P. Erdős, On the irrationality of certain series: problems and results, in *New Advances in Transcendence Theory*, Cambridge Univ. Press, 1988, pp. 102–109.

P. Erdős & M. Kac, Problem 4518, *Amer. Math. Monthly* **60**(1953) 47. Solution R. Breusch, **61**(1954) 264–265.

B15 Solutions of $\sigma(q) + \sigma(r) = \sigma(q + r)$.

Max Rumney (*Eureka*, **26**(1963) 12) asked if the equation $\sigma(q) + \sigma(r) = \sigma(q+r)$ has infinitely many solutions which are primitive in a sense similar to that used in **B11**. If $q + r$ is prime, the only solution is $(q, r) = (1, 2)$. If $q + r = p^2$ where p is prime, then one of q and r, say q, is prime, and $r = 2^n k^2$ where $n \geq 1$ and k is odd. If $k = 1$, there is a solution if $p = 2^n - 1$ is a and $q = p^2 - 2^n$ is prime; this is so for $n = 2$, 3, 5, 7, 13 and 19. For $k = 3$ there are no solutions, and none for $k = 5$ with $n < 189$. For $k = 7$, $n = 1$ and 3 give $(q, r, q + r) = (5231, 2 \cdot 7^2, 73^2)$ and $(213977, 2^3 \cdot 7^2, 463^2)$. Other solutions are $(k, n) = (11,1)$ (11,3), (19,5), (25,1), (25,9), (49,9), (53,1), (97,5), (107,5), (131,5), (137,1), (149,5), (257,5), (277,1), (313,3) and (421,3). Solutions with $q + r = p^3$ and p prime are $\sigma(2) + \sigma(6) = \sigma(8)$ and

$$\sigma(11638687) + \sigma(2^2 \cdot 13 \cdot 1123) = \sigma(227^3)$$

Erdős asks how many solutions (not necessarily primitive) are there with $q + r < x$; is it $cx + o(x)$ or is it of higher order? If $s_1 < s_2 < \cdots$ are the numbers for which $\sigma(s_i) = \sigma(q) + \sigma(s_i - q)$ has a solution with $q < s_i$, what is the density of the sequence $\{s_i\}$?

M. Sugunamma, PhD thesis, Sri Venkataswara Univ., 1969.

B16 Powerful numbers. Squarefree numbers.

Erdős & Szekeres studied numbers n such that if a prime p divides n, then p^i divides n where i is a given number greater than one. Golomb named these numbers **powerful** and exhibited infinitely many pairs of consecutive ones. In answer to his conjecture that 6 was not representable as the difference of two powerful numbers, Władysław Narkiewicz noted that $6 = 5^4 7^3 - 463^2$, and that there were infinitely many such representations. In fact in 1971 Richard P. Stanley (unpublished, since a simultaneous discovery was made by Peter Montgomery) used the theory of the Bhaskara (Pell) equation to show that every non-zero integer is the difference between two powerful numbers and that 1 is the difference between two non-square powerful numbers, each in infinitely many ways. A typical result of Stanley is that

if $a_1 = 39$, $b_1 = 1$, $a_n = 24335a_{n-1} + 7176b_{n-1}$ and
$b_n = 82524a_{n-1} + 24335b_{n-1}$, then $2^3(a_n)^2 - 23^3(b_n)^2 = 1$.

Many have investigated which numbers are the difference of two powers, $m^p - n^q$, with $m, n \geq 1$, $p, q \geq 2$. Can any of the following numbers be so expressed?

6, 14, 34, 42, 50, 58, 62, 66, 70, 78, 82, 86, 90, 102, 110, 114, 130, 134, 158, 178, 182, 202, 206, 210, 226, 230, 238, 246, 254, 258, 266, 274, 278, 302, 306, 310, 314, 322, ...

Erdős denotes by $u_1^{(k)} < u_2^{(k)} < \ldots$ the integers all of whose prime factors have exponents $\geq k$; sometimes called k-**full numbers**. He asks if the equation $u_{i+1}^{(2)} - u_i^{(2)} = 1$ has infinitely many solutions which do not come from Pell equations $x^2 - dy^2 = \pm 1$. Is there a constant c, such that the number of solutions with $u_i < x$ is less than $(\ln x)^c$? Does $u_{i+1}^{(3)} - u_i^{(3)} = 1$ have no solutions? Do the equations $u_{i+2}^{(2)} - u_{i+1}^{(2)} = 1$, $u_{i+1}^{(2)} - u_i^{(2)} = 1$ have no simultaneous solutions? And several other questions, some of which have been answered by Mąkowski.

For example, Mąkowski notes that $7^3 x^2 - 3^3 y^2 = 1$ has infinitely many solutions, and that this is not usually counted as a Bhaskara (Pell) equation. He also notes that

$$(2^{k+1} - 1)^k, \quad 2^k(2k + 1 - 1)^k \quad \text{and} \quad (2^{k+1} - 1)^{k+1}$$

are k-full numbers in A.P., and that if a_1, a_2, ..., a_s are k-full and in A.P. with common difference d then

$$a_1(a_s + d)^k, \quad a_2(a_s + d)^k, \quad \ldots, \quad a_s(a_s + d)^k, \quad (a_s + d)^{k+1}$$

are $s + 1$ such numbers. As

$$a^k(a^l + \ldots + 1)^k + a^{k+1}(a^l + \ldots + 1)^k + \ldots + a^{k+l}(a^l + \ldots + 1)^k = a^k(a^l + \ldots + 1)^{k+1},$$

the sum of $l + 1$ k-ful numbers can be k-full. He says that these last two questions become difficult when we require that the numbers be relatively prime. Heath-Brown has shown that every sufficiently large number is the sum of three powerful numbers; his proof would be much shortened if his conjecture could be proved that the quadratic form $x^2 + y^2 + 125z^2$ represents every sufficiently large $n \equiv 7 \bmod 8$. Erdős suggested that this may follow from work of Duke and Iwaniec: in fact see the paper by Moroz.

Are there only finitely many powerful numbers n such that $n^2 - 1$ is also powerful? (See the Granville reference at **D2**.)

If $A(x)$ is the number of squareful integers $\leq x$ and $\Delta(x) = A(x) - a_1 x^{1/2} - a_2 x^{1/3}$ with $a-1$, a_2 known constants, then, assuming the Riemann Hypothesis, Cao showed that $\Delta(x) = O(x^{5/33+\epsilon})$, and Cai replaced 5/33 by 4/27.

Liu Hong-Quan shows that the number of 3-full numbers in the interval $(x, x + x^{\frac{2}{3}+\mu})$ is asymptotic to Cx^μ if $\frac{11}{92} < \mu < \frac{1}{3}$.

Nitaj proves the conjecture of Erdős that $x + y = z$ has solutions in relatively prime 3-full numbers, with $|z| \to \infty$. Cohn further shows that this can be done infinitely often with none of x, y, z being a perfect cube.

Gang Yu improves Menzer's estimate for the number of 4-full numbers $\leq x$ from $x^{35/316}(\ln x)^3$ to $x^{3626/35461+\epsilon}$.

Huxley & Trifonov show that the number of square-full numbers among $N + 1$, ..., $N + h$ is, for N sufficiently large in terms of ϵ and $h \geq \frac{1}{\epsilon} N^{\frac{5}{8}} (\ln N)^{\frac{5}{16}}$ is

$$\frac{\zeta(\frac{3}{2})}{2\zeta(3)} \frac{h}{\sqrt{N}} + O\left(\frac{\epsilon h}{\sqrt{N}}\right)$$

Earlier results, with exponents $\frac{2}{3}$, 0.6318, 0.6308, and $\frac{49}{78}$ in place of $\frac{5}{8}$, were obtained by Bateman & Grosswald, Heath-Brown, Liu, and Filaseta & Trifonov.

If p is a prime, $p \equiv 1 \pmod{4}$, and $\frac{1}{2}(t + u\sqrt{p})$ is the fundamental unit of $Q(\sqrt{p})$ (i.e., (t, u) are the least positive integers satisfying the Bhaskara equation $t^2 - pu^2 = 1$),then the Ankeny-Artin-Chowla conjecture asserts that $p \nmid u$ for any p. It was proved for all $p < 10^{11}$ by van der Poorten, te Riele & Williams. The conjecture is false if p is not prime; Gerry Myerson believes that 46 and 430 are the two smallest counterexamples.

At the other end of the spectrum from the powerful numbers are the **squarefree** numbers, with no repeated prime divisors. If we denote the sequence of squarefree numbers by $\{f_n\} = \{1, 2, 3, 5, 6, 7, 10, \ldots\}$, then it is well known that $f_{n+1} - f_n = 1$ for infinitely many n and $\limsup(f_{n+1} - f_n) = \infty$. Panaitopol further shows that

$$\limsup(\min\{f_{n+1} - f_n, f_n - f_{n-1}\}) = \infty$$

and that if $e_n = f_{n+1} - 2f_n + f_{n-1}$ and $g_n = f_{n+1}^2 - 2f_n^2 + f_{n-1}^2$, then each of $e_n < 0$, $e_n = 0$, $e_n > 0$, $g_n < 0$, $g_n > 0$ holds for infinitely many n, and $g_n \neq 0$ for all $n > 1$. He asks if there exist, for each positive integer k, an index n such that $f_{n+1} - f_n = k$?

N. C. Ankeny, E. Artin & S. Chowla, The class-number of real quadratic number fields, *Ann. of Math.*(2), **56**(1952) 479–493; *MR* **14**, 251.

R. C. Baker & J. Brüdern, On sums of two squarefull numbers, *Math. Proc. Cambridge Philos. Society*, **116**(1994) 1–5; *MR* **95f**:11073.

P. T. Bateman & E. Grosswald, On a theorem of Erdős and Szekeres, *Illinois J. Math.*, **2**(1958) 88–98.

B. D. Beach, H. C. Williams & C. R. Zarnke, Some computer results on units in quadratic and cubic fields, *Proc. 25th Summer Meet. Canad. Math. Congress*, Lakehead, 1971, 609–648; *MR* **49** #2656.

Cai Ying-Chun, On the distribution of square-full integers, *Acta Math. Sinica* (*N.S.*) **13**(1997) 269–280; *MR* **98j**:11070.

Catalina Calderón & M. J. Velasco, Waring's problem on squarefull numbers, *An. Univ. Bucureşti Mat.*, **44**(1995) 3–12.

Cao Xiao-Dong, The distribution of square-full integers, *Period. Math. Hungar.*, **28**(1994) 43–54; *MR* **95k**:11119.

J. H. E. Cohn, A conjecture of Erdős on 3-powerful numbers. *Math. Comput.*, **67**(1998) 439–440; *MR* **98c**:11104.

David Drazin & Robert Gilmer, Complements and comments, *Amer. Math. Monthly*, **78**(1971) 1104–1106 (esp. p. 1106).

W. Duke, Hyperbolic distribution problems and half-integral weight Maass forms, *Invent. Math.*, **92**(1988) 73–90; *MR* **89d**:11033.

P. Erdős, Problems and results on consecutive integers, *Eureka*, **38**(1975–76) 3–8.

P. Erdős & G. Szekeres, Über die Anzahl der Abelschen Gruppen gegebener Ordnung und über ein verwandtes zahlentheoretisches Problem, *Acta Litt. Sci. Szeged*, **7**(1934) 95–102; *Zbl.* **10**, 294.

S. W. Golomb, Powerful numbers, *Amer. Math. Monthly*, **77**(1970) 848–852; *MR* **42** #1780.

Ryūta Hashimoto, Ankeny-Artin-Chowla conjecture and continued fraction expansion, *J. Number Theory*, **90**(2001) 143–153; *MR* **2002e**:11149.

D. R. Heath-Brown, Ternary quadratic forms and sums of three squarefull numbers, *Séminaire de Théorie des Nombres, Paris, 1986-87*, Birkhäuser, Boston, 1988; *MR* **91b**:11031.

D. R. Heath-Brown, Sums of three square-full numbers, in Number Theory, I (Budapest, 1987), *Colloq. Math. Soc. János Bolyai*, **51**(1990) 163–171; *MR* **91i**:11036.

D. R. Heath-Brown, Square-full numbers in short intervals, *Math. Proc. Cambridge Philos. Soc.*, **110**(1991) 1–3; *MR* **92c**:11090.

M.N. Huxley & O. Trifonov, The square-full numbers in an interval, *Math. Proc. Cambridge Philos. Soc.*, **119**(1996) 201–208; *MR* **96k**:11114.

Aleksander Ivić, On the asymptotic formulas for powerful numbers, *Publ. Math. Inst. Beograd (N.S.)*, **23(37)**(1978) 85–94; *MR* **58** #21977.

A. Ivić & P. Shiue, The distribution of powerful integers, *Illinois J. Math.*, **26**(1982) 576–590; *MR* **84a**:10047.

H. Iwaniec, Fourier coefficients of modular forms of half-integral weight, *Invent. Math.*, **87**(1987) 385–401; *MR* **88b**:11024.

C.-H. Jia, On square-full numbers in short intervals, *Acta Math. Sinica (N.S.)* **5**(1987) 614–621.

Ekkehard Krätzel, On the distribution of square-full and cube-full numbers, *Monatsh. Math.*, **120**(1995) 105–119; *MR* **96f**:11116.

Hendrik W. Lenstra, Solving the Pell equation, *Notices Amer. Math. Soc.*, **49**(2002) 182–192.

Liu Hong-Quan, On square-full numbers in short intervals, *Acta Math. Sinica (N.S.)*, **6**(1990) 148–164; *MR* **91g**:11105.

Liu Hong-Quan, The number of squarefull numbers in an interval, *Acta Arith.*, **64**(1993) 129–149.

Liu Hong-Quan, The number of cube-full numbers in an interval, *Acta Arith.*, **67**(1994) 1–12; *MR* **95h**:11100.

Liu Hong-Quan, The distribution of 4-full numbers, *Acta Arith.*, **67**(1994) 165–176.

Liu Hong-Quan, The distribution of square-full numbers, *Ark. Mat.*, **32**(1994) 449–454; *MR* **97a**:11146.

Andrzej Mąkowski, On a problem of Golomb on powerful numbers, *Amer. Math. Monthly*, **79**(1972) 761.

Andrzej Mąkowski, Remarks on some problems in the elementary theory of numbers, *Acta Math. Univ. Comenian.*, **50/51**(1987) 277–281; *MR* **90e**:11022.

Wayne L. McDaniel, Representations of every integer as the difference of powerful numbers, *Fibonacci Quart.*, **20**(1982) 85–87.

H. Menzer, On the distribution of powerful numbers, *Abh. Math. Sem. Univ. Hamburg*, **67**(1997) 221–237; *MR* **98h**:11115.

Richard A. Mollin, The power of powerful numbers, *Internat. J. Math. Math. Sci.*, **10**(1987) 125–130; *MR* **88e**:11008.

Richard A. Mollin & P. Gary Walsh, On non-square powerful numbers, *Fibonacci Quart.*, **25**(1987) 34–37; *MR* **88f**:11006.

Richard A. Mollin & P. Gary Walsh, On powerful numbers, *Internat. J. Math. Math. Sci.*, **9**(1986) 801–806; *MR* **88f**:11005.

Richard A. Mollin & P. Gary Walsh, A note on powerful numbers, quadratic fields and the Pellian, *CR Math. Rep. Acad. Sci. Canada*, **8**(1986) 109–114; *MR* **87g**:11020.

Richard A. Mollin & P. Gary Walsh, Proper differences of non-square powerful numbers, *CR Math. Rep. Acad. Sci. Canada*, **10**(1988) 71–76; *MR* **89e**:11003.

L. J. Mordell, On a pellian equation conjecture, *Acta Arith.*, **6**(1960) 137–144; *MR* **22** #9470.

B. Z. Moroz, On representation of large integers by integral ternary positive definite quadratic forms, *Journées Arithmétiques* (*Geneva* 1991, *Astérisque*, **209**(1992), 15, 275–278; *MR* **94a**:11051.

Abderrahmane Nitaj, On a conjecture of Erdős on 3-powerful numbers, *Bull. London Math. Soc.*, **27**(1995) 317–318; *MR* **96b**:11045.

R. W. K. Odoni, On a problem of Erdős on sums of two squarefull numbers, *Acta Arith.*, **39**(1981) 145–162; *MR* **83c**:10068.

Laurenţiu Panaitopol, On square free integers, *Bull. Math. Soc. Sci. Math. Roumanie (N.S.)*, **43(91)**(2000) 19–23; *MR*.**2002k**:11156.

A. J. van der Poorten, H. J. J. te Riele & H. C. Williams, Computer verification of the Ankeny-Artin-Chowla conjecture for all primes less than 100 000 000 000, *Math. Comput.*, **70**(2001) 1311–1328; *MR* **2001j**:11125; corrigenda and addition, **72**(2003) 521–523; *MR* **2003g**:11162.

A. V. M. Prasad & V. V. S. Sastri, An asymptotic formula for partitions into square full numbers, *Bull. Calcutta Math. Soc.*, **86**(1993) 403–408; *MR* **96b**:11138.

Peter Georg Schmidt, On the number of square-full integers in short intervals, *Acta Arith.*, **50**(1988) 195–201; corrigendum, **54**(1990) 251–254; *MR* **89f**:11131.

R. Seibold & E. Krätzel, Die Verteilung der k-vollen und l-freien Zahlen, *Abh. Math. Sem. Univ. Hamburg*, **68**(1998) 305–320.

W. A. Sentance, Occurrences of consecutive odd powerful numbers, *Amer. Math. Monthly*, **88**(1981) 272–274.

P. Shiue, On square-full integers in a short interval, *Glasgow Math. J.*, **25** (1984) 127–134.

P. Shiue, The distribution of cube-full numbers, *Glasgow Math. J.*, **33**(1991) 287–295. *MR* **92g**:11091.

P. Shiue, Cube-full numbers in short intervals, *Math. Proc. Cambridge Philos. Soc.*, **112** (1992) 1–5; *MR* **93d**:11097.

A. J. Stephens & H. C. Williams, Some computational results on a problem concerning powerful numbers, *Math. Comput.*, **50**(1988) 619–632.

B. Sury, On a conjecture of Chowla et al., *J. Number Theory*, **72**(1998) 137–139; it *MR* **99f**:11005.

D. Suryanarayana, On the distribution of some generalized square-full integers, *Pacific J. Math.*, **72**(1977) 547–555; *MR* **56** #11933.

D. Suryanarayana & R. Sitaramachandra Rao, The distribution of square-full integers, *Ark. Mat.*, **11**(1973) 195–201; *MR* **49** #8948.

Charles Vanden Eynden, Differences between squares and powerful numbers, *816-11-305, *Abstracts Amer. Math. Soc.*, **6**(1985) 20.

David T. Walker, Consecutive integer pairs of powerful numbers and related Diophantine equations, *Fibonacci Quart.*, **14**(1976) 111–116; *MR* **53** #13107.

Wu Jie[1], On the distribution of square-full and cube-full integers, *Monatsh. Math.*, **126**(1998) 339–358.

Wu Jie[1], On the distribution of square-full integers, *Arch. Math. (Basel)*, **77**(2001) 233–240; *MR* **2002g**:11136.

Yu Gang[1], The distribution of 4-full numbers, *Monatsh. Math.*, **118**(1994) 145–152; *MR* **95i**:11101.

Yuan Ping-Zhi, On a conjecture of Golomb on powerful numbers (Chinese. English summary), *J. Math. Res. Exposition*, **9**(1989) 453–456; *MR* **91c**:11009.

OEIS: A001694, A060355, A060859-060860, A062739.

B17 Exponential-perfect numbers

If $n = p_1^{a_1} p_2^{a_2} \cdots p_r^{a_r}$, then Straus & Subbarao call d an **exponential divisor** (e-divisor) of n if $d|n$ and $d = p_1^{b_1} p_2^{b_2} \cdots p_r^{b_r}$ where $b_j|a_j$ $(1 \le j \le r)$, and they call n **e-perfect** if $\sigma_e(n) = 2n$, where $\sigma_e(n)$ is the sum of the e-divisors of n. Some examples of e-perfect numbers are

$$2^2 \cdot 3^2, \quad 2^2 \cdot 3^3 \cdot 5^2, \quad 2^3 \cdot 3^2 \cdot 5^2, \quad 2^4 \cdot 3^2 \cdot 11^2, \quad 2^4 \cdot 3^3 \cdot 5^2 \cdot 11^2,$$

$$2^6 \cdot 3^2 \cdot 7^2 \cdot 13^2, \ 2^6 \cdot 3^3 \cdot 5^2 \cdot 7^2 \cdot 13^2, \ 2^7 \cdot 3^2 \cdot 5^2 \cdot 7^2 \cdot 13^2, \ 2^8 \cdot 3^2 \cdot 5^2 \cdot 7^2 \cdot 139^2$$

$$\text{and} \quad 2^{19} \cdot 3^2 \cdot 5^5 \cdot 7^2 \cdot 11^2 \cdot 13^2 \cdot 19^2 \cdot 37^2 \cdot 79^2 \cdot 109^2 \cdot 157^2 \cdot 313^2.$$

If m is squarefree, $\sigma_e(m) = m$, so if n is e-perfect and m is squarefree with $m \perp n$, then mn is e-perfect. So it suffices to consider only powerful (**B16**) e-perfect numbers.

Straus & Subbarao show that there are no odd e-perfect numbers, in fact no odd n which satisfy $\sigma_e(n) = kn$ for any integer $k > 1$. They also show that for each r the number of (powerful) e-perfect numbers with r prime factors is finite, and that the same holds for e-**multiperfect numbers** $(k > 2)$.

Is there an e-perfect number which is *not* divisible by 3 (equivalently, not divisible by 36)?

Straus & Subbarao conjecture that there is only a finite number of numbers *not* divisible by any given prime p.

Are there any e-multiperfect numbers?

E. G. Straus & M. V. Subbarao, On exponential divisors, *Duke Math. J.*, **41**(1974) 465–471; *MR* **50** #2053.

M. V. Subbarao, On some arithmetic convolutions, *Proc. Conf. Kalamazoo MI, 1971, Springer Lecture Notes in Math.*, **251**(1972) 247–271; *MR* **49** #2510.

M. V. Subbarao & D. Suryanarayana, Exponentially perfect and unitary perfect numbers, *Notices Amer. Math. Soc.*, **18**(1971) 798.

OEIS A051377, A054979-054980.

B18 Solutions of $d(n) = d(n+1)$.

Claudia Spiro has proved that $d(n) = d(n + 5040)$ has infinitely many solutions and Heath-Brown used her ideas to show that there are infinitely many numbers n such that $d(n) = d(n+1)$, and Pinner has extended this to $d(n) = d(n + a)$ for any integer a. Many examples arise from pairs of consecutive numbers which are products of just two distinct primes, and it has been conjectured that there is an infinity of *triples* of consecutive products of two primes, n, $n + 1$, $n + 2$. For example, $n = 33$, 85, 93, 141, 201, 213, 217, 301, 393, 445, 633, 697, 921, ... It is clearly not possible to have *four* such numbers, but it *is* possible to have longer sequences of consecutive numbers with the same number of divisors. For example,

$$d(242) = d(243) = d(244) = d(245) = 6 \quad \text{and}$$

$$d(40311) = d(40312) = d(40313) = d(40314) = d(40315) = 8.$$

How long can such sequences be?

For 5 and 6 consecutive numbers, Haukkanen (see ref. at **B13**) showed that the least n is respectively 11605 and 28374. In an 87-07-16 letter Stephane Vandemergel sent the sequence of seven numbers: $171893 = 19 \cdot 83 \cdot 109$, $171894 = 2 \cdot 3 \cdot 28649$, $171895 = 5 \cdot 31 \cdot 1109$, $171896 = 2^3 \cdot 21487$, $171897 = 3 \cdot 11 \cdot 5209$, $171898 = 2 \cdot 61 \cdot 1409$, $171899 = 7 \cdot 13 \cdot 1889$, each with 8 divisors. In 1990, Ivo Düntsch & Roger Eggleton discovered several such sequences of 7 numbers, two of 8 and one of 9, each with 48 divisors; the last example starts at 17796126877482329126044, presumably not the smallest of its kind. At the beginning of 2002 Jud McCranie gave 1043710445721 as the smallest first member of eight consecutive numbers with the same number of divisors.

Erdős believes that there are sequences of length k for every k, but does not see how to give an upper bound for k in terms of n.

Erdős, Pomerance & Sárközy showed that the number of $n \leq x$ with $d(n) = d(n+1)$ is $\ll x/(\ln\ln x)^{1/2}$, and Hildebrand showed that this number is $\gg x/(\ln\ln x)^3$. The former authors also showed that the number of $n \leq x$ with the ratio $d(n)/d(n+1)$ in the set $\{2^{-3}, 2^{-2}, 2^{-1}, 1, 2, 2^2, 2^3\}$ is $\asymp x/(\ln\ln x)^{1/2}$.

Erdős showed that the density of numbers n with $d(n+1) > d(n)$ is $\frac{1}{2}$. This, with the above results, settles a conjecture of S. Chowla. Fabrykowski & Subbarao extend this to the case with $n + h$ in place of $n + 1$.

Erdős also lets

$$1 = d_1 < d_2 < \cdots < d_\tau = n$$

be the set of all divisors of n, listed in order, defines

$$f(n) = \sum_{1}^{\tau-1} d_i/d_{i+1}$$

and asks us to prove that $\sum_{n=1}^{x} f(n) = (1 + o(1))x\ln x$.

Erdős & Mirsky ask for the largest k so that the numbers $d(n)$, $d(n+1)$, ..., $d(n + k)$ are all distinct. They only have trivial bounds; probably $k = (\ln n)^c$.

Farkas (see reference at **E16**) has shown that the set of rational numbers $\frac{d(n^2)}{d(n)}$ contains all positive odd integers, and that $\frac{d(n^3)}{d(n)}$ contains all integers which are not multiples of 3.

P. Erdős, Problem P. 307, *Canad. Math. Bull.*, **24**(1981) 252.

Paul Erdős & Hugh L. Montgomery, Sums of numbers with many divisors, *J. Number Theory*, **75**(1999) 1–6; *MR* **99k**:11153.

Anthony D. Forbes, Fifteen consecutive integers with exactly four prime factors, *Math. Comput.*, **71**(2002) 449–452; *MR* **2002g**:11012.

P. Erdős & L. Mirsky, The distribution of values of the divisor function $d(n)$, *Proc. London Math. Soc.*(3), **2**(1952) 257–271.

P. Erdős, C. Pomerance & A. Sárközy, On locally repeated values of certain arithmetic functions, II, *Acta Math. Hungarica*, **49**(1987) 251–259; *MR* **88c**:11008.

J. Fabrykowski & M. V. Subbarao, Extension of a result of Erdős concerning the divisor function, *Utilitas Math.*, **38**(1990) 175–181; *MR* **92d**:11101.

D. R. Heath-Brown, A parity problem from sieve theory, *Mathematika*, **29** (1982) 1–6 (esp. p. 6).

D. R. Heath-Brown, The divisor function at consecutive integers, *Mathematika*, **31**(1984) 141–149.

Adolf Hildebrand, The divisors function at consecutive integers, *Pacific J. Math.*, **129** (1987) 307–319; *MR* **88k**:11062.

A. J. Hildebrand, Erdős' problems on consecutive integers, *Paul Erdős and his mathematics, I* (*Budapest*, 1999) 305–317, Bolyai Soc. Math. Stud., **11**, János Bolyai Math. Soc., Budapest, 2002; *MR* **2004b**:11142.

Kan Jia-Hai & Shan Zun, On the divisor function $d(n)$, *Mathematika*, **43**(1997) 320–322; II **46**(1999) 419–423; *MR* **98b**:11101; **2003c**:11122.

M. Nair & P. Shiue, On some results of Erdős and Mirsky, *J. London Math. Soc.*(2), **22**(1980) 197–203; and see *ibid.*, **17**(1978) 228–230.

C. Pinner, M.Sc. thesis, Oxford, 1988.

A. Schinzel, Sur un problème concernant le nombre de diviseurs d'un nombre naturel, *Bull. Acad. Polon. Sci. Ser. sci. math. astr. phys.*, **6**(1958) 165–167.

A. Schinzel & W. Sierpiński, Sur certaines hypothèses concernant les nombres premiers, *Acta Arith.*, **4**(1958) 185–208.

W. Sierpiński, Sur une question concernant le nombre de diviseurs premiers d'un nombre naturel, *Colloq. Math.*, **6**(1958) 209–210.

OEIS: A000005, A005237-005238, A006558, A006601, A019273, A039665, A049051.

B19 $(m, n+1)$ and $(m+1, n)$ with same set of prime factors. The abc-conjecture.

Motzkin & Straus asked for all pairs of numbers m, n such that m and $n+1$ have the same set of distinct prime factors, and similarly for n and $m+1$. It was thought that such pairs were necessarily of the form $m = 2^k + 1$, $n = m^2 - 1$ ($k = 0, 1, 2, \ldots$) until Conway observed that if $m = 5 \cdot 7$, $n + 1 = 5^4 \cdot 7$, then $n = 2 \cdot 3^7$, $m + 1 = 2^2 \cdot 3^2$. Are there others?

Similarly, Erdős asks if there are numbers m, n $(m < n)$ other than $m = 2^k - 2$, $n = 2^k(2^k - 2)$ such that m and n have the same prime factors and similarly for $m+1$, $n+1$. Mąkowski found the pair $m = 3 \cdot 5^2$, $n = 3^5 \cdot 5$ for which $m + 1 = 2^2 \cdot 19$, $n + 1 = 2^6 \cdot 19$. Compare problem **B29**.

Pomerance has asked if there are any odd numbers $n > 1$ such that n and $\sigma(n)$ have the same prime factors. He conjectures that there are not.

The example $1 + 2 \cdot 3^7 = 5^4 7$ in the first paragraph is of interest in connexion with the **abc-conjecture@**abc**-conjecture**:

Many of the classical problems of number theory (Goldbach conjecture, twin primes, the Fermat problem, Waring's problem, the Catalan conjecture) owe their difficulty to a clash between multiplication and addition. Roughly, if there's an additive relation between three numbers, their prime factors can't all be small.

Suppose that $A + B = C$ with $\gcd(A, B, C) = 1$. Define the radical R to be the maximum squarefree integer dividing ABC and the power P by

$$P = \frac{\ln \max(|A|, |B|, |C|)}{\ln R}$$

then for a given $\eta > 1$ are there only finitely many triples $\{A, B, C\}$ with $P \geq \eta$? Another form of this conjecture is that $\limsup P = 1$; both forms of the conjecture seem to be hopelessly beyond reach.

Joe Kanapka, a student of Noam Elkies, has produced a list of all examples with $C < 2^{32}$ and $P > 1.2$. There are nearly 1000 of them. The

"top ten" according to
http://www.math.unicaen.fr/~nitaj/abc.html#Ten<i>abc</i>
(which has an extensive bibliography) are

P	A	B	C	author
1.62991	2	$3^{10} \cdot 109$	23^5	Reyssat
1.62599	11^2	$3^2 \cdot 5^6 \cdot 7^3$	$2^{21} \cdot 23$	de Weger (**D10**)
1.62349	$19 \cdot 1307$	$7 \cdot 29^2 \cdot 31^8$	$2^8 \cdot 3^{22} \cdot 5^4$	Browkin-Brzeziński
1.58076	283	$5^{11} \cdot 13^2$	$2^8 \cdot 3^8 \cdot 17^3$	Br-Br, Nitaj
1.56789	1	$2 \cdot 3^7$	$5^4 \cdot 7$	Lehmer (**B29**)
1.54708	7^3	3^{10}	$2^{11} \cdot 29$	de Weger
1.54443	$7^2 \cdot 41^2 \cdot 311^3$	$11^{16} \cdot 13^2 \cdot 79$	$2 \cdot 3^3 \cdot 5^{23} \cdot 953$	Nitaj
1.53671	5^3	$2^9 \cdot 3^{17} \cdot 13^2$	$11^5 \cdot 17 \cdot 31^3 \cdot 137$	Montgomery-teRiele
1.52700	$13 \cdot 19^6$	$2^{30} \cdot 5$	$3^{13} \cdot 11^2 \cdot 31$	Nitaj
1.52216	$3^{18} \cdot 23 \cdot 2269$	$17^3 \cdot 29 \cdot 31^8$	$2^{10} \cdot 5^2 \cdot 7^{15}$	Nitaj

Browkin & Brzeziński generalize the abc-conjecture (which is their case $n = 3$) to an "n-conjecture" on $a_1 + \cdots + a_n = 0$ in coprime integers with non-vanishing subsums. With R and P defined analogously, they conjecture that $\limsup P = 2n - 5$. They prove that $\limsup P \geq 2n - 5$. They give a lot of examples for the abc-conjecture with $P > 1.4$. Their method is to look for rational numbers approximating roots of integers (note that the best example above is connected to the good approximation $23/9$ for $109^{1/5}$). Abderrahmane Nitaj used a similar method. Some of these were found independently by Robert Styer (**D10**). The Catalan relation $1 + 2^3 = 3^2$ gives a comparatively poor $P \approx 1.22629$.

For connexions between the abc-conjecture and the Fermat problem, see the Granville references at **D2**. Indeed, if $A = a^p$, $B = b^p$, $C = c^p$ and the Fermat equation $A + B = C$ is satisfied, then Gerhard Frey's elliptic curve

$$y^2 = x(x - A)(x + B)$$

has discriminant $16(abc)^{2p}$.

This area has had several stimuli: two being the proof of Fermat's Last Theorem and the announcement of the Beal prize. I thank Andrew Granville for the following remarks.

The problem has been much studied recently by several authors. Darmon & Granville showed that if we fix integers x, y, z with $1/x + 1/y + 1/z < 1$ then there are only finitely many triples of coprime integers a, b, c satisfying $a^x + b^y = c^z$. This is proved independently of any assumption, and fits well with the conjecture that x, y, $z > 2$ imply that a, b, c have a common factor since in this case $1/x + 1/y + 1/z < 1$ unless $x = y = z = 3$, but of course Euler, and possibly Fermat, knew that there are no solutions in that case. Following this result there has been extensive computer searching and exactly ten solutions have been found with $1/x + 1/y + 1/z < 1$ and a, b, c coprime:

$$1 + 2^3 = 3^2 \qquad 17^7 + 76271^3 = 21063928^2$$
$$2^5 + 7^2 = 3^4 \qquad 1414^3 + 2213459^2 = 65^7$$
$$7^3 + 13^2 = 2^9 \qquad 9262^3 + 15312283^2 = 113^7$$
$$2^7 + 17^3 = 71^2 \qquad 43^8 + 96222^3 = 30042907^2$$
$$3^5 + 11^4 = 122^2 \qquad 33^8 + 1549034^2 = 15613^3$$

The five big solutions were found by clever computations by Beukers & Zagier. From this, several people have conjectured that the only solutions do have some exponent equal to 2, and wonder if this is always the case. Granville is loth to believe or disbelieve such a statement. For a year he and Cohen believed that the five small solutions above were the only five, and then were totally shocked by these computations — and he sees no good reason to suppose that we've seen the last of the solutions!

The technique used by Darmon & Granville was to reduce the problem to applications of Faltings's Theorem. This is why they always say 'at most finitely many solutions'. Recently Darmon & Merel, and also Poonen, have revisited these problems, and tried to reduce several examples of x, y, z to applications of Wiles's Theorem (Darmon and Granville had done a couple of examples of this in their paper, but it is done much more skillfully in the recent papers). Darmon & Merel, and Poonen, prove that there are no coprime solutions with exponents $(x, x, 3)$ with $x \geq 3$.

As Mauldin pointed out, the abc-conjecture is relevant to this. Gerald Tenenbaum has long suggested an explicit and plausible version of the abc-conjecture: If $a + b = c$ in coprime positive integers then $c \leq$ (product of $p|abc$)^2. Assuming then that $a^x + b^y = c^z$ with a, b, $c > 0$ we'd have $c^{(z/2)} \leq abc < c^{z(1/x+1/y+1/z)}$ and thus $1/x + 1/y + 1/z > 1/2$. This leaves us with a list of cases to consider if we insist that x, y, $z > 2$: $(3, 3, z > 3)$, $(3, 4, z > 3)$, $(3, 5, z > 4)$, $(3, 6, z > 6)$, $(4, 4, z > 4)$, and a finite list.

I am also indebted to Andrew Bremner and the AMS for permission to reproduce the review of Mauldin's paper:

This note announces the award of a substantial monetary prize (since this article was written, fixed at \$50,000) to any person who provides a solution to the "Beal Conjecture", stated as the following: *Let A, B, C, x, y, z be positive integers with*

$x, y, z > 2$. If $A^x + B^y = C^z$ (1), then A, B, C have a non-trivial common factor.

The story of this conjecture is an interesting one, and told at slightly greater length in the author's follow-up letter to the Notices in March, 1998. Andrew Beal is a successful Texas businessman, with enthusiasm for number theory. He has a particular interest in Fermat and his methods, and evidently formulated this conjecture after several years of study following the announcement in 1993 of Andrew Wiles's work on Fermat's Last Theorem. So often, the amateur number-theorist turns out to be a well-intentioned crank; what is remarkable here is how close the stated problem is to current research activity by leaders in the field. In fact, the problem is essentially many decades old, and Brun [1914] asks many similar questions. The formulation in the 1980s by Masser, Oesterle & Szpiro of the abc-conjecture has had great influence on the discipline, and in fact a corollary of the abc-conjecture is that there are no solutions to the Beal Prize problem when the exponents are sufficiently large. The prize problem itself was implicitly posed by Andrew Granville in the Unsolved Problems section of the West Coast Number Theory Meeting, Asilomar, 1993 ("Find examples of $x^p + y^q = z^r$ with $1/p + 1/q + 1/r < 1$ other than $2^3 + 1^7 = 3^2$ and $7^3 + 13^2 = 2^{9}$"), and is stated and discussed in van der Poorten's book "Notes on Fermat's Last Theorem" (1996). The resolution by Wiles of Fermat's Last Theorem disposes of a special case of the prize problem; and Darmon and Granville prove the deep result that if $1/x + 1/y + 1/z < 1$ then there can only be finitely many triples of coprime integers A, B, C satisfying $A^x + B^y = C^z$ (ten solutions are known). Recently, Darmon & Merel have shown there can exist no coprime solutions to (1) with the exponents $(x, x, 3), x \geq 3$.

There has been some feeling expressed that the Conjecture should best be referred to as the "Beal Prize problem", though there is no doubt that Andrew Beal after much computation independently arrived at and formulated the conjecture without knowledge of the current literature. With a nod to T. S. Eliot, the matter of naming Conjectures can be as difficult as the naming of Cats.

See also **D2**.

Alan Baker, Logarithmic forms and the abc-conjecture, *Number Theory (Eger)*, 1996, de Gruyter, Berlin, 1998, 37–44; *MR* **99e**:11101.

Frits Beukers, The Diophantine equation $Ax^p + By^q = Cz^r$, *Duke Math. J.*, **91**(1998) 61–88: *MR* **99f**:11039.

Niklas Broberg, Some examples related to the abc-conjecture for algebraic number fields, *Math. Comput.*, **69**(2000) 1707–1710; *MR* **2001a**:11117.

Jerzy Browkin & Juliusz Brzeziński, Some remarks on the abc-conjecture, *Math. Comput.*, **62**(1994) 931–939; *MR* **94g**:11021.

Jerzy Browkin, Michael Filaseta, G. Greaves & Andrzej Schinzel, Squarefree values of polynomials and the abc-conjecture, *Sieve methods, exponential sums, and their applications in number theory (Cardiff* 1995) 65–85, London Math. Soc. Lecture Note Ser., **237**, Cambridge Univ. Press, 1997.

Juliusz Brzeziński, *ABC* on the abc-conjecture, *Normat*, **42**(1994) 97–107; *MR* **95h**:11024.

Cao Zhen-Fu, A note on the Diophantine equation $a^x + b^y = c^z$, *Acta Arith.*, **91**(1999) 85–93; *MR* **2000m**:11029.

Cao Zhen-Fu & Dong Xiao-Lei, The Diophantine equation $x^2 + b^y = c^z$, *Proc. Japan Acad. Ser. A Math. Sci.*, **77**(2001) 1–4; *MR* **2002c**:11027.

Cao Zhen-Fu, Dong Xiao-Lei & Li Zhong, A new conjecture concerning the Diophantine equation $x^2 + b^y = c^z$, *Proc. Japan Acad. Ser. A Math. Sci.*, **78**(2002) 199–202.

Todd Cochran & Robert E. Dressler, Gaps between integers with the same prime factors, *Math. Comput.*, **68**(1999) 395–401; *MR* **99c**:11118.

Henri Darmon, Faltings plus epsilon, Wiles plus epsilon, and the generalized Fermat equation, *C. R. Math. Rep. Acad. Sci. Canada*, **19**(1997) 3–14; corrigendum, 64; *MR* **98h**:11034ab.

Henri Darmon & Andrew Granville, On the equations $z^m = F(x, y)$ and $Ax^p + By^q = Cz^r$. *Bull. London Math. Soc.*, **27**(1995) 513–543; *MR* **96e**:11042.

Henri Darmon & Loïc Merel, Winding quotients and some variants of Fermat's last theorem, *J. reine angew. Math.*, **490**(1997) 81–100; *MR* **98h**:11076.

Johnny Edwards, A complete solution to $X^2 + Y^3 + Z^5 = 0$, 2001-Nov preprint.

Noam D. Elkies, *ABC* implies Mordell, *Internat. Math. Res. Notices*, **1991** no. 7, 99–109; *MR* **93d**:11064.

Michael Filaseta & Sergeĭ Konyagin, On a limit point associated with the abc-conjecture, *Colloq. Math.*, **76**(1998) 265–268; *MR* **99b**:11029.

Andrew Granville, *ABC* allows us to count squarefrees, *Internat. Math. Res. Notices*, **1998** 991–1009.

Andrew Granville & Thomas J. Tucker, It's as easy as abc, *Notices Amer. Math. Soc.*, **49**(2002) 1224–1231; *MR* **2003f**:11044.

Alain Kraus, On the equation $x^p + y^q = z^r$: a survey. *Ramanujan J.*, **3**(1999), no. 3, 315–333; *MR* **2001f**:11046.

Serge Lang, Old and new conjectured diophantine inequalities, *Bull. Amer. Math. Soc.*, **23**(1990) 37–75.

Serge Lang, Die abc-Vermutung, *Elem. Math.*, **48** (1993) 89–99; *MR* **94g**:11044.

Michel Langevin, Cas d'égalité pour le théorème de Mason et applications de la conjecture (abc), *C. R. Acad. Sci. Paris Sér I math.*, **317**(1993) 441–444; *MR* **94k**:11035.

Michel Langevin, Sur quelques conséquences de la conjecture (abc) en arithmétique et en logique, *Rocky Mountain J. Math.*, **26**(1996) 1031–1042; *MR* **97k**:11052.

Le Mao-Hua, The Diophantine equation $x^2 + D^m = p^n$, *Acta Arith.*, **52**(1989) 255–265; *MR* **90j**:11029.

Le Mao-Hua, A note on the Diophantine equation $x^2 + b^y = c^z$, *Acta Arith.*, **71**(1995) 253–257; *MR* **96d**:11037.

Allan I. Liff, On solutions of the equation $x^a + y^b = z^c$, *Math. Mag.*, **41**(1968) 174–175; *MR* **38** #5711.

D. W. Masser, On *abc* and discriminants, *Proc. Amer. Math. Soc.*, **130**(2002) 3141–3150.

R. Daniel Mauldin, A generalization of Fermat's last theorem: the Beal conjecture and prize problem, *Notices Amer. Math. Soc.*, **44**(1997) 1436–1437; *MR* **98j**:11020 (quoted above).

Abderrahmane Nitaj, An algorithm for finding good *abc*-examples, *C. R. Acad. Sci. Paris Sér I math.*, **317**(1993) 811–815.

Abderrahmane Nitaj, Algorithms for finding good examples for the *abc* and Szpiro conjectures, *Experiment. Math.*, **2**(1993) 223–230; *MR* **95b**:11069.

Abderrahmane Nitaj, La conjecture *abc*, *Enseign. Math.*, (2) **42**(1996) 3–24; *MR* **97a**:11051.

Abderrahmane Nitaj, Aspects expérimentaux de la conjecture *abc*, *Number Theory* (*Paris* 1993-1994), 145–156, *London Math. Soc. Lecture Note Ser.*, **235**, Cambridge Univ. Press, 1996; *MR* **99f**:11041.

A. Mąkowski, On a problem of Erdős, *Enseignement Math.*(2), **14**(1968) 193.

J. Oesterlé, Nouvelles approches du "thre" de Fermat, *Sém. Bourbaki*, 2/88, exposé #694.

Bjorn Poonen, Some Diophantine equations of the form $x^n + y^n = z^m$, *Acta Arith.*, **86**(1998) 193–205; *MR* **99h**:11034.

Paulo Ribenboim, On square factors of terms in binary recurring sequences and the ABC-conjecture, *Publ. Math. Debrecen*, **59**(2001) 459–469.

Paulo Ribenboim & Peter Gary Walsh, The *ABC*-conjecture and the powerful part of terms in binary recurring sequences, *J. Number Theory*, **74**(1999) 134–147; *MR* **99k**:11047.

András Sárközy, On sums $a + b$ and numbers of the form $ab + 1$ with many prime factors, *Österreichisch-Ungarisch-Slowakisches Kolloquium über Zahlentheorie* (*Maria Trost*, 1992), 141–154, *Grazer Math. Ber.*, **318** *Karl-Franzens-Univ. Graz*, 1993.

C. L. Stewart & Yu Kun-Rui, On the *abc*-conjecture, *Math. Ann.*, **291**(1991) 225–230; *MR* **92k**:11037; II, *Duke Math. J.*, **108**(2001) 169–181; *MR* **2002e**:11046.

Nobuhiro Terai, The Diophantine equation $a^x + b^y = c^z$, I, II, III, *Proc. Japan Acad. Ser. A Math. Sci.*, **70**(1994) 22–26; **71**(1995) 109–110; **72**(1996) 20–22; *MR* **95b**:11033; **96m**:11022; **98a**:11038.

R. Tijdeman, The number of solutions of Diophantine equations, in Number Theory, II (Budapest, 1987), *Colloq. Math. Soc. János Bolyai*, **51**(1990) 671–696.

Paul Vojta, A more general *abc*-conjecture, *Internat. Math. Res. Notices*, **1998** 1103–1116; *MR* **99k**:11096.

Yuan Ping-Zhi & Wang Jia-Bao, On the Diophantine equation $x^2 + b^y = c^z$, *Acta Arith.*, **84**(1998) 145–147.

OEIS: A027598.

B20 Cullen and Woodall numbers.

Some interest has been shown in the **Cullen numbers**, $n \cdot 2^n + 1$, which are all composite for $2 \leq n \leq 1000$, except for $n = 141$. This is probably a good example of the Law of Small Numbers, because for small n, where the density of primes is large, the Cullen numbers are very likely to be composite because Fermats (little) theorem@Fermat's (little) theorem tells us that $(p-1)2^{p-1}+1$ and $(p-2)2^{p-2}+1$ are both divisible by p. Moreover, as John Conway observes, the Cullen numbers are divisible by $2n-1$ if that is a prime of shape $8k \pm 3$. He asks if p and $p \cdot 2^p + 1$ can both be prime. Wilfrid Keller notes that Conway's remark can be generalized as follows. Write $C_n = n \cdot 2^n + 1$, $W_n = n \cdot 2^n - 1$: then a prime p divides $C_{(p+1)/2}$ and $W_{(3p-1)/2}$ or it divides $C_{(3p-1)/2}$ and $W_{(p+1)/2}$ according as the Legendre symbol (see **F5**) $\left(\frac{2}{p}\right)$ is -1 or $+1$. Known Cullen numbers include $n = 1$, 141, 4713, 5795, 6611, 18496, 32292, 32469, 59656, 90825, 262419 and 361275.

The corresponding numbers (which have been called Woodall primes) $n \cdot 2^n - 1$ are prime for $n = 2, 3, 6, 30, 75, 81, 115, 123, 249, 362, 384, 462, 512$ (i.e. M_{521}), 751, 822, 5312, 7755, 9531, 12379, 15822, 18885, 22971, 23005, 98726, 143018, 151023, 667071. In parallel with Conway's question above, Keller notes that here 3, 751 and 12379 are primes.

Ingemar Jönsson, On certain primes of Mersenne-type, *Nordisk Tidskr. Informationsbehandling (BIT)*, **12** (1972) 117–118; *MR* **47** #120.

Wilfrid Keller, New Cullen primes, (92-11-20 preprint).

Hans Riesel, *En Bok om Primtal* (Swedish), Lund, 1968; supplement Stockholm, 1977; *MR* **42** #4507, **58** #10681.

Hans Riesel, *Prime Numbers and Computer Methods for Factorization*, Progress in Math., **126**, Birkhäuser, 2nd ed. 1994.

OEIS: A002064, A003261, A005849, A050914.

B21 $k \cdot 2^n + 1$ composite for all n.

Let $N(x)$ be the number of odd positive integers k, not exceeding x, such that $k \cdot 2^n + 1$ is prime for no positive integer n. Sierpiński used covering congruences (see **F13**) to show that $N(x)$ tends to infinity with x. For example, if

$$k \equiv 1 \bmod 641 \cdot (2^{32} - 1) \quad \text{and} \quad k \equiv -1 \bmod 6700417,$$

then every member of the sequence $k \cdot 2^n + 1$ ($n = 0, 1, 2, \ldots$) is divisible by just one of the primes 3, 5, 17, 257, 641, 65537 or 6700417. He also noted that at least one of 3, 5, 7, 13, 17, 241 will always divide $k \cdot 2^n + 1$ for certain other values of k.

Erdős & Odlyzko have shown that

$$(\frac{1}{2} - c_1)x \geq N(x) \geq c_2 x.$$

What is the least value of k such that $k \cdot 2^n + 1$ is composite for all values of n? Selfridge discovered that one of 3, 5, 7, 13, 19, 37, 73 always divides $78557 \cdot 2^n + 1$. He also noted that there is a prime of the form $k \cdot 2^n + 1$ for each $k < 383$ and Hugh Williams discovered the prime $383 \cdot 2^{6393} + 1$.

In the first edition we wrote that the determination of the least k may now be within computer reach, though Keller has expressed his doubts about this. Extensive calculations have been made by Baillie, Cormack & Williams, by Keller, and by Buell & Young. Continuing activities by these and many others, including seventeenorbust.com have reduced the 35 possibilities of the second edition to twelve. The answer seems almost certain to be $k = 78557$, but there remain the possibilities

4847	5359	10223	19249	21181	22699
24737	27653	28433	33661	55459	67607

Riesel (see references at **B20**) investigated the corresponding question for $k \cdot 2^n - 1$. For $k = 509203, 762701, 992077$, the covering set of divisors is $\{3, 5, 7, 13, 17, 241\}$; for $k = 777149, 790841$, the covering set of divisors is $\{3, 5, 7, 13, 19, 37, 73\}$. There are no other values of $k < 10^6$ covered by the following six covering sets given by Stanton:
$\{3, 5, 7, 13, 19, 37, 73\}$, $\{3, 5, 7, 13, 19, 37, 109\}$, $\{3, 5, 7, 11, 13, 31, 41, 61, 151\}$,
$\{3, 5, 7, 11, 13, 19, 31, 37, 41, 61, 181\}$, $\{3, 5, 7, 13, 17, 241\}$, $\{3, 5, 7, 13, 17, 97, 257\}$.
It seems that $k = 509203$ is the smallest number such that $k \cdot 2^n - 1$ are all composite for every integer $n > 0$. To show this, one needs a prime $k \cdot 2^n - 1$ for each $k < 509203$. Many have been found; Wilfrid Keller is organizing a search to find primes for the remaining 95 values (see http://www.prothsearch.net/rieselprob.html):

2293	9221	23669	26773	31859	38473	40597	46663	65531	67117
71009	74699	81041	93839	93997	97139	107347	110413	113983	114487
121889	123547	129007	141941	143047	146561	149797	150847	152713	161669
162941	170591	191249	192089	192971	196597	206039	206231	215443	220033
226153	234343	234847	245561	250027	252191	273809	275293	304207	309817
315929	319511	324011	325123	325627	327671	336839	342673	342847	344759
345067	350107	353159	357659	362609	363343	364903	365159	368411	371893
384539	386801	397027	398023	402539	409753	412717	415267	417643	428639
444637	450457	460139	467917	469949	470173	474491	477583	485557	485767
494743	500621	502541	502573	504613					

Robert Baillie, New primes of the form $k \cdot 2^n + 1$, *Math. Comput.*, **33**(1979) 1333–1336; *MR* **80h**:10009.

Robert Baillie, G. V. Cormack & H. C. Williams, The problem of Sierpiński concerning $k \cdot 2^n + 1$ *Math. Comput.*, **37**(1981) 229–231; corrigendum, **39**(1982) 308.

Wieb Bosma, Explicit primality criteria for $h \cdot 2^k \pm 1$, *Math. Comput.*, **61**(1993) 97–109.

D. A. Buell & J. Young, Some large primes and the Sierpiński problem, SRC Technical Report 88-004, Supercomputing Research Center, Lanham MD, May 1988.

G. V. Cormack & H. C. Williams, Some very large primes of the form $k \cdot 2^n + 1$, *Math. Comput.*, **35**(1980) 1419–1421; *MR* **81i**:10011; corrigendum, Wilfrid Keller, **38**(1982) 335; *MR* **82k**:10011.

Paul Erdős & Andrew M. Odlyzko, On the density of odd integers of the form $(p - 1)2^{-n}$ and related questions, *J. Number Theory*, **11**(1979) 257–263; *MR* **80i**:10077.

Michael Filaseta, Coverings of the integers associated with an irreducibility theorem of A. Schinzel, *Number Theory for the Millenium, II* (*Urbana IL*, 2000) 1–24, AKPeters, Natick MA, 2002; *MR* **2003k**:11015.

Anatoly S. Izotov, A note on Sierpiński numbers, *Fibonacci Quart.*, **33**(1995) 206–207; *MR* **96f**:11020.

G. Jaeschke, On the smallest k such that all $k \cdot 2^N + 1$ are composite, *Math. Comput.*, **40**(1983) 381–384; *MR* **84k**:10006; corrigendum, **45**(1985) 637; *MR* **87b**:11009.

Wilfrid Keller, Factors of Fermat numbers and large primes of the form $k \cdot 2^n + 1$, *Math. Comput.*, **41**(1983) 661–673; *MR* **85b**:11119; II (incomplete draft, 92-02-19).

Wilfrid Keller, Woher kommen die größten derzeit bekannten Primzahlen? *Mitt. Math. Ges. Hamburg*, **12**(1991) 211–229;*MR* **92j**:11006.

N. S. Mendelsohn, The equation $\phi(x) = k$, *Math. Mag.*, **49**(1976) 37–39; *MR* **53** #252.

Raphael M. Robinson, A report on primes of the form $k \cdot 2^n + 1$ and on factors of Fermat numbers, *Proc. Amer. Math. Soc.*, **9**(1958) 673–681; *MR* **20** #3097.

J. L. Selfridge, Solution of problem 4995, *Amer. Math. Monthly*, **70**(1963) 101.

W. Sierpiński, Sur un problème concernant les nombres $k \cdot 2^n + 1$, *Elem. Math.*, **15**(1960) 73–74; *MR* **22** #7983; corrigendum, **17**(1962) 85.

W. Sierpiński, *250 Problems in Elementary Number Theory*, Elsevier, New York, 1970, Problem 118, pp. 10 & 64.

R. G. Stanton, Further results on covering integers of the form $1 + k * 2^N$ by primes, *Combinatorial mathematics, VIII* (*Geelong*, 1980), *Springer Lecture Notes in Math.*, **884**(1981) 107–114; *MR* **84j**:10009.

R. G. Stanton & H. C. Williams, Further results on covering of the integers $1 + k2^n$ by primes, *Combinatorial Math. VIII, Lecture Notes in Math.*, **884**, Springer-Verlag, Berlin–New York, 1980, 107–114.

Yong Gao-Chen, On integers of the forms $k^r - 2^n$ and $k^r 2^n + 1$, *J. Number Theory*, **98**(2003) 310–319; *MR* bf2003m:11004.

OEIS: A076336, A076337.

B22 Factorial n as the product of n large factors.

Straus, Erdős & Selfridge have asked that $n!$ be expressed as the product of n factors, with the least one, l, as large as possible. For example, for $n = 56$, $l = 15$,

$$56! = 15 \cdot 16^3 \cdot 17^3 \cdot 18^8 \cdot 19^2 \cdot 20^{12} \cdot 21^9 \cdot 22^5 \cdot 23^2 \cdot 26^4 \cdot 29 \cdot 31 \cdot 37 \cdot 41 \cdot 43 \cdot 47 \cdot 53$$

Selfridge has two conjectures: (a) that, except for $n = 56$, $l \geq \lfloor 2n/7 \rfloor$; (b) that for $n \geq 300000$, $l \geq n/3$. If the latter is true, by how much can 300000 be reduced?

Straus was reputed to have shown that for $n > n_0 = n_0(\epsilon)$, $l > n/(e+\epsilon)$, but a proof was not found in his Nachlaß. It is clear from Stirling's formula that this is best possible. It is also clear that l is a monotonic, though not strictly monotonic, increasing function of n. On the other hand it does not take all integer values: for $n = 124$, 125, l is respectively 35 and 37. Erdős asks how large the gaps in the values of l can be, and can l be constant for arbitrarily long stretches?

Alladi & Grinstead write $n!$ as a product of prime powers, each as large as $n^{\delta(n)}$ and let $\alpha(n) = \max \delta(n)$ and show that $\lim_{n \to \infty} \alpha(n) = e^{c-1} = \alpha$, say, where

$$c = \sum_2^\infty \frac{1}{k} \ln \frac{k}{k-1} \quad \text{so that} \quad \alpha = 0.809394020534\ldots .$$

If $n! = p_1^{a_1(n)} p_2^{a_2(n)} \cdots p_l^{a_l(n)}$, where $l = \pi(n)$. Erdős & Graham asked if there exist, for every k, some $n > 1$ for which each of the exponents $a_1(n), a_2(n), \ldots, a_k(n)$ is even. Berend proved that there are infinitely many positive integers $1 = n_0 < n_1 < n_2 < \cdots$ such that for each j all the numbers $a_1(n_j), a_2(n_j), \cdots, a_k(n_j)$ are even. See also the papers of Chen & Zhu, of Luca & Stănică, and of Sander. Chen later proved Sander's conjecture that, given a set of primes, there are infinitely many $n!$ whose factorizations contain these primes with any specified pattern of parities.

See also **D25**.

K. Alladi & C. Grinstead, On the decomposition of $n!$ into prime powers, *J. Number Theory*, **9**(1977) 452–458; *MR* **56** #11934.

Daniel Berend, On the parity of exponents in the factorization of $n!$, *J. Number Theory*, **64**(1997) 13–19; *MR* **98g**:11019.

Chen Yong-Gao, On the parity of exponents in the standard factorization of $n!$, *J. Number Theory*, **100**(2003) 326–331; *MR* **2004b**:11136.

Chen Yong-Gao & Zhu Yao-Chen, On the prime power factorization of $n!$ *J. Number Theory*, **82**(2000) 1–11; *MR* **2001c**:11027.

P. Erdős, Some problems in number theory, *Computers in Number Theory*, Academic Press, London & New York, 1971, 405–414.

Paul Erdős, S. W. Graham, Aleksandar Ivić & Carl Pomerance, On the number of divisors of $n!$, *Analytic Number Theory, Vol.* 1(*Allerton Park IL*, 1995) 337–355, *Progr. Math.*, **138**, Birkhäuser Boston, 1996; *MR* **97d**:11142.

Florian Luca & Pantelimon Stănică, On the prime power factorization of $n!$ *J. Number Theory*, **102**(2003) 298–305.

J. W. Sander, On the parity of exponents in the prime factorization of factorials, *J. Number Theory*, **90**(2001) 316–328; *MR* **2002j**:11105.

B23 Equal products of factorials.

Suppose that $n! = a_1! a_2! \ldots a_r!$, $r \geq 2$, $a_1 \geq a_2 \geq \ldots \geq a_r \geq 2$. A trivial example is $a_1 = a_2! \ldots a_r! - 1$, $n = a_2! \ldots a_r!$ Dean Hickerson notes that the only nontrivial examples with $n \leq 410$ are $9! = 7!3!3!2!$, $10! = 7!6! = 7!5!3!$ and $16! = 14!5!2!$ and asks if there are any others. Jeffrey Shallit & Michael Easter have extended the search to $n = 18160$ and Chris Caldwell has shownthat any other n is greater than 10^6.

Erdős observes that if $P(n)$ is the largest prime factor of n and if it were known that $P(n(n+1))/\ln n$ tends to infinity with n, then it would follow that there are only finitely many nontrivial examples.

He & Graham have studied the equation $y^2 = a_1! a_2! \ldots a_r!$ They define the set F_k to be those m for which there is a set of integers $m = a_1 > a_2 > \ldots > a_r$ with $r \leq k$ which satisfies this equation for some y, and write D_k for $F_k - F_{k-1}$. They have various results, for example: for almost all primes p, $13p$ does not belong to F_5; and the least element of D_6 is 527. If $D_4(n)$ is the number of elements of D_4 which are $\leq n$, they do not know the order of growth of $D_4(n)$. They conjecture that $D_6(n) > cn$ but cannot prove this.

Chris Caldwell, The Diophantine equation $A!B! = C!$, *J. Recreational Math.*, **26**(1994) 128-133.

Donald I. Cartwright & Joseph Kupka, When factorial quotients are integers, *Austral. Math. Soc. Gaz.*, **29**(2002) 19–26.

Earl Ecklund & Roger Eggleton, Prime factors of consecutive integers, *Amer. Math. Monthly*, **79**(1972) 1082–1089.

E. Ecklund, R. Eggleton, P. Erdős & J. L. Selfridge, on the prime factorization of binomial coefficients, *J. Austral. Math. Soc. Ser. A*, **26**(1978) 257–269; *MR* **80e**:10009.

P. Erdős, Problems and results on number theoretic properties of consecutive integers and related questions, *Congressus Numerantium XVI* (Proc. 5th Manitoba Conf. Numer. Math. 1975), 25–44.

P. Erdős & R. L. Graham, On products of factorials, *Bull. Inst. Math. Acad. Sinica, Taiwan*, **4**(1976) 337–355.

T. N. Shorey, On a conjecture that a product of k consecutive positive integers is never equal to a product of mk consecutive positive integers except for $8 \cdot 9 \cdot 10 = 6!$ and related questions, *Number Theory* (*Paris*, 1992–1993), *L.M.S. Lect. Notes* **215**(1995) 231–244; *MR* **96g**:11028.

B24 The largest set with no member dividing two others.

Let $f(n)$ be the size of the largest subset of $[1, n]$ no member of which divides two others. Erdős asks how large can $f(n)$ be? By taking $[m + 1, 3m + 2]$ it is clear that one can have $\lceil 2n/3 \rceil$. D.J. Kleitman shows that $f(29) = 21$ by taking $[11,30]$ and omitting 18, 24 and 30, which then allows the inclusion of 6, 8, 9 and 10. However, this example does not seem to generalize. In fact Lebensold has shown that if n is large, then

$$0.6725n \leq f(n) \leq 0.6736n.$$

Erdős also asks if $\lim f(n)/n$ is irrational.

Dually, one can ask for the largest number of numbers $\leq n$, with no number a multiple of any two others. Kleitman's example serves this purpose also. More generally, Erdős asks for the largest number of numbers with no one divisible by k others, for $k > 2$. For $k = 1$, the answer is $\lceil n/2 \rceil$.

For some related problems, see **E2**.

Driss Abouabdillah & Jean M. Turgeon, On a 1937 problem of Paul Erdős concerning certain finite sequences of integers none divisible by another, Proc. 15th S.E. Conf. Combin. Graph Theory Comput., Baton Rouge, 1984, *Congr. Numer.*, **43**(1984) 19–22; *MR* **86h**:11020.

Neil J. Calkin & Andrew Granville, On the number of co-prime-free sets, *Number Theory (New York, 1991–1995)*, Springer, New York, 1996, 9–18; *MR 97j*:11006.

P. J. Cameron & P. Erdős, On the number of sets of integers with various properties, *Number Theory (Banff, 1988)*, de Gruyter, Berlin, 1990, 61–79; *MR* **92g**:11010.

P. Erdős, On a problem in elementary number theory and a combinatorial problem, *Math. Comput.*, (1964) 644–646; *MR* **30** #1087.

Kenneth Lebensold, A divisibility problem, *Studies in Appl. Math.*, **56**(1976–77) 291–294; *MR* **58** #21639.

Emma Lehmer, Solution to Problem 3820, *Amer. Math. Monthly*, **46**(1939) 240–241.

B25 Equal sums of geometric progressions with prime ratios.

Bateman asks if $31 = (2^5 - 1)/(2 - 1) = (5^3 - 1)/(5 - 1)$ is the only prime which is expressible in more than one way in the form $(p^r - 1)/(p - 1)$ where p is prime and $r \geq 3$ and $d \geq 1$ are integers. Trivially one has $7 = (2^3 - 1)/(2 - 1) = ((-3)^3 - 1)/(-3 - 1)$, but there are no others $< 10^{10}$. If the condition that p be prime is relaxed, the problem goes back to Goormaghtigh and we have the solution

$$8191 = (2^{13} - 1)/(2 - 1) = (90^3 - 1)/(90 - 1)$$

E. T. Parker observed that the very long proof by Feit & Thompson that every group of odd order is solvable would be shortened if it could be proved that $(p^q - 1)/(p - 1)$ never divides $(q^p - 1)/(q - 1)$ where p, q are distinct odd primes. In fact it has been conjectured that that these two expressions are relatively prime, but Nelson Stephens noticed that when $p = 17$, $q = 3313$ they have a common factor $2pq + 1 = 112643$. McKay has established that $p^2 + p + 1 \nmid 3^p - 1$ for $p < 53 \cdot 10^6$.

Karl Dilcher quotes Nelson Stephens to the effect that if $p^q - 1$ and $q^p - 1$ have a common factor r, then r is of shape $2\lambda pq + 1$ and has searched with all such $r < 8 \cdot 10^{10}$. Also with $1 \le \lambda \le 10$ and $p < q < 10^7$, and with $p = 3$ and $q < 10^{14}$.

P. T. Bateman & R. M. Stemmler, Waring's problem for algebraic number fields and primes of the form $(p^r - 1)/(p^d - 1)$, *Illinois J. Math.*, **6**(1962) 142–156; *MR* **25** #2059.

Ted Chinburg & Melvin Henriksen, Sums of kth powers in the ring of polynomials with integer coefficients, *Bull. Amer. Math. Soc.*, **81**(1975) 107–110; *MR* **51** #421; *Acta Arith.*, **29**(1976) 227–250; *MR* **53** #7942.

Karl Dilcher& Josh Knauer, On a conjecture of Feit and Thompson, (preprint, Williams60, Banff, May 2003).

A. Mąkowski & A. Schinzel, Sur l'équation indéterminée de R. Goormaghtigh, *Mathesis*, **68**(1959) 128–142; *MR* **22** # 9472; **70**(1965) 94–96.

N. M. Stephens, On the Feit-Thompson conjecture, *Math. Comput.*, **25**(1971) 625; *MR* **45** #6738.

B26 Densest set with no l pairwise coprime.

Erdős asks what is the maximum k so that the integers a_i, $1 \le a_1 < a_2 < \cdots < a_k \le n$ have no l among them which are pairwise relatively prime. He conjectures that this is the number of integers $\le n$ which have one of the first $l - 1$ primes as a divisor. He says that this is easy to prove for $l = 2$ and not difficult for $l = 3$; he offers $10.00 for a general solution.

Dually one can ask for the largest subset of $[1, n]$ whose members have pairwise least common multiples not exceeding n. If $g(n)$ is the cardinality of such a maximal subset, then Erdős showed that

$$\frac{3}{2\sqrt{2}} n^{1/2} - 2 < g(n) \le 2n^{1/2}$$

where the first inequality follows by taking the integers from 1 to $(n/2)^{1/2}$ together with the even integers from $(n/2)^{1/2}$ to $(2n)^{1/2}$. Choi improved the upper bound to $1.638 n^{1/2}$.

Rudolf F. Ahlswede & L. G. Khachatrian, Maximal sets of numbers not containing $k + 1$ pairwise coprime integers, *Acta Arith.*, **72**(1995) 77–100; *MR* **96k**:11020.

Neil J. Calkin & Andrew Granville, On the number of coprime-free sets, *Number Theory (New York, 1991–1995)* 9–18, Springer, New York, 1996; *MR* **97j**:11006.

S. L. G. Choi, The largest subset in $[1, n]$ whose integers have pairwise l.c.m. not exceeding n, *Mathematika*, **19**(1972) 221–230; **47** #8461.

S. L. G. Choi, On sequences containing at most three pairwise coprime integers, *Trans. Amer. Math. Soc.*, **183**(1973) 437–440; **48** #6052.

P. Erdős, Extremal problems in number theory, *Proc. Sympos. Pure Math. Amer. Math. Soc.*, **8**(1965) 181–189; *MR* **30** #4740.

B27 The number of prime factors of $n + k$ which don't divide $n + i$, $0 \leq i < k$.

Erdős & Selfridge define $v(n; k)$ as the number of prime factors of $n + k$ which do not divide $n + i$ for $0 \leq i < k$, and $v_0(n)$ as the maximum of $v(n; k)$ taken over all $k \geq 0$. Does $v_0(n) \to \infty$ with n? They show that $v_0(n) > 1$ for all n except 1, 2, 3, 4, 7, 8 and 16. More generally, define $v_l(n)$ as the maximum of $v(n; k)$ taken over $k \geq l$. Does $v_l(n) \to \infty$ with n? They are unable to prove even that $v_1(n) = 1$ has only a finite number of solutions. Probably the greatest n for which $v_1(n) = 1$ is 330.

They also denote by $V(n; k)$ the number of primes p for which p^α is the highest power of p dividing $n + k$, but p^α does not divide $n + i$ for $0 \leq i < k$, and by $V_l(n)$ the maximum of $V(n; k)$ taken over $k \geq l$. Does $V_1(n) = 1$ have only a finite number of solutions? Perhaps $n = 80$ is the largest solution. What is the largest n such that $V_0(n) = 2$?

Some further problems are given in their paper.

P. Erdős & J. L. Selfridge, Some problems on the prime factors of consecutive integers, *Illinois J. Math.*, **11**(1967) 428–430.

A. Schinzel, Unsolved problem 31, *Elem. Math.*, **14**(1959) 82–83.

OEIS: A059756-059757.

B28 Consecutive numbers with distinct prime factors.

Selfridge asked: do there exist n consecutive integers, each having either two distinct prime factors less than n or a repeated prime factor less than n? He gives two examples:

The numbers $a + 11 + i$ ($1 \leq i \leq n = 115$) where $a \equiv 0 \bmod 2^2 3^2 5^2 7^2 11^2$ and $a + p \equiv 0 \bmod p^2$ for each prime p, $13 \leq p \leq 113$, and

The numbers $a + 31 + i$ $(1 \leq i \leq n = 1329)$ where $a + p \equiv 0 \bmod p^2$ for each prime p, $37 \leq p \leq 1327$ and $a \equiv 0 \bmod 2^2 3^2 5^2 7^2 11^2 13^2 17^2 19^2 23^2 29^2 31^2$.

It is harder to find examples of n consecutive numbers, each one divisible by two distinct primes less than n or by the square of a prime $< n/2$, though he believes that they could be found by computer.

This is related to the problem: find n consecutive integers, each having a composite common factor with the product of the other $n - 1$. If the composite condition is relaxed, and one asks merely for a common factor greater than 1, then $2184 + i$ $(1 \leq i \leq n = 17)$ is a famous example.

Alfred Brauer, On a property of k consecutive integers, *Bull. Amer. Math. Soc.*, **47**(1941) 328–331; *MR* **2**, 248.

Ronald J. Evans, On blocks of N consecutive integers, *Amer. Math. Monthly* **76**(1969) 48–49.

Ronald J. Evans, On N consecutive integers in an arithmetic progression, *Acta Sci. Math. Univ. Szeged*, **33**(1972) 295–296; *MR* **47** #8408.

Heiko Harborth, Eine Eigenschaft aufeinanderfolgender Zahlen, *Arch. Math.* (*Basel*) **21**(1970) 50–51; *MR* **41** #6771.

Heiko Harborth, Sequenzen ganzer Zahlen, *Zahlentheorie* (*Tagung, Math. Forschungsinst. Oberwolfach*, 1970) 59–66; *MR* **51** #12775.

S. S. Pillai, On m consecutive integers I, *Proc. Indian Acad. Sci. Sect. A*, **11**(1940) 6–12; *MR* **1**, 199; II **11**(1940) 73–80; *MR* **1**, 291; III **13**(1941) 530–533; *MR* **3**, 66; IV *Bull. Calcutta Math. Soc.*, **36**(1944) 99–101; *MR* **6**, 170.

OEIS: A059756-059757.

B29 Is x determined by the prime divisors of $x + 1$, $x + 2$, ..., $x + k$?

Alan R. Woods asks if there is a positive integer k such that every x is uniquely determined by the (sets of) prime divisors of $x + 1$, $x + 2$, ..., $x + k$. Perhaps $k = 3$?

For primes less than 23 there are four ambiguous cases for $k = 2$: $(x+1, x+2) = (2,3)$ or $(8,9)$; $(6,7)$ or $(48,49)$; $(14,15)$ or $(224,225)$; $(75,76)$ or $(1215,1216)$. The first three of these are members of the infinite family $(2^n - 2, 2^n - 1)$, $(2^n(2^n - 2), (2^n - 1)^2)$. Compare **B19**.

D. H. Lehmer, On a problem of Størmer, *Illinois J. Math.*, **8**(1964) 57–79; *MR* **28** #2072.

OEIS: A059756-059757.

B30 A small set whose product is square.

Erdős, Graham & Selfridge want us to find the least value of t_n so that the integers $n + 1$, $n + 2$, ..., $n + t_n$ contain a subset the product of whose members with n is a square. The Thue-Siegel theorem implies that $t_n \to \infty$ with n, faster than a power of $\ln n$.

I was asked for a justification or reference for this last sentence. Andrew Granville kindly supplied the following comments:

> The point is that if you do have such a subset then there is an integer point (n, m) on some hyperelliptic curve
>
> $$y^2 = x(x + i_1)(x + i_2) \cdots (x + i_k)$$
>
> where $0 < i_i < i_2 < ... < i_k \le t_n$. If t_n were to be $< T$ for infinitely many n then some such curve would have infinitely many rational points (or even integer points), contradicting Faltings's Theorem if $k \ge 3$, and Thue's Theorem for $k \ge 0$. Thus $t_n \to \infty$.
>
> More difficult would be to estimate quite how fast we can prove $t_n \to \infty$. To do this one needs some effective version of Faltings's or Thue's Theorem. There is probably a pretty good effective version of Thue's Theorem, especially for hyperelliptic curves.
>
> It is amusing to note that the abc-conjecture is certainly applicable to this question, via Elkies's paper (see ref. at **B19**) or Langevin, though this would take some working out. Presumably $t_n > n^c$ for some $c > 0$ (assuming abc) though I have not proved this! Perhaps Silverman is interested in this question.

Joseph Silverman responded:

> Granville's argument that $t_n \to \infty$ is fine, but it depends on the fact that a hyperelliptic curve has only finitely many integer points (due to Siegel, I believe, not Thue). It seems to me that Theorem 1 of my paper with Evertse might be helpful. Let
>
> $$f(X) = X(X + i_1) \cdots (X + i_k) \quad \text{with} \quad 0 < i_1 < \cdots < i_k$$
>
> and assume that $k \ge 3$, since the case $k = 2$ can be dealt with separately. Then Theorem 1(b) can be applied with $K = \mathbb{Q}$, $m = 1$, S is the infinite place of $\mathbb{Q}$ together with the primes dividing $D(f)$, the discriminant of f, $s = |S| = 1 + \nu(D(f))$, R_S is the ring of S-integers in $\mathbb{Q}$, $L = K = \mathbb{Q}$, $M = 1$, $n = 2$, $\kappa_n(L) = 0$. This appears to give that the number of integer solutions to $Y^2 = f(X)$ is $\le 7^{4+9s} = 7^{13+9\nu(D(f))}$.

Selfridge (W. No. Theory problem 97:22) says that it is conjectured that 6 and 392 are the only numbers of shape $n = rs^2$ with $r > 1$ and squarefree for which there do not exist a, b with $n < a < b < r(s+1)^2$ and nab a square.

To revert to the opening paragraph, Selfridge has shown that $t_n \leq \max(P(n), 3\sqrt{n})$, where $P(n)$ is the largest prime factor of n.

Alternatively, is it true that for every c there is an n_0 so that for every $n > n_0$ the products $\prod a_i$, taken over $n < a_1 < \ldots < a_k < n + (\ln n)^c$ ($k = 1, 2, \ldots$) are all distinct? Erdős, Graham & Selfridge proved this for $c < 2$.

Selfridge conjectures that if n is not a square, and t is the next larger number than n such that nt is a square, then, unless $n = 8$ or 392, it is always possible to find r and s, $n < r < s < t$ such that nrs is a square. E.g., if $n = 240 = 2^4 3 \cdot 5$ then $t = 375 = 3 \cdot 5^3$ and we can find $r = 243 = 3^5$ and $s = 245 = 5 \cdot 7^2$. Selfridge and Meyerowitz have confirmed the conjecture for $n < 10^{30000}$.

Several of the papers referred to at **D10** are relevant here.

P. Erdős & Jan Turk, Products of integers in short intervals, *Acta Arith.*, **44**(1984) 147–174; *MR* **86d**:11073.

Paul Erdős, Janice Malouf, John Selfridge & Esther Szekeres, Subsets of an interval whose product is a power, Paul Erds memorial collection. *Discrete Math.*, **200**(1999) 137–147; *MR* **2000e**:11017.

Jan-Hendrik Evertse & J. H. Silverman, Uniform bounds for the number of solutions to $Y^n = f(X)$ Math. *Proc. Cambridge Philos. Soc.*, **100**(1986) 237–248; *MR* **87k**:11034.

L. Hajdu & Ákos Pintér, Square product of three integers in short intervals, *Math. Comput.*, **68**(1999) 1299–1301; **99j**:11027.

Michel Langevin, Cas d'égalité pour le théorème de Mason et applications de la conjecture (*abc*), *C.R. Acad. Sci. Paris Sér. I Math.*, **317**(1993) 441–444; *MR* **94j**:11027.

T. N. Shorey, Perfect powers in products of integers from a block of consecutive integers, *Acta Arith.*, **49**(1987) 71–79; *MR* **88m**:11002.

T. N. Shorey & Yu. V. Nesterenko, Perfect powers in products of integers from a block of consecutive integers, II *Acta Arith.*, **76**(1996) 191–198; *MR* **97d**:11005.

OEIS: A068568.

B31 Binomial coefficients.

Earl Ecklund, Roger Eggleton, Erdős & Selfridge (see **B23**) write the **binomial coefficient** $\binom{n}{k} = n!/k!(n-k)!$ as a product UV in which every prime factor of U is at most k and every prime factor of V is greater than k. There are only finitely many cases with $n \geq 2k$ for which $U > V$. They determine all such cases except when $k = 3$, 5 or 7.

S. P. Khare lists all cases with $n \le 551$: $k = 3$, $n = 8$, 9, 10, 18, 82, 162; $k = 5$, $n = 10$, 12, 28; and $k = 7$, $n = 21$, 30, 54.

Most binomial coefficients $\binom{n}{k}$ with $n \ge 2k$ have a prime factor $p \le n/k$. After some computing with Lacampagne & Erdős, Selfridge conjectured that this inequality is true whenever $n > 17.125k$. A slightly stronger conjecture is that any such binomial coefficient has least prime factor $p \le n/k$ or $p \le 17$ with just 4 exceptions: $\binom{62}{6}$, $\binom{959}{56}$, $\binom{474}{66}$, $\binom{284}{28}$ for which $p = 19$, 19, 23 and 29 respectively.

These authors define the **deficiency** of the binomial coefficient $\binom{n+k}{k}$, $k \le n$, as the number of i for which $b_i = 1$, where $n + i = a_i b_i$, $1 \le i \le k$, the prime factors of b_i are greater than k, and $\prod a_i = k!$ Then $\binom{44}{8}$, $\binom{74}{10}$, $\binom{174}{12}$, $\binom{239}{14}$, $\binom{5179}{27}$, $\binom{8413}{28}$, $\binom{8414}{28}$ and $\binom{96622}{42}$ each have deficiency 2; $\binom{46}{10}$, $\binom{47}{10}$, $\binom{241}{16}$, $\binom{2105}{25}$, $\binom{1119}{27}$ and $\binom{6459}{33}$ have deficiency 3; $\binom{47}{11}$ has deficiency 4; and $\binom{284}{28}$ has deficiency 9; and they conjecture that there are no others with deficiency greater than 1. Are there only finitely many binomial coefficients with deficiency 1?

Erdős & Selfridge noted that if $n \ge 2k \ge 4$, then there is at least one value of i, $0 \le i \le k - 1$, such that $n - i$ does not divide $\binom{n}{k}$, and asked for the least n_k for which there was only one such i. For example, $n_2 = 4$, $n_3 = 6$, $n_4 = 9$, $n_5 = 12$. $n_k \le k!$ for $k \ge 3$.

For a positive integer k, the Erdős-Selfridge function is the least integer $g(k) > k + 1$ such that all prime factors of $\binom{g(k)}{k}$ exceed k. The first few values are

$k =$	2	3	4	5	6	7	8	9	10	11	12	13
$g(k) =$	6	7	7	23	62	143	44	159	46	47	174	2239

They are far from being monotonic. Ecklund, Erdős & Selfridge conjecture that $\limsup$ & $\liminf$ of $g(k+1)/g(k)$ are ∞ & 0. They also conjecture that, for each n and for $k > k_0(n)$, $g(k) > k^n$, that $\lim g(k)^{1/k} = 1$ and that $g(k) < e^{c\pi(k)}$. For example, they conjecture that $g(k) > k^3$ for $k > 35$ and that $g(k) > k^5$ for $k > 100$. They show that $g(k) < k^2 L_k P_l$ with $l = \lfloor 6k/\ln k \rfloor$, where L_k is the l.c.m. of 1, 2, $\ldots$, k and P_l is the product of the primes not exceeding l and Erdős, Lacampagne & Selfridge show that $g(k) > ck^2/\ln k$. The tables of $g(k)$ have been extended by Scheidler & Williams (ref. at **B33**) to $k \le 140$ and, with the help of Lukes, to $k \le 200$.

$$g(200) = 520\,8783889271\,0191382732$$

Granville shows that the abc-conjecture implies that there are only finitely many powerful binomial coefficients $\binom{n}{k}$ with $3 \le k \le n/2$.

Harry Ruderman asks for a proof or disproof that for every pair (p, q) of nonnegative integers there is a positive integer n such that

$$\frac{(2n - p)!}{n!(n + q)!}$$

is an integer.

A problem which has briefly baffled good mathematicians is: is $\binom{n}{r}$ ever prime to $\binom{n}{s}$, $0 < r < s \le n/2$? The negative answer follows from the identity

$$\binom{n}{s}\binom{s}{r} = \binom{n}{r}\binom{n-r}{s-r}.$$

Erdős & Szekeres ask if the greatest prime factor of the g.c.d. is always greater than r; the only counterexample with $r > 3$ that they noticed is

$$\gcd\left(\binom{28}{5}, \binom{28}{14}\right) = 2^3 \cdot 3^3 \cdot 5$$

On 98-01-12 David Gale reported that a Berkeley student proposed a natural generalization, of the noncoprimality of two binomial coefficients, to trinomial coeffients, with the further generalization to k-nomial coefficients. If $n = a + b + c$, let $T(n; a, b, c) = n!/(a!b!c!)$. The conjecture, supported by fairly extensive computer evidence, is that for $0 < a, b, c < n$, every *three* $T(n; a, b, c)$ have a common factor. George Bergman and Hendrik Lenstra and others have some relevant partial results.

Wolstenholme's theorem states that if n is a prime > 3, then

$$\binom{2n-1}{n} \equiv 1 \bmod n^3.$$

There are no composite solutions for $n < 10^9$ and it is conjectured that there are none. Call a prime p satisfying $\binom{2p-1}{p-1} \equiv 1 \bmod p^4$ a **Wolstenholme prime**. McIntosh shows that p is Wolstenholme just if it divides the numerator of the Bernoulli number B_{p-3} (see **A17**). The two known such are 16843 and 2124679. He conjectures that there are infinitely many and that none satisfies $\binom{2p-1}{p-1} \equiv 1 \bmod p^5$.

For other problems and results on the divisors of binomial coefficients, see **B33**.

Emre Alkan, Variations on Wolstenholme's theorem, *Amer. Math. Monthly*, **101**(1994) 1001–1004.

D. F. Bailey, Two p^3 variations of Lucas's theorem, *J. Number Theory*, **35**(1990) 208–215; *MR* **90f**:11008.

M. Bayat, A generalization of Wolstenholme's theorem, *Amer. Math. Monthly*, **104**(1997) 557–560 (but see Gessel reference).

Daniel Berend & Jørgen E. Harmse, On some arithmetical properties of middle binomial coefficients, *Acta Arith.*, **84**(1998) 31–41.

Cai Tian-Xin & Andrew Granville, On the residues of binomisl coefficients and their residues modulo prime powers. *Acta Math. Sin.* (*Engl. Ser.*), **18**(2002) 277–288.

Chen Ke-Ying, Another equivalent form of Wolstenholme's theorem and its generalization (Chinese), *Math. Practice Theory*, **1995** 71–74; *MR* **97d**:11006.

Paul Erdős, C. B. Lacampagne & J. L. Selfridge, Estimates of the least prime factor of a binomial coefficient, *Math. Comput.*, **61**(1993) 215–224; *MR* **93k**:11013.

P. Erdős & J. L. Selfridge, Problem 6447, *Amer. Math. Monthly* **90**(1983) 710; **92**(1985) 435–436.

P. Erdős & G. Szekeres, Some number theoretic problems on binomial coefficients, *Austral. Math. Soc. Gaz.*, **5**(1978) 97–99; *MR* **80e**:10010 is uninformative.

Ira M. Gessel, Wolstenholme revisited, *Amer. Math. Monthly*, **105**(1998) 657–658; *MR* **99e**:11009.

Andrew Granville, Arithmetic properties of binomial coefficients. I. Binomial coefficients modulo prime powers, *Organic mathematics* (*Burnaby BC*, 1995), *CMS Conf. Proc.*, **20**(1997) 253–276; *MR* **99h**:11016.

Andrew Granville, On the scarcity of powerful binomial coefficients, *Mathematika* **46**(1999) 397–410; *MR* **2002b**:11029.

Andrew Granville & Olivier Ramaré, Explicit bounds on exponential sums and the scarcity of squarefree binomial coefficients, *Mathematika*, **43**(1996) 73–107; *MR* **99m**:11023.

A. Grytczuk, On a conjecture of Erdős on binomial coefficients, *Studia Sci. Math. Hungar.*, **29**(1994) 241–244.

Hong Shao-Fang, A generalization of Wolstenholme's theorem, *J. South China Normal Univ. Natur. Sci. Ed.*, **1995** 24–28; *MR* **99f**:11004.

Gerhard Larcher, On the number of odd binomial coefficients, *Acta Math. Hungar.*, **71**(1996 183–203.

Lee Dong-Hoon[1] & Hahn Sang-Geun, Some congruences for binomial coefficients, *Class field theory—its centenary and prospect* (*Tokyo*, 1998) 445–461, *Adv. Stud. Pure Math. 30 Math. Soc. Japan, Tokyo*, 2001; II *Proc. Japan Acad. Ser. A Math. Sci.*, **76**(2000) 104–107; *MR* **2002k**:11024, 11025.

Grigori Kolesnik, Prime power divisors of multinomial and q-multinomial coefficients, *J. Number Theory*, **89**(2001) 179–192; *MR* **2002i**:11020.

Richard F. Lukes, Renate Scheidler & Hugh C. Williams, Further tabulation of the Erdős-Selfridge function, *Math. Comput.*, **66**(1997) 1709–1717; *MR* **98a**:11191.

Richard J. McIntosh, A generalization of a congruential property of Lucas, *Amer. Math. Monthly*, **99**(1992) 231–238.

Richard J. McIntosh, On the converse of Wolstenholme's theorem, *Acta Arith.*, **71**(1995) 381–389; *MR* **96h**:11002.

Marko Razpet, On divisibility of binomial coefficients, *Discrete Math.*, **135** (1994) 377–379; *MR* **95j**:11014.

Harry D. Ruderman, Problem 714, *Crux Math.*, **8**(1982) 48; **9**(1983) 58.

J. W. Sander, On prime divisors of binomial coefficients. *Bull. London Math. Soc.*¡ **24**(1992) no. 2 140–142; *MR* **93g**:11019.

J. W. Sander, Prime power divisors of multinomial coefficients and Artin's conjecture, *J. Number Theory*, **46**(1994) 372–384; *MR* **95a**:11018.

J. W. Sander, On the order of prime powers dividing $\binom{2n}{n}$, *Acta Math.*, **174**(1995) 85–118; *MR* **96b**:11018.

J. W. Sander, A story of binomial coefficients and primes, *Amer. Math. Monthly*, **102**(1995) 802–807; *MR* **96m**:11015.

Renate Scheidler & Hugh C. Williams, A method of tabulating the number-theoretic function $g(k)$, *Math. Comput.*, **59**(1992) 199, 251–257; *MR* **92k**:11146.

David Segal, Problem E435, partial solution by H.W. Brinkman, *Amer. Math. Monthly*, **48**(1941) 269–271.

OEIS: A034602.

B32 Grimm's conjecture.

Grimm has conjectured that if $n + 1$, $n + 2$, ..., $n + k$ are all composite, then there are distinct primes p_{i_j} such that $p_{i_j} | (n + j)$ for $1 \le j \le k$. For example

$$1802\ 1803\ 1804\ 1805\ 1806\ 1807\ 1808\ 1809\ 1810$$

are respectively divisible by

$$53\ \ 601\ \ 41\ \ \ 19\ \ \ 43\ \ \ 139\ \ 113\ \ 67\ \ \ 181$$

and

$$114\ 115\ 116\ \ 117\ \ 118\ \ 119\ \ 120\ \ 121\ \ 122\ \ 123\ \ 124\ 125\ 126$$

by

$$19\ \ 23\ \ 29\ \ \ 13\ \ \ 59\ \ \ 17\ \ \ \ 2\ \ \ \ 11\ \ \ 61\ \ \ 41\ \ \ 31\ \ \ 5\ \ \ \ 7$$

Ramachandra, Shorey & Tijdeman proved, under the hypothesis of Schinzel mentioned in **A2**, that there are only finitely many exceptions to Grimm's conjecture.

Erdős & Selfridge asked for an estimate of $f(n)$, the least number such that for each m there are distinct integers a_1, a_2, ..., $a_{\pi(n)}$ in the interval $[m+1, m+f(n)]$ with $p_i | a_i$ where p_i is the ith prime. They and Pomerance show that, for large n,

$$(3 - \epsilon)n \le f(n) \ll n^{3/2}(\ln n)^{-1/2}$$

You & Zhang prove that for $k > 1$, $\binom{m+k}{k}$ can be written $\prod_{i=1}^{k} a_i$, where each a_i divides $m + i$ and $\gcd(a_i, a_j) = 1$ when $i \ne j$. They claim that Grimm's conjecture is a consequence of their own conjecture that if $m + 1, m + 2, \ldots, m + k$ are composite, then each of the $a_i > 1$.

P. Erdős, Problems and results in combinatorial analysis and combinatorial number theory, in *Proc. 9th S.E. Conf. Combin. Graph Theory, Comput., Boca Raton, Congressus Numerantium* XXI, Utilitas Math. Winnipeg, 1978, 29–40.

P. Erdős & C. Pomerance, Matching the natural numbers up to n with distinct multiples in another interval, *Nederl. Akad. Wetensch. Proc. Ser. A*, **83**(= *Indag. Math.*, **42**)(1980) 147–161; *MR* **81i**:10053.

Paul Erdős & Carl Pomerance, An analogue of Grimm's problem of finding distinct prime factors of consecutive integers, *Utilitas Math.*, **24**(1983) 45–46; *MR* **85b**:11072.

P. Erdős & J. L. Selfridge, Some problems on the prime factors of consecutive integers II, in *Proc. Washington State Univ. Conf. Number Theory*, Pullman, 1971, 13–21.

C. A. Grimm, A conjecture on consecutive composite numbers, *Amer. Math. Monthly*, **76**(1969) 1126–1128.

Michel Langevin, Plus grand facteur premier d'entiers en progression arithmétique, *Sém. Delange-Pisot-Poitou*, **18**(1976/77) *Théorie des nombres: Fasc.* 1, *Exp. No.* 3, Paris, 1977; *MR* **81a**:10011.

Carl Pomerance, Some number theoretic matching problems, in *Proc. Number Theory Conf.*, Queen's Univ., Kingston, 1979, 237–247.

Carl Pomerance & J. L. Selfridge, Proof of D.J. Newman's coprime mapping conjecture, *Mathematika*, **27**(1980) 69–83; *MR* **81i**:10008.

K. Ramachandra, T. N. Shorey & R. Tijdeman, On Grimm's problem relating to factorization of a block of consecutive integers, *J. reine angew. Math.*, **273**(1975) 109–124.

You Yong-Xing & Zhang Shao-Hua, A new theorem on the binomial coefficient $\binom{m+n}{n}$, *J. Math. (Wuhan)* **23**(2003) 146–148; *MR* **2004a**:11012.

B33 Largest divisor of a binomial coefficient.

What can one say about the largest divisor, less than n, of the binomial coefficient $\binom{n}{k} = n!/k!(n-k)!$? Erdős points out that it is easy to show that it is at least n/k and conjectures that there may be one between cn and n for any $c < 1$ and n sufficiently large. Marilyn Faulkner showed that if p is the least prime $> 2k$ and $n \geq p$, then $\binom{n}{k}$ has a prime divisor $\geq p$, except for $\binom{9}{2}$ and $\binom{10}{3}$. Earl Ecklund showed that if $n \geq 2k > 2$ then $\binom{n}{k}$ has a prime divisor $p \leq n/2$, except for $\binom{7}{3}$.

John Selfridge conjectures that if $n \geq k^2 - 1$, then, apart from the exception $\binom{62}{6}$, there is a prime divisor $\leq n/k$ of $\binom{n}{k}$. Among those binomial coefficients whose least prime factor p is $\geq n/k$ there may be only a finite number with $p \geq 13$, but there could be infinitely many with $p = 7$. That there are infinitely many with $p = 5$ was proved by Erdős, Lacampagne & Selfridge (**B31**).

A classical theorem, discovered independently by Sylvester and Schur, stated that the product of k consecutive integers, each greater than k, has a prime divisor greater than k. Leo Moser conjectured that the Sylvester-Schur theorem holds for primes $\equiv 1 \bmod 4$, in the sense that for n sufficiently large (and $\geq 2k$), $\binom{n}{k}$ has a prime divisor $\equiv 1 \bmod 4$ which is greater than k. However, Erdős does not think that this is true, but it may not be at all easy to settle. In this connexion John Leech notices that the fourteen integers $280213, \ldots, 280226$ have no prime factor of the form $4m + 1 > 13$.

Thanks to Ira Gessel and John Conway, we can say that the generalization of the **Catalan numbers** $\frac{1}{n+1}\binom{2n}{n}$, requested in the first edition by Neil Sloane, is $\frac{(n,r)}{n}\binom{n}{r}$, which is always an integer (multiply by n and by r and Euclid knew that (n,r) is a linear combination of n and r). These are also known as generalized ballot numbers and they occur when enumerating certain lattice paths.

If $f(n)$ is the sum of the reciprocals of those primes $< n$ which do not divide $\binom{2n}{n}$, then Erdős, Graham, Ruzsa & Straus conjectured that there is an absolute constant c so that $f(n) < c$ for all n. Erdős also conjectured that $\binom{2n}{n}$ is never squarefree for $n > 4$. Since $4 \,|\, \binom{2n}{n}$ unless $n = 2^k$, it suffices to consider

$$\binom{2^{k+1}}{2^k}.$$

Sárközy proved this for n sufficiently large and Sander has shown, in a precise sense, that binomial coefficients near the centre of the Pascal triangle are not squarefree. Granville & Ramaré completed Sárközy's proof by showing that $k > 300000$ was sufficiently large, and checking it computationally for $2 \leq k \leq 300000$. They also improved Sander's result by showing that there is a constant δ, $0 < \delta < 1$, such that if $\binom{n}{k}$ is squarefree then k or $n - k$ must be $< n^\delta$ for sufficiently large n. They conjecture that k or $n - k$ must in fact be $< (\ln n)^{2-\delta}$, and that this is best possible in the sense that there are infinitely many squarefree $\binom{n}{k}$ with $\frac{1}{2}n > k > c(\ln n)^2$ for some $c > 0$. They prove such a result for $\frac{1}{2}n > k > \frac{1}{5}\ln n$. They show that there is a constant $\rho_k > 0$ such that the number of $n \leq N$ with $\binom{n}{k}$ squarefree is $\sim \rho_k N$. Since $\rho_k < c/k^2$ for some $c > 0$, they conjecture that there is a constant $\gamma > 0$ such that the number of squarefree entries in the first N rows of Pascal's triangle is $\sim \gamma N$.

Erdős has also conjectured that for $k > 8$, 2^k is not the sum of distinct powers of 3 $[2^8 = 3^5 + 3^2 + 3 + 1]$. If that's true, then for $k \geq 9$,

$$3 \,\Big|\, \binom{2^{k+1}}{2^k}.$$

In answer to the question, is $\binom{342}{171}$ the largest $\binom{2n}{n}$ which is not divisible by the square of an odd prime, Eugene Levine gave the examples $n = 784$ and 786. Erdős feels sure that there are no larger such n.

Denote by $e = e(n)$ the largest exponent such that, for some prime p, p^e divides $\binom{2n}{n}$. It is not known whether $e \to \infty$ with n. On the other hand Erdős cannot disprove $e > c \ln n$.

Ron Graham offers \$100.00 for deciding if $(\binom{2n}{n}, 105) = 1$ infinitely often. Kummer knew that n, when written in base 3 or 5 or 7, would have to have only the digits 0, 1 or 0, 1, 2 or 0, 1, 2, 3 respectively. H. Gupta & S. P. Khare found the 14 values 1, 10, 756, 757, 3160, 3186, 3187, 3250, 7560, 7561, 7651, 20007, 59548377, 59548401 of n less than 7^{10}, while Peter Montgomery, Khare and others found many larger values.

Erdős, Graham, Ruzsa & Straus showed that for any *two* primes p, q there are infinitely many n for which $(\binom{2n}{n}, pq) = 1$. If $g(n)$ is the smallest odd prime factor of $\binom{2n}{n}$, then $g(n) \leq 11$ for $3160 < n < 10^{10000}$, while $g(3160) = 13$.

More complicated quotients of products of factorials which yield integers have been considered by Picon.

Gould repeated Hermite's 1889 observations that

$$\frac{m}{(m,n)} \left| \binom{m}{n} \right. \qquad \text{and} \qquad \frac{m-n+1}{(m+1,n)} \left| \binom{m}{n} \right.$$

and asked for more general a, b, c, r, s, u, v such that

$$\frac{am+bn+c}{(rm+s,un+v)} \left| \binom{m}{n} \right.$$

John McKay notes that $\binom{72-1}{72/2}$ is squarefree, and states that $\binom{2n-1}{n}$ is square-free only for $n = 1, 2, 3, 4, 6, 9, 10, 12, 36$ with $n \leq 500$.

E. F. Ecklund, On prime divisors of the binomial coefficient, *Pacific J. Math.*, **29**(1969) 267–270.

P. Erdős, A theorem of Sylvester and Schur, *J. London Math. Soc.*, **9**(1934) 282–288.

Paul Erdős, A mélange of simply posed conjectures with frustratingly elusive solutions, *Math. Mag.*, **52**(1979) 67–70.

P. Erdős & R. L. Graham, On the prime factors of $\binom{n}{k}$, *Fibonacci Quart.*, **14**(1976) 348–352.

P. Erdős, R. L. Graham, I. Z. Ruzsa & E. Straus, On the prime factors of $\binom{2n}{n}$, *Math. Comput.*, **29**(1975) 83–92.

M. Faulkner, On a theorem of Sylvester and Schur, *J. London Math. Soc.*, **41**(1966) 107–110.

Henry W. Gould, Advanced Problem 5777*, *Amer. Math. Monthly*, **78**(1971) 202.

Henry W. Gould & Paula Schlesinger, Extensions of the Hermite G.C.D. theorems for binomial coefficients, *Fibonacci Quart.*, **33**(1995) 386–391; *MR* **97g**:11015.

Hansraj Gupta, On the parity of $(n+m-1)!(n,m)/n!m!$, *Res. Bull. Panjab Univ.* (*N.S.*), **20**(1969) 571–575; *MR* **43** #3201.

L. Moser, Insolvability of $\binom{2n}{n} = \binom{2a}{a}\binom{2b}{b}$, *Canad. Math. Bull.*, **6**(1963)167–169.

P. A. Picon, Intégrité de certains produits-quotients de factorielles, *Séminaire Lotharingien de Combinatoire* (*Saint-Nabor,* 1992), *Publ. Inst. Rech. Math. Av.*, **498**, 109–113; *MR* **95j**:11013.

P. A. Picon, Conditions d'intégrité de certains coefficients hypergéometriques: généralisation d'un théorème de Landau, *Discrete Math.*, **135**(1994) 245–263; *MR* **96f**:11015.

J. W. Sander, Prime power divisors of $\binom{2n}{n}$, *J. Number Theory*, **39**(1991) 65–74; *MR* **92i**:11097.

J. W. Sander, Prime power divisors of binomial coefficients, *J. reine angew. Math.*, **430**(1992) 1–20; *MR* **93h**:11021; reprise **437**(1993) 217–220.

J. W. Sander, On primes not dividing binomial coefficients, *Math. Proc. Cambridge Philos. Soc.*, **113**(1993) 225–232; *MR* **93m**:11099.

J. W. Sander, An asymptotic formula for ath powers dividing binomial coefficients, *Mathematika*, **39**(1992) 25–36; *MR* **93i**:11110.

J. W. Sander, On primes not dividing binomial coefficients, *Math. Proc. Cambridge Philos. Soc.*, **113**(1993) 225–232.

A. Sárközy, On divisors of binomial coefficients I, *J. Number Theory*, **20**(1985) 70–80; *MR* **86c**:11002.

I. Schur, Einige Sätze über Primzahlen mit Anwendungen und Irreduzibilitätsfragen I, *S.-B. Preuss, Akad. Wiss. Phys.-Math. Kl.*, **14**(1929) 125–136.

Sun Zhi-Wei, Products of binomial coefficients modulo p^2, *Acta Arith.*, **97**(2001) 87–98; *MR* **2002m**:11013.

J. Sylvester, On arithmetical series, *Messenger of Math.*, **21**(1892) 1–19, 87–120.

W. Utz, A conjecture of Erdős concerning consecutive integers, *Amer. Math. Monthly*, **68**(1961) 896–897.

G. Velammal, Is the binomial coefficient $\binom{2n}{n}$ squarefree? *Hardy-Ramanujan J.*, **18**(1995) 23–45; *MR* **95j**:11016.

OEIS: A000984, A059097.

B34 If there's an i such that $n - i$ divides $\binom{n}{k}$.

If $H_{k,n}$ is the proposition: there is an i, $0 \leq i < k$ such that $n - i$ divides $\binom{n}{k}$, then Erdős asked if $H_{k,n}$ is true for all k when $n \geq 2k$. Schinzel gave the counterexample $n = 99215$, $k = 15$. If H_k is the proposition: $H_{k,n}$ is true for all n, then Schinzel showed that H_k is false for $k = 15, 21, 22, 33,$ 35 and thirteen other values of k. He showed that H_k is true for all other $k \leq 32$ and asked if there are infinitely many k, other than prime-powers, for which H_k is true: he conjectures not and later reported that it is true for $k = 34$, but for no other non-prime-powers between 34 and 201.

E. Burbacka & J. Piekarczyk, P. 217, R. 1, *Colloq. Math.*, **10**(1963) 365.

A. Schinzel, Sur un problème de P. Erdős, *Colloq. Math.*, **5**(1957–58) 198–204.

B35 Products of consecutive numbers with the same prime factors.

Let $f(n)$ be the the least integer such that at least one of the numbers n, $n + 1$, ..., $n + f(n)$ divides the product of the others. It is easy to see that $f(k!) = k$ and $f(n) > k$ for $n > k!$ Erdős has also shown that

$$f(n) > \exp((\ln n)^{1/2 - \epsilon})$$

for an infinity of values of n, but it seems difficult to find a good upper bound for $f(n)$.

Erdős asks if $(m+1)(m+2)\cdots(m+k)$ and $(n+1)(n+2)\cdots(n+l)$ with $k \geq l \geq 3$ can contain the same prime factors infinitely often. For example $(2 \cdot 3 \cdot 4 \cdot 5 \cdot 6) \cdot 7 \cdot 8 \cdot 9 \cdot 10$ and $14 \cdot 15 \cdot 16$ and $48 \cdot 49 \cdot 50$; also $(2 \cdot 3 \cdot 4 \cdot 5 \cdot 6) \cdot 7 \cdot 8 \cdot 9 \cdot 10 \cdot 11 \cdot 12$ and $98 \cdot 99 \cdot 100$. For $k = l \geq 3$ he conjectures that this happens only finitely many times.

If $L(n; k)$ is the l.c.m. of $n+1$, $n+2$, ..., $n+k$, then Erdős conjectures that for $l > 1$, $n \geq m + k$, $L(m; k) = L(n; l)$ has only a finite number of solutions. Examples are $L(4; 3) = L(13; 2)$ and $L(3; 4) = L(19; 2)$. He asks if there are infinitely many n such that for all k $(1 \leq k < n)$ we have $L(n; k) > L(n - k; k)$. What is the largest $k = k(n)$ for which this inequality can be reversed? He notes that it is easy to see that $k(n) = o(n)$, but he believes that much more is true. He expects that for every $\epsilon > 0$ and $n > n_0(\epsilon)$, $k(n) < n^{1/2+\epsilon}$ but cannot prove this.

P. Erdős, How many pairs of products of consecutive integers have the same prime factors? *Amer. Math. Monthly* **87**(1980) 391–392.

B36 Euler's totient function.

Euler's **totient function**, $\phi(n)$, is the number of numbers not greater than n and prime to n. For example $\phi(1) = \phi(2) = 1$, $\phi(3) = \phi(4) = \phi(6) = 2$, $\phi(5) = \phi(8) = \phi(10) = \phi(12) = 4$, $\phi(7) = \phi(9) = 6$. Are there infinitely many pairs of consecutive numbers, n, $n + 1$, such that $\phi(n) = \phi(n + 1)$? For example, $n = 1, 3, 15, 104, 164, 194, 255, 495, 584, 975$. It is not even known if $|\phi(n + 1) - \phi(n)| < n^\epsilon$ has an infinity of solutions for each $\epsilon > 0$. Jud McCranie found 1267 solutions of $\phi(n) = \phi(n + 1)$ with $n < 10^{10}$, the largest of which is $n = 9985705185$, $\phi(n) = \phi(n + 1) = 2^{11}3^5 7 \cdot 11$. He also looked for solutions of $\phi(n + k) = \phi(n)$. Schinzel conjectures that for every even k there are infinitely many solutions, but observes that the corresponding conjecture with k odd is implausible. Indeed, when k is an odd multiple of 3, in a search to $n = 10^{10}$, McCranie found only a few solutions. For $k = 3$ only $n = 3$ and $n = 5$. In fact, the only value of $k \equiv 3 \pmod 6$ which yielded as many as 13 solutions was $k = 141$ and the largest solution found was $n = 715$ with $k = 245$, except that $k = 27$ yielded a very atypical $n = 4135966808$. Other odd k were still yielding a steady supply of solutions; the most for $k < 101$ being 1673 solutions for $k = 47$, and the least being 278 for $k = 55$, except that $k = 35$ gave only 29 solutions, 12 of them $> 10^9$, including $n = 9423248800$.

Sierpiński has shown that there is at least one solution for each k and Schinzel & Wakulicz that there are at least two for each $k < 2 \cdot 10^{58}$. For even k Holt raises this to $k < 1.38 \cdot 10^{26595411}$.

Mąkowski has shown that $\phi(n + k) = 2\phi(n)$ has at least one solution for every k. For the equation $\phi(n+k) = 3\phi(n)$ see the solution to Problem E3215 in *Amer. Math. Monthly*, **96**(1989) 63–64.

McCranie found no solutions of $\phi(n+k) = \phi(n)+\phi(k)$ for $k = 3$, nor any of $\phi(n) = \phi(n+1) = \phi(n+2)$, for $n < 10^{10}$, apart from $\phi(5186) = \phi(5187) = \phi(5188) = 2^5 3^4$. Other curiosities are $\phi(25930) = \phi(25935) = \phi(25940) = \phi(25942) = 2^7 3^4$ and $\phi(404471) = \phi(404473) = \phi(404477) = 2^8 3^2 5^2 7$.

Graham, Holt & Pomerance can show that $|\phi(n) - \phi(n + 2)| < n^{9/10}$ infinitely often by looking at $n = 2(2p - 1)$, where p is a prime, and using a theorem of Chen that says that there are infinitely many primes p such that either $2p - 1$ is a prime or $2p - 1$ is a product of two prime factors, both of which exceed $p^{1/10}$. With a little work, they could improve the $9/10$ exponent. On the other hand, they don't see any similar argument for $|\phi(n) - \phi(n + 1)|$. This seems to be a fundamentally different problem.

Erdős lets $a_1, \dots, a_t$ be the longest sequence for which

$$a_1 < \cdots < a_t \le n \quad \text{and} \quad \phi(a_1) < \cdots < \phi(a_t)$$

and suggests that $t = \pi(n)$. Can one even prove $t < (1 + o(1))\pi(n)$ or at least $t = o(n)$? Similar questions can be asked about $\sigma(n)$.

Nontotients are positive *even* values of n for which $\phi(x) = n$ has no solution; for example, $n = 14$, 26, 34, 38, 50, 62, 68, 74, 76, 86, 90, 94, 98. The number, $\#(y)$, of these less than y has been calculated by the Lehmers.

y	10^3	10^4	$2 \cdot 10^4$	$3 \cdot 10^4$	$4 \cdot 10^4$	$5 \cdot 10^4$	$6 \cdot 10^4$	$7 \cdot 10^4$	$8 \cdot 10^4$	$9 \cdot 10^4$
$\#(y)$	210	2627	5515	8458	11438	14439	17486	20536	23606	26663

Zhang Ming-Zhi has shown that for every positive integer m, there is a prime p such that mp is a nontotient.

Browkin & Schinzel have proved the conjecture of Sierpiński and Erdős that there are infinitely many noncototients, by showing that none of the numbers $2^k \cdot 509203$, $k = 1, 2, \dots$ is of the form $x - \phi(x)$.

Erdős & Hall have shown that the number, $\Phi(y) = y - \#(n)$, of n for which $\phi(x) = n$ *has* a solution is $ye^{f(y)}/\ln y$, where $f(y)$ lies between $c(\ln \ln \ln y)^2$ and $c(\ln y)^{1/2}$. Maier & Pomerance more recently showed that the lower bound was correct, with $c \approx 0.8178$. Erdős conjectures that $\Phi(cy)/\Phi(y) \to c$, and that this, if true, may be the best substitute that one can find for an asymptotic formula for $\Phi(y)$.

Noncototients are positive values of n for which $x - \phi(x) = n$ has no solution; for example, $n = 10$, 26, 34, 50, 52, 58, 86, 100. Sierpiński and Erdős conjecture that there are infinitely many noncototients.

Erdős once asked if it was true that for every ϵ there is an n with $\phi(n) = m$, $m < \epsilon n$ and for no $t < n$ is $\phi(t) = m$; perhaps there are many such n.

Michael Ecker has asked for which values of x do each of the series $\sum_{n=1}^{\infty} \phi(n)/n^x$ and $\sum_{n=1}^{\infty} (-1)^{n+1} \phi(n)/n^x$ converge.

David Angell, in Western Number Theory problem 003:03, asks if there is a closed form for
$\sum_{n=1}^{\infty} \frac{\phi(n)}{2^n} = 1.36763080198502235079050814621308813907489199896\ldots$

Donald Newman has shown that if a, b, c, d are nonnegative integers with a, $c > 0$ and $ad - bc \neq 0$, then there exists a positive integer n such that $|phi(an + b) < \phi(cn + d)$, but there is no such $n < 2 \cdot 10^6$ such that $\phi(30n + 1) < \phi(30n)$. In fact Greg Martin showed that that the least such n has 1116 decimal digits.

$n =$ 232909 8101754967 9381404968 4205233780 0048598859 6605123536 3345311075
8883445287 2315452798 4260176895 8541826348 0290710927 1610432287 6529769074
6757436240 0134090318 3559621214 7678571289 1544538210 9667040369 9088529244
6155135679 7175658080 6376638384 6220120606 1438265094 3354025008 5111624970
4645413809 3448637568 8208918750 6406746299 4246549936 9036578640 3317590359
7936930268 5371156272 2454663962 2786562195 1101808240 6922599602 0309133058
9296656888 0117910114 1606263156 5320593772 2871189137 2860899790 1791216356
1086654763 0608074012 1528236888 6801201524 7913832745 1088404280 9290483149
1212278487 9758304016 8324367515 3225518564 0249324065 4924915110 7252158598
0547438748 6893071593 6348123396 5802331725 0336638626 1895716897 4043547448
8796632179 7108144561 9618789985 4720743031 0030363607 8827273695 5511620897
2543511024 6701964021 0458490818 1160442733 1227553783 5908215100 9160756717
8842569576 6995480382 1767317189 5383249326 8006674329 9353118643 7659910632
8654198923 7095772215 4266351039 8085481508 2886896882 0675198820 3811355236
4636120238 3915218571 0178014630 1149110878 4343253284 3935116502 5450659792
3969653616 8138977106 2175669382 7471154701 1512223204 4334740818 0047964860

J. M. Aldaz, A. Bravo, S. Gutiérrez & A. Ubis, A theorem of D. J. Newman on Euler's ϕ function and arithmetic progressions. *Amer. Math. Monthly*, **108**(2001) 364–367; *MR* **2002i**:11099.

Robert Baillie, Table of $\phi(n) = \phi(n + 1)$, *Math. Comput.*, **30**(1976) 189–190.

Roger C. Baker & Glyn Harman, Sparsely totient numbers, *Ann. Fac. Sci. Toulouse Math.*(6), **5**(1996) 183–190; *MR* **97k**:11129.

David Ballew, Janell Case & Robert N. Higgins, Table of $\phi(n) = \phi(n + 1)$, *Math. Comput.*, **29**(1975) 329–330.

Jerzy Browkin & Andrzej Schinzel, On integers not of the form $n - \phi(n)$, *Colloq. Math.*, **58**(1995) 55–58; *MR* **95m**:11106.

József Bukor & János T. Tóth, Estimation of the mean value of some arithmetical functions, *Octogon Math. Mag.*, **3**(1995) 31–32; *MR* **97a**:11013.

Cai Tian-Xin, On Euler's equation $\phi(x) = k$, *Adv. Math. (China)*, **27**(1998) 224–226.

Thomas Dence & Carl Pomerance, Euler's function in residue classes, *Ramanujan J.*, **2**(1998) 7–20; *MR* **99k**:11148.

Michael W. Ecker, Problem E-1, *The AMATYC Review*, **5**(1983) 55; comment **6**(1984)55.

N. El-Kassar, On the equations $k\phi(n) = \phi(n + 1)$ and $k\phi(n + 1) = \phi(n)$, *Number theory and related topics (Seoul 1998)* 95–109, Yonsei Univ. Inst. Math. Sci., Seoul 2000; *MR* **2003g**:11001.

P. Erdős, Über die Zahlen der Form $\sigma(n) - n$ und $n - \phi(n)$, *Elem. Math.*, **28**(1973) 83–86.

P. Erdős & R. R. Hall, Distinct values of Euler's ϕ-function, *Mathematika*, **23**(1976) 1–3.

S. W. Graham, Jeffrey J. Holt & Carl Pomerance, On the solutions to $\phi(n) = \phi(n + k)$. *Number theory in progress*, Vol. 2 (Zakopane-Kościelisko, 1997), 867–882, de Gruyter, Berlin, 1999; *MR* **2000h**:11102.

Jeffrey J. Holt, The minimal number of solutions to $\phi(n) = \phi(n + k)$, *Math. Comput.*, **72**(2003) 2059–2061; *MR* **2004c**:11171.

Patricia Jones, On the equation $\phi(x) + \phi(k) = \phi(x + k)$, *Fibonacci Quart.*, **28**(1990) 162–165; *MR* **91e**:11008.

M. Lal & P. Gillard, On the equation $\phi(n) = \phi(n + k)$, *Math. Comput.*, **26**(1972) 579–582.

Antanas Laurinčikas, On some problems related to the Euler ϕ-function, *Paul Erdős and his mathematics* (*Budapest*, 1999) 152–154, János Bolyai Math. Soc., Budapest 1999.

Florian Luca & Carl Pomerance, On some problems of Mąkowski-Schinzel and Erdős concerning the arithmetic functions ϕ and σ, *Colloq. Math.* **92**(2002) 111–130; *MR* **2003e**:11105.

Helmut Maier & Carl Pomerance, On the number of distinct values of Euler's ϕ-function, *Acta Arith.*, **49**(1988) 263–275.

Andrzej Mąkowski, On the equation $\phi(n+k) = 2\phi(n)$, *Elem. Math.*, **29**(1974) 13.

Greg Martin, The smallest solution of $\phi(30n + 1) < \phi(30n)$ is …, *Amer. Math. Monthly*, **106**(1999) 449–452; *MR* **2000d**:11011.

Kathryn Miller, UMT **25**, *Math. Comput.*, **27**(1973) 447-448.

Donald J. Newman, Euler's ϕ-function on arithmetic progressions, *Amer. Math. Monthly*, **104**(1997) 256–257; *MR* **97m**:11010.

Laurenţiu Panaitopol, On some properties concerning the function $a(n) = n - \phi(n)$, *Bull. Greek Math. Soc.*, **45**(2001) 71–77; *MR* **2003k**:11008.

Laurenţiu Panaitopol, On the equation $n - \phi(n) = m$, *Bull. Math. Soc. Sci. Math. Roumanie* (*N.S.*), **44(92)**(2001) 97–100.

Jan-Christoph Puchta, On the distribution of totients, *Fibonacci Quart.*, **40**(2002) 68–70.

HermanJ. J. te Riele, On the size of solutions of the inequality $\phi(ax + b) < \phi(ax)$, *Public-key cryptography and computational number theory* (*Warsaw*, 2000) 249–255, de Gruyter, Berlin, 2001; *MR* **2003c**:11007.

A. Schinzel, Sur l'équation $\phi(x + k) = \phi(x)$, *Acta Arith.*, **4**(1958) 181–184; *MR* **21** #5597.

A. Schinzel & A. Wakulicz, Sur l'équation $\phi(x + k) = \phi(x)$ II, *Acta Arith.*, **5**(1959) 425–426; *MR* **23** #A831.

W. Sierpiński, Sur un propriété de la fonction $\phi(n)$, *Publ. Math. Debrecen*, **4**(1956) 184–185.

Mladen Vassilev-Missana, The numbers which cannot be values of Euler's function ϕ, *Notes Number Theory Discrete Math.*, **2**(1996) 41–48; *MR* **97m**:11012.

Charles R. Wall, Density bounds for Euler's function, *Math. Comput.*, **26** (1972) 779-783 with microfiche supplement; *MR* **48** #6043.

Masataka Yorinaga, Numerical investigation of some equations involving Euler's ϕ-function, *Math. J. Okayama Univ.*, **20**(1978) 51–58.

Zhang Ming-Zhi, On nontotients, *J. Number Theory*, **43**(1993) 168–173; *MR* **94c**:11004.

OEIS: A000010, A001274, A001494, A003275, A005277-005278, A007015, A007617, A050472-050473, A051953.

B37 Does $\phi(n)$ properly divide $n - 1$?

D. H. Lehmer has conjectured that there is no composite value of n such that $\phi(n)$ is a divisor of $n - 1$, i.e., that for no value of n is $\phi(n)$ a *proper* divisor of $n-1$. Such an n must be a Carmichael number (**A13**). He showed that it would have to be the product of at least seven distinct primes, and Lieuwens has shown that if $3|n$, then $n > 5.5 \cdot 10^{571}$ and $\omega(n) \geq 212$; if the smallest prime factor of n is 5, then $\omega(n) \geq 11$; if the smallest prime factor of n is at least 7, then $\omega(n) \geq 13$. This supersedes and corrects the work of Schuh. Masao Kishore has shown that at least 13 primes are needed in any case, and Cohen & Hagis have improved this to 14. Siva Rama Prasad & Subbarao improve Lieuwens's 212 result to $\omega(n) \geq 1850$ and Hagis to $\omega(n) \geq 298848$. Siva Rama Prasad & Rangamma show that if $3|n$, n composite, $M\phi(n) = n - 1$, $M \neq 4$, then $\omega(n) \geq 5334$.

Pomerance has proved that the number of composite n less than x for which $\phi(n) \mid n - 1$ is

$$O(x^{1/2}(\ln x)^{3/4}(\ln \ln x)^{-1/2})$$

and Shan Zun improved the exponent $\frac{3}{4}$ to $\frac{1}{2}$.

Schinzel notes that if $n = p$ or $2p$, where p is prime, then $\phi(n) + 1$ divides n and asks if the converse is always true. Segal (see paper with Cohen) observes that Schinzel's question reduces to that of Lehmer, that it arises in group theory, and may have been raised by G. Hajós (see Miech's paper, though there it is attributed to Gordon).

Lehmer gives eight solutions to $\phi(n) \mid n+1$, namely $n = 2$, $n = 2^{2^k} - 1$ for $1 \leq k \leq 5$, $n = n_1 = 3 \cdot 5 \cdot 17 \cdot 353 \cdot 929$ and $n = n_1 \cdot 83623937$. [Note that $353 = 11 \cdot 2^5 + 1$, $929 = 29 \cdot 2^5 + 1$, $83623937 = 11 \cdot 29 \cdot 2^{18} + 1$ and $(353-2^8)(929-2^8) = 2^{16}-2^8+1$.] This exhausts the solutions with less than seven factors. Victor Meally notes that $n = n_1 \cdot 83623937 \cdot 699296672132097$ would be a solution were the largest factor a prime, but Peter Borwein notes that this is divisible by 73. The Borweins & Roland Girgensohn conjecture that there are no more solutions.

See Erick Wong's thesis, quoted at **A17**.

If n is prime, it divides $\phi(n)d(n) + 2$. Is this true for any composite n other than $n = 4$? Subbarao also notes that if n is prime, then $n\sigma(n) \equiv 2 \bmod \phi(n)$, and also only if $n = 4$, 6 or 22. Jud McCranie finds no others with $n < 10^{10}$ for either problem.

Subbarao has an analogous conjecture to Lehmer's, based on the function $\phi^{\cdot}(n) = \prod(p^\alpha - 1)$, where the product is taken over the maximal prime power divisors of n, $p^\alpha \| n$. He conjectures that $\phi^{\cdot}(n)|(n-1)$ if and only if n

is a power of a prime. He also has a 'dual' of Lehmer's conjecture, namely that $\psi(n) \equiv 1 \bmod n$ only when n is a prime, where $\psi(n)$ is Dedekind's function (see **B41**). Again, Jud McCranie finds no counterexamples $< 10^{10}$ to either of these conjectures.

Ron Graham makes the following conjecture

¿ For all k there are infinitely many n such that $\phi(n)|(n-k)$?

and he observes that it is true, for example, for $k = 0$, $k = 2^a$ $(a \geq 0)$ and for $k = 2^a 3^b$ $(a, b > 0)$. Pomerance (see *Acta Arith.* paper quoted at **B2**) has treated Graham's problem.

Ronald Alter, Can $\phi(n)$ properly divide $n - 1$? *Amer. Math. Monthly* **80** (1973) 192–193.

G. L. Cohen & P. Hagis, On the number of prime factors of n if $\phi(n)|n - 1$, *Nieuw Arch. Wisk.* (3), **28**(1980) 177–185.

G. L. Cohen & S. L. Segal, A note concerning those n for which $\phi(n) + 1$ divides n, *Fibonacci Quart.*, **27**(1989)285–286.

Aleksander Grytczuk & Marek Wójtowicz, On a Lehmer problem concerning Euler's totient function, *Proc. Japan Acad. Ser. A Math. Sci..* **79**(2003) 136–138.

Masao Kishore, On the equation $k\phi(M) = M - 1$, *Nieuw Arch. Wisk.* (3), **25**(1977) 48–53; see also *Notices Amer. Math. Soc.*, **22**(1975) A501–502.

D. H. Lehmer, On Euler's totient function, *Bull. Amer. Math. Soc.*, **38**(1932) 745–751.

E. Lieuwens, Do there exist composite numbers for which $k\phi(M) = M - 1$ holds? *Nieuw Arch. Wisk.* (3), **18**(1970) 165–169; *MR* **42** #1750.

R. J. Miech, An asymptotic property of the Euler function, *Pacific J. Math.*, **19**(1966) 95–107; *MR* **34** #2541.

Carl Pomerance, On composite n for which $\phi(n)|n-1$, *Acta Arith.*, **28**(1976) 387–389; II, *Pacific J. Math.*, **69**(1977) 177–186; *MR* **55** #7901; see also *Notices Amer. Math. Soc.*, **22**(1975) A542.

József Sándor, On the arithmetical functions $\sigma_k(n)$ and $\phi_k(n)$, *Math. Student*, **58**(1990) 49–54; *MR* **91h**:11005.

Fred. Schuh, Can $n - 1$ be divisible by $\phi(n)$ when n is composite? *Mathematica*, Zutphen B, **12**(1944) 102–107.

V. Siva Rama Prasad & M. Rangamma, On composite n satisfying a problem of Lehmer, *Indian J. Pure Appl. Math.*, **16**(1985) 1244–1248; *MR* **87g**:11017.

V. Siva Rama Prasad & M. Rangamma, On composite n for which $\phi(n)|n-1$, *Nieuw Arch. Wisk.* (4), **5**(1987) 77–81; *MR* **88k**:11008.

M. V. Subbarao, On two congruences for primality, *Pacific J. Math.*, **52**(1974) 261–268; *MR* **50** #2049.

M. V. Subbarao, On composite n satisfying $\psi(n) \equiv 1 \bmod n$,Abstract 882-11-60 *Abstracts Amer. Math. Soc.*, **14**(1993) 418.

M. V. Subbarao, The Lehmer problem on the Euler totient: a Pandora's box of unsolvable problems, *Number theory and discrete mathematics* (*Chandigarh*, 2000) 179–187.

David W. Wall, Conditions for $\phi(N)$ to properly divide $N - 1$, *A Collection of Manuscripts Related to the Fibonacci Sequence*, 18th Anniv. Vol., Fibonacci Assoc., 205–208.

Shan Zun, On composite n for which $\phi(n)|n-1$, *J. China Univ. Sci. Tech.*, **15**(1985) 109–112; *MR* **87h**:11007.

OEIS: A051948.

B38 Solutions of $\phi(m) = \sigma(n)$.

Are there infinitely many pairs of numbers m, n such that $\phi(m) = \sigma(n)$? Since for p prime $\phi(p) = p - 1$ and $\sigma(p) = p + 1$ this question would be answered affirmatively if there were infinitely many twin primes (**A7**). Also if there were infinitely many Mersenne primes (**A3**) $M_p = 2^p - 1$, since $\sigma(M_p) = 2^p = \phi(2^{p+1})$. However there are many solutions other than these, sometimes displaying little noticeable pattern, e.g., $\phi(780) = 192 = \sigma(105)$.

Erdős remarks that the equation $\phi(x) = n!$ is solvable, and (apart from $n = 2$) $\sigma(y) = n!$ is probably solvable also. Charles R. Wall can show that $\psi(n) = n!$ is solvable for $n \neq 2$, where ψ is Dedekind's function (see **B41**).

Jean-Marie De Koninck asks if R is the radical of n, i.e., the greatest squarefree divisor of n, then are $n = 1$ and 1782 the only solutions of $\sigma(n) = R^2$?

Jean-Marie De Koninck, Problem 10966(b), *Amer. Math. Monthly*, **109**(2002) 759.

Le Mao-Hua, A note on primes p with $\sigma(p^m) = z^n$, *Colloq. Math.*, **62**(1991) 193–196.

B39 Carmichael's conjecture.

Carmichael's conjecture. For every n it appears to be possible to find an m, not equal to n, such that $\phi(m) = \phi(n)$ and for a few years early in the last century it was thought that Carmichael had proved this. Klee verified the conjecture for $\phi(n) < 10^{400}$, and for all $\phi(n)$ not divisible by $2^{42} \cdot 3^{47}$. Masai & Valette have raised the bound to 10^{100000}, and Schlafly & Wagon to $10^{10900000}$. Pomerance has shown that if n is such that for every prime p for which $p - 1$ divides $\phi(n)$ we have p^2 divides n, then n is a counterexample. He can also show (unpublished) that if the first k primes $p \equiv 1 \pmod q$ (where q is prime) are all less than q^{k+1}, then there are no numbers n which satisfy his theorem. This also implies the truth of his conjecture that $p_k - 1 | \prod_{i<k} p_i(p_i - 1)$. The truth of this last conjecture for all k also implies that there are no numbers n which satisfy his theorem.

Define the **multiplicity** of an integer as the number of times it occurs as a value of $\phi(n)$. For example, 6 has multiplicity 4 because $\phi(n) = 6$ for $n = 7$, 9, 14, 18 and no other values of n. The multiplicity may be zero (for any odd $n > 1$, and $n = 14$, 26, 34, ...), but not, according to the Carmichael conjecture, equal to one. Sierpiński conjectured that all integers greater than 1 occur as multiplicities and Erdős has shown that if a multiplicity occurs once it occurs infinitely often. Kevin Ford has proved Sierpiński's conjecture. He also showed that if there is a counterexample to Carmichael's conjecture, then a positive proportion of totients are counterexamples.

There are examples of *even* numbers n such that there is no *odd* number m such that $\phi(m) = \phi(n)$. Lorraine Foster has given $n = 33817088 = 2^9 \cdot 257^2$ as the least such.

Erdős proved that if $\phi(x) = k$ has exactly s solutions, then there are infinitely many other k for which there are exactly s solutions, and that $s > k^c$ for infinitely many k. If C is the least upper bound of those c for which this is true, then Wooldridge showed that $C \geq 3 - 2\sqrt{2} > 0.17157$. Pomerance used Hooley's improvement on the Brun–Titchmarsh theorem to improve this to $C \geq 1 - 625/512e > 0.55092$ and notes that further improvements by Iwaniec enable him to get $C > 0.55655$ so that $s > k^{5/9}$ for infinitely many k. Erdős conjectures that $C = 1$. In the other direction Pomerance also shows that

$$s < k \exp\{-(1 + o(1)) \ln k \ln \ln \ln k / \ln \ln k\}$$

and gives a heuristic argument to support the belief that this is best possible.

R. D. Carmichael, Note on Euler's ϕ-function, *Bull. Amer. Math. Soc.*, **28** (1922) 109–110; and see **13**(1907) 241–243.

P. Erdős, On the normal number of prime factors of $p - 1$ and some other related problems concerning Euler's ϕ-function, *Quart. J. Math. Oxford Ser.*, **6**(1935) 205–213.

P. Erdős, Some remarks on Euler's ϕ-function and some related problems, *Bull. Amer. Math. Soc.*, **51**(1945) 540–544.

P. Erdős, Some remarks on Euler's ϕ-function, *Acta Arith.*, 4(1958) 10–19; *MR* **22**#1539.

Kevin Ford, The distribution of totients, *Ramanujan J.*, **2**(1998) 67–151; *MR* **99m**:11106.

Kevin Ford, The distribution of totients, *Electron. Res. Announc. Amer. Math. Soc.*, **4**(1998) 27–34; *MR* **99f**:11125.

Lorraine L. Foster, Solution to problem E3361, *Amer. Math. Monthly* **98** (1991) 443.

Peter Hagis, On Carmichael's conjecture concerning the Euler phi function (Italian summary), *Boll. Un. Mat. Ital.* (6), **A5**(1986) 409–412.

C. Hooley, On the greatest prime factor of $p + a$, *Mathematika*, **20**(1973) 135–143.

Henryk Iwaniec, On the Brun-Tichmarsh theorem and related questions, *Proc. Queen's Number Theory Conf., Kingston, Ont. 1979*, Queen's Papers Pure Appl. Math., **54**(1980) 67–78; *Zbl.* **446**.10036.

V. L. Klee, On a conjecture of Carmichael, *Bull. Amer. Math. Soc.*, **53**(1947) 1183–1186; *MR* **9**, 269.

P. Masai & A. Valette, A lower bound for a counterexample to Carmichael's conjecture, *Boll. Un. Mat. Ital. A* (6), **1** (1982) 313–316; *MR* **84b**:10008.

Carl Pomerance, On Carmichael's conjecture, *Proc. Amer. Math. Soc.*, **43** (1974) 297–298.

Carl Pomerance, Popular values of Euler's function, *Mathematika*, **27** (1980) 84–89; *MR* **81k**:10076.

Aaron Schlafly & Stan Wagon, Carmichael's conjecture on the Euler function is valid below $10^{10,000,000}$, *Math. Comput.*, **63**(1994) 415–419; *MR* **94i**:11008.

M. V. Subbarao & L.-W. Yip, Carmichael's conjecture and some analogues, in *Théorie des Nombres* (Québec, 1987), de Gruyter, Berlin–New York, 1989, 928–941 (and see *Canad. Math. Bull.*, **34**(1991) 401–404.

Alain Valette, Fonction d'Euler et conjecture de Carmichael, *Math. et Pédag.*, Bruxelles, **32**(1981) 13–18.

Stan Wagon, Carmichael's 'Empirical Theorem', *Math. Intelligencer*, **8**(1986) 61-63; *MR* **87d**:11012.

K. R. Wooldridge, Values taken many times by Euler's phi-function, *Proc. Amer. Math. Soc.*, **76**(1979) 229–234; *MR* **80g**:10008.

OEIS: A000010, A000142, A000203, A005100-005101, A014197, A054973, A055486-055489, A055506.

B40 Gaps between totatives.

If $a_1 < a_2 < \ldots < a_{\phi(n)}$ are the integers less than n and prime to it, then Erdős conjectured that $\sum(a_{i+1} - a_i)^2 < cn^2/\phi(n)$ and offered $500.00 for a proof. Hooley showed that, for $1 \le \alpha < 2$, $\sum(a_{i+1} - a_i)^\alpha \ll n(n/\phi(n))^{\alpha-1}$ and that $\sum(a_{i+1} - a_i)^2 \ll n(\ln\ln n)^2$, Vaughan established the conjecture "on the average" and he & Montgomery finally won the prize.

Jacobsthal asked what bounds can be placed on the maximum gap, $J(n) = \max(a_{i+1} - a_i)$. Erdős asks if, for infinitely many x, there are two integers n_1, n_2, $n_1 < n_2 < x$, $n_1 \perp n_2$, with $J(n_1) > \ln x$, $J(n_2) > \ln x$.

In a 95-03-30 letter Bernardo Recamán asks if the set of totatives of every sufficiently large n contains a Pythagorean triple, and whether, for each k, it contains an arithmetic progression of length k.

P. Erdős, On the integers relatively prime to n and on a number-theoretic function considered by Jacobsthal, *Math. Scand.*, **10**(1962) 163–170; *MR* **26** #3651.

R. R. Hall & P. Shiu, The distribution of totatives, *Cand. Math. Bull.*, **45**(2002); *MR* **2003a**:11005.

C. Hooley, On the difference of consecutive numbers prime to n, *Acta Arith.*, **8**(1962/63) 343–347; *MR* **27** #5741.

H. L. Montgomery & R. C. Vaughan, On the distribution of reduced residues, *Ann. of Math.* (2), **123**(1986) 311–333; *MR* **87g**:11119.

R. C. Vaughan, Some applications of Montgomery's sieve, *J. Number Theory*, **5**(1973) 64–79.

R. C. Vaughan, On the order of magnitude of Jacobsthal's function, *Proc. Edinburgh Math. Soc.*(2), **20**(1976/77) 329–331; *MR* **56** #11937.

B41 Iterations of ϕ and σ.

Pomerance asks if, for each positive integer n, there is a positive integer k such that $\sigma^k(n)/n$ is an integer. E.g., $(n, k) = (1,1)$, $(2,2)$, $(3,4)$, $(4,2)$, $(5,5)$, $(6,4)$, $(7,5)$, ...

There is a close relative to the sum of divisors and the sum of the unitary divisors function, which complements Euler's totient function and which is often named for Dedekind. If $n = p_1^{a_1} p_2^{a_2} \ldots p_k^{a_k}$, denote by $\psi(n)$ the product $\prod p_i^{a_i-1}(p_i + 1)$, i.e., $\psi(n) = n \prod(1 + p^{-1})$, where the product is taken over the distinct prime divisors of n. It is easy to see that iteration of the function leads eventually to terms of the form $2^a 3^b$ where b is fixed and a increases by one in successive terms. Given any value of b there are infinitely many values of n which lead to such terms, for example, $\psi^k(2^a 3^b 7^c) = 2^{a+4k} 3^b 7^{c-k}$ $(0 \le k \le c)$ and $\psi^k(2^a 3^b 7^c) = 2^{a+5k-c} 3^b$ $(k > c)$.

That there are values of n for which the iterates of the function $\psi(n) - n$ are unbounded as the number of iterations tends to infinity is the subject of te Riele's thesis (reference at **B8**); the least such n is 318.

If we average ψ with the ϕ-function, $\frac{1}{2}(\phi + \psi)$, and iterate, we produce sequences whose terms become constant whenever they are prime powers; for example 24, $\frac{1}{2}(8+48) = 28$, $\frac{1}{2}(12+48) = 30$, $\frac{1}{2}(8+72) = 40$, $\frac{1}{2}(16+72) = 44$, $\frac{1}{2}(20+72) = 46$, $\frac{1}{2}(22+72) = 47$, $\frac{1}{2}(46+48) = 47$, ... Charles R. Wall gives examples where iteration leads to an unbounded sequence: start with 45, 48, ... or 50, 55, ... and continue 56, 60, 80, 88, 92, 94, 95, 96, ... ; each term after the 35th is the double of the last but seven!

We can also average the σ- and ϕ-functions, and iterate. Since $\phi(n)$ is always even for $n > 2$ and $\sigma(n)$ is odd when n is a square or twice a square, we will sometimes get a noninteger value. For example, 54, 69, 70, 84, 124, 142, 143 ,144, $225\frac{1}{2}$; in this case we say that the sequence **fractures**. It is easy to show that $(\sigma(n)+\phi(n))/2 = n$ just if $n = 1$ or a prime, so sequences can become constant, for example, 60, 92, 106, 107, 107, ... Are there sequences which increase indefinitely without fracturing?

Of course, if we iterate the ϕ-function, it eventually arrives at 1. Call the least integer k for which $\phi^k(n) = 1$ the **class** of n.

k	n												
1						2							
2	3					4	6						
3	5	7				8	9	10	12	14	18		
4	11	13	15			16	19	20	21	22	24	26	...
5	17	23	25	29	31	32	33	34	35	37	39	40	43...
6	41	47	51	53	55	59	61	64	65	67	68	69	71 73 ...
7	83 85 89 97 101 103 107 113 115 119 121 122 123 125 128 ...												

The set of least values of the classes is $M = \{2, 3, 5, 11, 17, 41, 83, \dots\}$.
Shapiro conjectured that M contained only prime values, but Mills found
several composite members. If S is the union, for all k, of the members of
class k which are $< 2^k$, then
$$S = \{3; 5, 7; 11, 13, 15; 17, 23, 25, 29, 31; 41, 47, 51, 53, 55, 59, 61; 83, 85, \dots\}$$
and Shapiro showed that the factors of an element of S is also in S. Catlin
showed that if m is an odd element of M, then the factors of M are in M,
and that there are finitely many primes in M just if there are finitely many
odd numbers in M. Does S contain infinitely many odd numbers? Does
M contain infinitely many odd numbers?

Pillai showed that that the class, $k = k(n)$, of n satisfies

$$\left\lfloor \frac{\ln n}{\ln 3} \right\rfloor \le k(n) \le \left\lfloor \frac{\ln n}{\ln 2} \right\rfloor$$

and it's easy to see (look at $2^a 3^b$) that $k(n)/\ln n$ is dense in the interval
$[1/\ln 3, 1/\ln 2]$. What is the average and normal behavior of $k(n)$? Erdős,
Granville, Pomerance & Spiro conjecture that there is a constant α such
that the normal order of $k(n)$ is $\alpha \ln n$ and prove this under the assumption
of the Elliott-Halberstam conjecture. They also showed that the normal
order of $\phi^h(n)/\phi^{h+1}(n)$ is $h e^\gamma \ln \ln \ln n$ for each positive integer h, where
γ is Eulers constant@Euler's constant. See their paper for many unsolved
problems: for example, if $\sigma^k(n)$ is the kth iterate of the sum of divisors function, they are unable to prove or disprove any of the following statements.

¿   for every $n > 1$, $\sigma^{k+1}(n)/\sigma^k(n) \to 1$ as $k \to \infty$?

¿   for every $n > 1$, $\sigma^{k+1}(n)/\sigma^k(n) \to \infty$ as $k \to \infty$?

¿   for every $n > 1$, $\left(\sigma^k(n)\right)^{1/k} \to \infty$ as $k \to \infty$?

¿   for every $n > 1$, there is some k with $n | \sigma^k(n)$?

¿   for every $n, m > 1$, there is some k with $m | \sigma^k(n)$?

¿   for every $n, m > 1$, there are some k, l with $\sigma^k(m) = \sigma^l(n)$?

Miriam Hausman has characterized those integers n which are solutions
of the equation $n = m\phi^k(n)$; they are mainly of the form $2^a 3^b$.

Finucane iterated the function $\phi(n) + 1$ and asked: in how many steps
does one reach a prime? Also, given a prime p, what is the distribution
of the values of n whose sequences end in p? Are 5, 8, 10, 12 the only
numbers which lead to 5? And 7, 9, 14, 15, 16, 18, 20, 24, 30 the only ones
leading to 7?

Erdős similarly asked about the iteration of $\sigma(n) - 1$. Does it always end on a prime, or can it grow indefinitely? In none of the cases of iteration of $\sigma(n) - 1$, of $(\psi(n) + \phi(n))/2$, or of $(\phi(n) + \sigma(n))/2$ is he able to show that the growth is slower than exponential. For several results and conjectures, consult the quadruple paper cited below.

Atanassov defines some additive analogs of ϕ and σ, poses 17 questions and answers only three of them.

Iannucci, Moujie & Chen define **perfect totient numbers** as those n for which $n = \phi(n) + \phi^2(n) + \phi^3(n) + \cdots + 2 + 1$. All powers of 3 are perfect totient numbers. They find 30 others $< 5 \cdot 10^9$ of which the largest is 4764161215.

Krassimir T. Atanassov, New integer functions, related to ϕ and σ functions, *Bull. Number Theory Related Topics*, **11**(1987) 3–26; *MR* **90j**:11007.

P. A. Catlin, Concerning the iterated ϕ-function, *Amer. Math. Monthly*, **77**(1970) 60–61.

P. Erdős, A. Granville, C. Pomerance & C. Spiro, On the normal behavior of the iterates of some arithmetic functions, in Berndt, Diamond, Halberstam & Hildebrand (editors) Analytic Number Theory, *Proc. Conf. in honor P.T. Bateman, Allerton Park, 1989*, Birkhäuser, Boston, 1990, 165–204; *MR* **92a**:11113.

P. Erdős, Some remarks on the iterates of the ϕ and σ functions, *Colloq. Math.*, **17**(1967) 195–202; *MR* **36** #2573.

Paul Erdős & R. R. Hall, Euler's ϕ-function and its iterates, *Mathematika*, **24**(1977) 173–177; *MR* **57** #12356.

Miriam Hausman, The solution of a special arithmetic equation, *Canad. Math. Bull.*, **25**(1982) 114–117.

Douglas E. Iannucci, Deng Moujie & Graeme L. Cohen, On perfect totient numbers, preprint, 2003.

H. Maier, On the third iterates of the ϕ- and σ-functions, *Colloq. Math.*, **49**(1984) 123–130; *MR* **86d**:11006.

W. H. Mills, Iteration of the ϕ-function, *Amer. Math. Monthly* **50**(1943) 547–549; *MR* **5**, 90.

C. A. Nicol, Some diophantine equations involving arithmetic functions, *J. Math. Anal. Appl.*, **15**(1966) 154–161.

Ivan Niven, The iteration of certain arithmetic functions, *Canad. J. Math.*, **2**(1950) 406–408; *MR* **12**, 318.

S. S. Pillai, On a function connected with $\phi(n)$, *Bull. Amer. Math. Soc.*, **35**(1929) 837–841.

Carl Pomerance, On the composition of the arithmetic functions σ and ϕ, *Colloq. Math.*, **58**(1989) 11–15; *MR* **91c**:11003.

Harold N. Shapiro, An arithmetic function arising from the ϕ-function, *Amer. Math. Monthly* **50**(1943) 18–30; *MR* **4**, 188.

Charles R. Wall, Unbounded sequences of Euler-Dedekind means, *Amer. Math. Monthly* **92**(1985) 587.

Richard Warlimont, On the iterates of Euler's function, *Arch. Math. (Basel)*, **76**(2001) 345–349; *MR* **2002k**:11167.

OEIS: A000010, A003434, A005239, A007755, A019268, A040176, A049108.

B42 Behavior of $\phi(\sigma(n))$ and $\sigma(\phi(n))$.

Erdős asks us to prove that $\phi(n) > \phi(n - \phi(n))$ for almost all n, but that $\phi(n) < \phi(n - \phi(n))$ for infinitely many n.

Mąkowski & Schinzel prove that $\limsup \phi(\sigma(n))/n = \infty$,

$$\limsup \phi(\phi(n))/n = \frac{1}{2}, \quad \text{and} \quad \liminf \sigma(\phi(n))/n \leq \frac{1}{2} + \frac{1}{2^{34} - 4}$$

and they ask if $\sigma(\phi(n))/n \geq \frac{1}{2}$ for all n. They point out that even $\inf \sigma(\phi(n))/n > 0$ is not proved, but Pomerance has since established this, using Bruns method@Brun's method. Graeme Cohen and later Segal each thought that he had proved the main result, but it remains open.

John Selfridge, Fred Hoffman & Rich Schroeppel found 24 solutions of $\phi(\sigma(n)) = n$, namely

$$2^k \text{ for } k = 0, 1, 3, 7, 15, 31; \qquad 2^2 \cdot 3;\ 2^8 \cdot 3^3;\ 2^{10} \cdot 3^3 \cdot 11^2;$$
$$2^{12} \cdot 3^3 \cdot 5 \cdot 7 \cdot 13;\ 2^4 \cdot 3 \cdot 5;\ 2^4 \cdot 3^2 \cdot 5;\ 2^9 \cdot 3 \cdot 5^2 \cdot 31;\ 2^9 \cdot 3^2 \cdot 5^2 \cdot 31;$$
$$2^5 \cdot 3^4 \cdot 5 \cdot 11;\ 2^5 \cdot 3^4 \cdot 5^2 \cdot 11;\ 2^8 \cdot 3^4 \cdot 5 \cdot 11;\ 2^8 \cdot 3^4 \cdot 5^2 \cdot 11;\ 2^5 \cdot 3^6 \cdot 7^2 \cdot 13;$$
$$2^6 \cdot 3^6 \cdot 7^2 \cdot 13;\ 2^{13} \cdot 3^7 \cdot 5 \cdot 7^2;\ 2^{13} \cdot 3^7 \cdot 5^2 \cdot 7^2;\ 2^{21} \cdot 3^3 \cdot 5 \cdot 11^3 \cdot 31;$$
$2^{21} \cdot 3^3 \cdot 5^2 \cdot 11^3 \cdot 31$; and there are, of course, 24 corresponding solutions of $\sigma(\phi(m)) = m$.

Terry Raines, in January 1995, found ten further solutions:

$$2^8 \cdot 3^6 \cdot 7^2 \cdot 13;\ 2^9 \cdot 3^4 \cdot 5^2 \cdot 11^2 \cdot 31;\ 2^{13} \cdot 3^3 \cdot 5^4 \cdot 7^3;\ 2^{13} \cdot 3^8 \cdot 5 \cdot 7^3;$$
$$2^{13} \cdot 3^6 \cdot 5 \cdot 7^3 \cdot 13;\ 2^{13} \cdot 3^3 \cdot 5^4 \cdot 7^4;\ 2^{13} \cdot 3^8 \cdot 5^2 \cdot 7^3;\ 2^{13} \cdot 3^6 \cdot 5^2 \cdot 7^3 \cdot 13;$$
$$2^{21} \cdot 3^5 \cdot 5 \cdot 11^3 \cdot 31;\ 2^{21} \cdot 3^5 \cdot 5^2 \cdot 11^3 \cdot 31;$$

and on Independence Day, 1996, Graeme Cohen added

$$2^{13} \cdot 3^5 \cdot 5^4 \cdot 7^4 \quad \text{and} \quad 2^{24} \cdot 3^9 \cdot 5^7 \cdot 11 \cdot 13$$

Are there others? An infinite number?

Golomb observes that if $q > 3$ and $p = 2q - 1$ are primes and $m \in \{2, 3, 8, 9, 15\}$, then $n = pm$ is a solution of $\phi(\sigma(n)) = \phi(n)$. Undoubtedly there are infinitely many such and undoubtedly no one will prove this in the foreseeable future. There are other solutions, 1, 3, 15, 45, ... ; an infinite number? He gives the solutions 1, 87, 362, 1257, 1798, 5002, 9374 to $\sigma(\phi(n)) = \sigma(n)$. He also notes that if p and $(3^p - 1)/2$ are primes (e.g., $p = 3, 7, 13, 71, 103$), then $n = 3^{p-1}$ is a solution of $\sigma(\phi(n)) = \phi(\sigma(n))$; and shows that $\sigma(\phi(n)) - \phi(\sigma(n))$ is both positive and negative infinitely often and asks what is the proportion of each?

Walter Nissen found at least 14000 solutions of $\sigma(\phi(n)) = \phi(\sigma(n))$. The twenty values $< 10^5$ are 1, 9, 225, 242, 516, 729, 3872, 13932, 14406, 17672, 18225, 20124, 21780, 29262, 29616, 45996, 65025, 76832, 92778 and 95916.

There are numerous questions that one may ask about these two roughly dual functions. Zhang Ming-Zhi notes that if n is prime, then it divides $\phi(n) + \sigma(n)$, but if $n = p^\alpha$ with $\alpha > 1$ or $n = p^\alpha q$ with p, q distinct primes, then this is not so. He finds 17 composite n less than 10^7 which divide $\phi(n) + \sigma(n)$: 312, 560, 588, 1400, 23760, 59400, 85632, 147492, 153720, 556160, 569328, 1590816, 2013216 and four others.

U. Balakrishnan, Some remarks on $\sigma(\phi(n))$, *Fibonacci Quart.*, **32**(1994) 293–296; *MR* **95j**:11091.

Cao Fen-Jin, The composite number-theoretic function $\sigma(\phi(n))$ and its relation to n, *Fujian Shifan Daxue Xuebao Ziran Kexue Ban*, **10**(1994) 31–37; *MR* **96c**:11008.

Graeme L. Cohem, On a conjecture of Mąkowski and Schinzel, *Colloq. Math.*, **74**(1997) 1–8; *MR* **98e**:11006.

P. Erdős, Problem P. 294, *Canad. Math. Bull.*, **23**(1980) 505.

Kevin Ford & Sergei Konyagin, On two conjectures of Sierpiński concerning the arithmetic functions σ and ϕ, *Number Theory in Progress, Vol. 2* (*Zakopane-Kościelisko*, 1997) 795–803, de Gruyter, Berlin, 1999; *MR* **2000d**:11120.

Kevin Ford, Sergei Konyagin & Carl Pomerance, Residue classes free of values of Euler's function, *Number Theory in Progress, Vol. 2* (*Zakopane-Kościelisko*, 1997) 805–812, de Gruyter, Berlin, 1999; *MR* **2000f**:11120.

Solomon W. Golomb, Equality among number-theoretic functions, preprint, Oct 1992; Abstract 882-11-16, *Abstracts Amer. Math. Soc.*, **14**(1993) 415–416.

A. Grytczuk, F. Luca & M. Wójtowicz, A conjecture of Erdős concerning inequalities for the Euler totient function, *Publ. Math. Debrecen*, **59**(2001) 9–16; *MR* **2002f**:11005.

A. Grytczuk, F. Luca & M. Wójtowicz, Some results on $\sigma(\phi(n))$, *Indian J. Math.*, **43**(2001) 263–275; *MR* **2002k**:11166.

Richard K. Guy, Divisors and desires, *Amer. Math. Monthly*, **104**(1997) 359–360.

Pentti Haukkanen, On an inequality for $\sigma(n)\phi(n)$, *Octogon Math. Mag.*, **4**(1996) 3–5.

Lin Da-Zheng & Zhang Ming-Zhi, On the divisibility relation $n|(\phi(n)+\sigma(n))$, *Sichuan Daxue Xuebao*, **34**(1997) 121–123; *MR* **98d**:11010.

A. Mąkowski & A. Schinzel, On the functions $\phi(n)$ and $\sigma(n)$, *Colloq. Math.*, **13**(1964–65) 95–99; *MR* **30** #3870.

Carl Pomerance, On the composition of the arithmetic functions σ and ϕ, *Colloq. Math.*, **58**(1989) 11–15; *MR* **91c**:11003.

József Sándor, On Dedekind's arithmetical function, *Seminarul de Teoria Structurilor, Univ. Timişoara*, **51**(1988) 1–15.

József Sándor, On the composition of some arithmetic functions, *Studia Univ. Babeş-Bolyai Math.*, **34**(1989) 7–14; *MR* **91i**:11008.

József Sándor & R. Sivaramakrishnan, The many facets of Euler's totient. III. An assortment of miscellaneous topics, *Nieuw Arch. Wisk.*, **11**(1993) 97–130; *MR* **94i**:11007.

Robert C. Vaughan & Kevin L. Weis, On sigma-phi numbers, *Mathematika*, **48**(2001) 169–189 (2003).

Zhang Ming-Zhi, A divisibility problem (Chinese), *Sichuan Daxue Xuebao*, **32**(1995) 240–242; *MR* **97g**:11003.

OEIS: A000010, A000203, A001229, A006872, A033631, A051487-051488, A066831, A067382-0670385.

B43 Alternating sums of factorials.

The numbers
$$3! - 2! + 1! = 5,$$
$$4! - 3! + 2! - 1! = 19,$$
$$5! - 4! + 3! - 2! + 1! = 101,$$
$$6! - 5! + 4! - 3! + 2! - 1! = 619,$$
$$7! - 6! + 5! - 4! + 3! - 2! + 1! = 4421,$$
and
$$8! - 7! + 6! - 5! + 4! - 3! + 2! - 1! = 35899$$

are each prime. Are there infinitely many such? Here are the factors of $A_n = n! - (n-1)! + (n-2)! - + \ldots - (-1)^n 1!$ for the next few values of n :

n	A_n	n	A_n
9	$79 \cdot 4139$	19	15578717622022981 (prime)
10	3301819 (prime)	20	$8969 \cdot 210101 \cdot 1229743351$
11	$13 \cdot 2816537$	21	$113 \cdot 167 \cdot 4511191 \cdot 572926421$
12	$29 \cdot 15254711$	22	$79 \cdot 239 \cdot 569947572104043899$
13	$47 \cdot 1427 \cdot 86249$	23	$85439 \cdot 289993909455734779$
14	$211 \cdot 1679 \cdot 229751$	24	$12203 \cdot 24281 \cdot 2010359484638233$
15	1226280710981 (prime)	25	$59 \cdot 555307 \cdot 4552540056662640637$
16	$53 \cdot 6581 \cdot 56470483$	26	$1657 \cdot 23438498653915383253 8067$
17	$47 \cdot 7148742955723$	27	$127^2 \cdot 271 \cdot 1163 \cdot 2065633479970130593$
18	$2683 \cdot 2261044646593$	28	$61 \cdot 221171 \cdot 21820357757749410439949$

The example $n = 27$ shows that these numbers are not necessarily square-free. Wilfrid Keller has continued the calculations for $n \leq 335$; A_n is prime for $n = 41, 59, 61, 105$ and 160.

If there is a value of n such that $n+1$ divides A_n, then $n+1$ will divide A_m for all $m > n$, and there would be only a finite number of prime values. This problem has been answered by Miodrag Živkovič, who has shown that $p = 3612703$ divides $A(n)$ for all $n \geq p$.

There are questions if $0!$ is included. The numbers are now even, and only $2! - 1! + 0! = 2$ is prime. Kevin Buzzard reported that a teenage friend asked for which n does n divide $A_n + (-1)^n$? If $\mathcal{S}$ is the set of such n, then $a \in \mathcal{S}$, $b \in \mathcal{S}$, $a \perp b$ imply $ab \in \mathcal{S}$ and $c \in \mathcal{S}$, $d \mid c$ imply $d \in \mathcal{S}$. So one needs only search for prime powers. Buzzard found only $2, 4, 5, 13, 37,$ 463 less than 160000, and asks: is $\mathcal{S}$ just the set of divisors of 4454060?

Miodrag Živkovič, The number of primes $\sum_{i=1}^{n}(-1)^{n-i}i!$ is finite, *Math. Comput.*, **68**(1999) 403–409; *MR***99c**:11163.

OEIS: A000142, A001272, A002981-002982, A003422, A005165.

B44 Sums of factorials.

Ð. Kurepa defined $!n = 0! + 1! + 2! + \ldots + (n-1)!$ and asks if $!n \not\equiv 0 \bmod n$ for all $n > 2$. Slavić established this for $3 \leq n \leq 1000$. The conjecture is that $(!n, n!) = 2$. Wagstaff verified the conjecture for $n < 50000$, and Mijajlović and Gogić independently for $n \leq 10^6$. Mijajlović notes that for $K_n = !(n+1) - 1 = 1! + 2! + \ldots + n!$ we have $3|K_n$ for $n \geq 2$, $9|K_n$ for $n \geq 5$ and $99|K_n$ for $n \geq 10$. Wilfrid Keller has since extended this and found no new divisibilities for K_n with $n < 10^6$. In a 91-03-21 letter, and a February 1998 preprint, Reg. Bond offered an as yet unpublished proof of the conjecture.

Miodrag Živkovič (see ref. at **B43**) has shown that $(54503)^2$ divides $!26540$, so that $!n$ is not always squarefree for $n > 3$.

L. Carlitz, A note on the left factorial function, *Math. Balkanika*, **5**(1975) 37–42.

Goran Gogić, Kurepa's hypothesis on the left factorial (Serbian), *Zb. Rad. Mat. Inst. Beograd. (N.S.)* **5(13)**(1993) 41–45.

Aleksandar Ivić & Žarko Mijajlović, On Kurepa's problems in number theory, Ðuro Kurepa memorial volume, *Publ. Inst. Math. (Beograd) (N.S.)* **57(71)**(1995) 19–28; *MR* **97a**:11007.

Winfried Kohnen, A remark on the left-factorial hypothesis, *Univ. Beograd. Publ. Elektrotehn. Fak. Ser. Mat.*, **9**(1998) 51–53; *MR* **99m**:11005.

Ðuro Kurepa, On some new left factorial propositions, *Math. Balkanika*, **4**(1974) 383–386; *MR* **58** #10716.

Ž. Mijajlović, On some formulas involving $!n$ and the verification of the $!n$-hypothesis by use of computers, *Publ. Inst. Math. (Beograd) (N.S.)* **47(61)**(1990) 24–32; *MR* **92d**:11134.

Alexandar Petojević, On Kurepa's hypothesis for the left factorial, *Filomat* No. 12, part 1 (1998) 29–37.

Zoran Šami, On generalization of functions $n!$ and $!n$, *Publ. Inst. Math. (Beograd)(N.S.)* **60(74)**(1996) 5–14; *MR* **98a**:11006.

Zoran Šami, A sequence $u_{n,m}$ and Kurepa's hypothesis on left factorial, Sympos. dedicated to memory of Ðuro Kurepa, Belgrade 1996, *Sci. Rev. Ser. Sci. Eng.*, **19-20**(1996) 105–113; *MR* **98b**:11016.

V. S. Vladimirov, Left factorials, Bernoulli numbers, and the Kurepa conjecture, *Publ. Inst. Math. Beograd (N.S.)*, **72(86)**(2002) 11–22.

OEIS: A000142, A003422, A005165, A007489, A014144, A049782.

B45 Euler numbers.

The coefficients in the expansion of $\sec x = \sum E_n(ix)^n/n!$ are the **Euler numbers**, and arise in several combinatorial contexts. $E_0 = 1$, $E_2 = -1$, $E_4 = 5$, $E_6 = -61$, $E_8 = 1385$, $E_{10} = -50521$, $E_{12} = 2702765$, $E_{14} = -199360981$, $E_{16} = 19391512145$, $E_{18} = -2404879675441$, ... Is it true that for any prime $p \equiv 1 \bmod 8$, $E_{(p-1)/2} \not\equiv 0 \bmod p$? Is it true for $p \equiv 5 \bmod 8$?

V. I. Arnol'd, Bernoulli-Euler updown numbers associated with function singularities, their combinatorics and arithmetics, *Duke Math. J.*, **63**(1991) 537–555; *MR* **93b**:58020.

V. I. Arnol'd, Snake calculus and the combinatorics of the Bernoulli, Euler and Springer numbers of Coxeter groups, *Uspekhi Mat. Nauk*, **47**(1992) 3–45; transl. in *Russian Math. Surveys*, **47**(1992) 1–51; *MR* **93h**:20042.

E. Lehmer, On congruences involving Bernoulli numbers and the quotients of Fermat and Wilson, *Annals of Math.* **39**(1938) 350–360; *Zbl.* **19**, 5.

Barry J. Powell, Advanced problem 6325, *Amer. Math. Monthly* **87**(1980) 826.

B46 The largest prime factor of n.

Erdős denotes by $P(n)$ the largest prime factor of n and asks if there are infinitely many primes p such that $(p-1)/P(p-1) = 2^k$? Or $= 2^k \cdot 3^l$?

If $n > 2$, then $P(n)$, $P(n+1)$, $P(n+2)$ are all distinct. Show that each of the six permutations of {low, medium, high} occurs infinitely often, and that they occur with equal frequency. $2^k - 2$, $2^k - 1$, 2^k show that medium, high, low occurs for infinitely many k because $P(2^k - 1) \to \infty$ as $k \to \infty$ by a theorem of Bang (or Mahler). To see that low, medium high occurs infinitely often, ask if $p - 1$, p, $p + 1$ works for p prime. No! Try $p^2 - 1$, p^2, $p^2 + 1$. Maybe. If $P(p^2 + 1) < p$, try $p^4 - 1$, p^4, $p^4 + 1$. Eventually, for each prime p, there will be a value of k such that $P(p^{2^k} + 1) > p$.

Selfridge settled the low, high, medium case with 2^k, $2^k + 1$, $2^k + 2$ and Tijdeman gave the following argument for medium, low, high: consider the possibilities $2^k - 1$, 2^k, $2^k + 1$; $2^{2k} - 1$, 2^{2k}, $2^{2k} + 1$; $2^{4k} - 1$, 2^{4k}, $2^{4k} + 1$; ....

Mabkhout showed that $P(n^4 + 1) \geq 137$ for all $n > 3$; he used a classical result of Størmer, quoted at **D10**.

P. Erdős & Carl Pomerance, On the largest prime factors of n and $n + 1$, *Aequationes Math.*, **17**(1978) 311–321; *MR* **58** #476.

M. Mabkhout, Minoration de $P(x^4 + 1)$, *Rend. Sem. Fac. Sci. Univ. Cagliari*, **63**(1993) 135–148; *MR* **96e**:11039.

B47 When does $2^a - 2^b$ divide $n^a - n^b$?

Selfridge notices that $2^2 - 2$ divides $n^2 - n$ for all n, that $2^{2^2} - 2^2$ divides $n^{2^2} - n^2$ and $2^{2^{2^2}} - 2^{2^2}$ divides $n^{2^{2^2}} - n^{2^2}$ and asks for what a and b does $2^a - 2^b$ divide $n^a - n^b$ for all n. The case $n = 3$ was proposed as E2468*, *Amer. Math. Monthly*, **81**(1974) 405 by Harry Ruderman. In his solution (**83**(1976) 288–289) Bill Vélez omits $(b, a - b) = (0, 1)$ as trivial and gives 13 other solutions, (1,1), (1,2), (2,2), (3,2), (1,4), (2,4), (3,4), (4,4), (2,6), (3,6), (2,12), (3,12), (4,12). Remarks by Pomerance (**84**(1977) 59–60) show that results of Schinzel complete Vélez's solution. The problem was also solved by Sun Qi & Zhang Ming Zhi.

A. Schinzel, On primitive prime factors of $a^n - b^n$, *Proc. Cambridge Philos. Soc.*, **58**(1962) 555–562.

Sun Qi & Zhang Ming-Zhi, Pairs where $2^a - 2^b$ divides $n^a - n^b$ for all n, *Proc. Amer. Math. Soc.*, **93**(1985) 218–220; *MR* **86c**:11004.

Marian Vâjâitu & Alexandru Zaharescu, A finiteness theorem for a class of exponential congruences, *Proc. Amer. Math. Soc.*, **127**(1999) 2225–2232; *MR* **99j**:11003.

B48 Products taken over primes.

David Silverman noticed that if p_n is the n-th prime, then

$$\prod_{n=1}^{m} \frac{p_n + 1}{p_n - 1}$$

is an integer for $m = 1, 2, 3, 4$ and 8 and asked is it ever again an integer? Equivalently, as Mąkowski observes (reference at **B16**), for what $n = \prod_{r=1}^{m} p_r$ does $\phi(n)$ divide $\sigma(n)$? For example, if $\sigma(n) = 4\phi(n)$ then $2n$ is either perfect or abundant, $\sigma(2n) \geq 4n$. Jud McCranie checked that the product is not an integer for $8 < m \leq 98222287$, i.e for primes p, $23 \leq p < 2 \cdot 10^9$. He notes the connexion with Sophie Germain primes (primes p such that $2p + 1$ is also prime) and with Cunningham chains (**A7**).

Wagstaff asked for an elementary proof (e.g., without using properties of the Riemann zeta-function@Riemann ζ-function) that

$$\prod \frac{p^2 + 1}{p^2 - 1} = \frac{5}{2}$$

where the product is taken over all primes. It seems very unlikely that there is a proof which doesn't involve analytical methods. At first glance it might appear that the fractions might cancel, but none of the numerators are divisible by 3. Euler's proof is

$$\prod \frac{p^2+1}{p^2-1} = \prod \frac{p^4-1}{(p^2-1)^2} = \prod \frac{1-p^{-4}}{(1-p^{-2})^2} = \frac{\zeta^2(2)}{\zeta(4)} = \frac{(\pi^2/6)^2}{\pi^4/90} = \frac{5}{2}.$$

This uses $\sum n^{-k} = \prod(1-p^{-k})^{-1}$ and $\sum n^{-2} = \pi^2/6$ and $\sum n^{-4} = \pi^4/90$.
Wagstaff regards the first as elementary, but not the latter two. He would
like to see a direct proof of $2(\sum n^{-2})^2 = 5\sum n^{-4}$ or of

$$4\sum_{n=1}^{\infty} \frac{1}{n^2} \sum_{m=n+1}^{\infty} \frac{1}{m^2} = 3\sum \frac{1}{n^4}$$

David Borwein & Jonathan M. Borwein, On an intriguing integral and
some series related to $\zeta(4)$, *Proc. Amer. Math. Soc.*, **123**(1995) 1191–1198;
MR **95e**:11137.

B49 Smith numbers.

Albert Wilansky named **Smith numbers** from his brother-in-law's tele-
phone number

$$4937775 = 3 \cdot 5 \cdot 5 \cdot 65837,$$

the sum of whose digits is equal to the sum of the digits of its prime
factors, and they soon caught the public fancy. Trivially, any prime is a
Smith number: so are 4, 22, 27, 58, 85, 94, 121, ... Oltikar & Wayland
gave the examples $3304(10^{317}-1)/9$ and $2 \cdot 10^{45}(10^{317}-1)/9$ and the race
to find larger and larger Smith numbers was on. Yates has given

$$10^{3913210}(10^{1031}-1)(10^{4594}+3 \cdot 10^{2297}+1)^{1476}$$

with 10694985 decimal digits, but has since beaten his own record with a
13614513-digit Smith number.

Patrick Costello, A new largest Smith number, *Fibonacci Quart.*, **40**(2002)
369–371.
Patrick Costello & Kathy Lewis, Lots of Smiths, *Math. Mag.*, **75**(2002) 223–
226.
Stephen K. Doig, Math Whiz makes digital discovery, *The Miami Herald*,
1986-08-22; *Coll. Math. J.*, **18**(1987) 80.
Editorial, Smith numbers ring a bell? *Fort Lauderdale Sun Sentinel*, 86-09-16,
p. 8A.
Editorial, Start with 4,937,775, *New York Times*, 86-09-02.
Kathy Lewis, Smith numbers: an infinite subset of **N**, M.S. thesis, Eastern
Kentucky Univ., 1994.
Wayne L. McDaniel, The existence of infinitely many k-Smith numbers, *Fi-
bonacci Quart.*, **25**(1987) 76–80.

Wayne L. McDaniel, Powerful k-Smith numbers, *Fibonacci Quart.*, **25**(1987) 225–228.

Wayne L. McDaniel, Palindromic Smith numbers, *J. Recreational Math.*, **19**(1987) 34–37.

Wayne L. McDaniel, Difference of the digital sums of an integer base b and its prime factors, *J. Number Theory*, **31**(1989) 91–98; *MR* **90e**:11021.

Wayne L. McDaniel & Samuel Yates, The sum of digits function and its application to a generalization of the Smith number problem, *Nieuw Arch. Wisk.*(4), **7**(1989) 39–51.

Sham Oltikar & Keith Wayland, Construction of Smith numbers, *Math. Mag.*, **56**(1983) 36–37.

Ivars Peterson, In search of special Smiths, *Science News*, 86-08-16, p. 105.

Michael Smith, Cousins of Smith numbers: Monica and Suzanne sets, *Fibonacci Quart.*, **34**(1996) 102–104; *MR* **97a**:11020.

A. Wilansky, Smith numbers, *Two-Year Coll. Math. J.*, **13**(1982) 21.

Brad Wilson, For $b \geq 3$ there exist infinitely many base b k-Smith numbers, *Rocky Mountain J. Math.*, **29**(1999) 1531–1535; *MR* **2000k**:11014.

Samuel Yates, Special sets of Smith numbers, *Math. Mag.*, **59**(1986) 293–296.

Samuel Yates, Smith numbers congruent to 4 (mod 9), *J. Recreational Math.*, **19**(1987) 139–141.

Samuel Yates, How odd the Smiths are, *J. Recreational Math.*, **19**(1987) 168–174.

Samuel Yates, Digital sum sets, in R. A. Mollin (ed.), Number Theory, *Proc. 1st Canad. Number Theory Assoc. Conf., Banff, 1988*, de Gruyter, New York, 1990, pp. 627–634; *MR* **92c**:11008.

Samuel Yates, Tracking titanics, in R. K. Guy & R. E. Woodrow (eds.), The Lighter Side of Mathematics, *Proc. Strens Mem. Conf., Calgary*, 1986, Spectrum Series, Math. Assoc. of America, Washington DC, 1994.

OEIS: A006753, A019506, A059754.

B50 Ruth-Aaron numbers.

When Hank Aaron's 715th home run beat Babe Ruth's record of 714, Carl Pomerance noted that $714 = 2 \cdot 3 \cdot 7 \cdot 17$ and $715 = 5 \cdot 11 \cdot 13$ contained the first seven primes, and that the sums were each 29. He called n a **Ruth-Aaron number** if the sum of its prime divisors, counting multiplicity, was the same as that for $n + 1$. The first two dozen are 5, 8, 15, 77, 125, 714, 948, 1330, 1520, 1862, 2491, 3248, 4185, 4191, 5405, 5560, 5959, 6867, 8280, 8463, 10647, 12351, 14587, 16932,

The number of them that are less than x has been shown by Pomerance to be

$$O\left(x(\ln \ln x)^4/(\ln x)^2\right)$$

John L. Drost, Ruth/Aaron Pairs, *J. Recreational Math.*, **28** 120–122.

P. Erdős & C. Pomerance, On the largest prime factors of n and $n + 1$, *Aequationes Math.*, **17**(1978) 311–321; *MR* **58** #476.

C. Nelson, D. E. Penney & C. Pomerance, "714 and 715", *J. Recreational Math.*, **7**(1994) 87–89.

C. Pomerance, Ruth-Aaron numbers revisited, *Paul Erdős and his mathematics, I* (*Budapest* 1999) 567–579, *Bolyai Soc. Math. Stud.*, **11**, János Bolyai Math. Soc., Budapest, 2002.

David Wells, *The Penguin Dictionary of Curious and Interesting Numbers*, Penguin 1986, pp. 159–160.

OEIS: A006145, A006146, A039752, A039753, A054378

C. Additive Number Theory

C1 Goldbach's conjecture.

One of the most infamous problems is Goldbach's conjecture that every even number greater than 4 is expressible as the sum of two odd primes. Richstein has verified it up to $4 \cdot 10^{14}$, and on 2003-10-03 I learnt that Oliviera e Silva has extended this to $6 \cdot 10^{16}$. Vinogradov proved that every *odd* number greater than $3^{3^{15}}$ is the sum of *three* primes and Chen Jing-Run has shown that all large enough even numbers are the sum of a prime and the product of at most two primes. Wang & Chen have reduced the number $3^{3^{15}}$ to $e^{114} \approx 3.23274 \times 10^{49}$ under the assumption of the generalized Riemann hypothesis.

That every odd number greater than 5 can be expressed as $p + 2q$ with p, q both prime has been called Levy's conjecture. The number of such representations is expected to be similar to that given by "Conjecture A" of Hardy & Littlewood (cf. **A1**, **A8**): i.e., that the number, $N_2(n)$, of representations of an even number n as the sum of two primes, is given asymptotically by

$$ N_2(n) \sim \frac{2cn}{(\ln n)^2} \prod \left(\frac{p-1}{p-2} \right), $$

where, as in **A8**, $2c \approx 1.3203$ and the product is taken over all odd prime divisors of n.

Stein & Stein have calculated $N_2(n)$ for $n < 10^5$ and have found values of n for which $N_2(n) = k$ for all $k < 1911$. It is conjectured that $N_2(n)$ takes all positive integer values. They also verified the conjecture for $n < 10^8$. Granville, van de Lune & te Riele have extended this to $2 \cdot 10^{10}$.

Effinger summarizes his paper with Deshouillers, te Riele & Zinoviev as follows:

1. (Zinoviev) Under GRH, every odd number above 10^{20} is a sum of three primes.

2. (Effinger) If GRH holds and if n lies between 6 and 10^{20}, then there exists a prime p such that $n - p$ lies between 4 and 1.615×10^{12}

3. (Deshouillers & te Riele) Every even number 4 up to 10^{13} is a sum of two primes.

Kaniecki has shown, under the assumption of the Riemann hypothesis, that every odd integer is the sum of at most 5 primes, and 'with a plausible amount of computation this will be improved to at most 4 primes. Ramaré has shown unconditionally that every sufficiently large even integer is the sum of at most 6 primes.

Heath-Brown & J.-C. Puchta have shown that every sufficiently large even integer is a sum of two primes and exactly 13 powers of 2, and that under the Generalized Riemann Hypothesis 13 can be replaced by 7. The numbers 13 & 7 replace earlier values of 1906 by Li, & 200 by Liu & Wang.

If $E(x)$ is the number of even integers not exceeding x which are not the sum of two primes, then Mikawa has shown that

$$E(x + x^{\theta}) - E(x) \ll x^{\theta}(lnx)^{-c}$$

whenever $\theta > \frac{7}{48}$ and $c > 0$.

Let $\phi(n)$ be Euler's totient function (**B36**) so that if p is prime, $\phi(p) = p - 1$. If the Goldbach conjecture is true, then there are, for each number m, prime numbers p, q, such that

$$\phi(p) + \phi(q) = 2m.$$

If we relax the condition that p and q be prime, then it should be easier to show that there are always numbers p and q satisfying this equation. Erdős & Leo Moser asked if this can be done.

Antonio Filz defined a **prime circle** of order $2m$ to be a circular permutation of the numbers from 1 to $2m$ with each adjacent pair summing to a prime. There is essentially only one prime circle for $m = 1, 2$ and 3; two for $m = 4$ and 48 for $m = 5$.

Are there prime circles for all m? Give an asymptotic estimate of their number.

Similarly, Margaret Kenney proposed the **prime pyramid**

```
                        *
                   1        2
               1       2       3
           1       2       3       4
       1       4       3       2       5
   1       4       3       2       5       6
1      -       -       -       -       -       7
```

in which row n contains the numbers $1, 2, \ldots, n$, begins with 1, ends with n, and the sum of two consecutive entries is prime. How many ways are there of arranging the numbers in row n? This problem was also proposed by

Morris Wald; the solutions given are almost certain always to work, but a proof of this may be almost as difficult as proving the Goldbach conjecture itself. The slightly less restricted problem in which the end numbers are not prescribed was earlier asked by E. T. H. Wang.

Erdős asks if there are infinitely many primes p such that every even number $\leq p-3$ can be expressed as the difference between two primes each $\leq p$. For example, $p = 13$: $10 = 13 - 3$, $8 = 11 - 3$, $6 = 11 - 5$, $4 = 7 - 3$, $2 = 5 - 3$.

Zhang Ming-Zhi asked, 15 years ago, whether, for each odd $n > 1$, there were a, $b > 0$ with $a + b = n$ and $a^2 + b^2$ prime.

R. C. Baker & Glyn Harman, The three primes theorem with almost equal summands, *R. Soc. Lond. Philos. Trans. Ser. A Math. Phys. Eng. Sci.*, **356**(1998) 763–780; *MR* **99c**:11124.

R. C. Baker, Glyn Harman & J. Pintz, The exceptional set for Goldbach's problem in short intervals, *Sieve methods, exponential sums and their applications in number theory* (*Cardiff*, 1995) 87–100, London Math. Soc. Lecture Note Ser., **237**, Cambridge Univ. Press, 1997; *MR* **99g**:11121.

Claus Bauer, On Goldbach's conjecture in arithmetic progressions, *Studia Math. Sci. Hungar.*, **37**(2001) 1–20; *MR* **2002c**:11124.

J. Bohman & C.-E. Froberg, Numerical results on the Goldbach conjecture, *BIT*, **15**(1975) 239–243.

Cai Ying-Chun & Lu Ming-Gao, On Chen's theorem, *Analytic number theory* (*Beijing/Kyoto*, 1999), 99–119, Dev. Math., **6**(2002), Kluwer, Dordrecht; *MR* **2003a**:11130.

Chen Jing-Run, On the representation of a large even number as the sum of a prime and the product of at most two primes, *Sci. Sinica*, **16**(1973) 157–176; *MR* **55** #7959; II, **21**(1978) 421–430; *MR* **80e**:10037.

Chen Jing-Run & Wang Tian-Ze, On the Goldbach problem (Chinese), *Acta Math. Sinica*, **32**(1989) 702–718; *MR* **91e**:11108.

Chen Jing-Run & Wang Tian-Ze, The Goldbach problem for odd numbers (Chinese), *Acta Math. Sinica*, **39**(1996) 169–174; *MR* **97k**:11138.

J. G. van der Corput, Sur l'hypothèse de Goldbach pour presque tous les nombres pairs, *Acta Arith.*, **2**(1937) 266–290.

N. G. Čudakov, On the density of the set of even numbers which are not representable as the sum of two odd primes, *Izv. Akad. Nauk SSSR Ser. Mat.*, **2**(1938) 25–40.

N. G. Čudakov, On Goldbach-Vinogradov's theorem, *Ann. Math.*(2), **48** (1947) 515–545; *MR***9**, 11.

Jean-Marc Deshouillers, G. Effinger, H. T. T. te Riele & D. Zinoviev, A complete Vinogradov 3-primes theorem under the Riemann hypothesis, Electronic Research Announcements of the AMS (www.ams.org/era), page 99

Jean-Marc Deshouillers, Andrew Granville, Władysław Narkiewicz & Carl Pomerance, An upper bound in Goldbach's problem, *Math. Comput.*, **61**(1993) 209–213; *MR* **94b**:11101.

Gunter Dufner & Dieter Wolke, Einige Bemerkungen zum Goldbachschen Problem, *Monatsh. Math.*, **118**(1994) 75–82; *MR* **95h**:11106.

T. Estermann, On Goldbach's problem: proof that almost all even positive integers are sums of two primes, *Proc. London Math. Soc.*(2), **44**(1938) 307–314.

S. Yu. Fatkina, On a generalization of Goldbach's ternary problem for almost equal summands, *Vestnik Moskov. Univ. Ser. I Mat. Mekh.*, **2001** no.2, 22–28, 70; transl. *Moscow Univ. Math. Bull.*, **56**(2001) 22–29; *MR* **2002c**:11127.

Antonio Filz, Problem 1046, *J. Recreational Math.*, **14**(1982) 64; **15**(1983) 71.

D. A. Goldston, Linnik's theorem on Goldbach numbers in short intervals, *Glasgow Math. J.*, **32**(1990) 285–297; *MR* **91i**:11134.

Dan A. Goldston, On Hardy and Littlewood's contribution to the Goldbach conjecture, *Proc. Amalfi Conf. Anal. Number Theory (Maiori 1989)*, 115–155, Univ. Salerno, 1992; *MR* **94m**:11122.

A. Granville, J. van de Lune & H. J. J. te Riele, Checking the Goldbach conjecture on a vector computer, in R. A. Mollin (ed.), Number Theory and Applications, NATO ASI Series, Kluwer, Boston, 1989, pp. 423–433; *MR* **93c**:11085.

Richard K. Guy, Prime Pyramids, *Crux Mathematicorum*, **19**(1993) 97–99.

D. R. Heath-Brown & J.-C. Puchta, Integers represented as a sum of primes and powers of two, *Asian J. Math.*, **6**(2002) 535–565; *MR* **2003i**:11146.

Christopher Hooley, On an almost pure sieve, *Acta Arith.*, **66**(1994) 359–368; *MR* **95g**:11092.

Jia Chao-Hua, Goldbach numbers in short interval, I, *Sci. China Ser. A*, **38**(1995) 385–406; II, 513–523; *MR* **96f**:11130, 11131.

Jia Chao-Hua, On the exceptional set of Goldbach numbers in a short interval, *Acta Arith.*, **77**(1996) 207–287; *MR* **97e**:11126.

J. Kaczorowski, A. Perelli & J. Pintz, A note on the exceptional set for Goldbach's problem in short intervals, *Monatsh. Math.*, **116**(1993) 275–282; *MR* **95c**:11136; corrigendum, **119**(1995) 215–216.

Leszek Kaniecki, On Šnirelman's constant under the Riemann hypothesis, *Acta Arith.*, **72**(1995) 361–374; *MR* **96i**:11112.

Margaret J. Kenney, Student Math Notes, *NCTM News Bulletin*, Nov. 1986.

Li Hong-Ze, Goldbach numbers in short intervals, *Sci. China Ser. A*, **38**(1995) 641–652; *MR* **96h**:11100.

Liu Jian-Ya & Liu Ming-Chit, Representation of even integers as sums of squares of primes and powers of 2, *J. Number Theory*, **83**(2000) 202–225; *MR* **2001g**:11156.

Liu Jian-Ya & Liu Ming-Chit, Representation of even integers by cubes of primes and powers of 2, *Acta Math. Hungar.*, **91**(2001) 217–243; *MR* **2003c**:11131.

Liu Ming-Chit & Wang Tian-Ze, On the Vinogradov bound in the three primes Goldbach conjecture, *Acta Arith.*, **105**(2002) 133–175; *MR* **2003i**:11147.

Hiroshi Mikawa, On the exceptional set in Goldbach's problem, *Tsukuba J. Math.*, **16**(1992) 513–543; *MR* **94c**:11098.

Hiroshi Mikawa, Sums of two primes (Japanese), *Sūrikaisekikenkyūsho Kōkyūroku No.* **886**(1994) 149–169; *MR* **96g**:11123.

H. L. Montgomery & R. C. Vaughan, The exceptional set in Goldbach's problem, *Acta Arith.*, **27**(1975) 353–370.

Pan Cheng-Dong, Ding Xia-Xi & Wang Yuan, On the representation of every large even integer as the sum of a prime and an almost prime, *Sci. Sinica*, **18**(1975) 599–610; *MR* **57** #5897.

A. Perelli & J. Pintz, On the exceptional set for Goldbach's problem in short intervals, *J. London Math. Soc.*, **47**(1993) 41–49; *MR* **93m**:11104.

J. Pintz, & I. Z. Ruzsa, On Linnik's approximation to Goldbach's problem, I, *Acta Arith.*, **109**(2003) 169–194; *MR* **2004c**:11185.

Olivier Ramaré, On Šnirel'man's constant *Ann. Scuola Norm. Sup. Pisa Cl. Sci.*(2), **22**(1995) 645–706; *MR* **97a**:11167.

Jörg Richstein, Verifying the Goldbach conjecture up to $4 \cdot 10^{14}$, *Math. Comput.*, **70**(2001) 1745–1749 (electronic); *MR* **2002c**:11131.

Jörg Richstein, Computing the number of Goldbach partitions up to $5 \cdot 10^8$, *Algorithmic Number Theory (Leiden, 2000)*, *Springer Lecture Notes in Comput. Sci.*, **1838**(2000) 475–490; *MR* **2002g**:11142.

P. M. Ross, On Chen's theorem that every large even number has the form $p_1 + p_2$ or $p_1 + p_2 p_3$, *J. London Math. Soc.*(2), **10**(1975) 500–506.

Yannick Saouter, Checking the odd Goldbach conjecture up to 10^{20}, *Math. Comput.*, **67**(1998) 863–866; *MR* **98g**:11115.

Shan Zun & Kan Jia-Hai, On the representation of a large even number as the sum of a prime and an almost prime: the prime belongs to a fixed arithmetic progression, *Acta Math. Sinica*, **40**(1997) 625–638.

Matti K. Sinisalo, Checking the Goldbach conjecture up to $4 \cdot 10^{11}$, *Math. Comput.*, **61**(1993) 931–934.

Mels Sluyser, Parity of prime numbers and Goldbach's conjecture, *Math. Sci.*, **19**(1994) 64–65; *MR* **95f**:11004.

M. L. Stein & P. R. Stein, New experimental results on the Goldbach conjecture, *Math. Mag.*, **38**(1965) 72–80; *MR* **32** #4109.

M. L. Stein & P. R. Stein, Experimental results on additive 2-bases, *Math. Comput.*, **19**(1965) 427–434.

D. I. Tolev, On the number of representations of an odd integer as the sum of three primes, one of which belongs to an arithmetic progression, *Tr. Mat. Inst. Steklova* **218**(1997) *Anal. Teor. Chisel i Prilozh.*, 415–432.

Robert C. Vaughan, On Goldbach's problem, *Acta Arith.*, **22**(1972) 21–48.

Robert C. Vaughan, A new estimate for the exceptional set in Goldbach's problem, *Proc. Symp. Pure Math.*, (Analytic Number Theory, St. Louis), Amer. Math. Soc., **24**(1972) 315–319.

I. M. Vinogradov, Representation of an odd number as the sum of three primes, *Dokl. Akad. Nauk SSSR*, **15**(1937) 169–172.

I. M. Vinogradov, Some theorems concerning the theory of primes, *Mat. Sb. N.S.*, **2(44)** (1937) 179–195.

Morris Wald, Problem 1664, *J. Recreational Math.*, **20**(1987–88) 227–228; Solution **21**(1988–89) 236–237.

Edward T. H. Wang, Advanced Problem 6189, *Amer. Math. Monthly* **85** (1978) 54 [no solution has appeared].

Wang Tian-Ze, The odd Goldbach problem for odd numbers under the generalized Riemann hypothesis, II, *Adv. in Math.*, **25**(1996) 339–346; *MR* **98h**:11125.

Wang Tian-Ze & Chen Jing-Run, An odd Goldbach problem under the general Riemann hypothesis, *Sci. China Ser. A*, **36**(1993) 682–691; *MR* **95a**:11090.

Yan Song-Y., A simple verification method for the Goldbach conjecture, *Internat. J. Math. Ed. Sci.*, **25**(1994) 681-688; *MR* **95e**:11109.

Yu Xin-He, A new approach to Goldbach's conjecture, *Fujian Shifan Daxue Xuebao Ziran Kexue Ban*, **9**(1993) 1–8; *MR* **95e**:11110.

Zhang Zhen-Feng & Wang Tian-Ze, The ternary Goldbach problem with primes in arithmetic progressions, *Acta Math. Sin. (Engl. Ser.)*, **17**(2001) 679–696.

Dan Zwillinger, A Goldbach conjecture using twin primes, *Math. Comput.*, **33**(1979) 1071; *MR* **80b**:10071.

OEIS: A000341, A036440, A038133-038135, A039506-039507, A051237, A051239, A051252, A070897, A072184, A072325, A072616-072618, A072676.

C2 Sums of consecutive primes.

Let $f(n)$ be the number of ways of representing n as the sum of (one or more) *consecutive* primes. For example

$$5 = 2 + 3 \quad \text{and} \quad 41 = 11 + 13 + 17 = 2 + 3 + 5 + 7 + 11 + 13$$

so that $f(5) = 2$ and $f(41) = 3$. Leo Moser has shown that

$$\lim_{x \to \infty} \frac{1}{x} \sum_{n=1}^{x} f(n) = \ln 2$$

and he asks: is $f(n) = 1$ infinitely often? Is $f(n) = k$ solvable for every k? Do the numbers for which $f(n) = k$ have a density for every k? Is $\limsup f(n) = \infty$?

Erdős asks if there is an infinite sequence of integers $1 < a_1 < a_2 < \cdots$ such that $f(n)$ the number of solutions of $a_i + a_{i+1} + \cdots + a_k = n$, tends to infinity with n. He notes that if we insist that $k > i$, then it is not even known if $f(n) > 0$ for all but finitely many n. If $a_i = i$, $f(n)$ is the number of odd divisors of n.

Robert Silverman (W. No. Theory problem 97:10) allows powers of the primes and asks if 33 is the largest n not representable as $n = \sum_{j=1}^{L} p_j^{\alpha_j}$. He also asks for the number of representations as n becomes large, and for the behaviour of L and the α_j.

L. Moser, Notes on number theory III. On the sum of consecutive primes, *Canad. Math. Bull.*, **6**(1963) 159–161; *MR* **28** #75.

OEIS: A054845, A054859, A054996-055001.

C3 Lucky numbers.

Gardiner and others define **lucky numbers** by modifying the sieve of Eratosthenes in the following way. From the natural numbers strike out all even ones, leaving the odd numbers. Apart from 1, the first remaining number is 3. Strike out every third member (those of shape $6k - 1$) in the new sequence, leaving

$$1, 3, 7, 9, 13, 15, 19, 21, 25, 27, 31, 33, \ldots$$

The next number remaining is 7. Strike out every seventh term (numbers $42k - 23$, $42k - 3$) in this sequence. Next 9 remains: strike out every ninth term from what's left, and so on, until we are left with the lucky numbers

$$1, 3, 7, 9, 13, 15, 21, 25, 31, 33, 37, 43, 49, 51, 63, 67, 69,$$
$$73, 75, 79, 87, 93, 99, 105, 111, 115, 127, 129, 133, 135, 141,$$
$$151, 159, 163, 169, 171, 189, 193, 195, 201, 205, 211, 219, 223,$$
$$231, \ldots,$$

Many questions arise concerning lucky numbers, parallel to the classical ones asked about primes. For example, if $L_2(n)$ is the number of solutions of $l + m = n$, where n is even and l and m are lucky, then Stein & Stein find values of n such that $L_2(n) = k$ for all $k \leq 1769$, and there is a corresponding conjecture to that made in **C1**.

Hawkins discusses 'random primes', a generalization of lucky numbers, and Heyde shows that they almost always satisfy the prime number theorem and also the Riemann Hypothesis.

W. E. Briggs, Prime-like sequences generated by a sieve process, *Duke Math. J.*, **30**(1963) 297–312; *MR* **26** #6145.

R. G. Buschman & M. C. Wunderlich, Sieve-generated sequences with translated intervals, *Canad. J. Math.*, **19**(1967) 559–570; *MR* **35** #2855.

R. G. Buschman & M. C. Wunderlich, Sieves with generalized intervals, *Boll. Un. Mat. Ital.*(3), **21**(1966) 362–367.

Paul Erdős & Eri Jabotinsky, On sequences of integers generated by a sieving process, I, II, *Nederl. Akad. Wetensch. Proc. Ser. A*, **61** = *Indag. Math.*, **20**(1958) 115–128; *MR* **21** #2628.

Verna Gardiner, R. Lazarus, N. Metropolis & S. Ulam, On certain sequences of integers generated by sieves, *Math. Mag.*, **31**(1956) 117–122; **17**, 711.

David Hawkins & W. E. Briggs, The lucky number theorem, *Math. Mag.*, **31**(1957–58) 81–84, 277–280; *MR* **21** #2629, 2630.

D. Hawkins, The random sieve, *Math. Mag.*, **31**(1958) 1–3.

C. C. Heyde, *Ann. Probability*, **6**(1978) 850–875.

M. C. Wunderlich, Sieve generated sequences, *Canad. J. Math.*, **18**(1966) 291–299; *MR* **32** #5625.

M. C. Wunderlich, A general class of sieve-generated sequences, *Acta Arith.*, **16**(1969–70) 41–56; *MR* **39** #6852.

M. C. Wunderlich & W. E. Briggs, Second and third term approximations of sieve-generated sequences, *Illinois J. Math.*, **10**(1966) 694–700; *MR* **34** #153.

OEIS: A000959, A049781.

C4 Ulam numbers.

Ulam constructed increasing sequences of positive integers by starting from arbitrary u_1 and u_2 and continuing with those numbers which can be expressed in just one way as the sum of two distinct earlier members of the sequence. Recamán asked some of the questions which arise in connexion with the **U-numbers** ($u_1 = 1$, $u_2 = 2$).

> 1, 2, 3, 4, 6, 8, 11, 13, 16, 18, 26, 28, 36, 38, 47, 48, 53, 57,
> 62, 69, 72, 77, 82, 87, 97, 99, 102, 106, 114, 126, 131, 138, 145,
> 148, 155, 175, 177, 180, 182, 189, 197, 206, 209, 219, ...

(1) Can the sum of two consecutive U-numbers, apart from 1+2=3, be a U-number?

(2) Are there infinitely many numbers

$$23, \ 25, \ 33, \ 35, \ 43, \ 45, \ 67, \ 92, \ 94, \ 96, \ \ldots$$

which are *not* the sum of two U-numbers?

(3) (Ulam) Do the U-numbers have positive density?

(4) Are there infinitely many pairs

$$(1,2), \quad (2,3), \quad (3,4), \quad (47,48), \quad \ldots$$

of consecutive U-numbers?

(5) Are there arbitrarily large gaps in the sequence of U-numbers?

In answer to Question 1, Frank Owens noticed that $u_{19}+u_{20} = 62+69 = 131 = u_{31}$. In answer to Question 4, Muller calculated 20000 terms and found no further examples. On the other hand, more than 60% of these terms differed from another by exactly 2.

David Zeitlin and Stefan Burr noted that the sequence of U-numbers is **complete** in the sense that every positive number is expressible as the sum of distinct members of the sequence.

The reader is warned that the name "U-numbers" was also used by Mahler in the theory of algebraic numbers in connexion with Alan Baker's characterization of "S-numbers" and "T-numbers".

Such sequences can be defined with initial values other than (1,2). Queneau showed that the sequences initiated by (2,5), (2,7) and (2,9) are **regular** in the sense that their differences are ultimately periodic. Finch proved that if such a sequence has only finitely many even terms, then it is regular. Schmerl & Spiegel proved that sequences starting $(2, 2k+1)$ have just two even terms for any $k > 1$.

More generally, one can define s-**additive sequences** which are constructed in the same way, except that each term is the sum of two earlier terms in exactly s ways, the U-numbers corresponding to $s = 1$. If $s = 0$ the sequence is **sum-free**, i.e., constructed from numbers which are *not* the sum of two distinct earlier members. Compare problems **C9, C14, E12, E28, E32**. More generally still there are (s, t)-**additive sequences** where each term has exactly s representations as the sum of t distinct earlier members. In this notation, the U-numbers are the $(1,2)$-sequence initiated by $u_1 = 1$, $u_2 = 2$. Steven Finch has experimented in this area and has a number of conjectures. For example, that the sequences initiated by (u_1, u_2) with $u_1 < u_2$ and $u_1 \perp u_2$ contain only finitely many even terms in the cases (a) $(u_1, u_2) = (2, u_2)$ for $u_2 \geq 5$, (b) $(4, u_2)$, (c) $(5,6)$, (d) $u_1 \geq 6$ and even, and (e) $u_1 \geq 7$ odd with u_2 even; but infinitely many even terms otherwise.

Julien Cassaigne & Steven R. Finch, A class of 1-additive sequences and quadratic recurrences, *Experiment. Math.*, **4**(1995) 49–60; *MR* **96g**:11007.

Steven R. Finch, Conjectures about s-additive sequences, *Fibonacci Quart.*, **29**(1991) 209–214; *MR* **92j**:11009.

Steven R. Finch, Are 0-additive sequences always regular? *Amer. Math. Monthly*, **99** (1992) 671–673.

Steven R. Finch, On the regularity of certain 1-additive sequences, *J. Combin. Theory Ser. A*, **60**(1992) 123–130; *MR* **93c**:11009.

Steven R. Finch, Patterns in 1-additive sequences, *Experiment. Math.*, **1** (1992) 57–63; *MR* **93h**11014.

P. N. Muller, On properties of unique summation sequences, M.Sc. thesis, University of Buffalo, 1966.

Frank W. Owens, Sums of consecutive U-numbers, Abstract 74T-A74, *Notices Amer. Math. Soc.*, **21**(1974) A-298.

Raymond Queneau, Sur les suites s-additives, *J. Combin. Theory*, **12**(1972) 31–71; *MR* **46** #1741.

Bernardo Recamán, Questions on a sequence of Ulam, *Amer. Math. Monthly*, **80**(1973) 919–920.

James Schmerl & Eugene Spiegel, The regularity of some 1-additive sequences, *J. Combin. Theory Ser. A*, **66**(1994) 172–175; *MR* **95h**:11010.

S. M. Ulam, *Problems in Modern Mathematics*, Interscience, New York, 1964, p. ix.

Marvin C. Wunderlich, The improbable behaviour of Ulam's summation sequence, in *Computers and Number Theory*, Academic Press, 1971, 249–257.

David Zeitlin, Ulam's sequence $\{U_n\}$, $U_1 = 1$, $U_2 = 2$, is a complete sequence, Abstract 75T-A267, *Notices, Amer. Math. Soc.*, **22**(1975) A-707.

OEIS: A001857, A002858-002859, A003044-003045, A006844, A007300, A033627-033629, A054540, A072832.

C5 Sums determining members of a set.

Leo Moser asked, and Selfridge, Straus and others largely settled, to what extent the sums of all the pairs of numbers in a set determine the set. They show that if the cardinality is not a power of two, then the members are determined. Suppose that y_1, y_2, ..., y_s are the sums $x_i + x_j$ $(i \neq j)$ of the numbers x_1, x_2, ..., x_{2^k}, so that $s = 2^{k-1}(2^k - 1)$. Are there more than two sets $\{x_i\}$ which give rise to the same set $\{y_j\}$? If $k = 3$ there may be three such sets, for example.

$$\{\pm 1, \pm 9, \pm 15, \pm 19\}, \quad \{\pm 2, \pm 6, \pm 12, \pm 22\}, \quad \{\pm 3, \pm 7, \pm 13, \pm 21\}$$

but there can't be more than three. For $k > 3$ the problem is open.

The corresponding problem where sums of *triples* of elements of a set are given has been settled by Boman & Linusson. The exceptions are precisely 3, 6, 27, 486. For $n = 27$ they give five examples of which the simplest is $\{-4, -1^{10}, 2^{16}\}$ and its negative, where exponents denote repetitions. For $n = 486$ they give $\{-7, -4^{56}, -1^{231}, 2^{176}, 5^{22}\}$ and its negative.

The corresponding problem for sums of *four* distinct elements was settled by Ewell.

I. N. Baker, Solutions of the functional equation $(f(x))^2 - f(x^2) = h(x)$, *Canad. Math. Bull.*, **3**(1960) 113–120.

E. Bolker, The finite Radon transform, *Contemp. Math.*, **63**(1987) 27–50; *MR* **88b**:51009.

J. Boman, E. Bolker & P. O'Neil, The combinatorial Radon transform modulo the symmetric group, *Adv. Appl. Math.*, **12**(1991) 400–411; *MR* **92m**:05014.

Jan Boman & Svante Linusson, Examples of non-uniqueness for the combinatorial Radon transform modulo the symmetric group, *Math. Scand.*, **78**(1996) 207–212; *MR* **97g**:05014.

John A. Ewell, On the determination of sets by sets of sums of fixed order, *Canad. J. Math.*, **20**(1968) 596–611.

B. Gordon, A. S. Fraenkel & E. G. Straus, On the determination of sets by the sets of sums of a certain order, *Pacific J. Math.*, **12**(1962) 187–196; *MR* **27** #3576.

Ross A. Honsberger, A gem from combinatorics, *Bull. ICA*, 1(1991) 56–58.

J. Lambek & L. Moser, On some two way classifications of the integers, *Canad. Math. Bull.*, **2**(1959) 85–89.

B. Liu & X. Zhang, On harmonious labelings of graphs, *Ars Combin.*, **36**(1993) 315–326.

L. Moser, Problem E1248, *Amer. Math. Monthly*, **64**(1957) 507.

J. Ossowski, On a problem of Galvin, *Congressus Numerantium*, **96**(1993) 65–74.

D. G. Rogers, A functional equation: solution to Problem 89-19*, *SIAM Review*, **32**(1990) 684–686.

J. L. Selfridge & E. G. Straus, On the determination of numbers by their sums of a fixed order, *Pacific J. Math.*, **8**(1958) 847–856; *MR* **22** #4657.

C6 Addition chains. Brauer chains. Hansen chains.

An **addition chain** for n is a sequence $1 = a_0 < a_1 < \ldots < a_r = n$ with each member after the zeroth the sum of two earlier, not necessarily distinct, members. For example

$$1, 1+1, 2+2, 4+2, 6+2, 8+6 \quad \text{and} \quad 1, 1+1, 2+2, 4+2, 4+4, 8+6$$

are addition chains for 14 of **length** $r = 5$. The minimal length of an addition chain for n is denoted by $l(n)$.

The main unsolved problem is the Scholz conjecture

$$¿ \qquad l(2^n - 1) \leq n - 1 + l(n) \qquad ?$$

It has been proved for $n = 2^a$, $2^a + 2^b$, $2^a + 2^b + 2^c$, $2^a + 2^b + 2^c + 2^d$ by Utz, Gioia et al, and Knuth, and demonstrated for $1 \leq n \leq 18$ by Knuth and Thurber. Brauer proved the conjecture for those n for which a shortest chain exists which is a **Brauer chain**, that is one in which each member uses the previous member as a summand. The second of the examples is not a Brauer chain, because the term 4+4 does not use the summand 6. Such an n is called a **Brauer number**. Hansen proved that there are infinitely many non-Brauer numbers, but also that the Scholz conjecture still holds if n has a shortest chain which is a **Hansen chain**, that is one for which there is a subset H of the members such that each member of the chain uses the largest element of H which is less than the member. The second example is a Hansen chain, with $H = \{1, 2, 4, 8\}$. Knuth gives the example

$$1, 2, 4, 8, 16, 17, 32, 64, 128, 256, 512, 1024, 1041, 2082, 4164, 8328, 8345, 12509$$

of a Hansen chain ($H = \{1, 2, 4, 8, 16, 32, 64, 128, 256, 512, 1024, 1041, 2082, 4164, 8328, 8345\}$) for $n = 12509$ which is not a Brauer chain (32 does not use 17) and no such short Brauer chain exists for $n = 12509$.

Are there non-Hansen numbers?

It is clear that $l(2n) \leq l(n) + 1$. That strict inequality is possible was shown by Knuth with $l(382) = l(191) = 11$. The smallest even n with $l(2n) = l(n)$ is 13818, given by Thurber, who also noticed the odd adjacent pair 22453, 22455. Andrew Granville asks if there are n for which $l(4n) = l(2n) = l(n)$.

D. J. Newman considers a computer which costs 1 cent to perform each addition but nothing to perform multiplication. Then the addition chain for n costs maximally $(\log n)^{\frac{1}{2} + o(1)}$ instead of $\log n$, where log is to base 2.

An addition chain $1 = a_0 < a_1 < \ldots < a_r = n$ which satisfies the additional condition that the pair (i, j) in $a_k = a_i + a_j$ is either such that $i = j$ or there is an l such that $a_j - a_i = a_l$ is called a **Lucas chain**. Kutz has shown that most n do not have Lucas chains shorter than $(1 - \epsilon) \log n$, where the base of the log is the golden ratio.

For a good survey and list of problems, see Subbarao's article.

Walter Aiello & M. V. Subbarao, A conjecture in addition chains related to Scholz's conjecture, *Math. Comput.*, **61**(1993) 17–23; *MR* **93k**:11015.

Hatem M. Bahig, Mohamed H. El-Zahar & Ken Nakamula, Some results for some conjectures in addition chains, *Combinatorics, computability and logic* (*Constanţa*, 2001) 47–54, *Springer Ser. Discrete Math. Theor. Comput. Sci.*, 2001; *MR* **2003i**:11031.

F. Bergeron & J. Berstel, Efficient computation of addition chains, *J. Théor. Nombres Bordeaux*, **6**(1994) 21–38; *MR* **95m**:11144.

A. T. Brauer, On addition chains, *Bull. Amer. Math. Soc.*, **45**(1939) 736–739.

A. Cottrell, A lower bound for the Scholz-Brauer problem, Abstract 73T-A200, *Notices Amer. Math. Soc.*, **20**(1973) A-476.

Paul Erdős, Remarks on number theory III. On addition chains, *Acta Arith.*, **6**(1960) 77–81.

R. P. Giese, PhD thesis, University of Houston, 1972.

R. P. Giese, A sequence of counterexamples of Knuth's modification of Utz's conjecture, Abstract 72T-A257, *Notices Amer. Math. Soc.*, **19**(1972) A-688.

A. A. Gioia & M. V. Subbarao, The Scholz-Brauer problem in addition chains II, *Congr. Numer. XXII*, Proc. 8th Manitoba Conf. Numer. Math. Comput. 1978, 251–274; *MR* **80i**: 10078; *Zbl.* 408.10037.

A. A. Gioia, M. V. Subbarao & M. Sugunamma, The Scholz-Brauer problem in addition chains, *Duke Math. J.*, **29**(1962) 481–487; *MR* **25** #3898.

W. Hansen, Zum Scholz-Brauerschen Problem, *J. reine angew. Math.*, **202**(1959) 129–136; *MR* **25** #2027.

Kevin Hebb, Some problems on addition chains, thesis, Univ. of Alberta, 1974.

A. M. Il'in, On additive number chains (Russian), *Problemy Kibernet.*, **13** (1965) 245–248.

H. Kato, On addition chains, PhD dissertation, Univ. Southern California, June 1970.

Donald Knuth, *The Art of Computer Programming*, Vol. 2, Addison-Wesley, Reading MA, 1969, 398–422.

D. J. Newman, Computing when multiplications cost nothing, *Math. Comput.*, **46**(1986) 255–257.

Arnold Scholz, Aufgabe 253, *Jber. Deutsch. Math.-Verein. II*, **47**(1937) 41–42 (supplement).

Rolf Sonntag, Theorie der Addition Sketten, PhD Technische Universität Hannover, 1975.

K. B. Stolarsky, A lower bound for the Scholz-Brauer problem, *Canad. J. Math.*, **21**(1969) 675–683; *MR* **40** #114.

M. V. Subbarao, Addition chains – some results and problems, in R. A. Mollin (ed.) Number Theory and Applications, NATO ASI Series, Kluwer, Boston, 1989, pp. 555–574; *MR* **93a**:11105.

E. G. Straus, Addition chains of vectors, *Amer. Math. Monthly* **71**(1964) 806–808.

E. G. Thurber, The Scholz-Brauer problem on addition chains, *Pacific J. Math.*, **49**(1973) 229–242; *MR* **49** #7233.

E. G. Thurber, On addition chains $l(mn) \leq l(n) - b$ and lower bounds for $c(r)$, *Duke Math. J.*, **40**(1973) 907–913.

E. G. Thurber, Addition chains and solutions of $l(2n) = l(n)$ and $l(2^n - 1) = n + l(n) - 1$, *Discrete Math.*, **16**(1976) 279–289; *MR* **55** #5570; *Zbl.* **346**.10032.

W. R. Utz, A note on the Scholz-Brauer problem in addition chains, *Proc. Amer. Math. Soc.*, **4**(1953) 462–463; *MR* **14**, 949.

C. T. Wyburn, A note on addition chains, *Proc. Amer. Math. Soc.*, **16**(1965) 1134.

OEIS: A003064.

C7 The money-changing problem.

Given $n \geq 2$ integers $0 < a_1 < a_2 < \cdots < a_n$ with $(a_1, a_2, \ldots, a_n) = 1$, then $N = \sum_{i=1}^n a_i x_i$ has a solution in nonnegative integers x_i if N is large enough. The well known **coin problem** of Frobenius is to determine the greatest $N = g(a_1, a_2, \ldots, a_n)$ for which there is no solution. Sylvester showed that $g(a_1, a_2) = (a_1 - 1)(a_2 - 1) - 1$ and that the number of non-representable numbers is $(a_1 - 1)(a_2 - 1)/2$.

Denote by $S_m(a_1, a_2)$ the sum of the mth powers of the nonrepresentable integers. Sylvester's result is $S_0(a_1, a_2) = (a_1 - 1)(a_2 - 1)/2)$. Brown & Shiue evaluated $S_1(a, b)$ and Rødseth evaluated $S_m(a_1, a_2)$ for all nonnegative m.

The case $n = 3$ was first solved explicitly by Selmer & Beyer, using a continued fraction algorithm. Their result was simplified by Rödseth and later by Greenberg. No general formulas are known for $n \geq 4$. Roberts found the value of g if the a_i are in arithmetic progression.

Upper bounds for g are also sought. In 1942 Brauer showed that $g(a_1, a_2, \ldots, a_n) \leq \sum_{i=1}^n a_i(d_{i-1}/d_i - 1)$ where $d_i = (a_1, a_2, \ldots, a_i)$. Erdős & Graham showed that

$$g(a_1, a_2, \ldots, a_n) \leq 2a_{n-1}\lfloor a_n/n \rfloor - a_n$$

(which is best possible if $n = 2$ and a_2 is odd). They define

$$\gamma(n, t) = \max_{\{a_i\}} g(a_1, a_2, \ldots, a_n)$$

where the maximum is taken over all $0 < a_1 < a_2 < \cdots < a_n \leq t$ with $(a_1, a_2, \ldots, a_n) = 1$. Their theorem shows that $\gamma(n, t) < 2t^2/n$ and they proved that $\gamma(n, t) \geq t^2/(n - 1) - 5t$. Lewin showed that $\gamma(3, t) = \lfloor (t - 2)^2/2 \rfloor - 1$ and generally that $g(a_1, a_2, \ldots, a_n) \leq \lfloor (a_{n-1} - 1)(a_n - 2)/2 \rfloor - 1$ for $n \geq 3$.

In Conway's game of Sylver Coinage two players alternately mint a sufficient number of coins of a stated denomination subject to the condition that no denomination may be minted if it can be made up from existing coins. The player who names 1 loses. Hutchings used a strategy-stealing, and hence non-constructive, argument to prove that a prime $p \geq 5$ was a winning opening move. Almost nothing is known about the subsequent play, and it is not known if there are other winning opening moves. If there are any, they are 2- or 3-smooth and ≥ 16.

Matthias Beck, Ricardo Diaz & Sinai Robins, The Frobenius problem, rational polytopes, and Fourier-Dedekind sums, *J. Number Theory*, **96**(2002) 1–21; *MR* **2003j**:11117.

Matthias Beck, Ira M. Gessel & Takao Komatsu, The polynomial part of a restricted partition function related to the Frobenius problem, *Electron. J. Combin.*, **8**(2001) N7, 5 pp.; *MR* **2002f**:05017.

E. R. Berlekamp, J. H. Conway & R. K. Guy, The Emperor and his money, chap.18 of *Winning Ways for your Mathematical Plays*, 2nd ed., AKPeters, Natick MA, 2003.

Alfred Brauer, On a problem of partitions, *Amer. J. Math.*, **64**(1942) 299–312.

A. Brauer & J. E. Shockley, On a problem of Frobenius, *J. reine angew. Math.*, **211**(1962) 215–220; *MR* **26** #6113.

T. C. Brown & Peter Shiue Jau-Shyong, A remark related to the Frobenius problem, *Fibonacci Quart.*, **31**(1993) 32–36; *MR* **93k**:11018.

Chen Qing-Hua² & Liu Yu-Ji, A result on the Frobenius problem, *Fujian Shifan Daxue Xuebao Ziran Kexue Ban*, **11**(1995) 33–37; *MR* **97k**:11016.

Frank Curtis, On formulas for the Frobenius number of a numerical semigroup, *Math. Scand.*, **67**(1990) 190–192; *MR* **92e**:11019.

J. L. Davison, On the linear Diophantine problem of Frobenius, *J. Number Theory*, **48**(1994) 353–363; *MR* **95j**:11033.

J. Dixmier, Proof of a conjecture of Erdős and Graham concerning the problem of Frobenius, *J. Number Theory*, **34**(1990) 198–209; *MR* **91g**:11007.

P. Erdős & R. L. Graham, On a linear diophantine problem of Frobenius, *Acta Arith.*, **21**(1972) 399–408; *MR* **47** #127.

Harold Greenberg, Solution to a linear diophantine equation for nonnegative integers, *J. Algorithms*, **9**(1988) 343–353; *MR* **89j**:11122.

B. R. Heap & M. S. Lynn, On a linear Diophantine problem of Frobenius: An improved algorithm, *Numer. Math.*, **7**(1965) 226–231; *MR* **31** #1227.

Mihály Hujter & Béla Vizári, The exact solutions to the Frobenius problem with three variables, *J. Ramanujan Math. Soc.*, **2**(1987) 117–143; *MR* **89f**:11039.

S. M. Johnson, A linear diophantine problem, *Canad. J. Math.*, **12**(1960) 390–398; *MR* **22** #12074.

I. D. Kan, The Frobenius problem for classes of polynomial solvability, *Mat. Zametki*, **70**(2001) 845–853; *Math. Notes*, **70**(2001) 771–778; *MR* **2002k**:11034.

I. D. Kan, The Frobenius problem for classes of polynomial solvability. (Russian) *Mat. Zametki*, **70**(2001) 845–853; transl. *Math. Notes*, **70**(2001) 771–778; *MR* **2002k**:11034.

I. D. Kan, Boris S. Stechkin & I. V. Sharkov, On the Frobenius problem for three arguments, *Mat. Zametki*, **62**(1997) 626–629; transl. *Math. Notes*, **62**(1997) 521–523; *MR* **99h**:11022.

Ravi Kannan, The Frobenius problem, *Foundations of software technology and theoretical computer science* (Bangalore, 1989) 242–251, *Lecture Notes in Comput. Sci.*, **405** Springer, Berlin, 1989; *MR* **91e**:68063.

Ravi Kannan, Lattice translates of a polytope and the Frobenius problem, *Combinatorica*, **12**(1992) 161–177; *MR* **93k**:52015.

Géza Kiss, On the extremal Frobenius problem in a new aspect, *Ann. Univ. Sci. Budapest. Eötvös Sect. Math.*, **45**(2002) 139–142 (2003).

Vsevolod F. Lev, On the extremal aspect of the Frobenius problem, *J. Combin. Theory Ser. A*, **73**(1996) 111–119; *MR* **97d**:11045.

Vsevolod F. Lev, Structure theorem for multiple addition and the Frobenius problem, *J. Number Theory*, **58**(1996) 79–88; addendum **65**(1997) 96–100; *MR* **97d**:11056, **98e**:11045.

Vsevolod F. Lev & P. Y. Smeliansky, On addition of two distinct sets of integers, *Acta Arith.*, **70**(1995) 85–91.

Mordechai Lewin, On a linear diophantine problem, *Bull. London Math. Soc.*, **5**(1973) 75–78; *MR* **47** #3311.

Mordechai Lewin, An algorithm for a solution of a problem of Frobenius, *J. reine angew. Math.*, **276**(1975) 68–82; *MR* **52** #259.

Albert Nijenhuis, A minimal-path algorithm for the "money changing problem", *Amer. Math. Monthly*, **86**(1979) 832–835; correction, **87**(1980) 377; *MR* **82m**:68070ab.

Niu Xue-Feng, A formula for the largest integer which cannot be represented as $\sum_{i=1}^{s} a_i x_i$, *Heilongjiang Daxue Ziran Kexue Xuebao*, **9**(1992) 28–32; *MR* **93k**:11019.

Marek Raczunas & Piotr Chrząstowski-Wachtel, A Diophantine problem of Frobenius in terms of the least common multiple, *Discrete Math.*, **150**(1996) 347–357; *MR* **97d**:11057.

J. L. Ramírez-Alfonsín, Complexity of the Frobenius problem, *Combinatorica*, **16**(1996) 143–147; *MR* **97d**:11058.

Stefan Matthias Ritter, On a linear Diophantine problem of Frobenius: extending the basis, *J. Number Theory*, **69**(1998) 201–212; *MR* **99b**:11023.

J. B. Roberts, Note on linear forms, *Proc. Amer. Math. Soc.*, **7**(1956) 465–469.

Ø. J. Rødseth, On a linear Diophantine problem of Frobenius, *J. reine angew. Math.*, **301**(1978) 171–178; *MR* **58** #27741.

Øystein J. Rødseth, A note on T. C. Brown & P. J.-S. Shiue's paper: "A remark related to the Frobenius problem", *Fibonacci Quart.*, **32**(1994) 407–408; *MR* **95j**:11022.

Ernst S. Selmer, On the linear diophantine problem of Frobenius, *J. reine angew. Math.*, **293/294**(1977) 1–17; *MR* **56** #246.

E. S. Selmer & Ö. Beyer, On the linear diophantine problem of Frobenius in three variables, *J. reine angew. Math.*, **301**(1978) 161–170.

J. J. Sylvester, *Math. Quest. Educ. Times*, **37**(1884) 26; (not **41**(1884) 21 as in Dickson and elsewhere).

Calogero Tinaglia, Some results on a linear problem of Frobenius, *Geometry Seminars*, 1996–1997, *Univ. Stud. Bologna*, 1998, 231–244; *MR* **99e**:11026.

Amitabha Tripathi, On a variation of the coin exchange problem for arithmetic progressions, *Integers*, **3**(2003) A1; *MR* **2003k**:11018.

Herbert S. Wilf, A circle of lights algorithm for the "money changing problem", *Amer. Math. Monthly* **85**(1978) 562–565.

C8 Sets with distinct sums of subsets.

The set of integers $\{2^i : 0 \leq i \leq k\}$, of cardinality $k+1$, has the sums of all its 2^{k+1} subsets distinct. Erdős has asked for the maximum number, m, of positive integers $a_1 < a_2 < \cdots < a_m \leq 2^k$, with all sums of subsets distinct. With Leo Moser he showed that $k+1 \leq m < k + \frac{1}{2}\log k + 2$ where the logarithm is to base 2. Noam Elkies improved the constant 2 on the right to $\frac{1}{2}\log \pi < 0.826$.

Conway & Guy have given a Conway-Guy sequence, $u_0 = 0$, $u_1 = 1$, $u_{n+1} = 2u_n - u_{n-r}$ $(n \geq 1)$ where r is the nearest integer to $\sqrt{2n}$, from which may be derived the set of $k+2$ integers

$$A = \{a_i = u_{k+2} - u_{k+2-i} : 1 \leq i \leq k+2\}.$$

They conjecture that this set has subsets with distinct sums (established by Mike Guy for $k \leq 40$ and by Fred Lunnon for $n \leq 79$). Several claims of proof of the conjecture have all been found to be faulty. For $k \geq 21$, $u_{k+2} < 2^k$, so that $m \geq k+2$ for $k \geq 21$, since once a set with the desired cardinality is found, its cardinality may be increased by doubling the size of each member and adjoining the member 1 (or any odd number). Conway & Guy conjecture that A gives, in essence, the best possible solution, $m = k+2$, to the problem, though Lunnon defines a class of generalized Conway-Guy sequences, some of which give a smaller limit (e.g., 0.220963) than that of $u_n/2^n$ (≈ 0.23512531). Erdős offered \$500.00 for a proof or disproof of $m = k + O(1)$ (such prizes can still be negotiated via R. L. Graham).

M. D. Atkinson, A. Negro & N. Santoro, Sums of lexicographically ordered sets, *Discrete Math.*, **80**(1990) 115–122.

Jaegug Bae, On subset-sum-distinct sequences, *Analytic Number Theory*, Vol. 1 (*Allerton Park*, 1995) 31–37, *Progr. Math.*, **138**, Birkhäuser Boston MA, 1996; *MR* **97d**:11016.

Jaegug Bae, An extremal problem for subset-sum-distinct sequences with congruence conditions, *Discrete Math.*, **189**(1998) 1–20; *MR* **99h**:11018.

Jaegug Bae, A compactness result for a set of subset-sum-distinct sequences, *Bull. Korean Math. Soc.*, **35**(1998) 515–525; *MR* **99k**:11018.

Jaegug Bae, On maximal subset-sum-distinct sequences, *Commun. Korean Math. Soc.*, **17**(2002) 215–220; *MR* **2003b**:11014.

Jaegug Bae, On generalized subset-sum-distinct sequences, *Int. J. Pure Appl. Math.*, **1**(2002) 343–352; *MR* **2003c**:11022.

Tom Bohman, A sum packing problem of Erdős and the Conway-Guy sequence, *Proc. Amer. Math. Soc.*, **124**(1996) 3627–3636; *MR* **97b**:11027.

Thomas A. Bohman, A construction for sets of integers with distinct subset sums, *Electronic J. Combin.*, **5**(1998) no. 1, *Res. Paper* 3, 14pp.

Peter Borwein & Michael J. Mossinghoff, Newman polynomials with prescribed vanishing and integer sets with distinct subset sums, *Math. Comput.*, **72**(2003) 787–800.

Chen Yong-Gao, Distinct subset sums and an inequality for convex functions, *Proc. Amer. Math. Soc.*, **128**(2000) 2897–2898; *MR* **2000m**:11023.

J. H. Conway & R. K. Guy, Sets of natural numbers with distinct sums, *Notices Amer. Math. Soc.*, **15**(1968) 345.

J. H. Conway & R. K. Guy, Solution of a problem of P. Erdős, *Colloq. Math.*, **20**(1969) 307.

P. Erdős, Problems and results in additive number theory, *Colloq. Théorie des Nombres, Bruxelles*, 1955, Liège & Paris, 1956, 127–137, esp. p. 137.

Noam Elkies, An improved lower bound on the greatest element of a sum-distinct set of fixed order, *J. Combin. Theory Ser. A*, **41**(1986) 89–94; *MR* **87b**:05012.

P. E. Frenkel, Integer sets with distinct subset sums, *Proc. Amer. Math. Soc.*, **126**(1998) 3199–3200.

Martin Gardner, Number 5, *Science Fiction Puzzle Tales*, Penguin, 1981.

Hansraj Gupta, Some sequences with distinct sums, *Indian J. Pure Math.*, **5**(1974) 1093–1109; *MR* **57** #12440.

Richard K. Guy, Sets of integers whose subsets have distinct sums, *Ann. Discrete Math.* **12**(1982) 141–154

Y. O. Hamidoune, The representation of some integers as a subset sum, *Bull. London Math. Soc.*, **26**(1994) 557–563.

B. Lindström, On a combinatorial problem in number theory, *Canad. Math. Bull.*, **8**(1965) 477–490.

B. Lindström, Om et problem av Erdős for talfoljder, *Nordisk. Mat. Tidskrift*, **161**-2(1968) 29–30, 80.

W. Fred Lunnon, Integer sets with distinct subset-sums, *Math. Comput.*, **50**(1988) 297–320.

Roy Maltby, Bigger and better subset-sum-distinct sets, *Mathematika*, **44**(1997) 56–60; *MR* 98i:11012.

Melvyn B. Nathanson, Inverse theorems for subset sums, *Trans. Amer. Math. Soc.*, **347**(1995) 1409–1418.

Paul Smith, Problem E 2536*, *Amer. Math. Monthly* **82**(1975) 300. Solutions and comments, **83**(1976) 484.

Joshua Von Korff, Classification of greedy subset-sum-distinct-sequences, *Discrete Math.*, **271**(2003) 271–282.

C9 Packing sums of pairs.

Suppose that m is the maximum number of integers $1 \leq a_1 < a_2 < \ldots < a_m \leq n$ in a **Sidon sequence**, i.e., one in which the sums of pairs, $a_i + a_j$, are all different. It is known that

$$n^{1/2}(1 - \epsilon) < m \leq n^{1/2} + n^{1/4} + 1.$$

The upper bound is due to Lindstrom, improving a result of Erdős & Turán. The lower bound is due to Singer. Erdős & Turán ask, is $m = n^{1/2} + O(1)$? Erdős offers \$500 for settling this question. If there exists C such that $m < n^{1/2} + C$, then Zhang has shown that $C > 10.27$ and Lindström that $C > 13.71$.

Cameron & Erdős ask for an estimate of $F(n)$, the number of Sidon sequences whose members are at most n. With m as above, it is not even known if $F(n)/2^m \to \infty$, only that the upper limit is infinite. They believe that $F(n) < n^{\epsilon\sqrt{n}}$. Progress has been made by Alon and by Calkin & Thomson, who showed that $|F(n)| = O(2^{n/2+o(n)})$. For the corresponding problem, modulo a prime, p, Lev & Schoen show that

$$2^{\lfloor (p-2)/3 \rfloor}(p-1) \le |F(Z_p)| \le 2^{0.498p}$$

Cameron & Erdős would also like an estimate of the number of maximal Sidon sequences (those to which n further $a \le n$ can be adjoined).

Write $\mathrm{SF}(G)$ for the collection of all sum-free sets of a group G. Cameron & Erdős also conjecture that $|\mathrm{SF}(n)|$, the number of sum-free subsets of $\{1, 2, \ldots, n\}$, is $O(2^{n/2})$. See C9 for papers by Alon and by Calkin, who show $O(2^{n/2+o(n)})$, and by Lev & Schoen, who improve this to

$$2^{\lfloor \frac{p-2}{3} \rfloor}(p-1)(1 + O(2^{-\epsilon p})) \le |\mathrm{SF}(Z^p)| \le 2^{0.498p}$$

In April, 2003 the Cameron-Erdős conjecture was completely settled by Ben Green.

If $\{a_i\}$ continues as an infinite sequence, Erdős & Turán proved that $\limsup a_k/k^2 = \infty$ and gave a sequence with $\liminf a_k/k^2 < \infty$. Ajtai, Komlós & Szemerédi have shown that there is such a sequence with $a_k < ck^3/\ln n$.

Erdős & Rényi proved that there is a sequence satisfying $a_k < k^{2+\epsilon}$ for which the number of solutions of $a_i + a_j = t$ is $\le c$.

Erdős notes that $\sum_{i=1}^{x} a_i^{-1/2} < c(\ln x)^{1/2}$ and asks if this is best possible. He asks if it is true that

$$\frac{1}{\ln x} \sum_{a_i+a_j \le x} \frac{1}{a_i + a_j} \to 0$$

as $x \to \infty$ and suggests that perhaps

$$\sum_{a_i+a_j < x} \frac{1}{a_i + a_j} < c_1 \ln \ln x.$$

It is known that it can be $> c_2 \ln \ln x$.

Erdős also asks if a Sidon sequence $a_1 < a_2 < \ldots < a_k$ can be prolonged to a perfect difference set (see C10), i.e.,

$$a_1 < a_2 < \ldots < a_k < a_{k+1} < \ldots < a_{p+1} = p^2 + p + 1$$

with the differences $a_u - a_v$, $1 \leq u, v \leq p + 1$, $u \neq v$, representing every nonzero residue mod $p^2 + p + 1$ exactly once?

He could not even decide if it can be prolonged to

$$a_1 < a_2 < \ldots < a_k < a_{k+1} < \ldots < a_n, \quad a_n < (1 + o(1))n^2,$$

i.e., if it can be made as dense as possible asymptotically.

Let $a_1 < a_2 < \ldots < a_n$ be any sequence of integers. Is it true that it contains a Sidon subsequence $a_{i_1}, \ldots, a_{i_m}$ with $m = (1+o(1))n^{\frac{1}{2}}$? Komlós, Sulyok & Szemerédi (see **E11**) proved this with $m > cn^{\frac{1}{2}}$.

If $f(n)$ is the number of solutions of $n = a_i + a_j$, is there a sequence with

$$\lim f(n)/\ln n = c?$$

Erdős & Turán conjecture that if $f(n) > 0$ for all sufficiently large n, or if $a_k < ck^2$ for all k, then $\limsup f(n) = \infty$; Erdős also offers \$500 for settling this question.

Graham & Sloane rephrase the question in two more obviously packing forms:

Let $v_\alpha(k)$ [respectively $v_\beta(k)$] be the smallest v such that there is a k-element set $A = \{0 = a_1 < a_2 < \cdots < a_k\}$ of integers with the property that the sums $a_i + a_j$ for $i < j$ [respectively $i \leq j$] belong to $[0, v]$ and represent each element of $[0, v]$ at most once. The set A associated with v_β is often called a B_2-**sequence** (compare **E28**).

They give the values of v_α and v_β displayed in Table 3 and note that the bounds

$$2k^2 - O(k^{3/2}) < v_\alpha, v_\beta < 2k^2 + O(k^{36/23})$$

follow from a modification of the Erdős-Turán argument.

Table 3. Values of v_α, v_β and Exemplary Sets.

k	$v_\alpha(k)$	Example of A	$v_\beta(k)$	Example of A
2	1	{0,1}	2	{0,1}
3	3	{0,1,2}	6	{0,1,3}
4	6	{0,1,2,4}	12	{0,1,4,6}
5	11	{0,1,2,4,7}	22	{0,1,4,9,11}
6	19	{0,1,2,4,7,12}	34	{0,1,4,10,12,17}
7	31	{0,1,2,4,8,13,18}	50	{0,1,4,10,18,23,25}
8	43	{0,1,2,4,8,14,19,24}	68	{0,1,4,9,15,22,32,34}
9	63	{0,1,2,4,8,15,24,29,34}	88	{0,1,5,12,25,27,35,41,44}
10	80	{0,1,2,4,8,15,24,29,34,46}	110	{0,1,6,10,23,26,34,41,53,55}

Cilleruelo has shown that there is a sequence $\{a_k\}$, $a_k \ll k^2$ such that the sums $a_i^2 + a_j^2$ are all different.

If $g(m)$ is the largest integer n such that every set of integers of size m contains a subset of size n whose pairwise sums are distinct, then Abbott has shown that $g(m) > cm^{1/2}$ for any constant $c < \frac{2}{25}$ and all sufficiently large m.

In 1956 Erdős proved the existence of a sequence S such that all sufficiently large integers n are represented between $c_1 \ln n$ and $c_2 \ln n$ times as the sum of two members of S, and more recently Erdős & Tetali have obtained the corresponding result for the sum of k members of S.

See also **C14, C15, E12, E28, E32, F30**.

Harvey L. Abbott, Sidon sets, *Canad. Math. Bull.*, **33**(1990) 335–341; *MR* **91k**:11022.

Miklós Ajtai, János Komlós & Endre Szemerédi, A dense infinite Sidon sequence, *European J. Combin.*, **2**(1981) 1–11; *MR* **83f**:10056.

Noga Alon, Independent sets in regular graphs and sum-free subsets of finite groups, *Israel J. Math.*, **73**(1991) 247–256; *MR* **92k**:11024.

R. C. Bose & S. Chowla, Theorems in the additive theory of numbers, *Comment. Math. Helv.*, **37**(1962-63) 141–147.

John R. Burke, A remark on B_2-sequences in GF$[p,x]$, *Acta Arith.*, **70**(1995) 93–96; *MR* **96f**:11028.

Neil J. Calkin & Jan McDonald Thomson, Counting generalized sum-free sets, *J. Number Theory*, **68**(1998) 151–159; *MR* **98m**:11010.

Sheng Chen, A note on B_{2k}-sequences, *J. Number Theory*, **56**(1996) 1–3; *MR* **97a**:11035.

Fan Chung, Paul Erdős & Ronald Graham, On sparse sets hitting linear form, *Number Theory for the Millenium, I (Urbana IL, 2000)* 257–272, AKPeters, Natick MA, 2002; *MR* **2003k**:11012.

Javier Cilleruelo, B_2-sequences whose terms are squares, *Acta Arith.*, **55**(1990) 261–265; *MR* **91i**:11023.

J. Cilleruelo, $B_2[g]$ sequences whose terms are squares, *Acta Math. Hungar.*, **67**(1995) 79–83; *MR* **95m**:11032.

Javier Cilleruelo, New upper bounds for finite B_h sequences, *Adv. Math.*, **159**(2001) 1–17; *MR* **2002g**:11023.

Javier Cilleruelo & Antonio Córdoba, $B_2[\infty]$-sequences of square numbers *Acta Arith.*, **61**(1992) 265–270; *MR* **93g**:11014.

Ben Green, The Cameron-Erdős conjecture, *Bull. London Math. Soc.*, (to appear).

Javier Cilleruelo & Jorge Jiménez-Urroz, $B_h[g]$-sequences, *Mathematika*, **47**(2000) 109–115; *MR* **2003k**:11014.

Javier Cilleruelo, Imre Z. Ruzsa & Carlos Trujillo, Upper and lower bounds for finite $B_h[g]$ sequences, *J. Number Theory* **97**(2002) 26–34; *MR* **2003i**:11033.

Javier Cilleruelo & Carlos Trujillo, Infinite $B_2[g]$ sequences, *Israel J. Math.*, **126**(2001) 263–267; *MR* **2003d**:11032.

Martin Dowd, Questions related to the Erdős-Turán conjecture, *SIAM J. Discrete Math.*, **1**(1988) 142–150; *MR* **89h**:11006.

P. Erdős, Some of my forgotten problems in number theory, *Hardy-Ramanujan J.*, **15** (1992) 34–50.

P. Erdős & R. Freud, On sums of a Sidon-sequence, *J. Number Theory*, **38**(1991) 196–205.

P. Erdős & R. Freud, On Sidon-sequences and related problems (Hungarian), *Mat. Lapok*, **2**(1991) 1–44.

P. Erdős & W. H. J. Fuchs, On a problem of additive number theory, *J. London Math. Soc.*, **31**(1956) 67–73; *MR* **17**, 586d.

Paul Erdős, Melvyn B. Nathanson & Prasad Tetali, Independence of solution sets and minimal asymptotic bases, *Acta Arith.*, **69**(1995) 243–258; *MR* **96e**:11014.

Paul Erdős, A. Sárközy & V. T. Sós, On sum sets of Sidon sets, I, *J. Number Theory*, **47**(1994) 329–347; II, *Israel J. Math.*, **90**(1995) 221–233; *MR* **96f**:11034.

P. Erdős & E. Szemerédi, The number of solutions of $m = \sum_{i=1}^{k} x_i^k$, *Proc. Symp. Pure Math. Amer. Math. Soc.*, **24**(1973) 83–90; *MR* **49** #2529.

Paul Erdős & Prasad Tetali, Representations of integers as the sum of k terms, *Random Structures Algorithms*, **1**(1990) 245–261; *MR* **92c**:11012.

P. Erdős & P. Turán, On a problem of Sidon in additive number theory, and on some related problems, *J. London Math. Soc.*, **16**(1941) 212–215; *MR* **3**, 270. Addendum, **19**(1944) 208; *MR* **7**, 242.

H. Furstenberg, From the Erdős-Turán conjecture to ergodic theory—the contribution of combinatorial number theory to dynamics, *Paul Erdős and his mathematics* I (Budapest, 1999), 261–277, *Bolyai Soc. Math. Stud.*, **11**, János Bolyai Math. Soc., Budapest, 2002; *MR* **2003j**:37010.

S. Gouda & Mahmoud Abdel-Hamid Amer, A theorem on the h-range of B_h-sequences, *Ann. Univ. Sci. Budapest. Sect. Comput.*, **15**(1995) 65–69; *MR* **97k**:11026.

R. L. Graham & N. J. A. Sloane, On additive bases and harmonious graphs, *SIAM J. Alg. Discrete Math.*, **1**(1980) 382–404.

S.W. Graham, B_h-sequences, *Analytic Number Theory*, Vol. 1 (*Allerton Park*, 1995) 431–439, *Progr. Math.*, **138**, Birkhäuser Boston MA, 1996; *MR* **97h**:11019.

Ben Green, The number of squares and $B_h[g]$ sets, *Acta Arith.*, **100**(2001) 365–390; *MR* **2003d**:11033.

G. Grekos, L. Haddad, C. Helou & J. Pihko, On the Erdős-Turán conjecture, *J. Number Theory*, **102**(2003) 339–352.

Laurent Habsieger & Alain Plagne, Ensembles $B_2[2]$: l'étau se reserre, *Integers*, **2**(2002) Paper A2, 20pp.; *MR* **2002m**:11010.

H. Halberstam & K. F. Roth, *Sequences*, 2nd Edition, Springer, New York, 1982, Chapter 2.

Elmer K. Hayashi, Omega theorems for the iterated additive convolution of a nonnegative arithmetic function, *J. Number Theory*, **13**(1981) 176–191; *MR* **82k**:10067.

Martin Helm, Some remarks on the Erdős-Turán conjecture, *Acta Arith.*, **63** (1993) 373–378; *MR* **94c**:11012.

Gábor Horváth, On a theorem of Erdős and Fuchs, *Acta Arith.*, **103**(2002) 321–328; *MR* **2003k**:11019.

Gábor Horváth, On a generalization of a theorem of Erdős and Fuchs, *Acta Math. Hungar.*, **92**(2001) 83–110; *MR* **2003m**:11017.

Jia Xing-De, Some problems and results on subsets of asymptotic bases, *Qufu Shifan Daxue Xuebao Ziran Kexue Ban*, **13**(1987) 45–49; *MR* **88k**:11015.

Jia Xing-De, On the distribution of a B_2-sequence, *Qufu Shifan Daxue Xuebao Ziran Kexue Ban*, **14**(1988) 12–18; *MR* **89j**:11023.

Jia Xing-De, On finite Sidon sequences, *J. Number Thoery*, **44**(1993) 84–92.

F. Krückeberg, B_2-Folgen und verwandte Zahlenfolgen, *J. reine angew. Math.*, **206**(1961) 53–60.

Vsevolod Lev & Tomasz Schoen, Cameron-Erdős modulo a prime, *Finite Fields Appl.*, **8**(2002) 108–119; *MR* **2002k**:11029.

B. Lindström, An inequality for B_2-sequences, *J. Combin. Theory*, **6**(1969) 211–212; *MR* **38** #4436.

Bernt Lindström, $B_h[g]$-sequences from B_h-sequences. *Proc. Amer. Math. Soc.*, **128**(2000) 657–659; *MR* **2000e**:11022.

Bernt Lindström, Well distribution of Sidon sets in residue classes. *J. Number Theory*, **69**(1998) 197–200; *MR* **99c**:11021.

Bernt Lindström, Finding finite B_2-sequences faster, *Math. Comput.*, **67**(1998) 1173–1178; *MR* **98m**:11012.

Greg Martin & Kevin O'Bryant, Continuous Ramsey theory and Sidon sets, 70-page preprint, 2002.

M. B. Nathanson, Unique representation bases for the integers, *Acta Arith.*, **108**(2003) 1–8; *MR* **2004c**:11013.

K. G. Omel'yanov & A. A. Sapozhenko, On the number of sum-free sets in an interval of natural numbers, *Diskret. Mat.*, **14**(2002) 3–7; translation in *Discrete Math. Appl.*, **12**(2002) 319–323; *MR* **2003m**:11015.

Imre Z. Ruzsa, A just basis, *Monatsh. Math.*, **109**(1990) 145–151; *MR* **91e**:11016.

Imre Z. Ruzsa, Solving a linear equation in a set of integers I, *Acta Arith.*, **65**(1993) 259–282; *MR* **94k**:11112.

Imre Z. Ruzsa, Sumsets of Sidon sets, *Acta Arith.*, **77**(1996)353–359; *MR* **97j**:11013.

Imre Z. Ruzsa, An infinite Sidon sequence, *J. Number Theory*, **68**(1998) 63–71; *MR* **99a**:11014.

Imre Z. Ruzsa, A small maximal Sidon set, *Ramanujan J.*, **2**(1998) 55–58; *MR* **99g**:11026.

A. Sárközy, On a theorem of Erdős and Fuchs, *Acta Arith.*, **37**(1980) 333–338; *MR* **82a**:10066.

A. Sárközy, Finite addition theorems, II, *J. Number Theory*, **48**(1994) 197–218; *MR* **95k**:11010.

Tomasz Schoen, The distribution of dense Sidon subsets of $\mathbb{Z}_m$, *Arch. Math. (Basel)*, **79**(2002) 171–174; *MR* **2003f**:11022.

J. Singer, A theorem in finite projective geometry and some applications to number theory, *Trans. Amer. Math. Soc.*, **43**(1938) 377–385; *Zbl* **19**, 5.

Vera T. Sós, An additive problem in different structures, *Graph Theory, Combinatorics, Algorithms, and Applications*, (SIAM Conf., San Francisco, 1989), 1991, 486–510; *MR* **92k**:11026.

Zhang Zhen-Xiang, Finding finite B_2-sequences with larger $m - a_m^{1/2}$, *Math. Comput.*, **63**(1994) 403–414; *MR* **94i**:11109.

OEIS: A039836.

C10 Modular difference sets and error correcting codes.

Singer's result, mentioned in **C9**, is based on **perfect difference sets**, i.e., a set of residues $a_1, a_2, \ldots, a_{k+1}$ (mod n) such that every nonzero residue (mod n) can be expressed uniquely in the form $a_i - a_j$. For example, $\{1, 2, 4\}$ mod 7 and $\{1, 2, 5, 7\}$ mod 13. Perfect difference sets can exist only if $n = k^2 + k + 1$, and Singer proved that such a set exists whenever k is a prime power. Marshall Hall has shown that numerous non-prime-powers *cannot* serve as values of k and Evans & Mann that there is no such $k < 1600$ that is not a prime power. It is conjectured that no perfect difference set exists unless k is a prime power.

Can a given finite sequence, which contains no repeated differences, always be extended to form a perfect difference set?

Perfect difference sets may be used to make **Golomb rulers**. Subtract one from the elements of the difference set, e.g., $\{0,1,4,6\}$ and take these as marks on a ruler of length 6, which can be used to measure all the lengths 1, 2, 3, 4, 5, 6. More generally we may look for less than perfect rulers of length n with $k+1$ marks $\{0, a_1, \ldots, a_{k-1}, n\}$ subject to various conditions. E.g., (a) all $\binom{k+1}{2}$ distances distinct, (b) maximum number of distinct distances for given n and k, (c) all integer distances from 1 up to some maximum e to be measurable. We cannot satisfy all of these conditions if $k \geq 4$, but Leech has found examples of perfect 'jointed' rulers. The trees

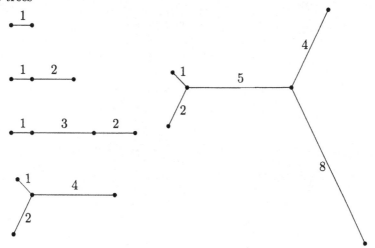

have edges with the lengths shown, and may be used to measure all lengths from 1 up to 1, 3, 6, 6, 15.

Gibbs & Slater, Herbert Taylor and Yang Yuan-Sheng have improved Leech's results for paths and for more general trees to

n	2	3	4	5	6	7	8	9	10	11	12
paths	1	3	6	9	13	18	24	29	37	45	(51)
trees	1	3	6	9	15	20	26	34	41	(48)	(55)

where the entries in parentheses are not necessarily best possible. There are connexions with the graceful labelling and harmonious labelling of graphs; see **C13** and a possibly forthcoming combinatorics volume in this series.

Graham & Sloane exhibit the problem of difference sets as the modular version of the packing problems of **C9**. They define $v_\gamma(k)$ [respectively $v_\delta(k)$] as the smallest number v such that there exists a subset $A = \{0 = a_1 < a_2 < \cdots < a_k\}$ of the integers (mod v) with the property that each r can be written in at most one way as $r \equiv a_i + a_j \bmod v$ with $i < j$ [respectively $i \leq j$].

Their interest in v_γ is in its application to **error-correcting codes**. If $A(k, 2d, w)$ is the maximum number of binary vectors with w ones and $k - w$ zeros (**words** of **length** k and **weight** w) such that any two vectors differ in at least $2d$ places, then (for $d = 3$)

$$A(k, 6, w) \geq \binom{k}{w} \Big/ v_\gamma(k)$$

(and the result for general d uses sets for which all sums of $d - 1$ distinct elements are distinct modulo v).

They note that $A(k, 2d, w)$ has been studied by Erdős & Hanani, by Schönheim, and by Stanton, Kalbfleisch & Mullin in the context of extremal set theory. Let $D(t, k, v)$ be the maximum number of k-element subsets of a v-element set S such that every t-element subset of S is contained in at most one of the k-element subsets. Then $D(t, k, v) = A(v, 2k - 2t + 2, k)$.

The values of v_δ in Table 4 are from Baumert's Table 6.1 and those of v_γ from Graham & Sloane who give the following bounds

$$k^2 - O(k) < v_\gamma(k) < k^2 + O(k^{36/23}),$$
$$k^2 - k + 1 \leq v_\delta(k) < k^2 + O(k^{36/23}).$$

Equality holds on the left of the latter whenever $k - 1$ is a prime power.

Table 4. Values of v_γ, v_δ and Exemplary Sets.

k	$v_\gamma(k)$	Example of A	$v_\delta(k)$	Example of A
2	2	{0,1}	3	{0,1}
3	3	{0,1,2}	7	{0,1,3}
4	6	{0,1,2,4}	13	{0,1,3,9}
5	11	{0,1,2,4,7}	21	{0,1,4,14,16}
6	19	{0,1,2,4,7,12}	31	{0,1,3,8,12,18}
7	28	{0,1,2,4,8,15,20}	48	{0,1,3,15,20,38,42}
8	40	{0,1,5,7,9,20,23,35}	57	{0,1,3,13,32,36,43,52}
9	56	{0,1,2,4,7,13,24,32,42}	73	{0,1,3,7,15,31,36,54,63}
10	72	{0,1,2,4,7,13,23,31,39,59}	91	{0,1,3,9,27,49,56,61,77,81}

L. D. Baumert, *Cyclic Difference Sets*, Springer Lect. Notes Math. **182**, New York, 1971.

M. R. Best, A. E. Brouwer, F. J. MacWilliams, A. M. Odlyzko & N. J. A. Sloane, Bounds for binary codes of length less than 25, *IEEE Trans. Inform. Theory*, **IT-24**(1978) 81–93.

F. T. Boesch & Li Xiao-Ming, On the length of Golomb's rulers, *Math. Appl.*, **2**(1989) 57–61; *MR* **91h**:11015.

J. H. Conway & N. J. A. Sloane, *Sphere Packings, Lattices and Groups*, Springer-Verlag, New York, 1988.

P. Erdős & H. Hanani, On a limit theorem in combinatorical analysis, *Publ. Math. Debrecen*, **10**(1963) 10–13.

T. A. Evans & H. Mann, On simple difference sets, *Sankhyā*, **11**(1951) 357–364; *MR* **13**, 899.

Richard A. Gibbs & Peter J. Slater, Distinct distance sets in a graph, *Discrete Math.*, **93**(1991) 155–165.

M. J. E. Golay, Note on the representation of 1, 2, ..., n by differences, *J. London Math. Soc.*(2), **4**(1972) 729–734; *MR* **45** #6784.

R. L. Graham & N. J. A. Sloane, Lower bounds for constant weight codes, *IEEE Trans. Inform. Theory*, **IT-26**(1980) 37–43; *MR* **81d**:94026.

R. L. Graham & N. J. A. Sloane, On additive bases and harmonious graphs, *SIAM J. Algebraic Discrete Methods*, **1**(1982) 382–404; *MR* **82f**:10067.

M. Hall, Cyclic projective planes, *Duke Math. J.*, **14**(1947) 1079–1090; *MR* **9**, 370.

S. V. Konyagin & A. P. Savin, Universal rulers (Russian), *Mat. Zametki*, **45**(1989) 136–137; *MR* **90k**:51038.

John Leech, On the representation of 1, 2, ..., n by differences, *J. London Math. Soc.*, **31**(1956) 160–169; *MR* **19**, 942f.

John Leech, Another tree labelling problem, *Amer. Math. Monthly* **82**(1975) 923–925.

F. J. MacWilliams & N. J. A. Sloane, *The Theory of Error-Correcting Codes*, North-Holland, 1977.

J. C. P. Miller, Difference bases: three problems in the additive theory of numbers, A. O. L. Atkin & B. J. Birch (editors) *Computers in Number Theory*, Academic Press, London, 1971, pp. 299–322; *MR* **47** #4817.

J. Schönheim, On maximal systems of k-tuples, *Stud. Sci. Math. Hungar.*, **1**(1966) 363–368.

R. G. Stanton, J. G. Kalbfleisch & R. C. Mullin, Covering and packing designs, *Proc. 2nd Conf. Combin. Math. Appl.*, Chapel Hill, 1970, 428–450.

Herbert Taylor, A distinct distance set of 9 nodes in a tree of diameter 36, *Discrete Math.*, **93**(1991) 167–168.

B. Wichmann, A note on restricted difference bases, *J. London Math. Soc.*, **38**(1963) 465–466; *MR* **28** #2080.

OEIS: A007187, A034470.

C11 Three-subsets with distinct sums.

One can generalize the ideas of **C9** and **E28** and define a B_h-**sequence** to be one in which the sums of h terms are distinct. Bose & Chowla showed that if $A_h(n)$ is the largest cardinality of a B_h-sequence in $[1, n]$, then

$$A_h(n) \geq n^{1/h}(1 + o(1)).$$

In the opposite direction, Jia has shown that if $h = 2k$, then

$$A_h(n) \leq k^{1/2k}(k!)^{1/k}n^{1/h}(1 + o(1)).$$

For $h = 2k - 1$, Chen and Graham independently proved the bound

$$A_h(n) \leq (k!)^{2/(2k-1)}n^{1/h}(1 + o(1)).$$

For $h = 3$ Graham obtained the further small improvement
    Martin Helm showed that no sequence containing $A(n) \sim \alpha n^{1/3}$ terms can be a B_3-sequence.

$$A_3(n) \leq \left(4 - \tfrac{1}{228}\right)^{1/3} n^{1/3}(1 + o(1)).$$

In the infinite case, Erdős offers \$500.00 for a proof or disproof of

$$\text{¿} \qquad \liminf \frac{A_h(n)}{n^{1/h}} = 0 \qquad ?$$

For $h = 2$ this was proved by Erdős himself, and for $h = 4$ by Nash. The case $h = 6$ was treated by Jia, and more generally Chen showed that if $h = 2k$ was even, then

$$\liminf A_h(n) \left(\frac{\ln n}{n}\right)^{\frac{1}{h}} < \infty$$

It remains an open problem to prove this when h odd.
    The paper of Chen & Kløve gives references to the electrical engineering literature on B_h-sequences.

W. C. Babcock, Intermodulation interference in radio systems, *Bell Systems Tech. J.*, **32**(1953) 63–73.

R. C. Bose & S. Chowla, *Report Inst. Theory of Numbers*, Univ. of Colorado, Boulder, 1959, p. 335.

R. C. Bose & S. Chowla, Theorems in the additive theory of numbers, *Comment. Math. Helvet.*, **37**(1962-63) 141–147.

Sheng Chen, On size of finite Sidon sequences, *Proc. Amer. Math. Soc.*, (to appear).

Sheng Chen, A note on B_{2k}-sequences, *J. Number Theory*, bf(1994)

Sheng Chen, On Sidon sequences of even orders, *Acta Arith.*, **64**(1993) 325–330.

Chen Wen-De & Torliev Kløve, Lower bounds on multiple difference sets, *Discrete Math.*, **98**(1991) 9–21; *MR* 93a:05028.

S. W. Graham, Upper bounds for B_3-sequences, Abstract 882-11-25, *Abstracts Amer. Math. Soc.*, **14**(1993) 416.

S. W. Graham, Upper bounds for Sidon sequences, (preprint 1993).

D. Hajela, Some remarks on $B_h[g]$-sequences, *J. Number Theory*, **29**(1988) 311–323; *MR* **90d**:11022.

Martin Helm, On B_{2k}-sequences, *Acta Arith.*, **63**(1993) 367–371; *MR* **95c**:11029.

Martin Helm, A remark on B_{2k}-sequences, *J. Number Theory*, **49**(1994) 246–249; *MR* **96b**:11024.

Martin Helm, On B_3-sequences, *Analytic Number Theory, Vol. 2 (Allerton Park, 1995)* 465–469, *Progr. Math.*, **139**, Birkhäuser, Boston, 1996; *MR* **97d**:11040.

Martin Helm, On the distribution of B_3-sequences, *J. Number Theory*, **58** (1996) 124–129; *MR* **97d**11041.

Jia Xing-De, On B_6-sequences, *Qufu Shifan Daxue Xuebao Ziran Kexue Ban*, **15**(1989) 7–11; *MR* **90j**:11022.

Jia Xing-De, on B_{2k}-sequences, *J. Number Theory*, (to appear).

T. Kløve, Constructions of $B_h[g]$-sequences, *Acta Arith.*, **58**(1991) 65–78; *MR* **92f**:11033.

Li An-Ping, On B_3-sequences, *Acta Math. Sinica*, **34**(1991) 67–71; *MR* **92f**: 11037.

B. Lindström, A remark on B_4-sequences, *J. Combin. Theory*, **7**(1969) 276–277.

John C. M. Nash, On B_4-sequences, *Canad. Math. Bull.*, **32**(1989) 446–449; *MR* **91e**: 11025.

Jukka Pihko, Proof of a conjecture of Selmer, *Acta Arith.*, **65**(1993) 117–135; *MR* **94k**:11010.

Alain Plagne, Removing one element from an exact additive basis, *J. Number Theory*, **87**(2001) 306–314; *MR* **2002d**:11014.

Imre Z. Ruzsa, An infinite Sidon sequence, *J. Number Theory*, **68**(1998) 63–71; *MR* **99a**:11014.

I. Z. Ruzsa, An almost polynomial Sidon sequence, *Studia Sci. Math. Hungar.*, **38**(2001) 367–375; *MR* **2002k**:11030.

Richard Warlimont, Über eine die Basen der natürlich Zahlen betreffende Ungleichung von Alfred Stöhr, *Arch. Math. (Basel)* **62**(1994) 227–229; *MR* **94m**: 11017.

R. Windecker, Eine Abschnittsbasis dritter Ordnung, *Kongel. Norske Vidensk. Selsk. Skrifter*, **9**(1976) 1–3.

C12 The postage stamp problem.

The covering problem which is dual to the packing problem **C9** goes back at least to Rohrbach. A popular form of it concerns the design of a set of integer denominations of postage stamp, $A_k = \{a_1, a_2, \ldots, a_k\}$ with $1 = a_1 < a_2 < \cdots < a_k$ to be used on envelopes with room for at most h

stamps, so that all integer amounts of postage up to a given bound can be affixed. What is the smallest integer $N(h, A_k)$ which is *not* representable by a linear combination $\sum_{i=1}^{k} x_i a_i$ with $x_i \geq 0$ and $\sum_{i=1}^{k} x_i \leq h$? The number of consecutive possible amounts of postage, $n(h, A_k) = N(h, A_k) - 1$ is called the h-**range** (German: h-Reichweite) of A_k. In this context A_k is called an **additive basis of order** h or h-**basis**. At first the main interest was in the 'global' problem: Given h and k, find an **extremal basis** A_k^* with largest possible h-range, $n(h, k) = n(h, A_k^*) = \max_{A_k} n(h, A_k)$. More recently the 'local' aspect has come more into focus: Find $n(h, A_k)$ when h and a particular basis A_k are given.

The local problem is completely solved only for $k = 2$ and $k = 3$. Trivially $n(h, A_2) = (h + 3 - a_2)a_2 - 2$ for $h \geq a_2 - 2$. Rødseth developed a general method, based on a continued fraction algorithm, for determining $n(h, A_3)$. From this, Selmer derived explicit formulas covering (asymptotically) about 99% of all A_3.

From the formula for $n(h, A_2)$ Stöhr concluded that

$$n(h, 2) = \lfloor (h^2 + 6h + 1)/4 \rfloor.$$

Shallit has shown that the local problem is NP-hard.

The global problem for $k = 3$ was solved by Hofmeister, who showed in particular that, for $h \geq 20$,

$$n(h, 3) = \tfrac{4}{3}(\tfrac{h}{3})^3 + 6(\tfrac{h}{3})^2 + Ah + B$$

where A and B depend on the residue of h modulo 9. Mossige showed that

$$n(h, 4) \geq 2.008 \left(\tfrac{h}{4}\right)^4 + O(h^3)$$

and, together with Kirfel (so far unpublished) that this bound is sharp. Kirfel has also shown that the limit

$$c_k = \lim_{h \to \infty} \frac{n(h, k)}{(h/k)^k}$$

exists for all $k \geq 1$. Kolsdorf showed that

$$n(h, 5) \geq 3.06 \left(\tfrac{h}{5}\right)^5 + O(h^4).$$

Mrose showed that

$$n(h_1 + h_2, k_1 + k_2) \geq (n(h_1, k_1) + 1) \cdot (n(h_2, k_2) + 1) \qquad (*)$$

for all positive integers $h - 1$, h_2, k_1, k_2 and deduced that if k_i $(i = 1, 2)$ are fixed and

$$n(h, k_i) \geq \alpha_i \left(\frac{h}{k_i}\right)^{k_i} + O(h^{k_i - 1})$$

then

$$n(h, k_1 + k_2) \geq \alpha_1 \alpha_2 \left(\frac{h}{k_1 + k_2}\right)^{k_1 + k_2} + O(h^{k_1 + k_2 - 1}).$$

Thus if x_i are fixed nonnegative integers with $k = \sum_{i=1}^{5} i x_i$ then

$$n(h,\, k) \geq (3.06)^{x_5} (2.008)^{x_4} \left(\frac{4}{3}\right)^{x_3} \left(\frac{h}{k}\right)^k + O(h^{k-1})$$

The best general upper bound for fixed k is due to Rødseth:

$$n(h,\, k) \leq \frac{(k-1)^{k-2}}{(k-2)!} \left(\frac{h}{k}\right)^k + O(h^{k-1}).$$

For fixed h, emphasis has been on the case $h = 2$. In 1937 Rohrbach showed that

$$c_1 \left(\frac{k}{2}\right)^2 + O(k) \leq n(2,\, k) \leq c_2 \left(\frac{k}{2}\right)^2 + O(k)$$

with $c_1 = 1$ and $c_2 = 1.9968$. After several improvements the best known results are $c_1 = \frac{8}{7}$ (Mrose) and $c_2 = 1.9208$ (Klotz). Windecker showed that

$$n(3,\, k) \geq \frac{4}{3} \left(\frac{k}{3}\right)^3 + \frac{16}{3} \left(\frac{k}{3}\right)^2 + O(k).$$

Again, with ($*$), if y_i are fixed nonnegative integers with $h = \sum_{i=1}^{3} i y_i$ then

$$n(h,\, k) \geq \left(\frac{4}{3}\right)^{y_3} \left(\frac{8}{7}\right)^{y_2} \left(\frac{k}{h}\right)^h + O(k^{h-1}).$$

Graham & Sloane (compare **C9**, **C10**) define $n_{\alpha(k)}$ [respectively $n_{\beta(k)}$] as the largest number n such that there is a k-element set $A = \{0 = a_1 < a_2 < \cdots < a_k\}$ of the integers with the property that each r in $[1, n]$ can be written in at least one way as $r = a_i + a_j$ with $i < j$ [respectively $i \leq j$], so that their $n_{\beta(k)}$ is here written $n(2, k-1)$, and their $n_{\alpha(k)}$ corresponds to the problem of two stamps of different denominations, with a zero denomination included.

They give the values for $n_{\alpha(k)}$ and $n_{\beta(k)}$ in Table 5.

Table 5. Values for $n_{\alpha(k)}$ and $n_{\beta(k)}$ and Exemplary Sets.

k	$n_{\alpha}(k)$	Example of A	$n_{\beta}(k)$	Example of A
2	1	{0,1}	2	{0,1}
3	3	{0,1,2}	4	{0,1,2}
4	6	{0,1,2,4}	8	{0,1,3,4}
5	9	{0,1,2,3,6}	12	{0,1,3,5,6}
6	13	{0,1,2,3,6,10}	16	{0,1,3,5,7,8}
7	17	{0,1,2,3,4,8,13}	20	{0,1,2,5,8,9,10}
8	22	{0,1,2,3,4,8,13,18}	26	{0,1,2,5,8,11,12,13}
9	27	{0,1,2,3,4,5,10,16,22}	32	{0,1,2,5,8,11,14,15,16}
10	33	{0,1,2,3,4,5,10,16,22,28}	40	{0,1,3,4,9,11,16,17,19,20}
11	40	{0,1,2,4,5,6,10,13,20,27,34}	46	{0,1,2,3,7,11,15,19,21,22,24}
12	47	{0,1,2,3,6,10,14,18,21,22,23,24}	54	{0,1,2,3,7,11,15,19,23,25,26,28}
13	56	{0,1,2,4,6,7,12,14,17,21,30,39,48}	64	{0,1,3,4,9,11,16,21,23,28,29,31,32}
14	65	{0,1,2,4,6,7,12,14,17,21,30,39,48,57}	72	{0,1,3,4,9,11,16,20,25,27,32,33,35,36}

Tables for more general $n(h, k)$ were computed by Lunnon and extended by Mossige and recently by Challis.

Serious students of these problems will consult the encyclopædic three volumes of Selmer, which contain 121 references. A useful summary, with 47 references, has been prepared by Djawadi & Hofmeister.

M. F. Challis, Two new techniques for computing extremal h-bases A_k, *Comput. J.*, **36**(1993) 117–126. [An updating of the appendix is available from the author.]

M. Djawadi & G. Hofmeister, The postage stamp problem, *Mainzer Seminarberichte, Additive Zahlentheorie*, **3**(1993) 187–195.

P. Erdős & R. L. Graham, On bases with an exact order, *Acta Arith.*, **37**(1980) 201–207; *MR* **82e**:10093.

P. Erdős & M. B. Nathanson, Partitions of bases into disjoint unions of bases, *J. Number Theory*, **29**(1988) 1–9; *MR* **89f**:11025.

Paul Erdős & Melvyn B. Nathanson, Additive bases with many representations, *Acta Arith.*, **52**(1989) 399–406; *MR* **91e**:11015.

Georges Grekos, Extremal problems about additive bases, *Acta Math. Inform. Univ. Ostraviensis*, **6**(1998) 87–92; *MR* **2002d**:11013.

N. Hämmerer & G. Hofmeister, Zu einer Vermutung von Rohrbach, *J. reine angew. Math.*, **286/287**(1976) 239–247; *MR* **54** #10181.

G. Hofmeister, Asymptotische Abschätzungen für dreielementige extremalbasen in natürlichen Zahlen, *J. reine angew. Math.*, **232**(1968) 77–101; *MR* **38** #1068.

G. Hofmeister, Die dreielementigen Extremalbasen, *J. reine angew. Math.*, **339**(1983) 207–214.

G. Hofmeister, C. Kirfel & H. Kolsdorf, Extremale Reichweitenbasen, No. **60**, Dept. Pure Math., Univ. Bergen, 1991.

Jia Xing-De, On a combinatorial problem of Erdős and Nathanson, *Chinese Ann. Math. Ser. A*, **9**(1988) 555–560; *MR* **90i**:11018.

Jia Xing-De, On the order of subsets of asymptotic bases, *J. Number Theory* **37**(1991) 37–46; *MR* **92d**:11006.

Jia Xing-De, Some remarks on minimal bases and maximal nonbases of integers, *Discrete Math.*, **122**(1993) 357–362; *MR* **94k**:11013.

Jia Xing-De & Melvyn B. Nathanson, A simple construction of minimal asymptotic bases, *Acta Arith.*, **52**(1989) 95–101; *MR* **90g**:11020.

Jia Xing-De & Melvyn B. Nathanson, Addition theorems for σ-finite groups, The Rademacher legacy to mathematics (University Park, 1992) *Contemp. Math.*, **166**, Amer. Math. Soc., 1994, 275–284.

Christoph Kirfel, Extremale asymptotische Reichweitenbasen, *Acta Arith.*, **60**(1992) 279–288; *MR* **92m**:11012.

W. Klotz, Extremalbasen mit fester Elementeanzahl, *J. reine andew. Math.*, **237**(1969) 194–220.

W. Klotz, Eine obere Schranke für die Reichweite einer Extremalbasis zweiter Ordnung, *J. reine angew. Math.*, **238**(1969) 161–168; *MR* **40** #117, 116.

W. F. Lunnon, A postage stamp problem, *Comput. J.*, **12**(1969) 377–380; *MR* **40** #6745.

L. Moser, On the representation of 1, 2, ..., n by sums, *Acta Arith.*, **6**(1960) 11–13; *MR* **23** #A133.

Robert W. Owens, An algorithm to solve the Frobenius problem, *Math. Mag.*, **76**(2003) 264–275.

L. Moser, J. R. Pounder & J. Riddell, On the cardinality of h-bases for n, *J. London Math. Soc.*, **44**(1969) 397–407; *MR* **39** #162.

S. Mossige, Algorithms for computing the h-range of the postage stamp problem, *Math. Comput.*, **36**(1981) 575–582; *MR* **82e**:10095.

S. Mossige, On extremal h-bases A_4, *Math. Scand.*, **61**(1987) 5–16; *MR* **89e**:11008.

Svein Mossige, The postage stamp problem: the extremal basis for $k = 4$, *J. Number Theory*, **90**(2001) 44–61; *MR* **2003c**:11009.

A. Mrose, Untere Schranken für die Reichweiten von Extremalbasen fester Ordnung, *Abh. Math. Sem. Univ. Hamburg*, **48**(1979) 118–124; *MR* **80g**:10058.

J. C. M. Nash, Some applications of a theorem of M. Kneser, *J. Number Theory*, **44**(1993) 1–8; *MR* **94k**:11015.

John C. M. Nash, Freiman's theorem answers a question of Erdős, *J. Number Theory*, **27**(1987) 7–8; *MR* **88k**:11014.

Melvyn B. Nathanson, Extremal properties for bases in additive number theory, Number Theory, Vol. I (Budapest, 1987), *Colloq. Math. Soc. János Bolyai* **51**(1990) 437–446; *MR* **91h**:11009.

H. O. Pollak, The postage stamp problem, *Consortium*, **69**(1999) 3–5.

J. Riddell & C. Chan, Some extremal 2-bases, *Math. Comput.*, **32**(1978) 630–634; *MR* **57** #16244.

Ø. Rødseth, On h-bases for n, *Math. Scand.*, **48**(1981) 165–183; *MR* **82m**: 10034.

Øystein J. Rødseth, An upper bound for the h-range of the postage stamp problem, *Acta Arith.* **54**(1990) 301–306; *MR* **91h**:11013.

H. Rohrbach, Ein Beitrag zur additiven Zahlentheorie, *Math. Z.*, **42**(1937) 1–30; *Zbl.* **15**, 200.

H. Rohrbach, Anwendung eines Satzes der additiven Zahlentheorie auf eine gruppentheoretische Frage, *Math. Z.*, **42**(1937) 538–542; *Zbl.* **16**, 156.

Ernst S. Selmer, On the postage stamp problem with three stamp denominations, *Math. Scand.*, **47**(1980) 29–71; *MR* **82d**:10046.

Ernst S. Selmer, The Local Postage Stamp Problem, Part I General Theory, Part II The Bases A_3 and A_4, Part III Supplementary Volume, No. 42, 44, 57, Dept. Pure Math., Univ. Bergen, (86-04-15, 86-09-15, 90-06-12) ISSN 0332-5047. (Available on request.)

Ernst S. Selmer, On Stöhr's recurrent h-bases for N, *Kongel. Norske Vidensk. Selsk.*, Skr. 3, 1986, 15 pp.

Ernst S. Selmer, Associate bases in the postage stamp problem, *J. Number Theory*, **42**(1992) 320–336; *MR* **94b**:11012.

Ernst S. Selmer & Arne Rödne, On the postage stamp problem with three stamp denominations, II, *Math. Scand.*, **53**(1983) 145–156; *MR* **85j**:11075.

Ernst S. Selmer & Björg Kristin Selvik, On Rödseth's h-bases $A_k = \{1, a_2, 2a_2, \ldots, (k-2)a_2, a_k\}$, *Math. Scand.*, **68**(1991) 180–186; *MR* **92k**:11010.

Jeffrey Shallit, The computational complexity of the local postage stamp problem, *SIGACT News*, **33**(2002) 90–94.

H. S. Shulz, The postage-stamp problem, number theory, and the programmable calculator, *Math. Teacher*, **92**(1999) 20–26.

Alfred Stöhr, Gelöste und ungelöste Fragen über Basen der natürlichen Zahlenreihe I, II, *J. reine angew. Math.*, **194**(1955) 40–65, 111–140; *MR* **17**, 713.

OEIS: A014616.

C13 The corresponding modular covering problem. Harmonious labelling of graphs.

Just as **C10** was the modular version of the packing problem **C9**, so we can propose the modular version of the corresponding covering problem **C12**.

Graham & Sloane complete their octad of definitions with $n_\gamma(k)$ [respectively $n_\delta(k)$] as the largest number n such that there is a subset $A = \{0 = a_1 < a_2 < \cdots < a_k\}$ of the residue classes modulo n with the property that each r can be written in at least one way as $r \equiv a_i + a_j \bmod n$ with $i < j$ [respectively $i \leq j$].

Table 6. Values of n_γ, n_δ and Exemplary Sets.

k	$n_\gamma(k)$	Example of A	$n_\delta(k)$	Example of A
2	1	—	3	{0,1}
3	3	{0,1,2}	5	{0,1,2}
4	6	{0,1,2,4}	9	{0,1,3,4}
5	9	{0,1,2,4,7}	13	{0,1,2,6,9}
6	13	{0,1,2,3,6,10}	19	{0,1,3,12,14,15}
7	17	{0,1,2,3,4,8,13}	21	{0,1,2,3,4,10,15}
8	24	{0,1,2,4,8,13,18,22}	30	{0,1,3,9,11,12,16,26}
9	30	{0,1,2,4,10,15,17,22,28}	35	{0,1,2,7,8,11,26,29,30}
10	36	{0,1,2,3,6,12,19,20,27,33}		

They call a connected graph with v vertices and $e \geq v$ edges **harmonious** if there is a labelling of the vertices x with distinct labels $l(x)$ so that when an edge xy is labelled with $l(x) + l(y)$, the edge labels form a complete system of residues (mod e). Trees (for which $e = v - 1$) are also called harmonious if just one vertex label is duplicated and the edge labels form a complete system (mod $v - 1$). The connexion with the present problem is that $n_\gamma(v)$ is the greatest number of edges in any harmonious graph on v vertices.

For example, from Table 6 we note that $n_\gamma(5) = 9$ is attained by the set {0,1,2,4,7} so that a maximum of 9 edges can occur in a harmonious graph on 5 vertices (Figure 6).

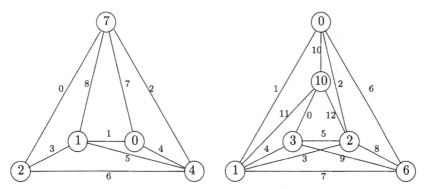

Figure 6. Maximal Harmonious Graphs.

Graham & Sloane compare and contrast harmonious graphs with graceful graphs, which will be discussed in the graph theory chapter of a later volume in this series. A graph is **graceful** if, when the vertex labels are chosen from $[0, e]$ and the edge labels are calculated by $|l(x) - l(y)|$, the latter are all distinct (i.e., take the values $[1, e]$).

Trees are conjectured to be both harmonious and graceful, but these are open questions. A cycle C_n is harmonious just if n is odd, and graceful just if $n \equiv 0$ or $3 \mod 4$. The friendship graph or windmill is harmonious just if $n \not\equiv 2 \mod 4$ and graceful just if $n \equiv 0$ or $1 \mod 4$. Fans and wheels are both harmonious and graceful, as is the Petersen graph. The graphs of the five Platonic solids are naturally graceful and one would expect them to be harmonious, but this is not so for the cube or octahedron. Joseph Gallian maintains a bibliography of graph labelling.

Joseph A. Gallian, A survey – recent results, conjectures and open problems in labeling graphs, *J. Graph Theory*, **13**(1989) 491–504; *MR* **90k:**05132.

B. Liu & X. Zhang, On harmonious labelings of graphs, *Ars Combin.*, **36**(1993) 315–326.

C14 Maximal sum-free sets.

Maximal sum-free sets. Denote by $l(n)$ the largest l so that if $a_1, a_2, \ldots,$ a_n are any distinct natural numbers, one can always find l of them so that $a_{i_j} + a_{i_k} \neq a_m$ for $1 \leq j < k \leq l$, $1 \leq m \leq n$. Note that $j \neq k$, else the set $\{a_i = 2^i \mid 1 \leq i \leq n\}$ would imply that $l(n) = 0$. A remark of Klarner shows that $l(n) > c \ln n$. On the other hand, the set $\{2^i + 0, \pm1 \mid 1 < i \leq s + 1\}$ implies that $l(3s) < s + 3$, so $l(n) < \frac{1}{3}n + 3$. Selfridge extends this by using the set $\{(3m + t)2^{m-i} \mid -i < t < i, 1 \leq i \leq m\}$ to show that $l(m^2) < 2m$. Choi, using sieve methods, has improved this to $l(n) \ll n^{0.4+\epsilon}$.

Neil Calkin notes the relevance of Peter Cameron's survey article and his own thesis to Steven Finch's 0-additive sequences problem (ref. at **C4**). Finch has calculated $1\frac{1}{2}$ million terms of the sequence beginning

$\{3, 4, 6, 9, 10, 17\}$ and continuing with the next least integer which is not the sum of two distinct earlier terms. He detected no regularity (ultimate periodicity of differences) and believes that this may be due to a massive initial segment of irregular values, while Calkin suspects that there may be counterexamples to Finch's conjecture.

The problem can be generalized to ask if, for every l, there is an $n_0 = n_0(l)$ so that if $n > n_0$ and a_1, a_2, ..., a_n are any n elements of a group with no product $a_{i_1} a_{i_2} = e$, the unit [here i_1, i_2 may be equal, so there is no a_i of order 1 or 2, and no a_i whose inverse is also an a_i] then there are l of the a_i such that $a_{i_j} a_{i_k} \neq a_m$, $1 \leq j < k \leq l$, $1 \leq m \leq n$. This has not even been proved for $l = 3$.

For the generalization to sets containing no solution of $a_1 x_1 + \ldots + a_k x_k = x_{k+1}$ see the papers of Funar and Moree.

Compare **E12**, **E32**.

Harvey L. Abbott, On a conjecture of Funar concerning generalized sum-free sets, *Niew Arch. Wisk.*(4), **4**(1991) 249–252.

Noga Alon & Daniel J. Kleitman, Sum-free subsets, *A Tribute to Paul Erdős*, 13–26, Cambridge Univ. Press, 1990; *MR* **92f**:11020.

Jean Bourgain, Estimates related to sumfree subsets of sets of integers, *Israel J. Math.*, **97**(1997) 71–92; *MR* **97m**:11026.

Neil J. Calkin, Sum-free sets and measure spaces, PhD thesis, Univ. of Waterloo, 1988.

Neil J. Calkin, On the number of sum-free sets, *Bull. London Math. Soc.*, **22**(1990) 141–144; *MR* **91b**:11015.

Neil J. Calkin, On the structure of a random sum-free set of positive integers, *Discrete Math.*, **190**(1998) 247–257. 99f:11015

Neil J. Calkin & Peter J. Cameron, Almost odd random sum-free sets, *Combin. Probab. Comput.*, **7**(1998) 27–32; *MR* **99c**:11012.

Neil J. Calkin & Paul Erdős, On a class of aperiodic sum-free sets, *Math. Proc. Cambridge Philos. Soc.*, **120**(1996) 1–5; *MR* **97b**:11030.

Neil J. Calkin & Steven R. Finch, Conditions on periodicity for sum-free sets, *Experiment. Math.*, **5**(1996) 131–137; *MR* **98c**:11010.

Neil J. Calkin & Jan MacDonald Thomson, Counting generalized sum-free sets, *J. Number Theory*, **68**(1998) 151–159; *MR* **98m**:11010.

Peter J. Cameron, Portrait of a typical sum-free set, Surveys in Combinatorics 1987, *London Math. Soc. Lecture Notes*, **123**(1987) Cambridge Univ. Press, 13–42.

S. L. G. Choi, On sequences not containing a large sum-free subsequence, *Proc. Amer. Math. Soc.*, **41**(1973) 415–418; *MR* **48** #3910.

S. L. G. Choi, On a combinatorial problem in number theory, *Proc. London Math. Soc.*,(3) **23**(1971) 629–642; *MR* **45** #1867.

Louis Funar, Generalized sum-free sets of integers, *Nieuw Arch. Wisk.*(4) **8**(1990) 49–54; *MR* **91e**:11012.

Vsevolod Lev, Sharp estimates for the number of sum-free sets, *J. reine angew. Math.*, **555**(2003) 1–25.

Vsevolod Lev, Tomasz Łuczak & Tomasz Schoen, Sum-free sets in abelian groups, *Israel J. Math.*, **125**(2001) 347–367; *MR* **2002i**:11023.

Tomasz Łuczak & Tomasz Schoen, On infinite sum-free sets of integers, *J. Number Theory*, **66**(1997) 211–224.

Tomasz Łuczak & Tomasz Schoen, On strongly sum-free subsets of abelian groups, *Colloq. Math.*, **71**(1996) 149–151.

Pieter Moree, On a conjecture of Funar, *Nieuw Arch. Wisk.*(4) **8**(1990) 55–60; *MR* **91e**:11013.

Tomasz Schoen, The number of (2,3)-sum-free subsets of $\{1, \ldots, n\}$, *Acta Arith.*, **98**(2001) 155–163; *MR* **2002m**11014.

Anne Penfold Street, A maximal sum-free set in A_5, *Utilitas Math.*, **5**(1974) 85–91; *MR* **49** #7156.

Anne Penfold Street, Maximal sum-free sets in abelian groups of order divisible by three, *Bull. Austral. Math. Soc.*, **6**(1972) 317–318; *MR* **47** #5147.

P. Varnavides, On certain sets of positive density, *J. London Math. Soc.*, **34**(1959) 358–360; *MR* **21** #5595.

Edward T. H. Wang, On double-free set of integers, *Ars Combin.*, **28**(1989) 97–100; *MR* **90d**:11011.

Yap Hian-Poh, Maximal sum-free sets in finite abelian groups, *Bull. Austral. Math. Soc.*, **4**(1971) 217–223; *MR* **43** #2081 [and see *ibid*, **5**(1971) 43–54; *MR* **45** #3574; *Nanta Math.*, **2**(1968) 68–71; *MR* **38** #3345; *Canad. J. Math.*, **22**(1970) 1185–1195; *MR* **42** #1897; *J. Number Theory*, **5**(1973) 293–300; *MR* **48** #11356].

C15 Maximal zero-sum-free sets.

Erdős & Heilbronn asked for the largest number $k = k(m)$ of distinct residue classes, modulo m, so that no subset has sum zero. For example, the set

$$1, \ -2, \ 3, \ 4, \ 5, \ 6$$

shows that $k(20) \geq 6$, and in fact equality holds. The pattern of this example shows that

$$k \geq \lfloor (-1 + \sqrt{8m+9})/2 \rfloor \qquad (m > 5)$$

Equality holds for $5 < m \leq 24$. However, Selfridge observes that if m is of the form $2(l^2 + l + 1)$, the set

$$1, \ 2, \ \ldots, \ l-1, \ l, \ \tfrac{1}{2}m, \ \tfrac{1}{2}m+1, \ \ldots, \ \tfrac{1}{2}m+l$$

implies that

$$k \geq 2l + 1 = \sqrt{2m - 3}$$

In fact he conjectures that, for any even m, this set or the set with l deleted always gives the best result. For example, $k(42) \geq 9$.

On the other hand, if p is a prime in the interval

$$\tfrac{1}{2}k(k+1) < p < \tfrac{1}{2}(k+1)(k+2)$$

he conjectures that $k(p) = k$, where the set can be simply

$$1, \ 2, \ \ldots, \ k$$

The case $k(43) = 8$ was confirmed by Clement Lam, so k is not a monotonic function of m.

The only case where a better inequality is known than $k \geq \lfloor \sqrt{2m-3} \rfloor$ is $k(25) \geq \sqrt{50-1} = 7$, as is shown by the set 1, 6, 11, 16, 21, 5, 10. If m is of the form $25l(l+1)/2$ and *odd*, then it is possible to improve on the set 1, -2, 3, 4, $\ldots$, but if m is of that form and *even*, then the construction already given for m even is always better.

Is $k = \lfloor (-1 + \sqrt{8m+9})/2 \rfloor$ for an infinity of values of m?

For which values of m are there realizing sets none of whose members are prime to m? For example, $m = 12$: $\{3,4,6,10\}$ or $\{4,6,9,10\}$. Is there a value of m for which *all* realizing sets are of this type?

Erdős & Heilbronn proved that if a_1, a_2, $\ldots$, a_k, $k \geq 3(6p)^{1/2}$, are distinct residues mod p, where p is prime, then every residue mod p can be written in the form $\sum_{i=1}^{k} \epsilon_i a_i$, $\epsilon_i = 0$ or 1. They conjectured that the same holds for $k > 2\sqrt{p}$ and that this is best possible and Olsen proved this. They further conjectured that the number, s, of distinct residues of the form $a_i + a_j$, $1 \leq i < j \leq k$, is at least $\min\{p, 2k-3\}$. Partial results have been obtained by Mansfield, by Rødseth and by Freiman, Low & Pitman. Dias da Silva & Hamidoune gave a complete proof of the Erdős-Heilbronn conjecture, and in fact proved that if A^h denotes the set of all sums of h distinct elements of A, $A \subseteq \mathbb{Z}/p\mathbb{Z}$, $|A| = k$, then $|A^h| \geq \min\{p, hk - h^2 + 1\}$. Nathanson simplifies their proof and, with Ruzsa, shows that if A, $B \subseteq \mathbf{Z}/p\mathbf{Z}$, $|A| = k > l = |B|$, then $|\{a+b : a \in A, b \in B, a \neq b\}| \geq \min\{p, k+l-2\}$.

One can ask the same question for members of more general groups. For example, if A is any sequence of size g of the group $G_n = Z_n \times Z_n$, and $s(G_n)$ is the least g such that there is always a zero-sum subsequence of A of size n, then it is easy to see that $s(G_n) > 4n-4$: take the sequence containing $n-1$ copies of each of (0,0), (0,1), (1,0) and (1,1). Kemnitz has conjectured that $s(G_n) = 4n-3$ for all n. Rónyai has proved that $s(G_p) \leq 4p-2$ for all primes p and Gao obtained the corresponding result with prime powers in place of primes. See also the paper of Thangadurai.

N. Alon, & M. Dubiner, Zero-sum sets of prescribed size, *Combinatorics, Paul Erdős is eighty*, Vol. 1, 33–50, *Bolyai Soc. Math. Stud., János Bolyai Math. Soc.*, Budapest, 1993; *MR* **94j**:11016.

Noga Alon, Melvyn B. Nathanson & Imre Ruzsa, Adding distinct congruence classes modulo a prime, *Amer. Math. Monthly*, **102**(1995) 250–255; *MR* **95k**:11009.

Noga Alon, Melvyn B. Nathanson & Imre Ruzsa, The polynomial method and restricted sums of congruence classes, *J. Number Theory*, **56**(1996) 404–417; *MR* **96k**:11011.

A. Bialostocki & P. Dierker, On the Erdős-Ginzburg-Ziv theorem and the Ramsey numbers for stars and matchings, *Discrete Math.*, **110**(1992) 1–8; *MR* **94a**:05147.

Arie Bialostocki, Paul Dierker & David Grynkiewicz, On some developments of the Erdős-Ginzburg-Ziv theorem, II, *Acta Arith.*, **110**(2003) 173–184.

A. Bialostocki & M. Lotspeich, Some developments of the Erdős-Ginzburg-Ziv theorem, I, *Sets, Graphs & Numbers (Budapest, 1991), Colloq. Math. Soc. János Bolyai*, **60**(1992), North-Holland, Amsterdam, 97–117; *MR* **94 k**:11021.

W. Brakemeier, *Ein Beitrag zur additiven Zahlentheorie*, Dissertation, Tech. Univ. Braunschweig 1973.

W. Brakemeier, Eine Anzahlformel von Zahlen modulo n, *Monatsh. Math.*, **85**(1978) 277–282.

Cao Hui-Qin & Sun Zhi-Wei, On sums of distinct representatives, *Acta Arith.*, **87**(1998) 159–169; *MR* **99k**:11021.

A. L. Cauchy, Recherches sur les nombres, *J. École Polytech.*, **9**(1813) 99–116.

H. Davenport, On the addition of residue classes, *J. London Math. Soc.*, **10**(1935) 30–32.

H. Davenport, A historical note, *J. London Math. Soc.*, **22**(1947) 100–101.

J. M. Deshouillers, P. Erdős $ A. Sárközi, On additive bases, *Acta Arith.*, **30**(1976) 121–132; *MR* **54** #5165.

Jean-Marc Deshouillers & Étienne Fouvry, On additive bases, II, *J. London Math. Soc.*(2), **14**(1976) 413–422; *MR* **56** #2951.

J. A. Dias da Silva & Y. O. Hamidoune, Cyclic spaces for Grassmann derivatives and additive theory, *Bull. London Math. Soc.*, **26**(1994) 140–146; see also *Linear Algebra Appl.*, **141**(1990) 283–287; *MR* **91i**:15004.

Shalom Eliahou & Michel Kervaire, Restricted sums of sets of cardinality $1+p$ in a vector space over F_p, *Discrete Math.*, **235**(2001) 199–213; *MR* **2002e**:11023.

R. B. Eggleton & P. Erdős, Two combinatorial problems in group theory, *Acta Arith.*, **21**(1972) 111–116; *MR* **46** #3643.

P. Erdős, Some problems in number theory, in *Computers in Number Theory*, Academic Press, London & New York, 1971, 405–413.

P. Erdős, A. Ginzburg & A. Ziv, A theorem in additive number theory, *Bull. Res. Council Israel Sect. F Math. Phys.*, **10F**(1961/62) 41–43.

P. Erdős & H. Heilbronn, On the addition of residue classes mod p, *Acta Arith.*, **9**(1964) 149–159; *MR* **29** #3463.

Carlos Flores & Oscar Ordaz, On the Erdős-Ginzburg-Ziv theorem, *Discrete Math.*, **152**(1996) 321–324; *MR* **97b**:05160.

G. A. Freiman, *Foundations of a Structural Theory of Set Addition*, Transl. Math. Monographs, **37**(1973), Amer. Math. Soc., Providence RI.

Gregory A. Freiman, Lewis Low & Jane Pitman, Sumsets with distinct summands and the Erdős-Heilbronn conjecture on sums of residues. Structure theory of set addition, *Astérisque*, **258**(1999) 163–172; *MR* **2000j**:11036.

Luis Gallardo & Georges Grekos, On Brakemeier's variant of the Erdős-Ginzberg-Ziz problem, *Tatra Mt. Math. Publ.*, **20**(2000) 91–98; *MR* **2002e**:11014.

L. Gallardo, G. Grekos, L. Habsieger, B. Landreau & A. Plagne, Restricted addition in $\mathbb{Z}/n\mathbb{Z}$ and an application to the Erdős-Ginzburg-Ziv theorem, *J. London Math. Soc.*(2), **65**(2002) 513–523; *MR* **2003b**:11011.

L. Gallardo, G. Grekos & J. Pihko, On a variant of the Erdős-Ginzburg-Ziv problem, *Acta Arith.*, **89**(1999) 331–336; *MR* **2000k**:11026.

Gao Wei-Dong, An addition theorem for finite cyclic groups, *Discrete Math.*, **163**(1997) 257–265; *MR* **97k**:20039.

Gao Wei-Dong, Addition theorems for finite abelian groups, *J. Number Theory*, **53**(1995) 241–246; *MR* **96i**:20070.

Gao Wei-Dong, On zero-sum subsequences of restricted size, *J. Number Theory*, **61**(1996) 97–102; II. *Discrete Math.* **271**(2003) 51–59; III. *Ars Combin.*, **61**(2001) 65–72; *MR* **97m**:11027; **2003e**:11019.

Gao Wei-Dong, On the Erdős-Ginsburg-Ziv theorem—a survey, *Paul Erdős and his mathematics* (*Budapest*, 1999) 76–78, János Bolyai Math. Soc., Budapest 1999.

Gao Wei-Dong, Note on a zero-sum problem, *J. Combin. Theory Ser. A*, **95**(2001) 387–389; *MR* **2002i**:11016.

Gao Wei-Dong & Alfred Geroldinger, On zero-sum sequences in $Z/nZ \oplus Z/nZ$, *Electr. J. Combin. Number Theory*, **3**(2003) A8.

Gao Wei-Dong & Y. O. Hamidoune, Zero sums in abelian groups, *Combin. Probab. Comput.*, **7**(1998) 261–263; *MR* **99h**:05118.

David J. Grynkiewicz, On four coloured sets with nondecreasing diameter and the Erdős-Ginzburg-Ziv theorem, *J. Combin. Theory Ser. A*, **100**(2002) 44–60; *MR* **2003i**:11032..

Y. O. Hamidoune, A. S. Lladó & O. Serra, On restricted sums, *Combin. Probab. Comput.*, **9**(2000) 513–518; *MR* **2002a**:20059.

Yahya Ould Hamidoune, Oscar Ordaz & Asdrubal Ortuño, On a combinatorial theorem of Erdős, Ginzburg and Ziv, *Combin. Probab. Comput.*, **7**(1998) 403–412; *MR* **2000c**:05146.

Yahya Ould Hamidoune & Gilles Zémor, On zero-free subset sums, *Acta Arith.*, **78**(1996) 143–152; *MR* **97k**:11023.

H. Harborth, Ein Extremalproblem für Gitterpunkte, *J. reine angew. Math.*, **262/263**(1973) 356–360; *MR* **48** #6008.

N. Hegyvári, F. Hennecart & A. Plagne, A proof of two Erdős' conjectures on restricted addition and further results, *J. reine angew. Math.*, **560**(2003) 199–220.

F. Hennecart, Additive bases and subadditive functions, *Acta Math. Hungar.*, **77**(1997) 29–40; *MR* **99a**:11013.

François Hennecart, Une propriété arithmétique des bases additives. Une critère de non-base, *Acta Arith.*, **108**(2003) 153–166; *MR* **2004c**:11012.

Hou Qing-Hu & Sun Zhi-Wei, Restricted sums in a field, *Acta Arith.*, **102**(2002) 239–249; *MR* **2003e**:11025.

A. Kemnitz, On a lattice point problem, *Ars Combin.*, **16**(1983) 151–160; *MR* **85c**:52022.

J. H. B. Kemperman, On small sumsets in an abelian group, *Acta Math.*, **103**(1960) 63–88; *MR* **22** #1615.

Vsevolod F. Lev, Restricted set addition in groups: I. The classical setting, *J. London Math. Soc.*(2), **62**(2000) 27–40; *MR* **2001g**:11024.

Li Zheng-Xue, Liu Hui-Qin & Gao Wei-Dong, A conjecture concerning the zero-sum problem, *J. Math. Res. Exposition*, **21**(2001) 619–622; *MR* **2002i**:11018.

Liu Jian-Xin & Sun Zhi-Wei, A note on the Erdős-Ginsburg-Ziv theorem, *Nanjing Daxue Xuebao Kiran Kexue Ban*, **37**(2001) 473–476; *MR* **2002j**:11014.

Henry B. Mann & John E. Olsen, Sums of sets in the elementary abelian group of type (p, p), *J. Combin. Theory*, **2**(1967) 275–284.

R. Mansfield, How many slopes in a polygon? *Israel J. Math.*, **39**(1981) 265–272; *MR* **84j**:03074.

Melvyn B. Nathanson, Ballot numbers, alternating products, and the Erdős-Heilbronn conjecture, *The Mathematics of Paul Erdős*, I 199–217, *Algorithms Combin.*, **13** Springer Berlin 1997; *MR* **97i**11008.

Melvyn B. Nathanson, *Additive number theory. Inverse problems and the geometry of sumsets*, Graduate Texts in Mathematics, **165**, Springer-Verlag, New York, 1996; *MR* **98f**:11011.

M. B. Nathanson & I. Z. Ruzsa, Sums of different residues modulo a prime, 1993 preprint.

John E. Olsen, An addition theorem modulo p, *J. Combin. Theory*, **5**(1968) 45–52.

John E. Olsen, An addition theorem for the elementary abelian group, *J. Combin. Theory*, **5**(1968) 53–58.

Pan Hao & Sun Zhi-Wei, A lower bound for $|\{a + b \colon a \in A, b \in B, P(a, b) \neq 0\}|$, *J. Combin. Theory Ser. A* **100**(2002) 387–393; *MR* **2003k**:11016.

J. M. Pollard, A generalisation of the theorem of Cauchy and Davenport, *J. London Math. Soc.*, **8**(1974) 460–462.

J. M. Pollard, Addition properties of residue classes, *J. London Math. Soc.*, **11**(1975) 147–152.

U.-W. Rickert, *Über eine Vermutung in der additiven Zahlentheorie*, Dissertation, Tech. Univ. Braunschweig 1976.

Øystein J. Rødseth, Sums of distinct residues mod p, *Acta Arith.*, **65**(1993) 181–184; *MR* **94j**:11004.

Øystein J. Rødseth, On the addition of residue classes mod p, *Monatsh. Math.*, **121**(1996) 139–143; *MR* **97a**:11009.

Lajos Rónyai, On a conjecture of Kemnitz, *Combinatorica*, **20**(2000) 569–573; *MR* **2001k**:11025.

C. Ryavec, The addition of residue classes modulo n, *Pacific J. Math.*, **26** (1968) 367–373.

Sun Zhi-Wei, Restricted sums of subsets of $\mathbb{Z}$, *Acta Arith.*, **99**(2001) 41–60; *MR* **2002j**:11016.

Sun Zhi-Wei, Unification of zero-sum problems, subset sums and covers of $\mathbb{Z}$, *Electr. Res. Announcements, Amer. Math. Soc.*, **9**(2003) 51–60.

B. Sury & R. Thangadurai, Gao's conjecture on zero-sum sequences, *Proc. Indian Acad. Sci. Math. Sci.*, **112**(2002) 399–414; *MR* **2003k**:11029.

E. Szemerédi, On a conjecture of Erdős and Heilbronn, *Acta Arith.*, **17**(1970-71) 227–229.

R. Thangadurai, On a conjecture of Kemnitz, *C.R. Math. Acad. Sci. Soc. R. Can.*, **23**(2001) 39–45; *MR* **2002i**:11017.

R. Thangadurai, Non-canonical extensions of Erdős-Ginzburg-Ziv theorem, *Integers*, **2**(2002) Paper A7, 14pp.; *MR* **2003d**:11025.

R. Thangadurai, Interplay between four conjectures on certain zero-sum problems, *Expo. Math.*, **20**(2002) 215–228; *MR* **2003f**:11023.

R. Thangadurai, On certain zero-sum problems in finite abelian groups, *Current trends in number theory* (*Allahabad*, 2000) 239–254, Hindustan Book Agency, 2002; *MR* **2003f**:11031.

Wang Chao[2], Note on a variant of the Erdős-Ginzburg-Ziv problem, *Acta Arith.*, **108**(2003) 53–59; *MR* **2004c**:11020.

OEIS: A034463.

C16 Nonaveraging sets. Nondividing sets.

A **nonaveraging set** A of integers $0 \le a_1 < a_2 < \cdots < a_n \le x$ was defined by Erdős & Straus by the property that no a_i shall be the arithmetic mean of any subset of A with more than one elent. Denote by $f(x)$ the maximum number of elements in such a set, and by $g(x)$ the maximum number of elements in a subset B of the integers $[0, x]$ such that no two distinct subsets of B have the same arithmetic mean, and by $h(x)$ the corresponding maximum where the subsets of B have different cardinality. Erdős & Straus and others cited below have shown that (note that $\log x = (\ln x)/(\ln 2)$ is the logarithm to base 2):

$$\frac{1}{4}\log x + O(1) < \log f(x) < \frac{1}{2}(\log x + \log \ln x) + O(1)$$

$$\frac{1}{2}\log x - 1 < g(x) < \log x + O(\ln \ln x)$$

and conjecture that $f(x) = \exp(c\sqrt{\ln x}) = o(x^{\epsilon})$ and that $h(x) = (1 + o(1))\log x$. The inequalities for $h(x)$ were misprinted in the second edition.

Erdős originally asked for the maximum number, $k(x)$, of integers in $[0, x]$ so that no one divides the sum of any others. Such **nondividing sets** are obviously nonaveraging, so $k(x) \le f(x)$. Straus showed that $k(x) \ge \max\{f(x/f(x)), f(\sqrt{x})\}$.

Abbott has shown that if $l(n)$ is the largest m such that *every* set of n integers contains a nonaveraging subset of size m, then $l(n) > n^{1/13 - \epsilon}$.

Compare **C14–16** with **E10–14**.

H. L. Abbott, On a conjecture of Erdős and Straus on non-averaging sets of integers, *Congr. Numer. XV, Proc. 5th Brit. Combin. Conf., Aberdeen*, 1975, 1–4; *MR* **53** #10752.

H. L. Abbott, Extremal problems on non-averaging and non-dividing sets, *Pacific J. Math.*, **91**(1980) 1–12; *MR* **82j**:10005.

H. L. Abbott, On nonaveraging sets of integers, *Acta Math. Acad. Sci. Hungar.*, **40**(1982) 197–200; *MR* **84g**:10090.

H. L. Abbott, On the Erdős–Straus non-averaging set problem, *Acta Math. Hungar.* **47**(1986) 117–119; *MR* **88e**:11010.

Á. P. Bosznay, On the lower estimation of non-averaging sets, *Acta Math. Hungar.*, **53**(1989) 155–157; *MR* **90d**:11016.

Fan R. K. Chung & John L. Goldwasser, Integer sets containing no solution to $x + y = 3z$, *The Mathematics of Paul Erdős, I* 218–227, *Algorithms Combin.*, **13** Springer Berlin 1997.

P. Erdős & A. Sárközy, On a problem of Straus, *Disorder in physical systems*, Oxford Univ. Press, New York, 1990, pp. 55–66; *MR* **91i**:11012.

P. Erdős & E. G. Straus, Non-averaging sets II, in *Combinatorial Theory and its Applications* II, *Colloq. Math. Soc. János Bolyai* **4**, North-Holland, 1970, 405–411; *MR* **47** #4804.

E. G. Straus, Non-averaging sets, *Proc. Symp. Pure Math.*, **19**, Amer. Math. Soc., Providence 1971, 215–222.

OEIS: A036779, A037021.

C17 The minimum overlap problem.

The minimum overlap problem. Let $\{a_i\}$ be an arbitrary set of n distinct integers, $1 \le a_i \le 2n$, and $\{b_j\}$ be the complementary set $1 \le b_j \le 2n$, with $b_j \ne a_i$. M_k is the number of solutions of $a_i - b_j = k$ ($-2n < k < 2n$) and $M = \min \max_k M_k$, where the minimum is taken over all sequences $\{a_i\}$. Erdős proved that $M > n/4$; Scherk improved this to $M > (1 - 2^{-1/2})n$ and Swierczkowski to $M > (4 - \sqrt{6})n/5$. Leo Moser obtained the further improvements $M > \sqrt{2}(n-1)/4$ and $M > \sqrt{4 - \sqrt{15}}(n-1)$. In the other direction, Motzkin, Ralston & Selfridge obtained examples to show that $M < 2n/5$, contrary to Erdős's conjecture that $M = \frac{1}{2}n$. Is there a number c such that $M \sim cn$?

The first few values of $M(n)$ are

n	1	2	3	4	5	6	7	8	9	10	11	12	13	14	15
$M(n)$	1	1	2	2	3	3	3	4	4	5	5	5	6	6	6

It's just the Law of Small Numbers that this is $\lfloor 5(n+3)/13 \rfloor$, because Haugland proves that if there is one value n_0 of n such that $M(n) \le tn_0$, then $\limsup M(i)/i \le t$, that 'sup' may be omitted from this statement, and that '-1' may be omitted from Leo Moser's result. He uses a theorem of Swinnerton-Dyer to improve the upper bound to $\lim M(n)/n \le 0.38200298812318988\ldots$.

Leo Moser asks the corresponding question where the cardinality of $\{a_i\}$ is not n, but k, where $k = \lfloor \alpha n \rfloor$ for some real α, $0 < \alpha < 1$.

A closely related problem is attributed to J. Czipszer: Let $A_k = \{a_1 + k, a_2 + k, \ldots, a_n + k\}$ where $a_1 < a_2 < \ldots < a_n$ are arbitrary integers and $k \ge 0$. Let M_k be the number of elements of A_k not in A_0 and $M = \min_{A_0} \max_{0 < k \le n} M_k$. Czipszer proved that $n/2 \le M \le 2n/3$ and conjectured that $M = 2n/3$. Katz & Schnitzer showed that $M > 0.6n$ for $n \ge 26$. Moser & Murdeshwar considered the continuous analog.

P. Erdős, Some remarks on number theory (Hebrew, English summary), *Riveon Lematematika*, **9**(1955) 45–48; *MR* **17**, 460.

Jan Kristian Haugland, Advances in the minimum overlap problem, *J. Number Theory*, **58**(1996) 71–78; *MR* **99c**:11031.

M. Katz & F. Schnitzer, On a problem of J. Czipszer, *Rend. Sem. Mat. Univ. Padova* **44**(1970) 85–90; *MR* **45** #8540.

L. Moser, On the minimum overlap problem of Erdős, *Acta Arith.*, **5**(1959) 117–119; *MR* **21** #5594.

L. Moser & M. G. Murdeshwar, On the overlap of a function with its translates, *Nieuw Arch. Wisk.*(3), **14**(1966) 15–18; *MR* **33** #4218.

T. S. Motzkin, K. E. Ralston & J. L. Selfridge, Minimum overlappings under translation, *Bull. Amer. Math. Soc.*, **62**(1956) 558.

S. Swierczkowski, On the intersection of a linear set with the translation of its complement, *Colloq. Math.*, **5**(1958) 185-197; *MR* **21** #1955.

C18 The n queens problem.

What is the minimum number of Queens which can be placed on an $n \times n$ chessboard so that every square is either occupied or attacked by a Queen? Berge noted that, in graph theory language, this is the same as finding the minimum externally stable set for a graph on n^2 vertices with two vertices joined just if they are on the same rank, file or diagonal. In his notation $\beta = 5$ for Queens (Figure 7a), 8 for Bishops (Figure 7b) and 12 for Knights (Figure 7c).

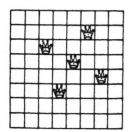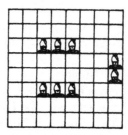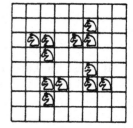

Figure 7. Minimum Covers of the Chessboard by Queens, Bishops and Knights.

Although there is no condition that a piece may not guard another piece, this condition is satisfied by the Queens and Bishops, but not by the Knights. Since in Chess a piece does not attack the square that it stands on, there are in fact two sets of problems. For example, Victor Meally notes that, if queens are allowed to guard one another, 3 queens (at $a6, c2, e4$) suffice on a 6×6 board, and 4 on a 7×7 board.

For the Queens, Kraitchik gave the following table for an $n \times n$ board.

n	5	6	7	8	9	10	11	12	13	14	15	16	17
number of Queens	3	4	5	5	5	5	5	6	7	8	9	9	9

Corresponding configurations for $n = 5, 6, 11$ are shown in Figure 8.

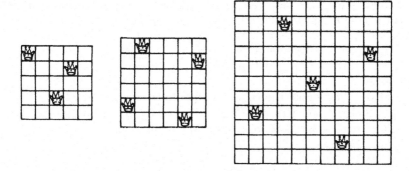

Figure 8. Queens Covering $n \times n$ Boards for $n = 5, 6$ and 11.

If we try to partition the numbers from 1 to $2n$ into n pairs a_i, b_i so that the $2n$ numbers $a_i \pm b_i$ fall one in each residue class modulo $2n$, then it is found to be impossible. Less restrictively, Shen & Shen asked that the $2n$ numbers $a_i \pm b_i$ be distinct. They gave examples for $n = 3$: 1, 5; 2, 3; 4, 6; for $n = 6$: 1, 10; 2, 6; 3, 9; 4, 11; 5, 8; 7, 12; and $n = 8$: 1, 10; 2, 14; 3, 16; 4, 11; 5, 9; 6, 12; 7, 15; 8, 13; and Selfridge showed that there was always a solution for $n \geq 3$. How many solutions are there for each n?

If the condition $b_i = i$ $(1 \leq i \leq n)$ is added, we have the reflecting Queens problem: place n queens on an $n \times n$ chessboard so that no two are on the same rank, file or diagonal, where, on a diagonal, we include reflexions in a mirror in the centre of the zero column (Fig. 9).

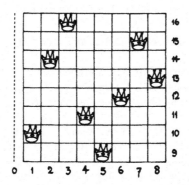

Figure 9. A Solution of the Reflecting Queens Problem.

We can again ask for the number of solutions for each n, both in the case where we distinguish between solutions obtained by rotation and reflexion and where we do not.

The classical form of the problem is to place n queens on an n by n board with no two attacking each other, and to ask in how many ways, $Q(n)$, can this be done. Also to ask the same question for $T(n)$, the number of ways on a toroidal board: Pólya (see Ahrens, pp. 363–374) proved that $T(n) > 0$ just if $n \perp 6$. Rivin, Vardi & Zimmermann conjecture that each of

$$\lim_{\substack{n \to \infty \\ n \perp 6}} \frac{\ln T(n)}{n \ln n} \quad \text{and} \quad \lim_{n \to \infty} \frac{\ln Q(n)}{n \ln n}$$

is positive. They quote the values of $Q(n)$ for $n = 1, 2, \ldots, 20$ as 1, 0, 0, 2, 10, 4, 40, 92, 352, 724, 2680, 14200, 73712, 365596, 2279184, 14772512, 95815104, 666090624, 4968057848, 39029188884.

They also give

n	5	7	11	13	17	19	23
$T(n)$	10	28	88	4524	140692	820496	128850048

For a **modular** $n \times n$ chessboard, on which each diagonal continues on the other side, Heden showed that $M(n)$, the maximum number of nonattacking queens, is n, $n - 1$, $n - 2$, $n - 4 \le M(n) \le n - 2$ or $n - 5 \le M(n) \le n - 2$ according as $\gcd(n, 6) = 1$, $\gcd(n, 12) = 2$, $\gcd(n, 12) = 3$ or 4, $\gcd(n, 12) = 6$ or $\gcd(n, 12) = 12$. He believes that $M(n) = n - 2$ in each of the last two cases and proves that $M(24) = 22$.

Mario Velucchi asks for the maximum number of **unattacked** squares when n queens are placed on an $n \times n$ chessboard. He gives

n	4	5	6	7	8	9	10	11	12	13	14	15	16	17	18	19	20	21	22	23	24	25	26	27	28	29	30
max	1	3	5	7	11	18	22	30	36	47	56	72	82	97	111	132	145	170	186	216	240	260	290	324	360	381	420

where, for $n \ge 14$, the values are described as 'probable'. He also asks the same question, but with the number of queens not necessarily equal to the side of the board.

W. Ahrens, *Mathematische Unterhaltungen und Spiele*, Vol. 1, B. G.Teubner, Leipzig 1921.

Claude Berge, *Theory of Graphs*, Methuen, 1962, p. 41.

Paul Berman, Problem 122, *Pi Mu Epsilon J.*, **3**(1959-64) 118, 412.

A. Bruen & R. Dixon, The n-queens problem, *Discrete Math.*, **12**(1975) 393–395; MR **51** #12555.

E. J. Cockayne & S. T. Hedetniemi, A note on the diagonal queens domination problem, DM-301-IR, Univ. of Victoria, BC, March 1984.

B. Gamble, B. Shepherd & E. T. Cockayne, Domination of chessboards by queens on a column, DM-318-IR, Univ. of Victoria, BC, July 1984.

Solomon W. Golomb & Herbert Taylor, Constructions and properties of Costas arrays, Dept. Elec. Eng., Univ. S. California, July 1981–Oct. 1983.

B. Hansche & W. Vucenic, On the n-queens problem, *Notices Amer. Math. Soc.*, **20**(1973) A-568.

Olof Heden, On the modular n-queens problem, *Discrete Math.*, **102**(1992) 155–161; *MR* **93d**:05041; II, **142**(1995) 97–106; *MR* **96e**:51007; III, **243**(2002) 135–150; *MR* **2002m**:51003.

G. B. Huff, On pairings of the first $2n$ natural numbers, *ActaArith.*, **23**(1973) 117–126.

D. A. Klarner, The problem of the reflecting queens, *Amer. Math. Monthly* **74**(1967) 953–955; *MR* **40** #7123.

Torleiv Kløve, The modular n-queen problem, *Discrete Math.*, **19**(1977) 289–291; *MR* **57** #2952; II, **36**(1981) 33–48; *MR* **82k**:05034.

Maurice Kraitchik, *Mathematical Recreations*, Norton, New York, 1942, 247–256.

Scott P. Nudelman, The modular n-queens problem in higher dimensions, *Discrete Math.*, **146**(1995) 159–167; *MR* **97b**:11028.

Patric R. J. Östergård & William D. Weakley, Values of domination numbers of the queen's graph, *Electron. J. Combin.*, **8**(2001) R29, 19pp.; *MR* **2002f**:05122.

Igor Rivin, Ilan Vardi & Paul Zimmermann, The n-queens problem, *Amer. Math. Monthly*, **101**(1994) 629–639; *MR* **95d**:05009.

J. D. Sebastian, Some computer solutions to the reflecting queens problem, *Amer. Math. Monthly*, **76**(1969) 399–400; *MR* **39** #4018.

J. L. Selfridge, Pairings of the first $2n$ integers so that sums and differences are all distinct, *Notices Amer. Math. Soc.*, **10**(1963) 195.

Shen Mok-Kong & Shen Tsen-Pao, Research Problem 39, *Bull. Amer. Math. Soc.*, **68** (1962) 557.

B. Shepherd, B. Gamble & E. T. Cockayne, Domination parameters for the bishops graph, DM-327-IR, Univ. of Victoria, BC, Oct. 1884.

M. Slater, Problem 1, *Bull. Amer. Math. Coc.*, **69**(1963) 333.

P. H. Spencer & E. J. Cockayne, An upper bound for the domination number of the queens graph, DM-376-IR, Univ. of Victoria, BC, June 1985.

Ilan Vardi, Computational Recreations in *Mathematica$^{©}$*, Addison-Wesley, Redwood City CA, 1991, Chap. 6.

OEIS: A000170, A002562, A065256.

C19 Is a weakly indedendent sequence the finite union of strongly independent ones?

Selfridge calls a set of positive integers $a_1 < a_2 < \cdots < a_k$ **independent** if $\sum c_i a_i = 0$ (where the c_i are integers, not all 0) implies that at least one of the c_i is < -1. By using the pigeonhole principle it is easy to show that if k positive integers are independent, then a_1 is at least 2^{k-1}. He offers \$10.00 for an answer to the question: is the set of k independent integers $a_i = 2^k - 2^{k-i}$ ($1 \leq i \leq k$) the only set with largest member less than 2^k? It is the only such set with $a_1 = 2^{k-1}$.

Call a(n infinite) sequence $\{a_i\}$ of positive integers **weakly indepen-dent** if any relation $\sum \epsilon_i a_i$ with $\epsilon_i = 0$ or ± 1 and $\epsilon_i = 0$ except finitely often, implies $\epsilon_i = 0$ for all i, and call it **strongly independent** if the same is true with $\epsilon_i = 0$. ± 1, or ± 2. Richard Hall asks if every weakly independent sequence is a finite union of strongly independent sequences.

J. L. Selfridge, Problem 123, *Pi Mu Epsilon J.*, **3**(1959-64) 118, 413–414.

C20 Sums of squares.

One would think that the last word had long been spoken on the classical problem of finding formulas for the number of ways of expressing an integer as the sum of s squares. But Shimura has recently given formulas for $2 \leq s \leq 8$, of which the ones for s odd are new, and Stephen Milne for $s = 4n^2$ or $s = 4n(n+1)$. See also the paper of Chan & Chua.

Paul Turán asks for a characterization of those positive integers n which can be expressed as the sum of four pairwise coprime squares, i.e., $n = x_1^2 + x_2^2 + x_3^2 + x_4^2$ with $x_i \perp x_j$ $(1 \leq i < j \leq 4)$. Leech notes that at most one of the x_i is even, so that $n \equiv 3, 4$ or $7 \bmod 8$. Similarly, numbers $n \equiv 2 \bmod 3$ are not so representable.

Turán also conjectured that all positive integers can be represented as the sum of at most five pairwise coprime squares, but Mąkowski (see reference at **B5**) notes that numbers $4^k(24l+15)$ with $k \geq 2$ can't be represented in this way. There are arbitrarily large integers that are not representable as the sum of *exactly* five coprime squares, since $3n = x_1^2 + \ldots + x_5^2$ implies that 3 divides two of the x_i. In fact, the product of two distinct primes of shape $24k + 7$ cannot be the sum of fewer than ten pairwise coprime squares, and Leech asks if this is the record.

Apart from 256 examples, the largest of which is 1167, every number can be expressed as the sum of at most five squares of *composite* numbers.

Chowla conjectures that every positive integer is the sum of at most four elements of the set $\{(p^2 - 1)/24\}$ where p is prime and $p \geq 5$. The smallest number which requires four such summands is 33.

Compare these problems with some earlier results of Wright, who showed, for example, that if $\lambda_1, \ldots, \lambda_4$ were given real numbers with sum 1, then every n with a sufficiently large odd factor is expressible as $n = m_1^2 + \ldots + m_4^2$ with $|m_i^2 - \lambda_i n| = o(n)$. He has similar results for five or more squares, and for three squares (provided, of course, that n is not of the form $4^k(8l + 7)$ in this last case).

For expression as the sum of *distinct* squares; if zero is allowed as one of the squares, then I correct and clarify earlier remarks by quoting Gordon Pall's Theorems 2 & 3, which are not as well known as they deserve.

The only integers $n > 0$ not sums of four squares ≥ 0 are $4^h a$, where $h = 0, 1, 2, \ldots$ and $a = 1, 3, 5, 7, 9, 11, 13, 15, 17, 19, 23, 25, 27, 31, 33,$

37, 43, 47, 55, 67, 73, 97, 103 or 2, 6, 10, 18, 22, 34, 58, 82.

The only integers $n > 0$ not sums of five unequal squares are: 1–29, 31–38, 40–45, 47–49, 52, 53, 56, 58–61, 64, 67–69, 72, 73, 76, 77, 80, 83, 89, 92, 96, 97, 101, 104, 108, 112, 124, 128, 136, 137, 188, 224.

If zero is not allowed, then Halter-Koch showed that every integer > 412 and not divisible by 8 is a sum of four distinct nonzero squares, and that every odd integer > 157 is so representable. He also showed that every integer > 245 is the sum of five distinct nonzero squares, every integer > 333 is the sum of six such, every integer > 390 the sum of seven such, every integer > 462 the sum of eight such, and so on, up to all > 1036 being the sum of twelve such squares. Bateman, Hildebrand & Purdy have produced a sequel to Halter-Koch's paper.

Štefan Porubský used a result of Cassels to give an affirmative answer to a question of R. E. Dressler: for each positive integer k, is every sufficiently large positive integer the sum of distinct k-th powers of primes?

One can also ask for every number to be expressible as the sum of as few as possible polygonal numbers of various kinds. For example there is Gauß's famous 1796-07-10 diary entry:

$$\text{EΥPHKA!} \qquad \text{num} = \triangle + \triangle + \triangle,$$

i.e., every number is expressible as the sum of three triangular numbers. For hexagonal numbers $r(2r-1)$, the answer is the same if you allow hexagonal numbers of negative rank, $r(2r+1)$, but if these are excluded, then 11 and 26 require six hexagonal numbers of positive rank to represent them. Is it possible that every sufficiently large number is expressible as the sum of of three such numbers? Equivalently, is every sufficiently large number $8n+3$ expressible as the sum of three squares of numbers of shape $4r-1$, with r positive? The corresponding question for pentagonal numbers $\frac{1}{2}r(3r-1)$ is to ask if every sufficiently large number of shape $24n+3$ is expressible as the sum of three squares of numbers of shape $6r-1$?

Schinzel and Sierpiński conjectured that there are infinitely many primes that are sums of two consecutive squares; and hence the existence of an infinity of triangular numbers that are not the sum of two positive triangular numbers.

Michael Hirschhorn proved Gosper's conjecture that every sum of four distinct odd squares is the sum of four distinct even squares. If $n \equiv 4 \bmod 8$ (so that a partition into four odd squares is at least possible), the number of partitions of n into four distinct odd squares is twice the number of partitions into four distinct even squares minus the number of partitions into three positive distinct even squares.

Kaplansky & Elkies showed that every number can be written as $x^2 + y^2 + z^3$, if z is allowed to be negative, and conjecture that there only a finite number of exceptions to $n = x^2 + y^3 + z^3$ where y and z are now restricted to being positive.

G. E. Andrews, EΥPHKA! num $= \triangle + \triangle + \triangle$, *J. Number Theory*, **23**(1986) 285–293.

Paul T. Bateman, Adolf Hildebrand & George B. Purdy, Expressing a positive integer as a sum of a given number of distinct squares of positive integers, *Acta Arith.*, **67**(1994)349–380; *MR* **95j**:11092.

Claus Bauer, A note on sums of five almost equal prime squares, *Arch. Math.* (*Basel*) **69**(1997) 20–30; *MR* **98c**:11110.

Claus Bauer, Liu Ming-Chit & Zhan Tao, On a sum of three prime squares, *J. Number Theory*), **85**(2000) 336–359; *MR* **2002e**:11135.

Jan Bohman, Carl-Erik Fröberg & Hans Riesel, Partitions in squares, *BIT*, **19**(1979) 297–301; see Gupta's *MR* **80k**:10043.

Jörg Brüdern & Étienne Fouvry, Lagrange's four squares theorem with almost prime variables, *J. reine angew. Math.*, **454**(1994) 59–96; *MR* **96e**:11125.

J. W. S. Cassels, On the representation of integers as the sums of distinct summands taken from a fixed set, *Acta Sci. Math. Szeged*, **21**(1960) 111–124; *MR* **24** #A103.

Chan Heng-Huat & Chua Kok-Seng, Representations of integers as sums of 32 squares, *Ramanujan J.*, (to appear)

Noam Elkies & Irving Kaplansky, Problem 10426, *Amer. Math. Monthly*, **102**(1995) 70; solution, Andrew Adler, **104**(1997) 574.

John A. Ewell, On sums of triangular numbers and sums of squares, *Amer. Math. Monthly*, **99**(1992) 752–757; *MR* **93j**:11021.

Andrew Granville & Zhu Yi-Liang, Representing binomial coefficients as sums of squares, *Amer. Math. Monthly* **97**(1990) 486–493; *MR* **92b**:11009.

Emil Grosswald & L. E. Mattics, Solutions to problem E 3262, *Amer. Math. Monthly*, **97**(1990) 240–242.

F. Halter-Koch, Darstellung natürliche Zahlen als Summe von Quadraten, *Acta Arith.*, **42**(1982) 11–20; *MR* **84b**:10025.

D. R. Heath-Brown & D. I. Tolev, Lagrange's four squares theorem with one prime and three almost-prime variables, *J. reine angew. Math.*, **558**(2003) 159–224; *MR* **2004c**:11180.

Michael D. Hirschhorn, On partitions into four distinct squares of equal parity, *Australas. J. Combin.*, **24**(2001) 285–291; *MR* **2002f**:05019.

Michael D. Hirschhorn, On sums of squares, *Austral. Math. Soc. Gaz.*, **29**(2002) 86–87.

Michael D. Hirschhorn, Comment on 'Dissecting squares', Note 87.33 *Math. Gaz.*, **87**(Jul 2003) 286–287.

Michael D. Hirschhorn & James A. Sellers, On representations of a number as a sum of three triangles, *Acta Arith.*, **77**(1996) 289–301; *MR* **97e**:11119.

Michael D. Hirschhorn & James A. Sellers, Some relations for partitions into four squares, *Special functions* (*Hong Kong*, 1999) 118–124, World Sci. Publishing, River Edge NJ, 2000; *MR* **2001k**:11201.

James G. Huard & Kenneth S. Williams, Sums of sixteen squares, *Far East J. Mat. Sci*, **7**(2002) 147–164; *MR* **2003m**:11059.

James G. Huard & Kenneth S. Williams, Sums of twelve squares, *Acta Arith.*, **109**(2003) 195–204; *MR* **2004a**:11029.

Lin Jia-Fu, The number of representations of an integer as a sum of eight triangular numbers, *Chinese Quart. J. Math.*, **15**(2000) 66–68; *MR* **2002e**:11049.

Liu Jian-Ya & Zhan Tao, On sums of five almost equal prime squares, *Acta Arith.*, **77**(1996) 369–383; *MR* **97g**:11113; II, *Sci. China Ser.*, **41**(1998) 710–722; *MR* **99h**:11115.

Stephen C. Milne, Infinite families of exact sums of squares formulas, Jacobi elliptic functions, continued fractions, and Schur functions. Reprinted from *Ramanujan J.*, **6**(2002) 7–149; *MR* **2003m**:11060. With a preface by George E. Andrews. *Developments in Mathematics*, **5**, Kluwer Academic Publishers, Boston MA, 2002; see also *Proc. Nat. Acad. Sci. U.S.A.*, **93**(1996) 15004–15008; *MR* **99k**:11162.

Hugh L. Montgomery & Ulrike M. A. Vorhauer, Greedy sums of distinct squares, *Math. Comput.*, **73**(2004) 493–513.

G. Pall, On sums of squares, *Amer. Math. Monthly*, **40**(1933) 10–18.

V. A. Plaksin, Representation of numbers as a sum of four squares of integers, two of which are prime, *Soviet Math. Dokl.*, **23**(1981) 421–424; transl. of *Dokl. Akad. Nauk SSSR*, **257**(1981) 1064–1066; *MR* **82h**:10027.

Štefan Porubský, Sums of prime powers, *Monatsh. Math.*, **86**(1979) 301–303.

A. Schinzel & W. Sierpiński, *Acta Arith.*, **4**(1958) 185–208; erratum, **5**(1959) 259; *MR* **21** #4936.

W. Sierpiński, Sur les nombres triangulaires qui sont sommes de deux nombres triangulaires, *Elem. Math.*, **17**(1962) 63–65.

Goro Shimura, The representation of integers as sums of squares, *Amer. J. Math.*, **124**(2002) 1059–1081; *MR* **20031**:11049.

R. P. Sprague, Über zerlegungen in ungleiche Quadratzahlen, *Math. Z.*, **51** (1949) 289–290; *MR* **10** 283d.

E. M. Wright, The representation of a number as a sum of five or more squares, *Quart. J. Math. Oxford*, **4**(1933) 37–51, 228–232.

E. M. Wright, The representation of a number as a sum of four 'almost proportional' squares, *Quart. J. Math. Oxford*, **7**(1936) 230–240.

E. M. Wright, Representation of a number as a sum of three or four squares, *Proc. London Math. Soc.*(2), **42**(1937) 481–500.

OEIS: A002808, A0224702, A055075, A055078.

C21 Sums of higher powers.

Denote by s_k the largest integer that is *not* the sum of distinct kth powers of positive integers. Sprague showed that $s_2 = 128$; Graham reported that $s_3 = 12758$ and Lin used his method to obtain $s_4 = 5134240$. Cam Patterson used his sieve, and a result of Richert, to obtain $s_5 = 67898771$.

Brüdern proved that all sufficiently large n can be represented as the sum of 17 distinct powers, $x_1^2 + x_2^3 + \cdots x_{17}^{18}$. Ford improves this to 15 and later to 14.

Brüdern also showed that almost all positive integers $m \equiv 4 \bmod 18$ can be expressed as the sum of three cubes of primes and a cube of P_4, where P_r has at most r prime factors. Kawada improved this to P_3.

Jörg Brüdern, Sums of squares and higher powers, I. II. *J. London Math. Soc.* (2), **35**(1987) 233–243, 244–250; *MR* **88f**:11096.

Jörg Brüdern, A problem in additive number theory, *Math. Proc. Cambridge Philos. Soc.*, **103**(1988) 27–33; *MR* **89c**:11150.

Jörg Brüdern, *Ann. Sci. École Norm. Sup.*, **28**(1995) 461–476; *MR* **96e**:11126.

Stephan Daniel, On gaps between numbers that are sums of three cubes, *Mathematika*, **44**(1997) 1–13; *MR* **98c**:11105.

Kevin B. Ford, The representation of numbers as sums of unlike powers, *J. London Math. Soc.* (2), **51**(1995) 14–26; *MR* **96c**:11115.

Kevin B. Ford, The representation of numbers as sums of unlike powers, II, *J. Amer. Math. Soc.*, **9**(1995) 919–940; *MR* **97g**:11111.

R. L. Graham, Complete sequences of polynomial values, *Duke Math. J.*, **31**(1964) 275–285; *MR* **29** #63.

Koichi Kawada, Note on the sums of cubes of primes and an almost prime, *Arch. Math. (Basel)* **69**(1997) 13–19; *MR* **98f**:11105.

Lin Shen, Computer experiments on sequences which form integral bases, in J. Leech (editor) *Computational Problems in Abstract Algebra*, 365–370, Pergamon, 1970.

Cameron Douglas Patterson, *The Derivation of a High Speed Sieve Device*, PhD thesis, The Univ. of Calgary, March 1992.

H. E. Richert, Über zerlegungen in paarweise verschiedene zahlen, *Norsk Mat. Tidskrift*, **31**(1949) 120–122.

Fernando Rodriguez Villegas & Don Zagier, Which primes are sums of two cubes? *Number Theory (Halifax NS, 1994)* 295–306, *CMS Conf. Proc.*, **15**, Amer. Math. Soc., 1995.

R. P. Sprague, Über zerlegungen in n-te Potenzen mit lauter verschiedenen Grundzahlen, *Math. Z.*, **51**(1948) 466–468; *MR* **10** 514h.

D. Diophantine Equations

"A subject which can be described briefly by saying that a great part of it is concerned with the discussion of the rational or integer solutions of a polynomial equation $f(x_1, x_2, \ldots, x_n) = 0$, with integer coefficients. It is well known that for many centuries, no other topic has engaged the attention of so many mathematicians, both professional and amateur, or has resulted in so many published papers."

This quotation from the preface of Mordell's book, *Diophantine Equations*, Academic Press, London, 1969, indicates that in this section we shall have to be even more eclectic than elsewhere. If you're interested in the subject, consult Mordell's book, which is a thoroughgoing but readable account of what is known, together with a great number of unsolved problems. There are well-developed theories of rational points on algebraic curves, so we mainly confine ourselves to higher dimensions, for which standard methods have not yet been developed.

D1 Sums of like powers. Euler's conjecture.

"It has seemed to many Geometers that this theorem [Fermat's Last Theorem] may be generalized. Just as there do not exist two cubes whose sum or difference is a cube, it is certain that it is impossible to exhibit three biquadrates whose sum is a biquadrate, but that at least four biquadrates are needed if their sum is to be a biquadrate, although no one has been able up to the present to assign four such biquadrates. In the same manner it would seem to be impossible to exhibit four fifth powers whose sum is a fifth power, and similarly for higher powers."

No advance was made on Euler's statement until 1911 when R. Norrie assigned four such biquadrates:

$$30^4 + 120^4 + 272^4 + 315^4 = 353^4$$

Fifty-five years later Lander & Parkin gave a counter-example to Euler's more general conjecture:

$$27^5 + 84^5 + 110^5 + 133^5 = 144^5$$

It's now well known, since his discovery hit the national newspapers, that Noam Elkies has disproved Euler's conjecture for fourth powers. The infinite family of solutions, of which the first member is

$$2682440^4 + 15365639^4 + 18796760^4 = 20615673^4,$$

comes from an elliptic curve given by $u = -5/8$ in the parametric solution of $x^4 - y^4 = z^4 + t^2$ given by Dem'janenko. The smallest solution,

$$95800^4 + 217519^4 + 414560^4 = 422481^4,$$

corresponding to $u = -9/20$, was subsequently found by Roger Frye. There are infinitely many values of u which give curves of positive rank, and further families of solutions.

The solution of Jan Kubiček, making the sum of three cubes a cube, co-incides with that of F. Vieta (see Dickson's *History* [**A17**], Vol. 2, pp. 550–551).

Simcha Brudno asks the following questions. Is there a parametric solution to $a^5 + b^5 + c^5 + d^5 = e^5$? [The above solution is the only one with $e \leq 765$.] Is there a parametric solution to $a^4 + b^4 + c^4 + d^4 = e^4$? Are there counterexamples to Euler's conjecture with powers higher than the fifth? Is there a solution of $a^6 + b^6 + c^6 + d^6 + e^6 = f^6$? Although there are solutions of $a_1^s + \ldots + a_{s-1}^s = b^s$ for $s = 4$ and 5, there is no known solution, even of $a_1^s + \ldots + a_s^s = b^s$, for $n \geq 6$.

Parametric solutions are known for equal sums of equal numbers of like powers,

$$\sum_{i=1}^{m} a_i^s = \sum_{i=1}^{m} b_i^s$$

with $a_i > 0$, $b_i > 0$, for $2 \leq s \leq 4$ and $m = 2$ and for $s = 5$, 6 and $m = 3$. Can a numerical solution be found for $s = 7$ and $m = 4$? For $s = 5$, $m = 2$, it is not known if there is any nontrivial solution of $a^5 + b^5 = c^5 + d^5$. Dick Lehmer once thought that there might be a solution with a sum of about 25 decimal digits, but a search by Blair Kelly yielded no nontrivial solution with sum $\leq 1.02 \times 10^{26}$.

Bob Scher & Ed Seidl discovered

$$14132^5 + 220^5 = 14068^5 + 6237^5 + 5027^5$$

in 1997 and recently Scott I. Chase found

$$966^8 + 539^8 + 81^8 = 954^8 + 725^8 + 481^8 + 310^8 + 158^8$$

The "Hardy-Ramanujan number" $1729 = 1^3 + 12^3 = 9^3 + 10^3$ was first found by Bernard Frénicle de Bessy in 1657. Conway writes

... that 1729 is the smallest discriminant of a pair of in-equivalent integer-valued quadratic forms in four variables that represent every integer the same number of times. This was discovered by Schiemann, who later proved that such pairs of isospectral forms can't exist with three or fewer variables.

The number 1729 is also the third Carmichael number (see **A13**) and Pomerance has observed that the second Carmichael number, 1105, is expressible as the sum of two squares in more ways than any smaller number. Granville invites the reader to make a corresponding statement about the first Carmichael number, 561.

But we digress. Leech, in 1957, found that

$$87539319 = 167^3 + 436^3 = 228^3 + 423^3 = 255^3 + 414^3$$

had three equal sums, and in 1991 Rosenstiel, Dardis & Rosenstiel found that 6963472309248 had four:

$$2421^3 + 19083^3 = 5436^3 + 18948^3 = 10200^3 + 18072^3 = 13322^3 + 16630^3.$$

Theorem 412 in Hardy & Wright shows that the number of such sums can be made arbitrarily large, but the least example is not known for five or more equal sums. If negative integers are allowed, then Randall Rathbun supplies the examples:

$$6017193 = 166^3 + 113^3 = 180^3 + 57^3 = 185^3 - (68)^3$$
$$= 209^3 - (146)^3 = 246^3 - (207)^3$$
$$1412774811 = 963^3 + 804^3 = 1134^3 - (357)^3 = 1155^3 - (504)^3$$
$$= 1246^3 - (805)^3 = 2115^3 - (2004)^3 = 4746^3 - (4725)^3$$
$$11302198488 = 1926^3 + 1608^3 = 1939^3 + 1589^3 = 2268^3 - (714)^3 = 2310^3 - (1008)^3$$
$$= 2492^3 - (1610)^3 = 4230^3 - (4008)^3 = 9492^3 - (9450)^3$$

Mordell and Mahler proved that the number of solutions of $n = x^3 + y^3$ can be $> c(\ln n)^\alpha$ and Silverman has improved their value of α from $\frac{1}{4}$ to $\frac{1}{3}$ and has shown that, if it is required that the pairs of cubes are to be mutually prime, then there is a constant c such that the number of such solutions is $< c^{r(n)}$, where $r(n)$ is the rank of the elliptic curve $x^3 + y^3 = n$. It is much harder to find these solutions. The largest number of representations found was three, by P. Vojta in 1983:

$$15170835645 = 517^3 + 2468^3 = 709^3 + 2456^3 = 1733^3 + 2152^3$$

but, as we predicted, this was soon beaten. Stuart Gascoigne and Duncan Moore independently found the specimen

$$1801049058342701083 = 12161023^3 + 1366353^3 = 11658843^3 + 6002593^3$$
$$= 12076023^3 + 3419953^3 = 12165003^3 + 922273^3$$

If negative integers are allowed, Rathbun gives the cubefree example

$$16776487 = 220^3 + 183^3 = 255^3 + 58^3 = 256^3 + (-9)^3 = 292^3 + (-201)^3$$

David Wilson, on 97-10-06, gave the smallest numbers which are sums of two cubes in four different ways (at least one pair coprime):

6963472309248, 12625136269928, 21131226514944, 26059452841000, 74213505639000, 95773976104625, 159380205560856, 174396242861568, 300656502205416, 376890885439488, 521932420691227, 573880096718136.

If the requirement of coprimality is removed, then there are the additional specimens:

55707778473984, 101001090159424, 169049812119552, 188013752349696,

He defines Taxicab(k) as the least integer expressible as the sum of two cubes in k different ways, so that Taxicab(1) $= 2 = 1^3 + 1^3$, Taxicab(2) $= 1729 = 9^3 + 10^3 = 1^3 + 12^3$, Taxicab(3) $= 87539319 = 167^3 + 436^3 = 228^3 + 423^3 = 255^3 + 414^3$, Taxicab(4) $= 6963472309248 = 2421^3 + 19083^3 = 5436^3 + 18948^3 = 10200^3 + 18072^3 = 13322^3 + 16630^3$ and comprise sequence A011541 in Sloane's Online Encyclopedia of Integer Sequences. He has found Taxicab(5) $= 48988659276962496 = 38787^3 + 365757^3 = 107839^3 + 362753^3 = 205292^3 + 342952^3 = 221424^3 + 336588^3 = 231518^3 + 331954^3$. Pending proof of minimality, Rathbun may have found Taxicab(6) by using the first three lines of

$$24153319581254312065344 = 28906206^3 + 582162^3 = 28894803^3 + 3064173^3$$
$$= 28657487^3 + 8519281^3 = 27093208^3 + 16218068^3$$
$$= 26590452^3 + 17492496^3 = 26224366^3 + 18289922^3$$
$$= 56461300^3 - 53813536^3 = 49162964^3 - 45576680^3$$

Andrew Bremner has computed the rational rank of the elliptic curve $x^3 + y^3 = $ Taxicab(n) as equal to $2, 4, 5, 4$ for $n = 2, 3, 4, 5$, respectively.

On 98-12-18 Daniel Bernstein reported that the smallest positive integer that can be written in eight ways as a sum of two (not necessarily positive) cubes is 137513849003496; and the smallest positive integer that can be written in five ways as a sum of two cubes of coprime integers is 506433677359393.

Euler knew that $635318657 = 133^4 + 134^4 = 59^4 + 158^4$, and Leech showed this to be the smallest example. No one knows of three such equal sums.

A method is known for generating parametric solutions of $a^4 + b^4 = c^4 + d^4$ which will generate all published solutions from the trivial one $(\lambda, 1, \lambda, 1)$; it will only produce solutions of degree $6n + 1$. Here, in answer to a question of Brudno, $6n + 1$ need not be prime. Although degree 25 does not appear, 49 does. More recently, Ajai Choudhry has found a parametric solution of degree 25.

Swinnerton-Dyer has a second method for generating new solutions from old and can show that the two methods, together with the symmetries, generate all *nonsingular* parametric solutions, i.e., all solutions which correspond to points on curves with no singular points. [A point on a curve with homogeneous equation $F(x, y, z) = 0$ is **singular** just if $\frac{\partial F}{\partial x} = \frac{\partial F}{\partial y} = \frac{\partial F}{\partial z} = 0$ there.] Moreover, the process is constructive in the sense that he can give a finite procedure for finding all nonsingular solutions of given degree. All nonsingular solutions have odd degree, and all sufficiently large odd degrees do occur. Unfortunately, singular solutions do exist. Swinnerton-Dyer has a process for generating them, but has no reason to believe that it gives them all. The problem of describing them all needs completely new ideas. Some of the singular solutions have even degree and he conjectures (and could probably prove) that all sufficiently large even degrees occur in this way.

In the same sense, Andrew Bremner can find "all" parametric solutions of $a^6 + b^6 + c^6 = d^6 + e^6 + f^6$ which also satisfy the equations

$$
\begin{aligned}
a^2 + ad - d^2 &= f^2 + fc - c^2 \\
b^2 + be - e^2 &= d^2 + da - a^2 \\
c^2 + cf - f^2 &= e^2 + eb - b^2
\end{aligned}
$$

(this is not such a restriction as might at first appear). Many solutions of $a^6 + b^6 + c^6 = d^6 + e^6 + f^6$ also satisfy $a^2 + b^2 + c^2 = d^2 + e^2 + f^2$ and all known simultaneous solutions (with appropriately chosen signs) of these two equations also satisfy the previous three equations, e.g., the smallest solution, found by Subba-Rao, $(a, b, c, d, e, f) = (3, 19, 22, -23, 10, -15)$. Is there a counter-example? Peter Montgomery has listed 18 equal sums of three sixth powers where the corresponding sums of squares are not equal. The least is

$$25^6 + 62^6 + 138^6 = 82^6 + 92^6 + 135^6.$$

Bremner can also find "all" parametric solutions of $a^5 + b^5 + c^5 = d^5 + e^5 + f^5$ which also satisfy $a + b + c = d + e + f$ and $a - b = d - e$.

Tong & Cao give integer solutions of $x^4 + y^4 = cz^4$ for $c = 2, 17, 82, 97, 257, 337, 626, 641, 706, 881$ and show that the only other $c < 1000$ for which there may be solutions are $c = 226, 562, 577, 977$.

Hardin & Sloane express $(a^2 + b^2 + c^2 + d^2)^3$ as a sum of 23 sixth powers of linear forms; it appears that no such 22-term identity exists.

Bob Scher calls a sum $\sum_{i=1}^{m} a_i^p = 0$ (where p is prime) "perfect" if, for any a_i, there is also a unique a_i' such that $a_i + a_i' \equiv 0 \bmod p$. If $a_i \equiv 0 \bmod p$, then $a_i' = a_i$. He shows that if $p = 3$ and $m < 9$ or if $p = 5$ and $m < 7$, then every such sum is perfect.

Randy Ekl has discovered that

$$149^7 + 123^7 + 14^7 + 10^7 = 146^7 + 129^7 + 90^7 + 15^7$$

(and $\quad 194^7 + 150^7 + 14^7 + 10^7 = 192^7 + 152^7 + 132^7 + 38^7$)

(and $\quad 354^7 + 112^7 + 52^7 + 19^7 = 343^7 + 281^7 + 46^7 + 35^7$)

in which the number of terms is one larger than the exponent. This was also found by Bob Scher together with

$$121^7 + 94^7 + 83^7 + 61^7 + 57^7 + 27^7 = 125^7 + 24^7$$

More spectacular is Bob Scher's discovery

$$14132^5 + 220^5 = 14068^5 + 6237^5 + 5027^5$$

which has no more terms than the size of the exponent.

On 97-08-11, Jim Sheen telephoned that

$$171611765^4 + 8674^8 + 9817^8 = 160565765^4 + 11567^8 + 5174^8.$$

He has a parametric solution based on Proth's identity

$$m^4 + n^4 + (m + n)^4 = 2(m^2 + mn + n^2)^2$$

There has been some interest in the simultaneous equations

$$
\begin{aligned}
a^n + b^n + c^n &= d^n + e^n + f^n \\
a^n b^n c^n &= d^n e^n f^n
\end{aligned}
$$

For $n = 2$ the problem goes back at least to Bini, with partial solutions by Dubouis and Mathieu; a neat general solution has recently been given by John B. Kelly. Stephane Vandemergel has found 62 solutions for $n = 3$ and three solutions for $n = 4$: (29,66,124;22,93,116), (54,61,196;28,122,189) and (19,217,657;9,511,589). He notes that if $r^n + s^n = u^n + v^n$, then $(ru, su, v^2; rv, sv, u^2)$ is a solution, which shows that there are infinitely many solutions for $n \leq 4$. Most of his solutions are not of this form.

There is a considerable history (see Dickson [**A17**], II, Ch. 24) of the Tarry–Escott problem, and a book by Gloden on (systems of) **multigrade equations**:

$$\sum_{i=1}^{l} n_i^j = \sum_{i=1}^{l} m_i^j \quad (j = 1, \ldots, k)$$

A spectacular example, with $k = 9$, $l = 10$, is due to Letac, with $(n_i; m_i) =$

±12, ±11881, ±20231, ±20885, ±23738; ±436, ±11857, ±20449, ±20667, ±23750.

Smyth has shown that this is a member of an infinite family of independent solutions.

Borwein, Lisoněk & Percival have found two other solutions of 'size 10':

±71, ±131, ±308, ±180, ±307; ±99, ±100, ±301. ± 188, ±313

±18, ±245, ±331, ±471, ±508; ±103, ±189, ±366. ± 452, ±515

Kuosa, Meyrignac & Chen Shuwen having earlier found the size 12 solution

±22, ±61, ±86, ±127, ±140, ±151; ±35, ±47, ±94, ±121. ± 146, ±148

Daniel J. Bernstein, Enumerating solutions to $p(a)+q(b) = r(c)+s(d)$. *Math. Comput.* **70**(2001) 389–394; *MR* **2001f**:11203.

U. Bini, Problem 3424, *L'Intermédiaire des Math.*, **15**(1908) 193.

B. J. Birch & H. P. F. Swinnerton-Dyer, Notes on elliptic curves, II, *J. reine angew. Math.*, **218**(1965) 79–108.

Peter Borwein, Petr Lisoněk & Colin Percival, Computational investigations of the Prouhet-Tarry-Escott problem, *Math. Comput.*, **72**(2003) 2063–2070; *MR* **2004c**:11042.

Andrew Bremner, A geometric approach to equal sums of sixth powers, *Proc. London Math. Soc.*(3), **43**(1981) 544–581; *MR* **83g**:14018.

Andrew Bremner, A geometric approach to equal sums of fifth powers, *J. Number Theory*, **13**(1981) 337–354; *MR* **83g**:14017.

Andrew Bremner & Richard K. Guy, A dozen difficult Diophantine dilemmas, *Amer. Math. Monthly*, **95**(1988) 31–36.

B. Brindza & Ákos Pintér, On equal values of power sums, *Acta Arith.*, **77**(1996) 97–101; *MR* **97m**:11033.

T. D. Browning, Equal sums of two kth powers, *J. Number Theory*, **96**(2002), 293–318; *MR* **2003m**:11163.

T. D. Browning, Sums of four biquadrates, *Math. Proc. Cambridge Philos. Soc.*, **134**(2003), 385–395.

S. Brudno, Some new results on equal sums of like powers, *Math. Comput.*, **23**(1969) 877–880.

S. Brudno, On generating infinitely many solutions of the diophantine equation $A^6 + B^6 + C^6 = D^6 + E^6 + F^6$, *Math. Comput.*, **24**(1970) 453–454.

S. Brudno, Problem 4, *Proc. Number Theory Conf. Univ. of Colorado*, Boulder, 1972, 256–257.

Simcha Brudno, Triples of sixth powers with equal sums, *Math. Comput.*, **30**(1976) 646–648.

S. Brudno & I. Kaplansky, Equal sums of sixth powers, *J. Number Theory*, **6**(1974) 401–403.

Chen Shuwen, The Prouhet-Tarry-Escott problem,
http://member.netease.com/~chin/eslp/~TarryPrb.htm

J. Choubey, Parametric solutions of Diophantine equation with equal sums of four fourth powers, *Acta Cienc. Indica Math.*, **20**(1994) 71–72; *MR* **96g**:11025.

Ajai Choudhry, The Diophantine equation $A^4 + B^4 = C^4 + D^4$, *Indian J. Pure Appl. Math.*, **22**(1991) 9–11; *MR* **92c**:11024.

Ajai Choudhry, Symmetric Diophantine systems, *Acta. Arith.*, **59**(1991) 291–307; *MR* **92g**:11030.

Ajai Choudhry, Multigrade chains with odd exponents, *J. Number Theory*, **42** (1992) 134–140; *MR* **94a**:11044.

Ajai Choudhry, On equal sums of sixth powers, *Indian J. Pure Appl. Math.*, **25**(1994) 834–841; *MR* **95k**:11036.

Ajai Choudhry, On the Diophantine equation $A^4 + hB^4 = C^4 + hD^4$, *Indian J. Pure Appl. Math.*, **26**(1995) 1057–1061; *MR***96k**:11032.

Ajai Choudhry, The Diophantine equation $X_1^5 + X_2^5 + 2X_3^5 = Y_1^5 + Y_2^5 + 2Y_3^5$, *Ganita*, **48**(1997) 115–116; *MR* **9**:110.

Ajai Choudhry, On equal sums of fifth powers, *Indian J. Pure Appl. Math.*, **28**(1997) 1443–1450; *MR* **99c**:11032.

Ajai Choudhry, The Diophantine equation $A^4 + 4B^4 = C^4 + 4D^4$, *Indian J. Pure Appl. Math.*, **29**(1998) 1127–1128; *MR* :110.

216   Unsolved Problems in Number Theory

Ajai Choudhry, A Diophantine equation involving fifth powers, *Enseign. Math.*, **44**(1998) 53–55; *MR* **99h**:11030.

Ajai Choudhry, On equal sums of cubes, *Rocky Mountain J. Math.*, **28**(1998) 1251–1257; *MR* :110.

Ajai Choudhry, On triads of squares with equal sums and equal products, *Ganita*, **49**(1998) 101–106; *MR* **99m**:11028.

Ajai Choudhry, The Diophantine equation $ax^5 + by^5 + cz^5 = au^5 + bv^5 + cw^5$, *Rocky Mountain J. Math.*, **29**(1999) 459–462; *MR* :110.

Ajai Choudhry, On equal sums of sixth powers, *Rocky Mountain J. Math.*, **30**(2000) 843–848; *MR* **2002e**:11036.

Ajai Choudhry, On equal sums of seventh powers, *Rocky Mountain J. Math.*, **30**(2000) 849–852; *MR* **2002e**:11037.

Ajai Choudhry, An elementary method of solving certain Diophantine equations, *Math. Student*, **68**(1999) 195–199; *MR* **2002e**:11038.

Ajai Choudhry, Ideal solutions of the Tarry-Escott problem of degree four and a related Diophantine system, *Enseign. Math.*(2) **46**(2000) 313–323; *MR* **2001m**:11033; and see **49**(2003) 101–108; *MR* **2004d**:11021 for degree five.

Ajai Choudhry, The system of simultaneous equations $\sum_{i=1}^{3} x_i^k = \sum_{i=1}^{3} y_i^k$, $k = 1$, 2 and 5 has no non-trivial solutions in integers, *Math. Student*, **70**(2001) 85–88.

Ajai Choudhry, Equal sums of like powers, *Rocky Mountain J. Math.*, **31**(2001) 115–129; *MR* **2001m**:11034.

Ajai Choudhry, Triads of cubes with equal sums and equal products, *Math. Student*, **70**(2001) 137–143.

Ajai Choudhry, Triads of biquadrates with equal sums and equal products, *Math. Student*, **70**(2001) 149–152.

Ajai Choudhry, On the number of representations of an integer as a sum of four integral fifth powers, *Math. Student*, **70**(2001) 215–217.

Ajai Choudhry, Triads of integers with equal sums of squares, cubes and fourth powers, *Bull. London Math. Soc.* **35**(2003) 821–824.

Jean-Joël Delorme, On the Diophantine equation $x_1^6 + x_2^6 + x_3^6 = y_1^6 + y_2^6 + y_3^6$, *Math. Comput.*, **59**(1992) 703–715; *MR* **93a**:11023.

V. A. Dem'janenko, L. Euler's conjecture (Russian), *Acta Arith.*, **25**(1973/74) 127–135; *MR* **50** #12912.

Dubouis & Mathieu, Réponse 3424, *L'Intermédiaire des Math.*, **16**(1909) 41–42, 112.

P. S. Dyer, A solution of $A^4 + B^4 = C^4 + D^4$, *J. London Math. Soc.*, **18**(1943) 2–4; *MR* **5**, 89e.

Randy L. Ekl, Equal sums of four seventh powers, *Math. Comput.*, **65**(1996) 1755–1756; *MR* **97a**:11050.

Randy L. Ekl, New results in equal sums of like powers, *Math. Comput.*, **67**(1998) 1309–1315; *MR* **98m**:11023.

Noam Elkies, On $A^4 + B^4 + C^4 = D^4$, *Math. Comput.*, **51**(1988) 825–835; and see Ivars Peterson, *Science News*, **133**, #5(88-01-30) 70; Barry Cipra, *Science*, **239**(1988) 464; and James Glieck, *New York Times*, 88-04-17.

Rachel Gar-el & Leonid Vaserstein, On the Diophantine equation $a^3 + b^3 + c^3 + d^3 = 0$, *J. Number Theory*, **94**(2002) 219–223; *MR* **200d**:11039.

A. Gérardin, *L'Intermédiare des Math.*, **15**(1908) 182; *Sphinx-Œdipe*, 1906-07 80, 128.

A. Gloden, *Mehrgradige Gleichungen*, Noordhoff, Groningen, 1944.

G. R. H. Greaves, Representation of a number by the sum of two fourth powers, *Mat. Zametki*, **55**(1994) 47–58, 188; transl. *Math. Notes* **55**(1994) 134–141; *MR* **95e**:11101.

R. H. Hardin & N. J. A. Sloane, Expressing $(a^2 + b^2 + c^2 + d^2)^3$ as a sum of 23 sixth powers, /JCA **68**(1994) 481–485; *MR* **96e**:11048.

Michael D. Hirschhorn, An amazing identity of Ramanujan, *Math. Mag.*, **68**(1995) 199–201; *MR* **96f**:11044.

M. A. Ivanov, The Diophantine equation $x_1^n + x_2^n + \cdots x_{r_1}^n = y_1^n + y_2^n + \cdots y_{r_2}^n$, *Rend. Sem. Mat. Univ. Politec. Torino*, **54**(1996) 25–33; *MR* **99d**:11033.

William C. Jagy & Irving Kaplansky, Sums of squares, cubes and higher powers, *Experiment. Math.*, **4**(1995) 169–173; *MR* **99a**:11046.

John B. Kelly, Two equal sums of three squares with equal products, *Amer. Math. Monthly*, **98**(1991) 527–529; *MR* **92j**:11025.

Jan Kubiček, A simple new solution to the diophantine equation $A^3 + B^3 + C^3 = D^3$ (Czech, German summary), *Časopis Pěst. Mat.*, **99**(1974) 177–178.

L. J. Lander, Geometric aspects of diophantine equations involving equal sums of like powers, *Amer. Math. Monthly*, **75**(1968) 1061–1073.

L. J. Lander & T. R. Parkin, Counterexample to Euler's conjecture on sums of like powers, *Bull. Amer. Math. Soc.*, **72**(1966) 1079; *MR* **33** #5554.

L. J. Lander, T. R. Parkin & J. L. Selfridge, A survey of equal sums of like powers, *Math. Comput.*, **21**(1967) 446–459; *MR* **36** #5060.

Leon J. Lander, Equal sums of unlike powers, *Fibonacci Quart.*, **28**(1990) 141–150; *MR* **91e**:11033.

John Leech, Some solutions of Diophantine equations, *Proc. Cambridge Philos. Soc.* **53**(1957) 778–780; *MR* **19** 837f.

Li Yuan, On the Diophantine equation $x^4 + y^4 \pm z^4 = 2c^4$, *Sichuan Daxue Xuebao*, **33**(1996) 366–368; *MR* **97f**:11023.

R. Norrie, in University of Saint Andrews 500th Anniversary Memorial Volume of Scientific Papers, published by the University of St. Andrews, 1911, 89.

Giovanni Resta & Jean-Charles Meyrignac, The smallest solutions to the Diophantine equation $x^6 + y^6 = a^6 + b^6 + c^6 + d^6 + e^6$, *Math. Comput.*, **72**(2003) 1051–1054; *MR* **2004b**:11174.

E. Rosenstiel, J. A. Dardis & C. R. Rosenstiel, The four least solutions in distinct positive integers of the Diophantine equation $s = x^3 + y^3 = z^3 + w^3 = u^3 + v^3 = m^3 + n^3$, *Bull. Inst. Math. Appl.*, **27**(1991) 155–157; *MR* **92i**:11134.

Csaba Sándor, On the equation $a^3 + b^3 + c^3 = d^3$, *Period. Math. Hungar.*, **33**(1996) 121–134; *MR* **99c**:11031.

Bob Scher, General-perfect sums, (95-11-22 preprint)

Joseph H. Silverman, Integer points on curves of genus 1, *J. London Math. Soc.*(2), **28**(1983) 1–7; *MR* **84g**:10033.

Joseph H. Silverman, Taxicabs and sums of two cubes, *Amer. Math. Monthly* **100**(1993) 331–340; *MR* **93m**:11025.

C. M. Skinner & T. D. Wooley, Sums of two kth powers, *J. reine angew. Math.*, **462**(1995) 57–68; *MR* **96b**:11132.

C. J. Smyth, Ideal 9th-order multigrades and Letac's elliptic curve, *Math. Comput.*, **57**(1991) 817–823.

K. Subba-Rao, On sums of sixth powers, *J. London Math. Soc.*, **9**(1934) 172–173.

Tong Rui-Zhou & Cao Yu-Shu, The Diophantine equation $x^4 + y^4 = cz^4$, *Heilongjiang Daxue Ziran Kexue Xuebo*, **11**(1994) 6–10; *MR* **96f**:11045.

Ju. D. Trusov, New series of solutions of the thirty second problem of the fifth book of Diophantus (Russian). *Ivanov. Gos. Ped. Inst. Učen. Zap.*, **44**(1969) 119–121 (and see 122–123); *MR* **47** #4925(-6).

Morgan Ward, Euler's three biquadrate problem, *Proc. Nat. Acad. Sci. U.S.A.*, **31**(1945) 125–127; *MR* **6**, 259.

Morgan Ward, Euler's problem on sums of three fourth powers, *Duke Math. J.*, **15**(1948) 827–837; *MR* **10**, 283.

A. S. Werebrusow, *L'Intermédiaire des Math.*, **12**(1905) 268; **25**(1918) 139.

David W. Wilson, The fifth taxicab number is 48 988 659 276 962 496, *J. Integer Seq.*, **2**(1999), Article 99.1.9, HTML document (electronic); *MR* **2000i**:11195.

Trevor D. Wooley, Sums of two cubes, *Internat. Math. Res. Notices*, **1995** 181–184 (electronic); *MR* **96a**:11103.

Aurel J. Zajta, Solutions of the diophantine equation $A^4 + B^4 = C^4 + D^4$, *Math. Comput.*, **41**(1983) 635–659.

OEIS: A001235, A003824-003828, A005130, A011541, A018786-018787, A018850, A023050-023051, A047696-047697, A049405-049406, A051055, A080642.

D2 The Fermat problem.

Fermat's Last Theorem, that the equation

$$x^p + y^p = z^p$$

is impossible in positive integers for an odd prime p, is at last indeed a theorem. Ribet had shown that the theorem follows from the Taniyama-Weil conjectures on elliptic curves. For as much of the proof as our margin will hold, see **B19**.

Kummer proved the theorem for all regular primes, where a prime p is **regular** if it doesn't divide h_p, the number of equivalence classes of ideals in the cyclotomic field $\mathbb{Q}(\zeta_p)$. He also showed that a prime was regular just if it did not divide the numerators of the **Bernoulli numbers** B_2, B_4, ..., B_{p-3}, where

$$B_{2k} = (-1)^{k-1} \frac{2(2k)!}{(2\pi)^{2k}} \zeta(2k)$$

($\zeta(2k) = \sum_{n=1}^{\infty} n^{-2k}$ and see **A17**). Of the 78497 primes less than 10^6, 47627 are regular, agreeing well with their conjectured density, $e^{-1/2}$. However, it has not even been proved that there are infinitely many. On the other hand, Jensen has proved that there are infinitely many irregular primes. The prime $p = 16843$ divides B_{p-3} and Richard McIntosh, as well as Buhler, Crandall, Ernvall & Metsänkylä, observes that this is also true for $p = 2124679$.

Some of J. M. Gandhi's subsidiary problems, mentioned in the first edition, are covered by Wiles's method, which shows the unsolvability of

$$x^p + y^p + \gamma z^p = 0$$

if $p \geq 11$ and γ is a power of one of the primes 3, 5, 7, 11, 13, 17, 19, 23, 29, 53, 59, ...

It follows from the work of Ribet via Mazur & Kamienny and Darmon & Merel that the equation $x^n + y^n = 2z^n$ has no solutions for $n > 2$ apart from the trivial $x = y = z$.

For the equation $x^3 + y^3 = z^t$ with $t \geq 4$, Darmon & Granville showed that there are finitely many primitive solutions. For $t \geq 17$, Kraus gave criteria for the nonexistence of primitive solutions and verified them for $17 \leq t \leq 9999$. Bruin has shown nonexistence for $t = 4$ and 5.

Here are some of Noam Elkies's near misses for a counterexample to Fermat's theorem:

$$3472073^7 + 4627011^7 = (1 + \epsilon)4710868^7 \quad \text{with} \quad \epsilon < 3.63615 \times 10^{-22}$$

$$280^{10} + 305^{10} = (1 - \epsilon)316^{10} \quad \text{with} \quad \epsilon < 0.00000000232$$

$13^5 + 16^5 = 17^5 + 12$ Noam notes 'as it happens also 13+16=17+12'

Elkies has also examined 'twists' of the Fermat equation, $x^3 + y^3 = nz^3$. For $n = 8837$ the smallest solution is

(9478937992673335356289864181224522415627973189452224987008153367976553669071 2886433241,
98903593562573785800455574280919162748021995450989744962454595882114614463951 42633010,
45831090648933774822827862041103878922298626345431860617524151571465321225266 512578677)

The prime $n = 5849$ comes in a poor second, with smallest solution

(4728618748476229625617988838113772096021241206255932896773863201719 59834,
47232460457097857774354275261178167019299454389106806971083849881632 703,
39482305224952708961137431710040143320650499109248727618241797267508 97)

though its smallest *positive* solution has values with more than 10^4 decimal digits!!

For what integers c are there integer solutions of $x^4 + y^4 = cz^4$ with $x \perp y$ and $z > 1$? Leech gives a method for finding non-trivial solutions for any $z = a^4 + b^4$: the smallest he has found is $25^4 + 149^4 = 5906 \cdot 17^4$. Bremner & Morton show that 5906 is the least integer that is the sum of two rational fourth powers, but not the sum of two integer fourth powers.

Prove that $x^n + y^n = n!z^n$ has no integer solutions with $n > 2$. Erdős & Obláth showed that $x^p \pm y^p = n!$ has none with $p > 2$, and Erdős states that $x^4 + y^4 = n!$ has no solutions with $x \perp y$. Indeed, even without this last condition, Leech notes that for $n > 3$ there's a prime $\equiv 3 \bmod 4$ in the interval $[n+1, 2n]$ and hence a simple (i.e. not repeated) such prime divisor of $n!$ for $n > 6$, so $n!$ is not even the sum of two squares for $n > 6$ (apart from $n = 0$, 1, 2, the only solution is $6! = 24^2 + 12^2$). [Compare **D25**.]

Granville's paper, in which the reviewer felt "a new powerful wave", had already related the Fermat problem to numerous other conjectures, including the abc-conjecture (**B19**) and Erdős's powerful numbers conjecture (**B16**).

M. A. Bennett & C. Skinner, Ternary Diophantine equations via Galois representations and modular forms, *Canad. J. Math.*, (to appear)

Frits Beukers, The Diophantine equation $Ax^p + By^q = Cz^r$, *Duke Math. J.*, **91**(1998) 61–88; *MR* **99f**:11039.

Andrew Bremner & Patrick Morton, A new characterization of the integer 5906, *Manuscripta Math.*, **44**(1983) 187–229; *MR* **84i**:10016.

Nils Bruin, The Diophantine equations $x^2 \pm y^4 = \pm z^6$ and $x^2 + y^8 = z^3$, *Compositio Math.*, **118**(1999) 305-321; *MR* **2001d**:11035.

Nils Bruin, On powers as sums of two cubes, *Algorithmic number theory* (*Leiden* 2000) 169–184, *Springer Lecture Notes in Comput. Sci.*, **1838**; *MR* **2002f**:11029.

J. P. Buhler, R. E. Crandall & R. W. Sompolski, Irregular primes to one million, *Math. Comput.*, **59**(1992) 717–722; *MR* **93a**:11106.

J. P. Buhler, R. E. Crandall, R. Ernvall & T. Metsänkylä, Irregular primes and cyclotomic invariants to four million, *Math. Comput.*, **61**(1993) 151–153; *MR* **93k**:11014.

M. Chellali, Accélération de calcul de nombres de Bernoulli, *J. Number Theory*, **28**(1988) 347–362.

Chen Wen-Jing, On the Diophantine equations $a^x + b^y = c^z$, $a^2 + b^2 = c^2$, *J. Changsha Comm. Inst.*, **10**(1994) 1–4; *MR* **95j**:11030.

Don Coppersmith, Fermat's last theorem (case 1) and the Wieferich criterion, *Math. Comput.*, **54**(1990) 895–902; *MR* **90h**:11024.

Henri Darmon & Andrew J. Granville, On the equations $z^m = F(x, y)$ and $Ax^p + By^p = Cz^r$, *Bull. London Math. Soc.*, **17**(1995) 513–543; *MR* **96e**:11042.

Henri Darmon & Claude Levesque, Infinite sums, Diophantine equations and Fermat's last theorem, *Number theoretic methods* (*Iizuka*, 2001), 73–95, *Dev. Math.*, **8**, Kluwer Acad. Publ., Dordrecht, 2002; *MR* **2004c**:11086.

Henri Darmon & Loïc Merel, Winding quotients and some variants of Fermat's last theorem, *J. reine angew. Math.*, **490**(1997) 81–100; *MR* **98h**:11076.

Karl Dilcher & Ladislav Skula, A new criterion for the first case of Fermat's Last Theorem, *Math. Comput.*, **64**(1995) 363–392; *MR* **93c**:11034.

Harold M. Edwards, *Fermat's Last Theorem, a Genetic Introduction to Algebraic Number Theory*, Springer-Verlag, New York, 1977.

Noam D. Elkies, Wiles minus epsilon implies Fermat, in *Elliptic Curves, Modular Forms & Fermat's Last Theorem* (*Hong Kong* 1993) 79–98, *Ser. Number Theory, I, Internat. Press, Cambridge MA* 1995; *MR* **97b**:11043.

P. Erdős & R. Obláth, Über diophantische Gleichungen der form $n! = x^p \pm y^p$ and $n! \pm m! = x^p$, *Acta Litt. Sci. Szeged*, **8**(1937) 241–255; *Zbl.* **17**.004.

Gerd Faltings, The proof of Fermat's last theorem by R. Taylor and A. Wiles, *Notices Amer. Math. Soc.*, **42**(1995) 743–746; *MR* **96i**:11030.

Andrew Granville, Some conjectures related to Fermat's last theorem, *Number Theory* (*Banff*), 1988, de Gruyter, 1990, 177–192; *MR* **92k**:11036.

Andrew Granville, Some conjectures in analytic number theory and their connection with Fermat's last theorem, in *Analytic Number Theory (Proc. Conf. Honor Bateman*, 1989), Birkhäuser, 1990, 311–326; *MR* **92a**:11031.

Andrew Granville, On the number of solutions to the generalized Fermat equation, *Number Theory (Halifax NS* 1994) 197–207, *CMS Conf. Proc.*, **15**, Amer. Math. Soc., 1995; *MR* **96j**:11035.

A. Granville & M. B. Monagan, The first case of Fermat's last theoerem is true for all exponents up to 714,591,416,091,389, *Trans. Amer. Math. Soc.*, **306**(1988) 329–359; *MR* **89g**:11025.

A. Grelak & Aleksander Grytczuk, *Comment. Math. Prace Mat.*, **24**(1984) 269–275; *MR* **86d**:11027.

K. Inkeri & A. J. van der Poorten, Some remarks on Fermat's conjecture, *Acta Arith.*, **36**(1980) 107-111.

Wilfrid Ivorra, Sur les équations $x^p + 2^\beta y^p = z^2$ et $x^p + 2^\beta y^p = 2z^2$, *Acta Arith.*, **108**(2003) 327–338; *MR* **2004b**:11036.

Wells Johnson, Irregular primes and cyclotomic invariants, *Math. Comput.*, **29**(1975) 113–120; *MR* **51** #12781.

Alain Kraus, Sur l'équation $a^3 + b^3 = c^p$, *Experiment. Math.*, **7**(1998) 1–13; *MR* **99f**:11040.

Le Mao-Hua, A note on Jeśmanowicz' conjecture, *Colloq. Math.*, **69**(1995) 47–51; *MR* **96g**:11029.

Le Mao-Hua, A note on the Diophantine equation $(m^3 - 3m)^x + (3m^2 - 1)^y = (m^2 + 1)^z$, *Proc. Japan Acad. Ser. A Math. Sci.*, **73**(1997) 148–149; *MR* **98m**:11021; *Publ. Math. Debrecen*, **58**(2001) 461–466; *MR* **2002b**:11044.

Le Mao-Hua, A conjecture concerning the exponential Diophantine equation $a^x + b^y = c^z$, *Heilongjiang Daxue Ziran Kexue Xuebao*, **20**(2003) 10–14.

Le Mao-Hua, On Terai's conjecture concerning the exponential Diophantine equation $a^x + b^y = c^z$, *Acta Math. Sinica*, **46**(2003) 245–250; *MR* **2004c**:11038.

D. H. Lehmer, E. Lehmer & H. S. Vandiver, An application of high-speed computing to Fermat's Last Theorem, *Proc. Nat. Acad. Sci. U.S.A.*, **40**(1954), 25–33.

R. Daniel Mauldin, A generalization of Fermat's last theorem: the Beal conjecture and prize problem, *Notices Amer. Math. Soc.*, **44**(1997) 1436–1437; *MR* 98j:11020.

M. Mignotte, On the diophantine equation $D_1 x^2 + D_2^m = 4y^n$, *Portugal. Math.*, **54**(1997) 457-460; *MR* **99a**:11039.

Satya Mohit & M. Ram Murty, Wieferich primes and Hall's conjecture, *C. R. Math. Acad. Sci. Soc. R. Can.*, **20**(1998) 29–32; *MR* **98m**:11004.

Fadwa S. Abu Muriefah, On the Diophantine equation $x^3 = dy^2 \pm q^6$, *Int. J. Math. Math. Sci.*, **28**(2001) 493–497; *MR* **2003b**:11026.

Bjorn Poonen, Some Diophantine equations of the form $x^n + y^n = z^m$, *Acta Arith.*, **86**(1998) 193–205; *MR* **99h**:11034.

Paulo Ribenboim, *13 Lectures on Fermat's Last Theorem*, Springer-Verlag, New York, Heidelberg, Berlin, 1979; see *Bull. Amer. Math. Soc.*, **4**(1981) 218–222; *MR* **81f**:10023.

Paulo Ribenboim, Fermat's last theorem, before June 23, 1993, *Number Theory (Halifax NS* 1994) 279–294, *CMS Conf. Proc.*, **15**, Amer. Math. Soc., 1995; *MR* **96i**:11032.

K. Ribet, On modular representations of $\mathrm{Gal}(^{-}\mathbf{Q}/\mathbf{Q})$ arising from modular forms, *Invent. Math.*, **100**(1990) 431–476.

Kenneth A. Ribet, On the equation $a^p + 2^\alpha b^p + c^p = 0$, *Acta Arith.*, **79**(1997) 7–16; *MR* **98e**:11035.

J. L. Selfridge, C. A. Nicol & H. S. Vandiver, Proof of Fermat's last theorem for all prime exponents less than 4002, *Proc. Nat. Acad. Sci., U.S.A.* **41**(1955) 970–973; *MR* **17**, 348.

Daniel Shanks & H.C. Williams, Gunderson's function in Fermat's last theorem, *Math. Comput.*, **36**(1981) 291–295.

R. W. Sompolski, The second case of Fermat's last theorem for fixed irregular prime exponents, Ph.D. thesis, Univ. Illinois Chicago, 1991.

Jiro Suzuki, On the generalized Wieferich criteria, *Proc. Japn Acad. Ser. A Math. Sci.*, **70**(1994) 230–234.

Jonathan W. Tanner & Samuel S. Wagstaff, New bound for the first case of Fermat's last theorem, *Math. Comput.*, **53**(1989) 743–750; *MR* **90h**:11028.

Richard Taylor & Andrew Wiles, Ring-theoretic properties of certain Hecke algebras, *Ann. of Math.*(2), **141**(1995) 553–572.

Samuel S. Wagstaff, The irregular primes to 125000, *Math. Comput.*, **32**(1978) 583–591; *MR* **58** #10711.

Andrew Wiles, Modular elliptic curves and Fermat's last theorem, *Ann. of Math.*(2), **141**(1995) 443–551.

John A. Zuehlke, Fermat's last theorem for Gaussian integer exponents, *Amer. Math. Monthly*, **106**(1999) 49.

D3 Figurate numbers.

Every number is expressible as the sum of three pentagonal numbers, $\frac{1}{2}r(3r-1)$, or of three hexagonal numbers, $r(2r-1)$. But if we restrict ourselves to those of positive rank, $r > 0$, this is not true. Is it true for all sufficiently large numbers? Equivalently, is every sufficiently large is every sufficiently large number of shape $24k + 3$ expressible as the sum of three squares of numbers of shape $6r - 1$ with $r > 0$ and is every number $8k + 3$ expressible as the sum of three squares of numbers of shape $4r - 1$ with $r > 0$?

Richard Blecksmith & John Selfridge found six numbers among the first million, namely

$$9, \ 21, \ 31, \ 43, \ 55 \text{ and } 89,$$

which require five pentagonal numbers of positive rank, and two hundred and four others, the largest of which is 33066, which require four. They believe that they have found them all.

Many numbers (what fraction of the whole, or are they of zero density?) require four hexagonal numbers of positive rank; several, e.g.,

$$5, 10, 20, 25, 38, 39, 54, 65, 70, 114, 130, \ldots,$$

require five, and 11 and 26 require six. Which numbers require five?

Corresponding questions can be asked for k-gonal numbers,

$$\frac{1}{2}\left((k-2)r - (k-4)\right),$$

with $k \geq 7$, in both the cases $r > 0$ and r unrestricted. E.g. is every number $40n + 27$ expressible as the sum of three squares of shape $(10r \pm 3)^2$?

10 and 16 are not the sum of three heptagons. Nor is 76; are there others?

Chou & Deng believe that all numbers > 343867 are expressible as the sum of four tetrahedral numbers; they have verified this for numbers $< 4 \cdot 10^9$.

Mordell, on p. 259 of his book, asks if the only integer solutions of

$$6y^2 = (x+1)(x^2 - x + 6)$$

are given by $x = -1$, 0, 2, 7, 15 and 74? By Theorem 1 of Mordell's Chapter 27 there are only finitely many. The equation arises from

$$y^2 = \binom{x}{0} + \binom{x}{1} + \binom{x}{2} + \binom{x}{3}.$$

Andrew Bremner gave the additional solution $(x, y) = (767, 8672)$ and showed that that is all, but the result was already given by Ljunggren in 1971.

Similarly, Martin Gardner took the figurate numbers: triangle, square, tetrahedron and square pyramid; and equated them in pairs. Of the six resulting problems, he noted that they were all solved except "triangle = square pyramid", which leads to the equation

$$3(2y+1)^2 = 8x^3 + 12x^2 + 4x + 3.$$

The number of solutions is again finite. Are they all given by $x = -1$, 0, 1, 5, 6 and 85? Schinzel sends Avanesov's affirmative answer, rediscovered by Uchiyama.

The "triangle = tetrahedron" problem is a special case of a more general question about equality of binomial coefficients (see **B31**) — the only nontrivial examples of $\binom{n}{2} = \binom{m}{3}$ are $(m, n) = (10,16)$, $(22,56)$ and $(36,120)$. Are there nontrivial examples of $\binom{n}{2} = \binom{m}{4}$ other than $(10,21)$?

The case "square pyramid = square" is Lucass problem@Lucas's problem. Is $x = 24$, $y = 70$ the only nontrivial solution of the diophantine equation

$$y^2 = x(x+1)(2x+1)/6?$$

This was solved affirmatively by Watson, using elliptic functions, and by Ljunggren, using a Bhaskara equation (often called a Pell equation) in a quadratic field. Mordell asked if there was an elementary proof, and affirmative answers have been given by Ma, by Xu & Cao, by Anglin and by Pintér.

The same equation in disguise is to ask if (48, 140) is the unique nontrivial solution to the case "square = tetrahedron", since the previous equation may be written

$$(2y)^2 = 2x(2x + 1)(2x + 2)/6,$$

though, as Peter Montgomery notes, this doesn't eliminate the possibility of an odd square. A more modern treatment is to put $12x = X - 6$, $72y = Y$ and note that $Y^2 = X^3 - 36X$ is curve 576H2 in John Cremona's tables. The point (12,36) (which gives an odd square) serves as a generator. There's an infinity of rational solutions, but the only nontrivial integer solution to the original problem is given by the point (294, 5040).

More general than asking for the sum of the first n squares to be square, we can ask for the sum of any n consecutive squares to be square. If S is the set of n for which is possible, then it is known that S is infinite, but has density zero, and that if n is a nonsquare member of S, then there are infinitely many solutions for such an n. If $N(x)$ is the number of members of S less than x, then the best that seems to be known is that

$$c\sqrt{x} < N(x) = O\left(\frac{x}{\ln x}\right).$$

The elements of S, $1 < n < 73$, and the corresponding least values of a for which the sum of n squares starting with a is square, are

n	2	11	23	24	26	33	47	49	50	59
a	3	18	7	1	25	7	539	25	7	22

More generally still, one can ask that the sums of the squares of the members of an arbitrary arithmetic progression should be square. K. R. S. Sastry notes that this can occur if the number of terms in the progression is square.

In answer to the question: which triangular numbers are the product of three consecutive integers, Tzanakis & de Weger gave the (only) answers 6, 120, 210, 990, 185136 and 258474216. Unfortunately, Mohanty's elementary proof of the same result is erroneous.

Richard Bruce noticed that 6, 66, and 666 are all triangular numbers. Dean Hickerson wonders if there are other examples; i.e., integers $1 \le d < b$ such that d, $d(b+1)$, and $d(b^2+b+1)$ are all triangular and answers 'Yes': If d is triangular and $b = 8d+1$ then it's easy to see that $d(b^n+b^{n-1}+\ldots b+1)$ is triangular for all $n \ge 0$. The only other solutions (d, b) that he knows are (6,10) and (6,2040). If we allow $d \ge b$, then there's also (15,2).

For three triangular numbers in geometric progression, see **D23**.

de Weger confirms that there are no nontrivial examples of $\binom{n}{2} = \binom{m}{4}$ apart from (10,21).

For $\binom{n}{k} = \binom{m}{l}$, de Weger notes that $(k,l) = (3,4)$ is essentially exercise 9.13 in Silverman's *The Arithmetic of Elliptic Curves* (put $X = n - 1$, $2Y = m^2 - 3m$). The only positive solutions are $\binom{3}{3} = \binom{4}{4}$ and $\binom{7}{3} = \binom{7}{4}$. The corresponding problem of making the product of two consecutive positive integers equal to the product of three consecutive integers has just the solutions $2 \cdot 3 = 1 \cdot 2 \cdot 3$ and $14 \cdot 15 = 5 \cdot 6 \cdot 7$. The cases $(k,l) = (2,6)$, (2,8), (3,6), (4,6), (4,8) also lead to elliptic curves.

Other examples of elliptic curves were treated by Bremner & Tzanakis who showed that there are just 26 integer points on $y^2 = x^3 - 7x + 10$ and they also examined $y^2 = x^3 - bx + c$ for $(b,c) = (172,505)$, (172,820) and (112,2320).

There are infinitely many solutions of $\binom{a}{k} - \binom{b}{k} = c^k$ for $k = 3$. Are there any for $k = 4$? And is $(a,b,c) = (18,12,6)$ an isolated example for $k = 5$?

Pintér & de Weger note that

$$210 = 14 \times 15 = 5 \times 6 \times 7 = \binom{21}{2} = \binom{10}{4}$$

and give complete solutions to the $\binom{4}{2}$ implied diophantine equations.

Pintér shows that the only nontrivial solutions of $\binom{x}{2} = y(y-1)(y-2)(y-3)$ are $(-15,-2)$, $(16,-2)$, $(-15,5)$, $(16,5)$.

Hajdu & Pintér survey and complete the solution of two dozen equations of the type considered here.

Győry completed the solution of $\binom{n}{k} = x^l$ (Erdős had settled $l \geq 4$) by showing that the only other solutions are with $k = l = 2$ or $(n,k,x,l) = (50,3,140,2)$.

Luo Ming gave a simpler proof that the only nontrivial triangle which is the square of a triangle is 36.

Grytczuk has studied the equation

$$2\binom{x+n-1}{n} = \binom{y+n-1}{n}$$

and solves some special cases. This is related to a conjecture of Erdős.

H. L. Abbott, P. Erdős & D. Hanson, On the number of times an integer occurs as a binomial coefficient, *Amer. Math. Monthly* **81**(1974) 256–261.

S. C. Althoen & C. Lacampagne, Tetrahedral numbers as sums of square numbers, *Math. Mag.*, **64**(1991) 104–108.

W. S. Anglin, The square pyramid puzzle, *Amer. Math. Monthly* **97**(1990) 120–124; *MR* **91e**:11026.

Krassimir T. Atanassov, Extension of one Sierpiński's problem related to the triangular numbers, *Adv. Stud. Contemp. Math. (Kyungshang)*, **5**(2002) 33–36.

È. T. Avanesov, The Diophantine equation $3y(y+1) = x(x+1)(2x+1)$ (Russian), *Volž. Mat. Sb. Vyp.*, **8**(1971) 3–6; *MR* **46** #8967.

È. T. Avanesov, Solution of a problem on figurate numbers (Russian), *Acta Arith.*, **12**(1966/67) 409–420; *MR* **35** #6619.

E. Barbette, Les sommes de p-ièmes puissances distinctes égales à une p-ième puissance, Liège, 1910, 77–104.

J. M. Barja, J. M. Molinelli & L. Blanco Ferro, Finiteness of the number of solutions of the Diophantine equation $\binom{x}{n} = \binom{y}{2}$, *Bull. Soc. Math. Belg. Sér. A*, **45**(1993) 39–43; *MR* **96a**:11027.

Laurent Beeckmans, Squares expressible as the sum of consecutive squares, *Amer. Math. Monthly*, **101**(1994) 437–442; *MR* **95c**:11033.

Michael A. Bennett, Lucas's square pyramid problem revisited, *Acta Arith.*, **105**(2002) 341–347; *MR* **2003g**:11028.

David W. Boyd & Hershey H. Kisilevsky, The diophantine equation $u(u + 1)(u + 2)(u + 3) = v(v + 1)(v + 2)$, *Pacific J. Math.*, **40**(1972) 23–32; *MR* **46** #5238.

Andrew Bremner, An equation of Mordell, *Math. Comput.*, **29**(1975) 925–928; *MR* **51** #10219.

Andrew Bremner & Nicholas Tzanakis, Integer points on $y^2 = x^3 - 7x + 10$, *Math. Comput.*, **41**(1983) 731–741.

A. Bremner, R. J. Stroeker & N. Tzanakis, On sums of consecutive squares. *J. Number Theory*, **62**(1997) 39–70; *MR* **97k**:11082.

B. Brindza, Ákos Pintér & S. Turjányi, On equal values of pyramidal and polygonal numbers, *Indag. Math. (N.S.)*, **9**(1998) 183–185.

Jasbir S. Chahal & Jaap Top, Triangular numbers and elliptic curves, *Rocky Mountain J. Math.*, **26**(1996) 937–949.

Chou Chung-Chiang & Deng Yue-Fan, Decomposing 40 billion integers by four tetrahedral numbers, *Math. Comput.*, **66**(1997) 893–901; *MR* **97i**:11123.

J. E. Cremona, *Algorithms for Modular Elliptic Curves*, Cambridge Univ. Press, 1992. [Information for conductor 702, inadvertently omitted from the tables, is obtainable from the author.]

Ion Cucurezeanu, An elementary solution of Lucas' problem, *J. Number Theory*, **44**(1993) 9-12.

H. E. Dudeney, *Amusements in Mathematics*, Nelson, 1917, 26, 167.

William Duke, Some old problems and new results about quadratic forms, *Notices Amer. Math. Soc.*, **44**(1997) 190–195.

P. Erdős, On a Diophantine equation, *J. London Math. Soc.*, **26**(1951) 176–178; *MR* **12**, 804d.

Raphael Finkelstein, On a Diophantine equation with no non-trivial integral solution, *Amer. Math. Monthly*, **73**(1966) 471–477; *MR* **33** #4004.

Martin Gardner, Mathematical games, On the patterns and the unusual properties of figurate numbers, *Sci. Amer.*, **231** No. 1 (July, 1974) 116–120.

P. Goetgheluck, Infinite families of solutions of the equation $\binom{n}{k} = 2\binom{a}{b}$, *Math. Comput.*, **67**(1998) 1727–1733; *MR* **99a**:11019.

Charles M. Grinstead, On a method of solving a class of Diophantine equations, *Math. Comput.*, **32**(1978) 936–940; *MR* **58** #10724.

A. Grytczuk, On a conjecture of Erdős on binomial coefficients, *Studia Sci. Math. Hungar.*, **29**(1994) 241–244; *MR* **95g**:11013.

Richard K. Guy, Every number is expressible as the sum of how many polygonal numbers? *Amer. Math. Monthly*, **101**(1994) 169–172; *MR* **95k**:11041.

K. Győry, On the Diophantine equation $\binom{n}{k} = x^l$, *Acta Arith.*, **80**(1997) 289–295; *MR* **98f**:11028.

L. Hajdu & Ákos Pintér, Combinatorial diophantine equations, (preprint late 1997?)

Heiko Harborth, Fermat-like binomial equations, in Applications of Fibonacci Numbers, Kluwer, 1988, 1–5.

Michael D. Hirschhorn & James A. Sellers, On representations of a number as a sum of three triangles, *Acta Arith.*, **77**(1996) 289–301.

Masanobu Kaneko & Katsuichi Tachibana, When is a polygonal pyramid number again polygonal? *Rocky Mountain J. Math.*, **32**(2002) 149–165; *MR* **2003b**:11025..

T. Krausz, A note on equal values of polygonal numbers, *Publ. Math. Debrecen*, **54**(1999) 321–325.

Masato Kuwata & Jaap Top, An elliptic surface related to sums of consecutive squares, *Exposition. Math.*, **12**(1994) 181–192.

Moshe Laub, O. P. Lossers & L. E. Mattics, Problem 6552 and solution, *Amer. Math. Monthly* **97**(1990) 622–625.

Le Mao-Hua, A note on the Diophantine equation $\binom{k}{2} - 1 = q^n + 1$, *Colloq. Math.*, **76**(1998) 31–34; *MR* **99a**:11038.

D. A. Lind, The quadratic field $Q(\sqrt{5})$ and a certain Diophantine equation, *Fibonacci Quart.*, **6**(1968) 86–93.

W. Ljunggren, New solution of a problem proposed by E. Lucas, *Norsk Mat. Tidskr.*, **34**(1952) 65–72; *MR* **14**, 353h.

W. Ljunggren, A diophantine problem, *J. London Math. Soc.* (2), **3**(1971) 385–391; *MR* **45** #171.

E. Lucas, Problem 1180, *Nouv. Ann. Math.*(2), **14**(1875) 336.

Ma De-Gang, An elementary proof of the solution to the Diophantine equation $6y^2 = x(x + 1)(2x + 1)$, *Sichuan Daxue Xuebao*, **4**(1985) 107–116; *MR* **87e**:11039.

R. A. MacLeod & I. Barrodale, On equal products of consecutive integers, *Canad. Math. Bull.*, **13**(1970) 255–259.

Maurice Mignotte, Complete resolution of some families of Diophantine equations, *Number theory* (*Eger* 1996) 383–399, de Gruyter, Berlin, 1998; *MR* **99d**:11026.

Luo Ming, On the Diophantine equation $(x(x-1)/2)^2 = y(y-1)/2$, *Fibonacci Quart.*, **34**(1996) 277-279; *MR* **97a**:11048.

S. P. Mohanty, Integer points of $y^2 = x^3 - 4x + 1$, *J. Number Theory*, **30**(1988) 86-93; *MR* **90e**:11041a; but see A. Bremner, **31**(1989) 373; **90e**:11041b.

Louis Joel Mordell, On the integer solutions of $y(y + 1) = x(x + 1)(x + 2)$, *Pacific J. Math.*, **13**(1963) 1347–1351.

Ken Ono, Sinai Robins & Patrick T. Wahl, On the representations of integers as sums of triangular numbers, *Aequationes Math.*, **50**(1995) 73–94; *MR* **96i**:11044.

Ákos Pintér, A note on the Diophantine equation $\binom{x}{4} = \binom{y}{2}$, *Publ. Math. Debrecen*, **47**(1995) 411–415; *MR* **961**:11027.

Ákos Pintér, On the equation $\binom{x}{2} = y(y - 1)(y - 2)(y - 3)$, *Publ. Math. Debrecen*, (submitted) (199) –; *MR* :110.

Ákos Pintér & B. M. M. de Weger, $210 = 14 \times 15 = 5 \times 6 \times 7 = \binom{21}{2} = \binom{10}{4}$, *Publ. Math. Debrecen*, **51**(1997) 175–189; *MR* **98k**:11032.

H. E. Saltzer & N. Levine, Tables of integers not exceeding 1000000 that are not expressible as the sum of four tetrahedral numbers, *Math. Tables Aids Comput.*, **12**(1958) 141–144; *MR* **20** #6194.

David Singmaster, How often does an integer occur as a binomial coefficient? *Amer. Math. Monthly* **78**(1971) 385–386.

David Singmaster, Repeated binomial coefficients and Fibonacci numbers, *Fibonacci Quart.*, **13** (1975) 295–298.

V. Siva Rama Prasad & B. Srinivasa Rao, Triangular numbers in the associated Pell sequence and Diophantine equations, $x^2(x+1)^2 = 8y^2 \pm 4$, *Indian J. Pure Appl. Math.*, **33**(2002) 1643–1648; *MR* **2004c**:11033.

C. M. Skinner & T. D. Wooley, Sums of two k-th powers, *J. reine angew. Math.*, **462**(1995) 57–68. 96b 11132

R. J. Stroeker, On the sum of consecutive cubes being a perfect square, *Compositio Math.*, **97**(1995) 295–307; *MR* **96i**:11038.

Roelof J. Stroeker & B. M. M. de Weger, Elliptic binomial diophantine equations, MaT **68**(1999) 1257–1281; *MR* **99i**:11122.

Craig A. Tovey, Multiple occurrences of binomial coefficients, *Fibonacci Quart.*, **23**(1985) 356–358.

Nikos Tzanakis, Explicit solution of a class of quartic Thue equations, *Acta Arith.*, **64**(1993) 271–283; *MR* **94e**:11022.

N. Tzanakis & B. M. M. de Weger, How to explicitly solve a Thue-Mahler equation, *Compositio Math.*, **84**(1992) 223–288; *MR* **93k**:11025; corrections, **89** (1993) 241–242.

Nikos Tzanakis, Explicit solution of a class of quartic Thue equations, *Acta Arith.*, **64**(1993) 271–283; *MR* **94e**:11022.

N. Tzanakis & B. M. M. de Weger, On the practical solution of the Thue equation, *J. Number Theory*, **31**(1989) 99-132; *MR* **90c**:11018.

Saburô Uchiyama, Solution of a Diophantine problem, *Tsukuba J. Math.*, **8**(1984) 131–137; *MR* **86i**:11010.

Richard Warlimont, On natural numbers as sums of consecutive hth powers, *J. Number Theory*, **68**(1998) 87–98.

G. N. Watson, The problem of the square pyramid, *Messenger of Math.*, **48**(1918/19) 1–22.

B. M. M. de Weger, A binomial diophantine equation, Report 9476/B, Econometric Institue, Erasmus Univerity Rotterdam, 94-12-13; *Quart. J. Math. Oxford Ser.*,

Benjamin M. M. de Weger, Equal binomial coefficients: some elementary considerations, *J. Number Theory*, **63**(1997) 373–386; *MR* **98b**:11027.

Z. Y. Xu & Cao Zhen-Fu, On a problem of Mordell, *Kexue Tongbao*, **30**(1985) 558–559.

OEIS: A001219, A047694-047695.

D4 Waring's problem. Sums of l kth Powers.

If all numbers are representable as the sum of l k-th powers, then it is usual to denote the least such l by $g(k)$. For example

$k =$	2	3	4	5	6	7	8	9	10	11	12	13	14	15	16	17
$g(k) =$	4	9	19	37	73	143	279	548	1079	2132	4223	8384	16673	33203	66190	132055

For $k > 2$, with a finite number of exceptions, it is possible to represent n as the sum of a smaller number, $G(k)$, of k-th powers. Much less is known about $G(k)$: $G(2) = 4$, $4 \leq G(3) \leq 7$, $G(4) = 16$, $G(5) \leq 18$. Probably $G(3) = 4$; see the opening paragraph of **D5**. In 1996 Li Hong-Ze showed that $G(16) \leq 111$.

Let $r_{k,l}(n)$ be the number of solutions of $n = \sum_{i=1}^{l} x_i^k$ in *positive* integers x_i. Hardy & Littlewood's Hypothesis K is that $\epsilon > 0$ implies that $r_{k,k}(n) = O(n^\epsilon)$. This is well-known for $k = 2$; in fact, for sufficiently large n,

$$r_{2,2}(n) < n(1 + \epsilon) \ln 2 / \ln \ln n$$

and this does not hold if $\ln 2$ is replaced by anything smaller. Mahler disproved the hypothesis for $k = 3$ by showing that $r_{3,3} > c_1 n^{1/12}$ for infinitely many n.

Erdős thinks it possible that for all n, $r_{3,3} < c_2 n^{1/12}$ but nothing is known. Probably Hypothesis K fails for every $k \geq 3$, but also it's probable that $\sum_{n=1}^{x} (r_{k,k}(n))^2 < x^{1+\epsilon}$ for sufficiently large x.

S. Chowla proved that for $k \geq 5$, $r_{k,k}(n) \neq O(1)$ and, with Erdős, that for every $k \geq 2$ and for infinitely many n,

$$r_{k,k} > \exp(c_k \ln n / \ln \ln n).$$

Mordell proved that $r_{3,2}(n) \neq O(1)$ and Mahler that $r_{3,2}(n) > (\ln n)^{1/4}$ for infinitely many n. No nontrivial upper bound for $r_{3,2}(n)$ is known. Jean Lagrange has shown that $\limsup r_{4,2}(n) \geq 2$ and that $\limsup r_{4,3}(n) = \infty$.

Another tough problem is to estimate $A_{k,l}(x)$, the number of $n \leq x$ which are expressible as the sum of l k-th powers. Landau showed that

$$A_{2,2}(x) = (c + o(1))x/(\ln x)^{1/2},$$

Erdős & Mahler proved that if $k > 2$, then $A_{k,2} > c_k x^{2/k}$, and Hooley that $A_{k,2} > (c_k + o(1))x^{2/k}$. It seems certain that if $l < k$, then $A_{k,l} > c_{k,l} x^{l/k}$ and that $A_{k,k} > x^{1-\epsilon}$ for every ϵ, but these have not been established.

It follows from the Chowla-Erdős result that for all k there is an n_k such that the number of solutions of $n_k = p^3 + q^3 + r^3$ is greater than k. No corresponding result is known for more than three summands.

The survey article of Vaughan & Wooley contains 162 references.

David Wilson gave a list of smallest positive k-th powers which are a sum of distinct smaller positive kth powers for $1 \leq k \leq 10$.

$3^1 = 1^1 + 2^1$, $5^2 = 3^2 + 4^2$, $6^3 = 3^3 + 4^3 + 5^3$, $15^4 = 4^4 + 6^4 + 8^4 + 9^4 + 14^4$,
$12^5 = 4^5 + 5^5 + 6^5 + 7^5 + 9^5 + 11^5$, $25^6 = 1^6 + 2^6 + 3^6 + 5^6 + 6^6 + 7^6 + 8^6 +$
$9^6 + 10^6 + 12^6 + 13^6 + 15^6 + 16^6 + 17^6 + 18^6 + 23^6$, $40^7 = 1^7 + 3^7 + 5^7 +$
$9^7 + 12^7 + 14^7 + 16^7 + 17^7 + 18^7 + 20^7 + 21^7 + 22^7 + 25^7 + 28^7 + 39^7$, $84^8 =$
$1^8 + 2^8 + 3^8 + 5^8 + 7^8 + 9^8 + 10^8 + 11^8 + 12^8 + 13^8 + 14^8 + 15^8 + 16^8 + 17^8 + 18^8 + 19^8 + 21^8 +$
$23^8 + 24^8 + 25^8 + 26^8 + 27^8 + 29^8 + 32^8 + 33^8 + 35^8 + 37^8 + 38^8 + 39^8 + 41^8 + 42^8 + 43^8 +$
$45^8 + 46^8 + 47^8 + 48^8 + 49^8 + 51^8 + 52^8 + 53^8 + 57^8 + 58^8 + 59^8 + 61^8 + 63^8 + 69^8 + 73^8$,
$47^9 = 1^9 + 2^9 + 4^9 + 7^9 + 11^9 + 14^9 + 15^9 + 18^9 + 26^9 + 27^9 + 30^9 + 31^9 + 32^9 + 33^9 + 36^9 +$
$38^9 + 39^9 + 43^9$, $63^{10} = 1^{10} + 2^{10} + 4^{10} + 5^{10} + 6^{10} + 8^{10} + 12^{10} + 15^{10} + 16^{10} + 17^{10} +$
$20^{10} + 21^{10} + 25^{10} + + 26^{10} + 27^{10} + 28^{10} + 30^{10} + 36^{10} + 37^{10} + 38^{10} + 40^{10} + 51^{10} + 62^{10}$.

Ramachandran Balasubramanian, Jean-Marc Deshouillers & François Dress, Problème de Waring pour les bicarrés II, Résultats auxiliaires pour le théorème asymptotique, *C. R. Acad. Sci. Paris Sér. I Math.*, **303**(1986) 161–163; *MR* **88e**:11095.

F. Bertault, Oliver Ramaré & Paul Zimmermann, On sums of seven cubes, *Math. Comput.*, **68**(1999) 1303–1310; *MR* **99j**:11116.

Jörg Brüdern, A sieve approach to the Waring-Goldbach problem II: on the seven cubes theorem, *Acta Arith.*, **72**(1995) 211–227; *MR* **97j**:11047.

Jörg Brüdern & Nigel Watt, On Waring's problem for four cubes, *Duke Math. J.*, **77**(1995) 583–606; *MR* **96e**:11121.

Jörg Brüdern & Trevor D. Wooley, On Waring's problem: three cubes and a sixth power, *Nagoya Math. J.*, **163**(2001) 13–53; *MR* **2002e**:11133.

François Castella & Alain Plagne, A distribution result for slices of sums of squares, *Math. Proc. Cambridge Philos. Soc.*, **132**(2002) 1–22.

Ajai Choudhry, On representing 1 as the sum or difference of kth powers of integers, *Math. Student*, **70**(2001) 205–208.

S. Chowla, The number of representations of a large number as a sum of non-negative nth powers, *Indian Phys.-Math. J.*, **6**(1935) 65–68; *Zbl.* **12**.339.

H. Davenport, Sums of three positive cubes, *J. London Math. Soc.*, **25**(1950) 339–343; *MR* **12**, 393.

Jean-Marc Deshouillers, François Hennecart & Bernard Landreau, Waring's problem for sixteen biquadrates – numerical results, *J. Théor. Nombres Bordeaux*, **12**(2000) 411–422; *MR* **2002b**:11133.

P. Erdős, On the representation of an integer as the sum of k kth powers, *J. London Math. Soc.*, **11**(1936) 133–136; *Zbl.* **13**.390.

P. Erdős, On the sum and difference of squares of primes, I, II, *J. London Math. Soc.*, **12**(1937) 133–136, 168–171; *Zbl.* **16**.201, **17**.103.

P. Erdős & K. Mahler, On the number of integers that can be represented by a binary form, *J. London Math. Soc.*, **13**(1938) 134–139; *Zbl.* **13**.390.

P. Erdős & E. Szemerédi, On the number of solutions of $m = \sum_{i=1}^{k} x_i^k$, *Proc. Symp. Pure Math.*, **24** Amer. Math. Soc., Providence, 1972, 83–90; *MR* **49** #2529.

W. Gorzkowski, On the equation $x_0^2 + x_1^2 + \ldots + x_n^2 = x_{n+1}^k$, *Ann. Soc. Math. Polon. Ser. I Comment. Math. Prace Mat.*, **10**(1966) 75–79; *MR* **32** #7495.

G. H. Hardy & J. E.Littlewood, Partitio Numerorum VI: Further researches in Waring's problem, *Math. Z.*, **23**(1925) 1–37.

Koichi Kawada & Trevor D. Wooley, On the Waring-Goldbach problem for fourth and fifth powers, *Proc. London Math. Soc.*(3) **83**(2001) 1–50; *MR* **2002b**:11134.

Koichi Kawada & Trevor D. Wooley, Slim exceptional sets for sums of fourth and fifth powers, *Acta Arith.*, **103**(2002) 225–248; *MR* **2003a**:11125.

Alain Kraus, Sur l'équation $a^3 + b^3 = c^p$, *Experiment. Math.*, **7**(1998) 1–13; *MR* **99f**:11040.

Jean Lagrange, Thèse d'État de l'Université de Reims, 1976.

Li Hong-Ze, Waring's problem for sixteenth powers, *Sci. China Ser. A*, **39**(1996) 56–64; *MR* **97g**:11112

K. Mahler, On the lattice points on curves of genus 1, *Proc. London Math. Soc.*, **39**(1935) 431–466.

K. Mahler, Note on Hypothesis K of Hardy and Littlewood, *J. London Math. Soc.*, **11**(1936) 136–138.

Meng Xian-Meng, The Waring-Goldbach problem in short intervals, *Shandong Daxue Xuebao Ziran Kexue Ban* **32**(1997) 255–264; *MR* **98k**:11134.

Robert C. Vaughan & Trevor D. Wooley, Further improvements in Waring's problem, *Acta Math.*, **174**(1995) 147–240; II. Sixth powers, *Duke Math. J.*, **76**(1994) 683–710; *MR* **96j**:11129ab; III. Eighth powers, *Philos. Trans. Roy. Soc. London Ser. A* **345**(1993) 385–396; *MR* **94m**:11118; IV. Higher powers, *Acta Arith.*, **94**(2000) 203–285; *MR* **2001g**:11154.

R. C. Vaughan & T. D. Wooley, Waring's problem: a survey, *Number Theory for the Millennium III* (*Urbana IL*, 2000, 301–340; *MR* **2003j**:11116.

Van H. Vu, On a refinement of Waring's problem, *Duke Math. J.*, **105**(2000) 107–134; *MR* **2002e**:11134.

Trevor D. Wooley, Slim exceptional sets for sums of cubes, *Canad. J. Math.*, **54**(2002) 417–448; *MR* **2003a**:11131.

Trevor D. Wooley, Slim exceptional sets and the asymptotic formula in Waring's problem, *Math. Proc. Cambridge Philos. Soc.*, **134**(2003) 193–206; *MR* **2004c**:11181.

OEIS: A002376, A002377, A002804, A018889, A079611.

D5 Sum of four cubes.

Is every number the sum of four cubes? Dem'janenko has proved this for all numbers except possibly those of the form $9n \pm 4$. Richard Lukes has found representations of all $n \leq 10^7$ as sums of four positive or negative cubes. The only 'difficult' number was 82562; since numbers appear to get 'easier' with increasing size, it is most plausible that all numbers can be represented as the sum of four cubes.

More demanding is to ask if every number is the sum of four cubes with two of them equal. In earlier editions we asked specifically if there is a solution of $76 = x^3 + y^3 + 2z^3$ and on 96-01-18 J. H. E. Cohn emailed the representation $(k, x, y, z) = (76, -122171, -21167, 97135)$ of $k = x^3 + y^3 + 2z^3$ together with

$(230, 27293, -14101, -20617)$, $(356, 1048469, 129521, -832693)$,
$(418, 91705, 15961, -72914)$, $(428, -117091, -111433, 114332)$,
$(445, 150439, -19178, -119321)$, $(482, -11878, -2254, 9449)$,
$(580, 89845, 85111, -87542)$, $(967, 641263, 380698, -542246)$.

Kenji Koyama sent the solutions $(491, 13476659, 13584908, -13531000)$, $(183, 4170061, -4494438, 2090533)$, $(931, -6942368, -23115371, 18510883)$, so that now the missing $k < 1000$ are at most 148, 671 and 788. There are 59 more, less than 10000, listed on the website quoted below.

1084	1121	1247	1444	1462	1588	1975	2246	2300	2372
2822	3047	3268	3307	3335	3380	3641	3676	3956	4036
4108	4369	4388	4819	4883	4990	5188	5279	5468	5540
5620	5629	6707	6980	7097	7106	7132	7177	7323	7519
7708	7727	7799	7853	7862	7988	8114	8380	8572	8588
8644	8779	8887	8968	9274	9463	9589	9724	9850	

Are all numbers which are not of the form $9n \pm 4$ the sum of *three* cubes? From the list of unknowns given in the second edition, Andrew Bremner has deleted 75 and 600; Conn & Vaserstein 84; Richard Lukes 110, 435 and 478; Kenyi Koyama 444, 501, 618, 912 and 969. Don Reble quoted $30 = 2220422932^3 - 283059965^3 - 2218888517^3$ and the website
http://www.asahi-net.or.jp/~KC2H-MSM/mathland/math04/cube01.htm
which implies that only 25 remain:

$$33 \quad 42 \quad 52 \quad 74 \quad 114 \quad 156 \quad 165 \quad 318 \quad 366 \quad 390 \quad 420 \quad 564 \quad 579$$
$$627 \quad 633 \quad 732 \quad 758 \quad 789 \quad 795 \quad 894 \quad 906 \quad 921 \quad 933 \quad 948 \quad 975$$

Noam Elkies sent 462 and many smaller representations of earlier unknowns, and he wrote

... is the representation $12 = 9730705^3 - 9019406^3 - 5725013^3$ known? Miller & Woolett found only $12 = 10^3 + 7^3 - 11^3$ and asked about the existence of further solutions, of which the above is the first. Likewise for $2 = 1214928^3 + 3480205^3 - 3528875^3$, the first instance not accounted for by the identity $2 = (6t^3 + 1)^3 - (6t^3 - 1)^3 - (6t^2)^3$.

The equation $3 = x^3 + y^3 + z^3$ has the solutions $(1, 1, 1)$ and $(4, 4, -5)$. Are there any others?

Antal Balog & Jörg Brüdern, Sums of three cubes in three linked three-progressions, *J. reine angew. Math.*, **466**(1995) 45–85; *MR* **96j**:11133.

Andrew Bremner, On sums of three cubes, *Number theory (Halifax NS, 1994)* 87–91, *CMS Conf. Proc.*, **15** Amer. Math. Soc., Providence RI, 1995; *MR* **96g**:11024.

Jörg Brüdern, Sums of four cubes, *Monatsh. Math.*, **107**(1989) 179–188; *MR* **90g**:11136.

Jörg Brüdern, On Waring's problem for cubes, *Math. Proc. Cambridge Philos. Soc.*, **109**(1991) 229–256; *MR* **91m**:11081.

J. W. S. Cassels, A note on the Diophantine equation $x^3 + y^3 + z^3 = 3$, *Math. Comput.*, **44**(1985) 265–266; *MR* **86d**:11021.

J. H. E. Cohn & L. J. Mordell, On sums of four cubes of polynomials, *J. London Math. Soc.*(2), **5**(1972) 74–78; *MR* **45** #3367.

W. Conn & L. N. Vaserstein, On sums of three integral cubes, The Rademacher legacy to mathematics (University Park, 1992), *Contemp. Math.*, **166**, Amer. Math. Soc., 1994, 285–294; *MR* **95g**:11128.

V. A. Dem'janenko, Sums of four cubes, *Izv. Vysš. Učebn. Zaved. Matematika*, **1966**(1966) 64–69; *MR* **34** #2525. [Thanks to Henri Cohen, an English translation is at http://www.math.u-bordeaux.fr/~cohen]

Leonard Eugene Dickson, Simpler proofs of Waring's Theorem on cubes with various generalizations, *Trans. Amer. Math. Soc.*, **30**(1928) 1–18.

S. W. Dolan, On expressing numbers as the sum of two cubes, *Math. Gaz.*, **66**(1982) 31–38.

W. J. Ellison, Waring's problem, *Amer. Math. Monthly*, **78**(1971) 10–36.

V. L. Gardiner, R. B. Lazarus & P. R. Stein, Solutions of the diophantine equation $x^3 + y^3 = z^3 - d$, *Math. Comput.*, **18**(1964) 408–413; *MR* **31** #119.

D. R. Heath-Brown, Searching for solutions of $x^3 + y^3 + z^3 = k$, *Séminaire de Théorie des Nombres, Paris*, 1989–90, 71–76, *Progr. Math.*, **102**(1992) Birkhäuser, Boston; *MR* **98f**:11025.

D. R. Heath-Brown, W. M. Lioen & H. J. J. te Riele, On solving the Diophantine equation $x^3 + y^3 + z^3 = k$ on a vector computer, *Math. Comput.*, **61**(1993) 235–244; *MR* **94f**:11132.

C. Hooley, On the representations of a number as the sum of four cubes, I, *Proc. London Math. Soc.*(3). **36**(1978) 117–140; II, **16**(1977) 424–428; *MR* **58** #21932ab.

Chao Ko, Decompositions into four cubes, *J. London Math. Soc.*, **11**(1936) 218–219.

Koichi Kawada, On the sum of four cubes, *Mathematika*, **43**(1996) 323–348; *MR* **97m**:11125.

Kenji Koyama, Tables of solutions of the Diophantine equation $x^3 + y^3 + z^3 = n$, *Math. Comput.*, **62**(1994) 941–942.

Kenji Koyama, On the solutions of the Diophantine equation $x^3 + y^3 + z^3 = n$, *Trans. Inst. Electron. Inform. Commun. Eng.* (*IEICE in Japan*), **E78-A**(1994) 444–449.

Kenji Koyama, On searching for solutions of the Diophantine equation $x^3 + y^3 + 2z^3 = n$, *Math. Comput.*, **69**(2000) 1735–1742; *MR* **2001a**:11206.

Kenji Koyama, Yukio Tsuruoka & Hiroshi Sekigawa, On searching for solutions of the Diophantine equation $x^3 + y^3 + z^3 = n$, *Math. Comput.*, **66**(1997) 841–851; *MR* **97m**:11041.

M. Lal, W. Russell & W. J. Blundon, A note on sums of four cubes, *Math. Comput.*, **23**(1969) 423–424; *MR* **39** #6819.

A. Mąkowski, Sur quelques problèmes concernant les sommes de quatre cubes, *Acta Arith.*, **5**(1959) 121–123; *MR* **21** #5609.

J. C. P. Miller & M. F. C. Woollett, Solutions of the diophantine equation $x^3 + y^3 + z^3 = k$, *J. London Math. Soc.*, **30**(1955) 101–110; *MR* **16**, 797e.

L. J. Mordell, On sums of four cubes of polynomials, *Acta Arith.*, **16**(1969/1970) 365–369; *MR* **42** #230.

G. Payne & L. Vaserstein, Sums of three cubes, in *The Arithmetic of Function Fields*, de Gruyter, 1992, 443–454; *MR* **93k**:11090.

Ren Xiu-Min, Sums of four cubes of primes, *J. Number Theory*, **98**(2003) 156–171; *MR* **2003j**:11122 (and see **2002c**:11130).

Puilippe Revoy, Sur les sommes de quatre cubes, *Enseign. Math.*(2), **29**(1983) 209–220; *MR* **85m**:11058.

H. W. Richmond, *Messenger Math.*(2), **51**(1921) 177–186.

Fernando Rodríguez Villegas & Don Zagier, Which primes are sums of two cubes? *Number Theory (Halifax NS* 1994) 295–306, *CMS Conf. Proc.*, **15**, Amer. Math. Soc., Providence RI, 1995; *MR* **96g**:11049.

W. Scarowsky & A. Boyarsky, A note on the Diophantine equation $x^n + y^n + z^n = 3$, *Math. Comput.*, **42**(1984) 235–237; *MR* **85c**:11029.

A. Schinzel, On sums of four cubes of polynominals, *J. London Math. Soc.*, **43**(1968) 143–145; *MR* **36** #6388.

A. Schinzel & W. Sierpiński, Sur les sommes de quatre cubes, *Acta Arith.*, **4**(1958) 20–30; *MR* **20**, #1664.

Sun Qi, On Diophantine equation $x^3 + y^3 + z^3 = n$, *Kexue Tongbao*, **33**(1988) 2007–2010; *MR* **90g**:11033 (see also **32**(1987) 1285–1287).

R. C. Vaughan, On Waring's problem for cubes, *J. reine angew. Math.*, **365**(1986) 122–170; *MR* **87j**:11103.

Edward Maitland Wright, An easier Waring's problem, *J. London Math. Soc.*, **9**(1934) 267–272.

OEIS: A003072, A046041, A060464-060467.

D6 An elementary solution of $x^2 = 2y^4 - 1$.

Ljunggren has shown that the only solutions of $y^2 = 2x^4 - 1$ in positive integers are (1,1) and (239,13) but his proof is difficult. Mordell asks if it is possible to find a simple or elementary proof. Whether Steiner & Tzanakis have simplified the solution may be a matter of taste; they use the theory of linear forms in logarithms of algebraic numbers.

Chen Jian-Hua gives an unconventional solution.

Ljunggren and others have made considerable investigations into equations of similar type. For references see the first edition. Cohn has considered the equation $y^2 = Dx^4 + 1$ for all $D \leq 400$. This will have rational solutions just when there are rational points on the curves $y^2 = x(x^2 - 4D)$ and $Y^2 = X(X^2 + 16D)$, which are nonsingular provided $\pm D$ is not square.

Chen Jian-Hua, A new solution of the Diophantine equation $X^2 + 1 = 2Y^4$, *J. Number Theory*, **48**(1994) 62–74; *MR* **95i**:11019.

Chen Jian-Hua, A note on the Diophantine equation $x^2 + 1 = dy^4$, *Abh. Math. Sem. Univ. Hamburg*, **64**(1994) 1–10; *MR* **95k**:11034.

J. H. E. Cohn, The diophantine equation $y^2 = Dx^4 + 1$, I, *J. London Math. Soc.*, **42**(1967) 475–476; *MR* **35** #4158; II, *Acta Arith.*, **28**(1975/76) 273–275; *MR* **52** #8029; III, *Math. Scand.*, **42**(1978) 180–188; *MR* **80a**:10031.

W. Ljunggren, Zur Theorie der Gleichung $x^2 + 1 = Dy^4$, *Avh. Norsk. Vid. Akad. Oslo*, (1942) 1–27.

W. Ljunggren, Some remarks on the diophantine equations $x^2 - \mathcal{D}y^4 = 1$ and $x^4 - \mathcal{D}y^2 = 1$, *J. London Math. Soc.*, **41**(1966) 542–544; *MR* **33** #5555.

L. J. Mordell, The diophantine equation $y^2 = Dx^4 + 1$, *J. London Math. Soc.*, **39**(1964) 161–164; *MR* **29** #65.

Ray Steiner & Nikos Tzanakis, Simplifying the solution of Ljunggren's equation $X^2 + 1 = 2Y^4$, *J. Number Theory*, **37**(1991) 123–132; *MR* **91m**:11018.

Sun Qi & Yuan Ping-Zhi, A note on the Diophantine equation $x^4 - Dy^2 = 1$. *Sichuan Daxue Xuebao*, **34**(1997) 265–268; *MR* **98g**:11034.

D7 Sum of consecutive powers made a power.

Rufus Bowen conjectured that the equation

$$1^n + 2^n + \cdots + m^n = (m+1)^n$$

has no nontrivial solutions, and Leo Moser showed that there were none with $m \le 10^{1000000}$ and none with n odd. Zhou & Kang raised the bound to $m \le 10^{2000000}$. Van de Lune & te Riele showed that the equation is almost never solvable. Note that $n \sim m \ln 2$.

Moree corrects and generalizes a result of Carlitz and applies it to Bowen's equation. He also shows that the equation

$$1^n + 2^n + \cdots + m^n = a(m+1)^n$$

has no integer solutions (a, m, n) with $m < \max(10^{10^6}, a \cdot 10^{22})$, $n > 1$.

Iseki & Sandor find all sums of consecutive squares which are square, where the largest square is less than 10000, and Iseki finds similar sums from $i = 0$ to n of $(x + 2i)^2$.

Tijdeman observes that general results on the equation

$$1^n + 2^n + \cdots + k^n = m^n$$

do not appear to have any implications for the special equation.

Erdős proposed the problem to prove that if m, n are integers satisfying (K), then (L$'$) and (M$'$) are true, where

$$(\text{K}) \qquad \left(1 - \frac{1}{m}\right)^n > \frac{1}{2} > \left(1 - \frac{1}{m-1}\right)^n,$$

$$(\text{L}') \qquad 1^n + 2^n + \cdots + (m-2)^n < (m-1)^n,$$

$$(\text{M}') \qquad 1^n + 2^n + \cdots + m^n > (m+1)^n,$$

and that (L) and (M) are each true infinitely often, where

(L) $1^n + 2^n + \cdots + (m-1)^n < m^n$,

(M) $1^n + 2^n + \cdots + (m-1)^n > m^n$.

Van de Lune proved that (K) implies (L') and Best & te Riele proved that (K) implies (M') and that (M) holds for at most $c \ln x$ values of $m \le x$. Van de Lune & te Riele proved that (L) is true for almost all pairs (m, n). Best & te Riele computed 33 pairs (m, n) for which (K) and (M) both hold, the smallest being

$$m = 1121626023352385, \qquad n = 777451915729368.$$

Are there infinitely many such pairs?

More recently Pieter Moree, te Riele & Urbanowicz have shown that, in the original equation, n must be divisible by $2^8 \cdot 3^5 \cdot 5^4 \cdot 7^3 \cdot 11^2 \cdot 13^2 \cdot 17^2 \cdot 19^2 \cdot 23 \cdot 29 \cdots \cdot 197 \cdot 199$ and that m is not divisible by any regular prime (see **D2**), nor by any irregular prime < 1000. Even more recently, Moree points to Kellner's substantially improved result that n must be divisible by all primes < 1000, so that n is divisible by an integer $> 5.7462 \times 10^{427}$.

M. R. Best & H. J. J. te Riele, On a conjecture of Erdős concerning sums of powers of integers, *Report NW 23/76*, Mathematisch Centrum Amsterdam, 1976.

A. Bremner, R. J. Stroeker & N. Tzanakis, On sums of consecutive squares, *J. Number Theory*, **62**(1997) 39–70; *MR* **97k**:11082.

P. Erdős, Advanced problem 4347, *Amer. Math. Monthly* **56**(1949) 343.

K. Győry, R. Tijdeman & M. Voorhoeve, On the equation $1^k + 2^k + \cdots + x^k = y^z$, *Acta Arith.*, **37**(1980) 233–240.

Kiyoshi Iseki, Positive integral solutions of $x^2 + (x+2)^2 + \cdots + (x+2n)^2 = y^2$ ($1 \le n \le 10000$), *Math. Japonica*, **35**(1990) 1003–1012.

Kiyoshi Iseki & Lajos Sandor, Positive integral solutions of the equation $x^2 + (x+1)^2 + (x+2)^2 + \cdots + (x+n)^2 = y^2$ ($1 \le x + n \le 10000$), *Math. Japonica*, *35*(1990) 817–830.

M. J. Jacobson, Á. Pintér & P. G. Walsh, A computational approach for solving $y^2 = 1^k + 2^k + \cdots x^k$, *Math. Comput.*, **72**(2003) 2099–2110.

Bernd Christian Kellner, Über irreguläre Paare höherer Ordnungen [und die Vermutungen von Giuga-Agoh und Erdős-Moser], Diplomarbeit, Göttingen, 2002.

J. van de Lune, On a conjecture of Erdős (I), *Report ZW 54/75*, Mathematisch Centrum Amsterdam, 1975.

J. van de Lune & H. J. J. te Riele, On a conjecture of Erdős (II), *Report ZW 56/75*, Mathematisch Centrum Amsterdam, 1975.

J. van de Lune & H. J. J. te Riele, A note on the solvability of the diophantine equation $1^n + 2^n + \cdots + m^n = G(m+1)^n$, *Report ZW 59/75*, Mathematisch Centrum Amsterdam, 1975.

Pieter Moree, On a theorem of Carlitz-von Staudt, *C.R. Math. Rep. Acad. Sci. Canada*, **16**(1994) 166–170; *MR* **95i**:11002.

Pieter Moree, Diophantine equations od Erdős-Moser type, *Bull. Austral. Math. Soc.*, **53**(1996) 281–292; *MR* **97b**:11048.

Pieter Moree, Herman J. J. te Riele & Jerzy Urbanowicz, Divisibility properties of integers x, k satisfying $1^k + \cdots + (x-1)^k = x^k$, Math. Comput., **63**(1994) 799–815; MR **94m**:11041.

L. Moser, On the diophantine equation $1^n + 2^n + \cdots + (m-1)^n = m^n$, Scripta Math., **19**(1953) 84–88; MR **14**, 950.

J. J. Schäffer, The equation $1^p + 2^p + \cdots + n^p = m^q$, Acta Math., **95**(1956) 155–189; MR **17**, 1187.

Jerzy Urbanowicz, Remarks on the equation $1^k + 2^k + \cdots + (x-1)^k = x^k$, Nederl. Akad. Wetensch. Indag. Math., **50**(1988) 343–348; MR **90b**:11026.

M. Voorhoeve, K. Győry & R. Tijdeman, On the diophantine equation $1^k + 2^k + \cdots + x^k + R(x) = y^z$, Acta Math., **143**(1979) 1–8; MR **80e**:10020.

Zhou Guo-Fu & Kang Ji-Ding, On the diophantine equation $\sum_{k=1}^{m} k^n = (m+1)^n$, J. Math. Res. Exposition, **3**(1983) 47–48; MR **85m**:11020.

D8 A pyramidal diophantine equation.

Wunderlich asks for (a parametric representation of) *all* solutions of the equation $x^3 + y^3 + z^3 = x + y + z$. Bernstein, S. Chowla, Edgar, Fraenkel, Oppenheim, Segal, and Sierpiński have given solutions, some of them parametric, so there are certainly infinitely many. Eighty-eight of them have unknowns less than 13000. Bremner has effectively determined all parametric solutions.

Leon Bernstein, Explicit solutions of pyramidal Diophantine equations, Canad. Math. Bull., **15**(1972) 177–184; MR **46** #3442.

Andrew Bremner, Integer points on a special cubic surface, Duke Math. J., **44**(1977) 757-765; MR **58** #27745.

Hugh Maxwell Edgar, Some remarks on the Diophantine equation $x^3 + y^3 + z^3 = x + y + z$, Proc. Amer. Math. Soc., **16**(1965) 148–153; MR **30** #1094.

A. S. Fraenkel, Diophantine equations involving generalized triangular and tetrahedral numbers, in *Computers in Number Theory, Proc. Atlas Symp. No. 2*, Oxford, 1969, Academic Press, 1971, 99–114; MR **48** #231.

A. Oppenheim, On the Diophantine equation $x^3 + y^3 + z^3 = x + y + z$, Proc. Amer. Math. Soc., **17**(1966) 493–496; MR **32** #5590.

A. Oppenheim, On the diophantine equation $x^3 + y^3 + z^3 = px + py - qz$, Univ. Beograd Publ. Elektrotehn. Fac. Ser. #235(1968); MR **39** #126.

S. L. Segal, A note on pyramidal numbers, Amer. Math. Monthly **69**(1962) 637–638; Zbl. **105**, 36.

W. Sierpiński, Sur une propriété des nombres tétraédraux, Elem. Math., **17**(1962) 29–30; MR **24** #A3118.

W. Sierpiński, Trois nombres tétraédraux en progression arithmetique, Elem. Math., **18**(1963) 54–55; MR **26** #4957.

M. Wunderlich, Certain properties of pyramidal and figurate numbers, Math. Comput., **16**(1962) 482–486; MR **26** #6115.

D9 Catalan conjecture. Difference of two powers.

Except that there remains a finite amount of computation, Tijdeman has settled the old conjecture of Catalan, that the only consecutive powers, higher than the first, are 2^3 and 3^2. We earlier said that this finite amount of computation is far beyond computer range and will not be achieved without some additional theoretical ideas. However, Predha Mihailescu announced a complete proof in May 2002.

Bennett has shown that $4 \leq N \leq k \cdot 3^k$ implies that

$$\left\| \left(\frac{N+1}{N} \right)^k \right\| > 3^{-k}$$

where $\|x\|$ is the distance from x to the nearest integer.

Leech asks if there are any solutions of $|a^m - b^n| < |a-b|$ with m, $n \geq 3$. With equality he notes $|5^3 - 2^7| = 5 - 2$ and $|13^3 - 3^7| = 13 - 3$. Are these all? are the shared exponents 3, 7 significant?

I quote from Noam Elkies's 'List of integers x,y with $x < 10^{18}$, $0 < |x^3 - y^2| < x^{1/2}$':

> Halls conjecture@Hall's conjecture asserts that if x and y are positive integers such that $k = x^3 - y^2$ is nonzero then $|k| > x^{1/2} - o(1)$. It is now recognized as a special case of the Masser-Oesterlé abc-conjecture (see **B19**). Known theoretical results in the direction of Hall's conjecture are far from satisfactory, but the question has been subject to considerable experimental work, starting with Hall's original paper.
>
> I have a new algorithm that finds all solutions of $|k| \ll x^{1/2}$ (or indeed of $|k| \ll x$) with $x < N$ in time $O(N^{1/2} + o(1))$. I found 10 new cases of $0 < |k| < x^{1/2}$, including two that improved on the previous record for $x^{1/2}/|k|$, one of which breaks the old record by a factor of nearly 10.

Elkies decorates the first two solutions with '!!' and '!' respectively:

$$5853886516781223^3 - 447884928428402042307918^2 = 1641843$$

$$381159910678861271^3 - 7441505802879036345061579^2 = 30032270$$

If $a_1 = 4$, $a_2 = 8$, $a_3 = 9$, ... is the sequence of powers higher than the first, Chudnovsky claims to have proved that $a_{n+1} - a_n$ tends to infinity with n. Erdős conjectures that $a_{n+1} - a_n > c'n^c$, but says that there is no present hope of proof.

Erdős asks if there are infinitely many numbers not of the form $x^k - y^l$, $k > 1$, $l > 1$. Let me know if any of the following nuts from **OEIS** A074981 have been cracked: 6,14,34,42,50,58,62,66,70,78, 82,86,90,102,110,114,130,

134,158,178,182,202,206,210,226,230,238,246,254,258,266,274,278, 290,302,
306,310,314,322,326,330,358,374,378,390,394,398,402,410, 418,422,426.

Carl Rudnick denotes by $N(r)$ the number of positive solutions of
$x^4 - y^4 = r$, and asks if $N(r)$ is bounded. Hansraj Gupta observes that
Hardy & Wright (p. 201) give Swinnerton-Dyer's version of Euler's parametric solution of $x^4 - y^4 = u^4 - v^4$, which establishes that $N(r)$ is 0, 1 or
2 infinitely often. For example $133^4 - 59^4 = 158^4 - 134^4 = 300783360$. For
an example with $N(r) = 3$, Zajta gives

$$401168^4 - 17228^4 = 415137^4 - 248289^4 = 421296^4 - 273588^4$$

There can hardly be any doubt that $N(r)$ is bounded.

Hugh Edgar asks how many solutions (m, n) does $p^m - q^n = 2^h$ have, for
primes p and q and h an integer? Examples are $3^2 - 2^3 = 2^0$; $3^3 - 5^2 = 2^1$;
$5^3 - 11^2 = 2^2$; $5^2 - 3^2 = 2^4$; $3^4 - 7^2 = 2^5$; are there others? Andrzej
Schinzel writes that work of Gelfond and of Rumsey & Posner implies
that the equation has only finitely many solutions. Reese Scott goes a fair
way towards settling the question. He observes that the finiteness of the
number of solutions for given $(p, q, c(= 2^h))$ follows from a result of Pillai,
and proves that this number is often at most one, with a small number
of specifically listed exceptional cases, where it is two or possibly three.
Later, in an 02-01-15 letter, he proves that the equation at the beginning
of this paragraph has at most one solution.

The Delaunay-Nagell theorem states that if $d > 1$ is cube-free, then the
equation $x^3 + dy^3 = 1$ has at most one nontrivial solution (x, y) and if there
is such, then $x + y\sqrt[3]{d}$ is the fundamental unit of $\mathbf{Q}(\sqrt[3]{d})$ or its square, the
latter in only finitely many cases. Adongo shows that these are just $d = 19$,
20, 28.

M. Aaltonen & K. Inkeri, Catalan's equation $x^p - y^q = 1$ and related congruences, *Math. Comput.*, **56**(1991) 359–370; *MR* **91g**:11025.

Harun P. K. Adongo, The Diophantine equation $3u^2 - 2 = v^6$, *J. Number
Theory*, **59**(1996) 302–208; *MR* **97k**:11038.

David M. Battany, Advanced Problem 6110*, *Amer. Math. Monthly*, **83**
(1976) 661.

M. Bennett, Fractional parts of powers of rational numbers, *Math. Proc.
Cambridge Philos. Soc.*, **114**(1993) 191–201.

Michael A. Bennett, On consecutive integers of the form ax^2, by^2 and cz^2,
Acta Arith., **88**(1999) 363–370.

Michael A. Bennett, Rational approximation to algebraic numbers of small
height: The diophantine equation $|ax^n - by^n| = 1$, *J. reine angew. Math.*,
535(2001) 1–49; *MR* **2002d**:11079.

Michael A. Bennett & Benjamin M. M. de Weger, On the Diophantine equa-
tion $|ax^n - by^n| = 1$, *Math. Comput.*, **67**(1998) 413–438; *MR* **98c**:11024.

Frits Beukers, The Diophantine equation $Ax^p + By^q = Cz^r$, *Duke Math. J.*,
91(1998) 61–88; *MR* **99f**:11039.

Yu. F. Bilu, Catalan's conjecture (after Mihailescu). See
http://www.ufr-mi.u-bordeaux.fr/~yuri/.

B. Brindza, J.-H. Evertse & K. Győry, *J. Austral. Math. Soc. Ser. A*, **51**(1991) 8–26; *MR* **92e**:11024.

Yann Bugeaud, Sur la distance entre deux puissances pures, *C.R. Acad. Sci. Paris Sér. I Math.*, **322**(1996) 1119–1121; *MR* **97i**:11030.

Yann Bugeaud & K. Győry, *Acta Arith.*, **74**(1996) 273–292; *MR* **97b**:11046.

Cao Zhen-Fu, Hugh Edgar's problem on exponential Diophantine equations (Chinese), *J. Math. (Wuhan)*, **9**(1989) 173–178; *MR* **90i**:11035.

Cao Zhen-Fu & Wang Du-Zheng, On the Diophantine equation $a^x - b^y = (2p)^z$ (Chinese). *Yangzhou Shiyuan Xuebao Ziran Kexue Ban*, **1987** no. 4 25–30; *MR* **90c**:11020.

Todd Cochran & Robert E. Dressler, Gaps between integers with the same prime factors, *Math. Comput.*, **68**(1999) 395–401.

Henri Darmon & Andrew Granville, On the equations $z^m = F(x, y)$ and $Ax^p + By^q = Cz^r$, *Bull. London Math. Soc.*, **27**(1995) 513–543; *MR* **96e**:11042.

Y. Domar, On the diophantine equation $|Ax^n - By^n| = 1$, $n \geq 5$, *Math. Scand.*, **2**(1954) 29–32.

P. S. Dyer, A solution of $A^4 + B^4 = C^4 + D^4$, *J. London Math. Soc.*, **18**(1943) 2–4; *MR* **5**, 89e.

Noam D. Elkies, Rational points near curves and small nonzero $|x^3 - y^2|$ via lattice reduction, *Algorithmic number theory (Leiden, 2000),,* Springer Lect. Notes Comput. Sci., **1838** 33–63; *MR* **2002g**:11035.

Gerhard Frey, Der Satz von Preda Mihăilescu: die Vermutung von Catalan ist richtig! *Mitt. Dtsch. Math.-Ver.*, **2002** 8–13; *MR* **2003j**:11034.

J. Gebel, A. Pethö & H. G. Zimmer, On Mordell's equation, *Compositio Math.*, **110**(1998) 335–367; *MR* **98m**:11049.

A. O. Gelfond, Sur la divisibilité de la différence des puissances de deux nombres entiers par une puissance d'un idéal premier, *Mat. Sbornik*, **7**(1940) 724.

A. M. W. Glass, David B. Meronk, Taihei Okada & Ray P. Steiner, A small contribution to Catalan's equation, *J. Number Theory*, **47**(1994) 131–137; *MR* **95g**:11021.

A. Granville, The latest on Catalan's conjecture, *Focus*, Math. Assoc. America, May/June 2001, 4–5.

Marshall Hall, The Diophantine equation $x^3 - y^2 = k$, in A. Atkin & B. Birch, eds., *Computers in Number Theory*, Academic Press, 1971, pp. 173–198; *MR* **48** #2061.

Aaron Herschfeld, The equation $2^x - 3^y = d$, *Bull. Amer. Math. Soc.*, **42** (1936) 231–234; Zbl. **14** 8a.

K. Inkeri, On Catalan's conjecture, *J. Number Theory*, **34** (1990) 142–152; *MR* **91e**:11030.

I. Jiménez Calvo, & G. Saez Moreno, Approximate power roots in Z_m, in George I. Davida & Yair Frankel (eds), *ISC 2001*, *Springer Lect. Notes Comput. Sci.*, **2200** 310–323.

Alain Kraus, Sur l'équation $a^3 + c^3 = c^p$, *Experiment. Math.*, **7**(1998) 1–13; *MR* **99f**:11040.

Michel Langevin, Quelques applications de nouveaux résultats de Van der Poorten, *Sém. Delange-Pisot-Poitou*, (1975/76). *Théorie des nombres: Fasc. 2, Exp. No. G12*, Paris, 1977; *MR* **58** #16550.

Le Mao-Hua, A note on the diophantine equation $ax^m - by^n = k$, *Indag. Math.*(*N.S.*), **3**(1992) 185–191; *MR* **93c**:11016.

W. J. Leveque, On the equation $a^x - b^y = 1$, *Amer. J. Math.*, **74**(1952) 325–331; *MR* **30** #56.

Florian Luca, On the equation $x^2 + 2^a \cdot 3^b = y^n$, *Int. J. Math. Math. Sci.*, **29**(2002) 239–244.

Maurice Mignotte, Sur l'équation de Catalan, *C.R. Acad. Sci. Paris Sér. I Math.*, **314**(1992) 165–168; *MR* **93e**:11044.

Maurice Mignotte, Une critère élémentaire pour l'équation de Catalan, *Math. Reports Acad. Sci. Royal Soc. Canada*, **15**(1993) 199–200.

M. Mignotte, Sur l'équation de Catalan, II, *Theor. Comput. Sci.*, **123**(1994) 145–149; *MR* **95b**:11032.

Maurice Mignotte, A criterion on Catalan's equation, *J. Number Theory*, **52**(1995) 280–283; *MR* **96b**:11042.

Maurice Mignotte, A note on the equation $ax^n - bx^n = c$, *Acta Arith.*, **75**(1996) 287–295; *MR* **9??**:110??.

M. Mignotte, Sur l'équation $x^p - y^q = 1$ lorsque $p \equiv 5 \bmod 8$, *C.R. Math. Rep. Acad. Sci. Canada*, **18**(1996) 228-232; *MR* **98e**:11041.

Maurice Mignotte, Arithmetical properties of the exponents of Catalan's equation, *Bull. Greek Math. Soc.*, **42**(1999) 85–87; *MR* **2002d**:11037.

Maurice Mignotte & Yves Roy, Catalan's equation has no new solution with either exponent less than 10651, *Experiment. Math.*, **4**(1995) 259–268; *MR* **97g**:11030.

Maurice Mignotte & Yves Roy, Minorations pour l'équation de Catalan, *C.R. Acad. Sci. Paris Sér. I Math.*, **324**(1997) 377–380; *MR* **98e**:11042.

Maurice Mignotte & Yves Roy, Lower bounds for Catalan's equation, *Ramanujan J.*, **1**(1997) 351–356; *MR* **99g**:11042.

Preda Mihăilescu, A clss number free criterion for Catalan's conjecture, *J. Number Theory*, **99**(2003) 225–231.

Trygve Nagell, Sur une classe d'équations exponentielles, *Ark. Mat.*, **3**(1958) 569–582; *MR* **21** #2621.

S. Sivasankaranarayana Pillai, On the inequality "$0 < a^x - b^y \leq n$", *J. Indian Math. Soc.*, **19**(1931) 1–11; *Zbl.* **1** 268b.

S. S. Pillai, On $a^x - b^y = c$, *ibid.* (N.S.) **2**(1936) 119–122; Zbl. **14** 392e.

S. S. Pillai, A correction to the paper "On $A^x - B^y = C$", *ibid.* (N.S.) **2**(1937) 215; Zbl. **16** 348b.

Paulo Ribenboim, *Catalan's Conjecture. Are 8 and 9 the only consecutive powers?* Academic Press, Boston MA 1994; *MR* **95a**:11029.

Paulo Ribenboim, Catalan's conjecture, *Amer. Math. Monthly*, **103**(1996) 529–538.

Howard Rumsey & Edward C. Posner, On a class of exponential equations, *Proc. Amer. Math. Soc.*, **15**(1964) 974–978.

Wolfgang Schwarz, A note on Catalan's equation, *Acta Arith.*, **72**(1995) 277–279; it MR **96f**:11048.

Reese Scott, On the equations $p^x - b^y = c$ and $a^x + b^y = c^z$, *J. Number Theory*, **44**(1993) 153–165; *MR* **94k**:11038.

T. N. Shorey, *Acta Arith.*, **36**(1980) 21–25; *MR* **81h**:10023.

Ray P. Steiner, Class number bounds and Catalan's equation, *Math. Comput.*, **67**(1998) 1317–1322; *MR* **98j**:11021.

R. Tijdeman, On the equation of Catalan, *Acta Arith.*, **29**(1976) 197–209; *MR* **53** #7941.

P. Gary Walsh, On Diophantine equations of the form $(x^n - 1)(y^m - 1) = z^2$, *Tatra Mt. Math. Publ.*, **20**(2000) 87–89; *MR* **2002d**:11038.

Aurel J. Zajta, Solutions of the Diophantine equation $A^4 + B^4 = C^4 + D^4$, *Math. Comput.*, **41**(1983) 635–659, esp. p. 652; *MR* **85d**:11025.

See also references at **D10**.

OEIS: A068583, A074981.

D10 Exponential diophantine equations.

Brenner & Foster pose the following general problem. Let $\{p_i\}$ be a finite set of primes and $\epsilon_i = \pm 1$. When can the exponential diophantine equation $\sum \epsilon_i p_i^{x_i} = 0$ be solved by elementary methods (e.g., by modular arithmetic)? More exactly, given p_i, ϵ_i, what criteria determine whether there exists a modulus M such that the given equation is equivalent to the congruence $\sum \epsilon_i p_i^{x_i} \equiv 0 \bmod M$? They solve many particular cases, mostly where the p_i are four in number and less than 108. In a few cases elementary methods avail, even if two of the primes are equal, but in general they do not. In fact, neither $3^a = 1 + 2^b + 2^c$ nor $2^a + 3^b = 2^c + 3^d$ can be reduced to a single congruence. Tijdeman notes that another approach to these purely exponential diophantine equations (which play a role in group theory) is by Baker's method (compare **F23**). This makes it possible to solve these last two equations.

Hugh Edgar asks if there is a solution, other than $1 + 3 + 3^2 + 3^3 + 3^4 = 11^2$, of the equation $1 + q + q^2 + \ldots + q^{x-1} = p^y$ with p, q odd primes and $x \geq 5$, $y \geq 2$. An important breakthrough in this area, is the paper of Reese Scott (see **D9**). Scott & Styer have since shown that, given primes p and q and positive integer h, there is at most one solution (x, y) to $p^x - q^y = 2^h$, and, with a few listed exceptions, at most two solutions (x, y) to $|p^x \pm q^y| = h$, and at most two solutions (x, y, z) to $p^x \pm q^y \pm 2^z = 0$.

The conjecture of Goormaghtigh, that the only solutions of

$$\frac{x^m - 1}{x - 1} = \frac{y^n - 1}{y - 1}$$

with $x, y > 1$ and $n > m > 2$ are $\{x, y, m, n\} = \{5, 2, 3, 5\}$ and $\{90, 2, 3, 13\}$ is still open; some results have been obtained by Le Mao-Hua, who also shows that the equation $(x^m - 1)/(x - 1) = y^n + 1$ has no solution in positive integers and that $(x^m - 1)/(x - 1) = y^n$ has no solution with x an nth perfect power, verifying a conjecture of Edgar.

Mignotte observed that the equation $n^2 = 2^a + 2^b + 1$ is completely solved by combining the results of Le Mao-Hua on $x^2 = 4q^n + 4q^m + 1$ with those of Tzanakis & Wolfskill, which in turn rest on those of Beukers, who showed that $x^2 + D = 2^n$ has at most one solution unless $D = 7$, 23 or 2^{k-1}, in which cases it has respectively 5, 2 or 2 solutions.

Compare sections **B16, B19, D2, D9, D23**

Ana-Maria Acu & Mugur Acu, On the exponential Diophantine equations of the form $a^x - b^y c^z = \pm 1$ with a, b, c prime numbers, *Gen. Math.*, **8**(2000) 73–77; *MR* **2003f**:11041.

Dumitru Acu, On some exponential Diophantine equations of the form $a^x - b^y c^z = \pm 1$, *An. Univ. Timişoara Ser. Mat.-Inform.*, **34**(1996) 167–171.

Norio Adachi, An application of Frey's idea to exponential Diophantine equations, *Proc. Japan Acad. Ser. A Math. Sci.*, **70**(1994) 261–263; *MR* **95i**:11022.

Leo J. Alex, Diophantine equationsrelated to finite groups, *Comm. Algebra*, **4**(1976) 77–100; *MR* **54** #12634.

Leo J. Alex, Problem E2880, *Amer. Math. Monthly*, **88**(1981) 291.

Leo J. Alex, Problem 6411, *Amer. Math. Monthly*, **89**(1982) 788.

Leo J. Alex, The diophantine equation $3^a + 5^b = 7^c + 11^d$, *Math. Student*, **48**(1980) 134–138 (1984); *MR* **86e**:11022.

Leo J. Alex & Lorraine L. Foster, Exponential Diophantine equations, in *Théorie des nombres, Québec 1987*, de Gruyter, Berlin – New York (1989) 1–6; *MR* **90d**:11030.

Leo J. Alex & Lorraine L. Foster, On the Diophantine equation $1 + x + y = z$ with $xyz = 2^r 3^s 7^t$, *Forum Math.*, **7**(1995) 645–663; *MR* **961**:11036.

Leo J. Alex & Lorraine L. Foster, On the Diophantine equation $w + x + y = z$ with $wxyz = 2^r 3^s 5^t$, *Rev. Mat. Univ. Complut. Madrid*, **8**(1995) 13–48; *MR* **961**:11035.

S. Akhtar Arif & Amal S. Al-Ali, On the Diophantine equation $ax^2 + b^m = 4y^n$, *Acta Arith.*, **103**(2002) 343–346; *MR* **2003f**:11042.

S. Akhtar Arif & Amal S. Al-Ali, On the Diophantine equation $x^2 + p^{2k+1} = 4y^n$, *Int. J. Math. Math. Sci.*, **31**(2002) 695–699; *MR* **2003g**:11030.

S. Akhtar Arif & Fadwa S. Abu Muriefah, On the Diophantine equation $x^2 + 2^k = y^n$, *Internat. J. Math. Math. Sci.*, **20**(1997) 299–304; *MR* **98b**:11032; II, *Arab J. Math. Sci.*, **7**(2001) 67–71; *MR* **2003g**:11038.

S. Akhtar Arif & Fadwa S. Abu Muriefah, The Diophantine equation $x^2 + 3^m = y^n$, *Internat. J. Math. Math. Sci.*, **21**(1998) 619–620; *MR* **99b**:11026.

S. Akhtar Arif & Fadwa S. Abu Muriefah, On the Diophantine equation $x^2 + q^{2k+1} = y^n$, *J. Number Theory*, **95**(2002) 95–100; *MR* **2003d**:11046.

M. Bennett, Effective measures of irrationality for certain algebraic numbers, *J. Austral. Math. Soc.*, **62**(1997) 329–344; *MR* **98c**:11070.

M. A. Bennett, On some exponential equations of S. S. Pillai, *Canad. J. Math.*, **53**(2001) 897–922; *MR* **2002g**:11032.

Michael A. Bennett, Pillai's conjecture revisited, *J. Number Theory*, **98**(2003) 228–235; *MR* **2003j**:11030.

Michael A. Bennett & Bruce Reznick, Positive rational solutions to $x^y = y^{mx}$: a number-theoretic excursion, *Amer. Math. Monthly*, **111**(2004) 13–21.

M. A. Bennett & C. M. Skinner, Diophantine equations via Galois representations and modular forms, *Canad. J. Math.*, (to appear)

Frits Beukers, On the generalized Ramanujan-Nagell equation, I *Acta Arith.*, **38**(1980/81) 389–410; II **39**(1981) 113–123; *MR* **83a**:10028a.

Frits Beukers, The Diophantine equation $Ax^p + By^q = Cz^r$, *Duke Math. J.*, **91**(1998) 61–88; *MR* **99f**:11039.

Yuri Bilu, On Le's and Bugeaud's papers about the equation $ax^2 + b^{2m-1} = 4c^p$, *Monatsh. Math.*, **137**(2002) 1–3; *MR* **2004a**:11024.

Yu. F. Bilu, B. Brindza, P. Kirschenhofer, Ákos Pintér & R. F. Tichy, Diophantine euquations and Bernoulli polynomials, with an appendix by A. Schinzel, *Compositio Math.*, **131**(2002) 173–188. (filed under Pintér)

J. L. Brenner & Lorraine L. Foster, Exponential Diophantine equations, *Pacific J. Math.*, **101**(1982) 263–301; *MR* **83k**:10035.

B. Brindza, On a generalization of the Ramanujan-Nagell equation, *Publ. Math. Debrecen*, **53**(1998) 225–233; *MR* **99j**:11032.

Yann Bugeaud, On the Diophantine equation $x^2 - 2^m = \pm y^n$, *Proc. Amer. Math. Soc.*, **125**(1997) 3203–3208; *MR* **97m**:11045.

Yann Bugeaud, On the Diophantine equation $x^2 - p^m = \pm y^n$, *Acta Arith.*, **79**(1997) 213–223.

Yann Bugeaud, Linear forms in p-adic logarithms and the Diophantine equation $\frac{x^n-1}{x-1} = y^q$, *Math. Proc. Cambridge Philos. Soc.*, **127**(1999) 373–381; *MR* **2000h**:11029.

Yann Bugeaud, On the Diophantine equation $a(x^n - 1)/(x - 1) = y^q$, *Number Theory* (*Turku*, 1999) 19–24, de Gruyter, 2001; *MR* **2002b**:11042.

Yann Bugeaud, On some exponential Diophantine equations, *Monatsh. Math.*, **132**(2001) 93–97; *MR* **2002c**:11026.

Yann Bugeaud, Guillaume Hanrot & Maurice Mignotte, Sur l'équation diophantienne $(x^n - 1)/(x - 1) = y^q$ III, *Proc. London Math. Soc.*(3), **84**(2002) 59–78; *MR* **2002h**:11027.

Yann Bugeaud & Maurice Mignotte, On integers with identical digits, *Mathematika*, **46**(1999) 411–417; *MR* **2002d**:11035.

Yann Bugeaud & Maurice Mignotte, Sur l'équation diophantienne $(x^n - 1)/(x - 1) = y^q$ II, *C.R. Acad. Sci. Paris Sér. I Math.*, **328**(1999) 741–744; *MR* **2001f**:11050.

Yann Bugeaud & Maurice Mignotte, On the Diophantine equation $(x^n - 1)/(x - 1) = y^q$ with negative x, *Number theory for the millennium* I (Urbana IL, 2000), 145–151, A K Peters, Natick MA, 2002; *MR* **2003k**:11053.

Yann Bugeaud & Maurice Mignotte, L'quation de Nagell-Ljunggren $\frac{x^n-1}{x-1} = y^q$. *Enseign. Math.*(2) **48**(2002) 147–168; *MR* **2003f**:11043.

Yann Bugeaud, Maurice Mignotte & Yves Roy, On the Diophantine equation $\frac{x^n-1}{x-1} = y^q$, *Pacific J. Math.*, **193**(2000) 257–268; *MR* **2001f**:11049.

Yann Bugeaud, Maurice Mignotte, Yves Roy & T. N. Shorey, The equation $(x^n - 1)/(x - 1) = y^q$ has no solution with x square, *Math. Proc. Cambridge Philos. Soc.*, **127**(1999) 353–372; *MR* **2000h**:11027.

Yann Bugeaud & T. N. Shorey, On the number of solutions of the generalized Ramanujan-Nagell equation, **J. reine angew. Math.**, **539**(2001) 55–74; *MR* **2002k**:11041. But see Leu & Li, below.

Yann Bugeaud & T. N. Shorey, On the Diophantine equation $\frac{x^m-1}{x-1} = \frac{y^n-1}{y-1}$. *Pacific J. Math.*, **207**(2002) 61–75; *MR* **2003m**:11053.

Cao Zhen-Fu, The equation $a^x - b^y = (2p^s)^z$ and Hugh Edgar's problem (Chinese), *J. Harbin Inst. Tech.*, **1986**7–11; *MR* **88a**:11029.

Cao Zhen-Fu, Exponential Diophantine equations of the form $a^x + b^y = c^z$ (Chinese), *J. Harbin Inst. Tech.*, **1987** 113–121; *MR* **89k**:11016.

Cao Zhen-Fu, Hugh Edgar's problem on exponential Diophantine equations (Chinese), *J. Math. (Wuhan)*, **9**(1989) 173–178; *MR* **90i**:11035.

Cao Zhen-Fu, On the Diophantine equation $ax^2 + by^2 = p^z$, *J. Harbin Inst. Tech.*, **23**(1991) 108–111; *MR* **94a**:11041.

Cao Zhen-Fu, The Diophantine equation $cx^4 + dy^4 = z^p$, *C. R. Math. Rep. Acad. Sci. Canada*, **14**(1992) 231–234; *MR* **93i**:11036.

Cao Zhen-Fu, The Diophantine equation $Cx^2 + 2^{2m}D = k^n$ (Chinese), *Chinese Ann. Math. Ser. A*, **15**(1994) 235–240; *MR* **95h**:11028.

Cao Zhen-Fu, On the Diophantine equation $x^4 - py^2 = z^p$, *C. R. Math. Rep. Acad. Sci. Canada*, **17**(1995) 61–66; *MR* **96h**:11020.

Cao Zhen-Fu, On the Diophantine equation $x^4 - py^2 = x^p$, *C.R. Math. Rep. Acad. Sci. Canada*, **17**(1995) 61-66; *MR* **96h**:11020; corrigendum **18**(1996) 233–234; *MR* **97i**:11026.

Cao Zhen-Fu, The Diophantine equations $x^4 - y^4 = z^p$ and $x^4 - 1 = dy^q$, *C.R. Math. Rep. Acad. Sci. Canada*, **21**(1999) 23–27; *MR* **99m**:11030.

Cao Zhen-Fu, On the Diophantine equation $x^p + 2^{2m} = py^2$, *Proc. Amer. Math. Soc.*, **128**(2000) 1927–1931; *MR* **2000m**:11028.

Cao Zhen-Fu, The Diophantine equations $x^4 + y^4 = z^p$ (Chinese), *J. Ningxia Univ. Nat. Sci. Ed.*, **20**(1999) 18–21; *MR* **2000h**:11025.

Cao Zhen-Fu & Cao Yu-Shu, Solutions to the Diophantine equation $y^2 + D^n = x^3$ (Chinese), *Heilongjiang Daxue Ziran Kexue Xuebao*, **10**(1993) 5–10; *MR* **94a**:11034.

Cao Zhen-Fu & Dong Xiao-Lei, The Diophantine equation $x^2 + b^y = c^z$, *Proc. Acad. Japan Ser. A Math. Sci.*, **77**(2001) 1–4; *MR* **2002c**:11027.

Cao Zhen-Fu & Dong Xiao-Lei, An application of a lower bound for linear forms in two logarithms to the Terai-Jeśmanowicz conjecture, *Acta Arith.*, **110**(2003) 153–164.

Cao Zhen-Fu & Pan Jia-Yu, The Diophantine equation $x^{2p} - Dy^2 = 1$ and the Fermat quotient $Q_p(m)$, *J. Harbin Inst. Tech.*, **25**(1993) 119–121; *MR* **95d**:11034.

Cao Chen-Fu & Wang Du-Zheng, On the Diophantine equation $a^x - b^y = (2p)^z$ (Chinese), *Yangzhou Shiyuan Xuebao Ziran Kexue Ban*, **1987** 25–30; *MR* **90c**:11020.

Cao Zhen-Fu & Wang Yan-Bin, On the Diophantine equation $ax^m - by^n = 4$, *J. Harbin Inst. Tech.*, **25**(1993) 111–113; *MR* **95b**:11031.

J. W. S. Cassels, On the equation $a^x - b^y = 1$ II, *Proc. Cambridge Philos. Soc.*, **56**(1960) 97–103.

Chen Xi-Geng, Guo Yong-Dong & Le Mao-Hua, On the number of solutions of the generalized Ramanujan-Nagell equation $x^2 + D = k^n$, *Acta Math. Sinica*, **41**(1998) 1249–1254; *MR* **2000b**:11030.

Chen Xi-Geng & Le Mao-Hua, On the number of solutions of the generalized Ramanujan-Nagell equation $x^2 - D = k^n$, *Publ. Math. Debrecen*, **49**(1996) 85–92; *MR* **97k**:11045.

Chen Xi-Geng & Le Mao-Hua, The Diophantine equation $(x^m - 1)(x^{mn} - 1) = y^2$, *Huaihua Shizhuan Xuebao*, **30**(2001) 11–12; *MR* **97k**:11045.

Mihai Cipu, A bound for the solutions of the Diophantine equation $D_1 x^2 + D_2^m = 4y^n$, *Proc. Japan Acad. Ser. A Math. Sci.*, **78**(2002) 179–180; *MR* **2003k**: 11054.

J. H. E. Cohn, The Diophantine equation $x^2 + 19 = y^n$, *Acta Arith.*, **61**(1992) 193–197; *MR* **93e**:11041.

J. H. E. Cohn, The Diophantine equation $x^2 + 2^k = y^n$, *Arch. Math. (Basel)* **59**(1992) 341–344; *MR* **93f**:11030.

J. H. E. Cohn, The Diophantine equation $x^2 + 3 = y^n$, *Glasgow Math. J.*, **35**(1993) 203–206; *MR* **94i**:11027.

J. H. E. Cohn, The Diophantine equation $x^2 + C = y^n$, *Acta Arith.*, **65**(1993) 367–381; *MR* **94k**:11037.

J. H. E. Cohn, The Diophantine equation $x^2 + 2^k = y^n$ II, *Int. J. Math. Math. Sci.*, **22**(1999) 459–462; *MR* **2000i**:11049.

J. H. E. Cohn, The Diophantine equation $(a^n - 1)(b^n - 1) = x^2$, *Period. Math. Hungar.*, **44**(2002) 169–175; *MR*.**2003e**:11035.

J. H. E. Cohn, The Diophantine equation $x^p + 1 = py^2$, *Proc. Amer. Math. Soc.*, **131**(2003) 13–15; *MR* **2003g**:11031.

J. H. E. Cohn, The Diophantine equation $x^n = Dy^2 + 1$, *Acta Arith.*, **106**(2003) 73–83; *MR* **2003j**:11033.

Ion Cucurezeanu, The Diophantine equation $(x^n - 1)/(x - 1) = y^2$, *An. Univ. "Ovidius" Constanţa Ser. Mat.*, **1**(1993) 101–104.

Ion Cucurezeanu, On the Diophantine equation $x^2 = 4y^{m+n} + 4y^n + 1$, *Stud. Circ. Mat.*, **49**(1997) 33–34; *MR* **99m**:11033.

Deng Mou-Jie, A note on the Diophantine equation $(a^2 - b^2)^x + (2ab)^y = (a^2 + b^2)^z$, *Heilongjiang Daxue Ziran Kexue Xuebao*, **19**(2002) 8–10.

Erik Dofs, Two exponential Diophantine equations, *Glasgow Math. J.*, **39**(1997) 231–232; *MR* **98e**:11039.

Dong Xiao-Lei & Cao Zhen-Fu, The Terai-Jeśmanowicz conjecture on the equation $a^x + b^y = c^z$ (Chinese), *Chinese Ann. Math. Ser. A*, **21**(2000) 709–714; *MR* **2001i**:11035; see also *Publ. Math. Debrecen*, **61**(2002) 253–265; *MR* **2003j**:11032.

Hugh M. Edgar, Problems and some results concerning the Diophantine equation $1 + A + A^2 + \cdots + A^{x-1} = P^y$, *Rocky Mountain J. Math.*, **15**(1985) 327–329; *MR* **87e**:11047.

P. Filakovszky & L. Hajdu, The resolution of the Diophantine equation $x(x + d) \cdots (x + (k - 1)d) = by^2$ for fixed d, *Acta Arith.*, **98**(2001) 151–154; *MR* **2002c**:11028.

A. Flechsenhaar, Über die Gleichung $x^y = y^x$, *Unterrichts. für Math.*, **17**(1911) 70–73.

Lorraine L. Foster, Solution to Problem S31 [1980, 403], *Amer. Math. Monthly*, **89**(1982) 62.

A. F. Gilman, The Diophantine equation $x^3 + y^3 + z^3 - 3xyz = p^r$, *Internat. J. Math. Ed. Sci. Tech.*, **28**(1997) 468–472; *MR* **98i**:11015.

Guo Zhi-Tang & Wu Yun-Fei, On the equation $1 + q + q^2 + \cdots + q^{n-1} = p^m$ (Chinese), *J. Harbin Inst. Tech.*, **24**(1992) 6–9; *MR* **93k**:11024.

K. Győry, On the Diophantine equation $n(n + 1) \cdots (n + k - 1) = bx^l$, *Acta Arith.*, **83**(1998) 87–92.

K. Győry, & Ákos Pintér, On the equation $1^k + 2^k + \cdots + x^k = y^n$, Dedicated to Professor Lajos Tamássy on the occasion of his 80th birthday, *Publ. Math. Debrecen*, **62**(2003) 403–414.

T. Hadano, On the diophantine equation $a^x + b^y = c^z$, *Math. J. Okayama Univ.*, **19**(1976/77) 25–29; *MR* **55** #5526

A. Hausner, Algebraic number fields and the Diophantine equation $m^n = n^m$, *Amer. Math. Monthly*, **68**(1961) 856–861.

Christoph Hering, On the Diophantine equations $ax^2 + bx + c = c_0 c_1^{y_1} \cdots c_r^{y_r}$, *Appl. Algebra Engrg. Comm. Comput.*, **7**(1996) 251–262.

Noriko Hiata-Kohno & Tarlok N. Shorey, On the equation $(x^m - 1)/(x - 1) = y^q$ with x power, *Analytic Number Theory (Kyoto,1996)*, 119–125, London Math. Soc. Lecture Ser., **247**(1997), Cambridge Univ. Press; *MR* bf2000c:11051.

S. Hurwitz, On the rational solutions to $m^n = n^m$ with $m \neq n$, *Amer. Math. Monthly*, **74**(1967) 298–300.

K. Inkeri, On the diophantine equation $a(x^n - 1)/(x - 1) = y^m$, *Acta Arith.*, **21**(1972) 299–311.

M. J. Jacobson, Ákos Pintér & P. Gary Walsh, A computational approach for solving $y^2 = 1^k + 2^k + \cdots + x^k$, *Math. Comput.*, (200); *MR* **2004c**:11241.

Alain Kraus, Sur les équations $a^p + b^p + 15c^p = 0$ et $a^p + 3b^p + 5c^p = 0$, *C. R. Acad. Sci. Paris Sér. I Math.*, **322**(1996) 809–812.

Alain Kraus, Une question sur les équations $x^m - y^m = Rz^n$, *Compositio Math.*, **132**(2002) 1–26; *MR* **2003k**:11043.

Klaus Langmann, Unlösbarkeit der Gleichung $x^n - y^2 = az^m$ bei kleinem ggT$(n, h(-a))$, *Arch. Math. (Basel)* **70**(1998) 197–203.

Le Mao-Hua, A note on the Diophantine equation $x^{2p} - Dy^2 = 1$, *Proc. Amer. Math. Soc.*, **107**(1989) 27–34; *MR* **90a**:11033.

Le Mao-Hua, The Diophantine equation $x^2 = 4q^n + 4q^m + 1$, *Proc. Amer. Math. Soc.*, **106**(1989) 599–604; *MR* **90b**:11024.

Le Mao-Hua, On the Diophantine equation $x^2 + D = 4p^n$, *J. Number Theory*, **41**(1992) 87–97; *MR* **93b**:11037.

Le Mao-Hua, A note on the Diophantine equation $ax^m - by^n = k$, *Indag. Math. (N.S.)*, **3**(1992) 185–191; *MR* **93c**:11016.

Le Mao-Hua, On the Diophantine equations $d_1 x^2 + 2^{2m} d_2 = y^n$ and $d_1 x^2 + d_2 = 4y^n$, *Proc. Amer. Math. Soc.*, **118**(1993) 67–70; *MR* **93f**:11031.

Le Mao-Hua, On the Diophantine equation $(x^m - 1)/(x - 1) = y^n$, *Acta Math. Sinica*, **36**(1993) 590–599; *MR* **95a**:11028.

Le Mao-Hua, A note on the Diophantine equation $(x^m - 1)/(x - 1) = y^n$, *Acta Arith.*, **64**(1993) 19–28; *MR* **94e**:11029.

Le Mao-Hua, On the Diophantine equation $D_1 x^2 + D_2 = 2^{n+2}$, *Acta Arith.*, **64**(1993) 29–41; *MR* **94e**:11030.

Le Mao-Hua, A note on the diophantine equation $(x^m - 1)/(x - 1) = y^n + 1$, *Math. Proc. Cambridge Philos. Soc.*, **116**(1994) 385–389; *MR* **95h**:11030.

Le Mao-Hua, Integer solutions of exponential Diophantine equations (Chinese), *Adv. in Math. (China)*, **23**(1994) 385–395; *MR* **95i**:11023.

Le Mao-Hua, The Diophantine equation $x^2 + D^m = 2^{n+2}$, *Comment. Math. Univ. St. Paul.*, **43**(1994) 127–133.

248 Unsolved Problems in Number Theory

Le Mao-Hua, On the number of solutions of the generalized Ramanujan-Nagell equation $x^2 - D = p^n$, Publ. Math. Debrecen, **45**(1994) 239–254; MR **96b**:11041.

Le Mao-Hua, On the Diophantine equation $D_1 x^2 + D_2^m = 4y^n$, Monatsh. Math., **120**(1995) 121–125; MR **96h**:11023. But see Bilu, above.

Le Mao-Hua, On the Diophantine equation $2^n + px^2 = y^p$, Proc. Amer. Math. Soc., **123**(1995) 321–326; MR **95c**:11038.

Le Mao-Hua, A note on the Diophantine equation $x^2 + b^y = c^z$, Acta Arith., **71**(1995) 253–257; MR **96d**:11037.

Le Mao-Hua, A note on the generalized Ramanujan-Nagell equation, J. Number Theory, **50**(1995) 193–201; MR **96f**:11051.

Le Mao-Hua, Some exponential Diophantine equations I. The equation $D_1 x^2 - D_2 y^2 = \lambda k^z$, J. Number Theory, **55**(1995) 209–221; MR **96j**:11036.

Le Mao-Hua, A note on perfect powers of the form $x^{m-1} + \cdots + x + 1$, Acta Arith, **69**(1995) 91–98; MR **96b**:11034 [but see **83**(1998); MR **98k**:11030].

Le Mao-Hua, The Diophantine equations $x^2 \pm 2^m = y^n$ (Chinese), Adv. in Math. (China), **25**(1996) 328–333; MR **98m**:11020.

Le Mao-Hua, On the Diophantine equation $D_1 x^4 - D_2 y^2 = 1$, Acta Arith., **76**(1996) 1–9; MR **97b**:11040.

Le Mao-Hua, A note on the number of solutions of the generalized Ramanujan-Nagell equation $x^2 - D = k^n$, Acta Arith., **78**(1996) 11–18; MR **97j**:11015.

Le Mao-Hua, A note on the number of solutions of the generalized Ramanujan-Nagell equation $D_1 x^2 + D_2 = 4p^n$, J. Number Theory, **62**(1997) 100–106; MR **97k**:11048.

Le Mao-Hua, A note on the Diophantine equation $x^2 + 7 = y^n$, Glasgow Math. J., **39**(1997) 59–63; MR **98e**:11040.

Le Mao-Hua, On the Diophantine equation $(x^m + 1)(x^n + 1) = y^2$, Acta Arith., **82**(1997) 17–26; MR **98h**:11036.

Le Mao-Hua, A note on the Diophantine equation $D_1 x^2 + D_2 = 2y^n$, Publ. Math. Debrecen, **51**(1997) 191–198; MR **98j**:11022.

Le Mao-Hua, On the Diophantine equation $(x^3 - 1)/(x-1) = (y^n - 1)/(y-1)$, Trans. Amer. Math. Soc., **351**(1999) 1063–1074; MR **99e**:11033. But see Leu & Li, below.

Le Mao-Hua, The number of solutions of the Diophantine equation $D_1 x^2 + 2^m D_2 = y^n$ (Chinese), Adv. in Math. (China), **26**(1997) 43–49; MR **98g**:11038.

Le Mao-Hua, On the Diophantine equation $x^2 + D = y^n$, Acta Math. Sinica, **40**(1997) 839–844; MR **99f**:11043.

Le Mao-Hua, On the diophantine equation $(x^m + 1)(x^n + 1) = y^2$, Acta Arith., **82**(1997) 17–26; MR **98h**:11036.

Le Mao-Hua, The Diophantine equation $x^2 + 2^m = y^n$, Kexue Tongbao, **42**(1997) 1255–1257.

Le Mao-Hua, A note on the Diophantine equation $x^2 + 7 = y^n$, Glasgow Math. J., **39**(1997) 59–63; MR **98e**:11040.

Le Mao-Hua, Diophantine equation $x^2 + 2^m = y^n$, Chinese Sci. Bull., **42**(1997) 1515–1517; MR **99h**:11036.

Le Mao-Hua, The number of solutions for a class of exponential Diophantine equations (Chinese), J. Jishou Univ. Nat. Sci. Edu., **21**(2000) 27–32; MR

2001m:11039.

Le Mao-Hua, The exceptional solutions of Goormaghtigh's equation $\frac{x^3-1}{x-1} = \frac{y^n-1}{y-1}$, *J. Jishou Univ. Nat. Sci. Ed.*, **22**(2001) 29–32; *MR* **2002d**:11036.

Le Mao-Hua, On Goormaghtigh's equation $\frac{x^3-1}{x-1} = \frac{y^n-1}{y-1}$, *Acta Math. Sinica*, **45**(2001) 505–508; *MR* **2003f**:11045.

Le Mao-Hua, Exeptional solutions of the exponential Diophantine equation $\frac{x^3-1}{x-1} = \frac{y^n-1}{y-1}$, *J. reine angew. Math.*, **543**(2002) 187–192; *MR* **2002k**:11042.

Le Mao-Hua, An exponential Diophantine equation, *J. Bull. Austral. Math. Soc.*, **64**(2001) 99–105; *MR* **2002e**:11040.

Le Mao-Hua, A note on the exponential Diophantine equations $a^x + db^y = c^z$, *Tokyo J. Math.*, **24**(2001) 169–171; *MR* **2002e**:11041.

Le Mao-Hua, On Cohn's conjecture concerning the Diophantine equation $x^2 + 2^m = y^n$, *Arch. Math.* (*Basel*), **78**(2002) 26–35; *MR* **2002m**:11022.

Le Mao-Hua, Exceptional solutions of the exponential Diophantine equation $(x^3 - 1)/(x - 1) = (y^n - 1)/(y - 1)$, *J. reine angew. Math.*, **543**(2002) 187–192; *MR* **2002k**:11042.

Le Mao-Hua, On the number of solutions of the generalized Ramanujan-Nagell equation $x^2 + D = 4p^n$, *J. Jishou Univ. Nat. Sci. Ed.*, **23**(2002) 44–46.

Le Mao-Hua, On the Diophantine equation $x^2 + p^2 = y^n$, *Publ. Math. Debrecen*, **63**(2003) 67–78; *MR* **2004c**:11039.

Leu Ming-Guang & Li Guan-Wei, The Diophantine equation $2x^2 + 1 = 3^n$, *Proc. Amer. Math. Soc.*, **131**(2003) 3643–3645 (electronic).

Li Fu-Zhong, On the family of diophantine equations $x^3 \pm (5^k)^3 = Dy^2$, *Dongbei Shida Xuebao*, **1998** 16–19.

W. Ljunggren, Noen setninger om ubestemte likninger av formen $(x^n - 1)/(x - 1) = y^q$ [New propositions about the indeterminate equation $(x^n - 1)/(x - 1) = y^q$], *Norske Mat. Tidsskrift*, **25**(1943) 17–20.

Florian Luca, On the equation $x^2 = 4y^{m+n} + 4y^n + 1$, *Bull. Math. Soc. Sci. Math. Roumanie* (*N.S.*), **42(90)**(1999) 231–235; *MR* **2003b**:11030.

Florian Luca, On the equation $1!^k + 2!^k + \cdots + n!^k = x^2$, *Period. Math. Hungar.*, **44**(2002) 219–224.

Florian Luca, On the equation $x^2 + 2^a \cdot 3^b = y^n$, *Int. J. Math. Math. Sci.*, **29**(2002) 239–244; *MR* **2003a**:11028.

Florian Luca, On the Diophantine equation $x^2 = 4q^m - 4q^n + 1$, *Proc. Amer. Math. Soc.*, **131**(2003) 1339–1345; *MR* **2004c**:11041.

Florian Luca & Maurice Mignotte, On the equation $y^x \pm x^y = z^2$, *Rocky Mountain J. Math.*, **30**(2000) 651–661; *MR* **2001h**:11040.

Florian Luca, Maurice Mignotte & Yves Roy, On the equation $x^y \pm y^x = \prod n_i!$, *Glasg. Math. J.*, **42**(2000) 351–357; *MR* **2001i**:11037.

Mustapha Mabkhout, *Équations diophantiennes exponentielles*, Thèse, Université Louis Pasteur, Strasbourg, 1993.

A. Mąkowski & A. Schinzel, Sur l'équation indéterminée de R. Goormaghtigh, *Mathesis*, **68**(1959) 128–142; *MR* **22** #9472.

Maurice Mignotte, A note on the equation $ax^n - by^n = c$, *Acta Arith.*, **75**(1996) 287–295; *MR* **97c**:11042.

Maurice Mignotte, On the Diophantine equation $\frac{x^n-1}{x-1} = y^q$, *Algebraic number theory and Diophantine analysis* (Graz 1998) 305–310, de Gruyter, Berlin;

MR **2001g**:11043.

Maurice Mignotte & A Pethö, On the Diophantine equation $x^p - x = y^q - y$, *PublMat.*, **43**(1999) 207–216; *MR* **2000d**:11044.

Maurice Mignotte & T. N. Shorey, The equations $(x + 1) \cdots (x + k) = (y+1) \cdots (y+mk)$, $m = 5, 6$, *Indag. Math. (N.S.)*, **7**(1996) 215–225; *MR* **99h**:11032.

Günter Mitas, Über die Lösungen der Gleichung $a^b = b^a$ in rationalen und in algebraischen Zahlen, *Mitt. Math. Gesellsch. Hamburg*, **10**(1976) 249–254; *MR* **58** #21930.

Mo De-Ze, Exponential Diophantine equations with four terms, II, *Acta Math. Sinica*, **37**(1994) 482–490; *MR* **96f**:11053.

Mo De-Ze & R. Tijdeman, Exponential Diophantine equations with four terms, *Indag. Math.(N.S.)*, **3**(1992) 47–57; *MR* **93d**:11035.

Fadwa S. Abu Muriefah & S. Akhtar Arif, On a Diophantine equation. *Bull. Autral. Math. Soc.*, **57**(1998) 189–198; *MR* **99a**:11035. [$x^2 + 3^{2k} = y^n$]

Fadwa S. Abu Muriefah & S. Akhtar Arif, The Diophantine equation, $x^2 + q^{2k} = y^n$ *Arab. J. Sci. Eng. Sect. A Sci.*, **26**(2001) 53–62; *MR* **2002e**:11042.

Trygve Nagell, Note sur l'équation indéterminée $(x^n - 1)/(x - 1) = y^q$, *Norsk Mat. Tidsskr.*, **1**(1920) 75–78.

S. S. Pillai, On $a^X - b^Y = b^y \pm a^x$, *J. Indian Math. Soc.*, **8**(1944) 10–13; *MR* **7** 145i.

Ákos Pintér, A note on the equation $1^k + 2^k + \cdots + (x - 1)^k = y^m$, *Indag. Math. (N.S.)*, **8**(1997) 119–123; *MR* **99a**:11040.

Stanley Rabinowitz, The solution of $3y^2 \pm 2^n = x^3$, *Proc. Amer. Math. Soc.*, **69**(1978) 213–218; *MR* **58** #499.

A. Rotkiewicz & W. Złotkowski, On the Diophantine equation $1 + p^{\alpha_1} + p^{\alpha_2} + \cdots + p^{\alpha_k} = y^2$, in Number Theory, Vol. II (Budapest, 1987), North-Holland, *Colloq. Math. Soc. János Bolyai*, **51**(1990) 917–937; *MR* **91e**:11032.

N. Saradha & T. N. Shorey, The equation $(x^n - 1)/(x - 1) = y^q$ with x square, *Math. Proc. Cambridge Philos. Soc.*, **125**(1999) 1–19; *MR* **99h**:11037.

Daihachiro Sato, Algebraic solution of $x^y = y^x$ $(0 < x < y)$, *Proc. Amer. Math. Soc.*, **31**(1972) 316; *MR* **44** #5272.

Daihachiro Sato, A characterization of two real algebraic numbers x, y such that $x^y = y^x$, *Sûgaku*, **24**(1972) 223–226; *MR* **58** #27891.

Reese Scott, On the generalized Pillai equation $\pm a^x \pm b^y = c$, (submitted).

Reese Scott & Robert Styer, On $p^x - q^y = c$ and related three term exponential Diophantine equations with prime bases, *J. Number Theory*, **105**(2004) 212–234.

T. N. Shorey, On the equation $z^q = (x^n - 1)/(x - 1)$, *Indag. Math.*, **48**(1986) 345–351; *MR* **87m**:11028.

T. N. Shorey, Integers with identical digits, *Acta Arith.*, **53** (1989) 187–205; *MR* **90j**:11027.

T. N. Shorey, Some exponential Diophantine equations, II, *Number Theory and Related Topics (Bombay, 1988)* 217–229, Tata Inst. Fund. Res. Stud. Math. **12**, Tata Inst. Fund. Res., Bombay, 1989; *MR* **98d**:11038.

T. N. Shorey, Some conjectures in the theory of exponential Diophantine equations, *Publ. Math. Debrecen*, **56**(2000) 631–641; *MR* **2001i**:11038.

Samir Siksek & John E. Cremona, On the Diophantine equation $x^2 + 7 = y^m$, *Acta Arith.*, **109**(2003) 143–149; *MR* **2004c**:11109.

Christopher M. Skinner, On the Diophantine equation $ap^x + bq^v = c + dp^z q^w$, *J. Number Theory*, **35**(1990) 194–207; *MR* **91h**:11021.

Jörg Stiller, The Diophantine equation $x^2 + 119 = 15 \cdot 2^n$ has exactly six solutions, *Rocky Mountain J. Math.*, **26**(1996)295–298; *MR* **97a**:11053.

C. Størmer, Quelques théorèmes sur l'equation de Pell $x^2 - Dy^2 = 1$ et leurs applications, *Skrifter Videnskabs-seskabet (Christiana), I, Mat. Naturv. Kl.*, (1897) no. 2 (48pp.).

Robert Styer, Small two-variable exponential Diophantine equations, *Math. Comput.*, **60**(1993) 811–816.

Sun Qi & Cao Zhen-Fu On the Diophantine equation $x^p - y^p = z^2$ (Chinese), An English summary appears in *Chinese Ann. Math. Ser. B*, **7**(1986) 527; *Chinese Ann. Math. Ser. A*, **7**(1986) 514–518; *MR* **88c**:11022.

Sun Qi & Cao Zhen-Fu, On the Diophantine equations $x^2 + D^m = p^n$ and $x^2 + 2^m = y^n$ (Chinese), *Sichuan Daxue Xuebao*, **25**(1988) 164–169; *MR* **89j**:11029.

Marta Sved, On the rational solutions of $x^y = y^x$, *Math. Mag.*, **63**(1990) 30–33; *MR* **91a**:11018.

László Szalay, The equations $2^n \pm 2^m \pm 2^l = z^2$, *Indag. Math. (N.S.)*, **13**(2002) 131-142.

Nobuhiro Terai (& Kei Takakuwa), The Diophantine equation $a^x + b^y = c^z$, *Proc. Japan Acad. Ser. A Math. Sci.*, **70**(1994) 22–26; II **71**(1995) 109–110; III **72**(1996) 20–22; **73**(1997) 161–164; *MR* **95b**:11033; **96m**:11022; **98a**:11038; **99f**:11045.

Nobuhiro Terai, A note on certain exponential Diophantine equations, *Number Theory (Eger)*, 1996, de Gruyter, Berlin, 1998, 479–487; *MR* **99d**:11032.

Nobuhiro Terai, On the exponential Diophantine equation $a^x + l^y = c^z$, *Proc. Japan Acad. Ser. A Math. Sci.*, **77**(2001) 151–154; *MR* **2002j**:11029.

Nobuhiro Terai & Tei Takakuwa, A note on the Diophantine equation $a^x + b^y = c^z$, *Proc. Japan Acad. Ser. A Math. Sci.*, **73**(1997) 161–164; *MR* **99f**:11045.

R. Tijdeman, Exponential Diophantine equations, *Number Theory (Eger)*, 1996, de Gruyter Berlin, 1998, 479–487; *MR* **99f**:11046.

Nikos Tzanakis & John Wolfskill, The Diophantine equation $x^2 = 4q^{a/2} + 4q + 1$, *J. Number Theory*, **26**(1987) 96–116; *MR* **88g**:11009.

S. Uchiyama, On the Diophantine equation $2^x = 3^y + 13^z$, *Math. J. Okayama Univ.*, **19**(1976/77) 31–38; *MR* **55** #5527.

Wang Du-Zheng & Cao Zhen-Fu, On the Diophantine equations $x^{2n} - Dy^2 = 1$ and $x^2 - Dy^{2n} = 1$ (Chinese), *Yangzhou Shiyuan Ziran Kexue Xuebao*, **1985** 16–18; *MR* **89f**:11048.

B. M. M. de Weger, Solving exponential Diophantine equations using lattice basis reduction algorithms, *J. Number Theory*, **26**(1987 325–367; erratum **31**(1989) 88-89; *MR* **88k**:11097 **90a**:11040.

Wu Ke-Jian & Le Mao-Hua, A note on the diophantine equation $x^4 - y^4 = z^p$, *C.R. Math. Rep. Acad. Sci. Canada*, **17**(1995) 197–200; *MR* **96i**:11033.

Xu Tai-Jin & Le Mao-Hua, On the Diophantine equation $D_1 x^2 + D_2 = k^n$, *Publ. Math. Debrecen*, **47**(1995) 293–297; *MR* **96h**:11025.

Yu Li & Le Mao-Hua, On the diophantine equation $(x^m - 1)/(x - 1) = y^n$, *Acta Arith.*, **73**(1995) 363–366; *MR* **97m**:11023 [but see **83**(1998); *MR* **98k**:11030].

Yuan Ping-Zhi, A note on the Diophantine equation $(x_m - 1)/(x - 1) = y_n$ (Chinese), *Acta Math. Sinica*, **39**(1996) 184–189.

Yuan Ping-Zhi, Comment: "A note on perfect powers of the form $x^{m-1} + \cdots + x + 1$, [*Acta Arith*, **69**(1995) 91–98; *MR* **96b**:11034] by M. H. Le and "On the diophantine equation $(x^m - 1)/(x - 1) = y^n$, [*ibid.*, **73**(1995) 363–366; *MR* **97m**:11023] by L. Yu and Le, *Acta Arith.*, **83**(1998) 199.

Yuan Ping-Zhi, The Diophantine equation $x^2 + b^y = c^z$ (Chinese), *Sichuan Daxue Xuebao*, **35**(1998) 5–7; *MR* **99f**:11047.

Yuan Ping-Zhi, On the number of the solutions of $x^2 - D = p^n$, *Sichuan Daxue Xuebao*, **35**(1998) 311–316; *MR* **99k**:11048.

Yuan Ping-Zhi & Wang Jia-Bao, On the Diophantine equation $x^2 + b^y = c^z$, *Acta Arith.*, **84**(1998) 145–147; *MR* **99b**:11028.

See also references at **D9**.

OEIS: A000961, A006549, A037087, A072633.

D11 Egyptian fractions.

Some wit has suggested that **Egyptian fractions**, or unit fractions, are so called from the Hungarian for 'one', namely *egy*. In fact the Rhind papyrus is amongst the oldest written mathematics that has come down to us; it concerns the representation of rational numbers as the sum of unit fractions,

$$\frac{m}{n} = \frac{1}{x_1} + \frac{1}{x_2} + \cdots + \frac{1}{x_k}$$

This has suggested numerous problems, many of which are unsolved, and continues to suggest new problems, so the interest in Egyptian fractions is as great as it has ever been. Our bibliography shows only a fraction of what has been written. Bleicher has given a careful survey of the subject and draws attention to the various algorithms that have been proposed for constructing representations of the given type: the Fibonacci-Sylvester algorithm, Erdős's algorithm, Golomb's algorithm and two of his own, the Farey series algorithm and the continued fraction algorithm. See also the extensive Section 4 of the collection of problems by Erdős & Graham mentioned at the beginning of this volume, and the bibliography obtainable from Paul Campbell.

Erdős & Straus conjectured that the equation

$$\frac{4}{n} = \frac{1}{x} + \frac{1}{y} + \frac{1}{z}$$

could be solved in positive integers for all $n > 1$. There is a good account of the problem in Mordell's book, where it is shown that the conjecture is true, except possibly in cases where n is congruent to 1^2, 11^2, 13^2, 17^2, 19^2 or 23^2 mod 840. Several have worked on the problem, including Bernstein,

Obláth, Rosati, Shapiro, Straus, Yamamoto and Allan Swett, who has verified the conjecture for all $n \leq 1003162753$, the 51-millionth prime.

Schinzel has observed that one can express

$$\frac{4}{at+b} = \frac{1}{x(t)} + \frac{1}{y(t)} + \frac{1}{z(t)}$$

where $x(t)$, $y(t)$, $z(t)$ are integer polynomials in t with positive leading coefficients and $a \perp b$, only if b is *not* a quadratic residue of a.

Sierpiński made the corresponding conjecture concerning

$$\frac{5}{n} = \frac{1}{x} + \frac{1}{y} + \frac{1}{z}.$$

Palamà confirmed it for all $n \leq 922321$ and Stewart has extended this to $n \leq 1057438801$ and for all n not of the form $278468k + 1$.

Schinzel relaxed the condition that the integers x, y, z should be positive, replaced the numerators 4 and 5 by a general m and required the truth only for $n > n_m$. That n_m may be greater than m is exemplified ny $n_{18} = 23$. The conjecture has been established for successively larger values of m by Schinzel, Sierpiński, Sedláček, Palamà and Stewart & Webb, who prove it for $m < 36$. Breusch and Stewart independently showed that if $m/n > 0$ and n is odd, then m/n is the sum of a finite number of reciprocals of odd integers. See also Graham's papers. Vaughan has shown that if $E_m(N)$ is the number of $n \leq N$ for which $m/n = 1/x + 1/y + 1/z$ has no solution, then

$$E_m(N) \ll N \exp\{-c(\ln N)^{2/3}\}$$

where c depends only on m. Hofmeister & Stoll have shown that if $F_m(N)$ is the number of $n \leq N$ for which $m/n = 1/x + 1/y$ has no solution, then

$$F_m(N) \ll N(\ln N)^{-1/\phi(m)}$$

where $\phi(m)$ is Euler's totient function (**B36**).

Hofmeister notes that this implies that $A_m(N)/N \to 1$ as $N \to \infty$ where $A_m(N)$ is the number of b, $1 \leq b \leq N$ for which there's a representation $m/b = 1/n_1 + 1/n_2$, so that almost *all* lattice points on the *line* $y = m$, $x \geq 1$ have such a representation. Paradoxically, Mittelbach lets $B(N)$ be the number of lattice points (a, b), $1 \leq a \leq b \leq N$ for which there's a representation $a/b = 1/n_1 + 1/n_2$ and proves that $B(N)/\frac{1}{2}N(N+1) \to 0$. I.e., almost *no* lattice points in the *triangle* $(1,1)$, $(1,N)$, (N,N) have a representation for large N.

In contrast to the result of Breusch and Stewart, the following problem, asked by Stein, Selfridge, Graham and others, has not been solved. If m/n, a rational number (n odd), is expressed as $\sum 1/x_i$, where the x_i

are successively chosen to be the least possible odd integers which leave a nonnegative remainder, is the sum always finite? For example,

$$\frac{2}{7} = \frac{1}{5} + \frac{1}{13} + \frac{1}{115} + \frac{1}{10465}$$

Kertesz checked the conjecture for $2 < m < n < 120$. His most recalcitrant fraction was $94/101$ with 16 terms, the last with 4561 digits.

Wagon noted that $3/179$ produces 19 terms, the last of which has 439492 decimal digits! The sequence of numerators of remainders is somewhat amazing: 3, 4, 5, 6, 7, 8, 9, 10, 11, 12, 13, 14, 15, 16, 17, 2, 3, 4, 1.

David Bailey beat this with the odd greedy algorithm for $3/2879$: this halts, but the last of the 21 terms has 3018195 digits. Later, $5/5809$ stumped his program; the representation has at least 22 terms and the last has over 60000000 decimal digits. Then David Eppstein, working modulo 37943838567204570000 was able to prove that the sequence does indeed halt, and that the numerators of the remainders are 5, 6, 7, ..., 29, 30, 1.

Broadhurst found the example $2/24631$; the numerator pattern appears as 2, 3, 4, 5, 6, ..., 25, 2, 3, 4, 5, 2, 1.

Ron Hardin and Neil Sloane conjectured that for every odd number n not a multiple of 3, one can write $3/n = 1/a + 1/b + 1/c$ with a, b, c all odd, positive and distinct. Note that $3/(4n+1) = 1/(2n+1) + 1/(4n+1) + 1/(2n+1)(4n+1)$, $3/(6n+1) = 1/(2n+1) + 1/(2n+1)(4n+1) + 1/(4n+1)(6n+1)$ and $3/(18k+11) = 1/(8k+5) + 1/3(8k+5) + 1/3(8k+5)(18k+11)$. Gary Mulkey and Thomas Hagedorn each proved the Hardin-Sloane conjecture.

John Leech, in a 77-03-14 letter, asked what is known about sets of unequal odd integers whose reciprocals add to 1, such as

$$\frac{1}{3} + \frac{1}{5} + \frac{1}{7} + \frac{1}{9} + \frac{1}{15} + \frac{1}{21} + \frac{1}{27} + \frac{1}{35} + \frac{1}{63} + \frac{1}{105} + \frac{1}{135} = 1$$

He says that you need at least nine in the set, while on the other hand the largest denominator must be at least 105. Notice the connexion with Sierpiński's pseudoperfect numbers (**B2**).

$$945 = 315 + 189 + 135 + 105 + 63 + 45 + 35 + 27 + 15 + 9 + 7$$

Peter Shiu confirms the optimality of Leech's observations with the sets $\{3, 5, 7, 9, 11, 15, 21, 135, 10395\}$ and $\{3, 5, 7, 9, 11, 33, 35, 45, 55, 77, 105\}$.

It is known that if n is odd, then m/n is always expressible as a sum of distinct odd unit fractions.

Beeckmans answers a question of B. M. Stewart by showing that, starting from $\frac{m}{n} = \frac{1}{n} + \cdots + \frac{1}{n}$ and successively replacing fractions $\frac{1}{x}$ by $\frac{1}{x+1} + \frac{1}{x(x+1)}$ until all fractions are distinct, is a finite process.

Tenenbaum & Yokota show that m/n can be expressed as the sum of r unit fractions each of whose denominators is at most $4n(\ln n)^2 \log_2 n$ where $r \leq (1 + \epsilon) \ln n / \log_2 n$ but that $1 + \epsilon$ cannot be replaced by $1 - \epsilon$.

Victor Meally ordered the rationals a/b, $a \perp b$, between 0 and 1 by size of $a + b$ and of a: $\frac{1}{2}$, $\frac{1}{3}$, $\frac{1}{4}$, $\frac{2}{3}$, $\frac{1}{5}$, $\frac{1}{6}$, $\frac{2}{5}$, $\frac{3}{4}$, ... and noted that $\frac{2}{3}$,

$\frac{4}{5}$ and $\frac{8}{11}$ are the earliest members of the sequence that need 2, 3 and 4 unitary fractions to represent them. Which are the earliest that need 5? 6? 7? Stephane Vandemergel, in a 93-04-28 letter, states that $\frac{16}{17}$ requires 5 unitary fractions, and $\frac{77}{79}$ needs 6.

Barbeau expressed 1 as the sum of the reciprocals of 101 distinct positive integers, no one dividing another. Erdős & Graham showed that if n is squarefree, then m/n can always be written as a finite sum of reciprocals of squarefree integers each having exactly ω distinct prime factors, for $\omega \geq 3$. Often ω can be taken as 2. For $m = n = 1$ at least 38 integers are required: Allan Johnson manages it with $\omega = 2$ and the 48 numbers

6	21	34	46	58	77	87	115	155	215	287	391
10	22	35	51	62	82	91	119	187	221	299	689
14	26	38	55	65	85	93	123	203	247	319	731
15	33	39	57	69	86	95	133	209	265	323	901

Is this the smallest possible set? Richard Stong also solved this problem, but used a larger set.

Erdős, in a 72-01-14 letter, sets $\frac{1}{2} + \frac{1}{3} + \ldots + \frac{1}{n} = \frac{a}{b}$, where $b = [2, 3, \ldots, n]$, the l.c.m. of $2, 3, \ldots, n$. He observes that $\frac{1}{2} + \frac{1}{3} = \frac{5}{6}$ and $\frac{1}{2} + \frac{1}{3} + \frac{1}{4} = \frac{13}{12}$ are such that $a \pm 1 \equiv 0 \bmod b$ and asks if this occurs again: he conjectures not. Is $a \perp b$ infinitely often?

If $\sum_{i=1}^{t} 1/x_i = 1$ with $x_1 < x_2 < x_3 < \ldots$ *distinct* positive integers Erdős & Graham ask what is $m(t)$, the min max x_i, where the minimum is taken over all sets $\{x_i\}$. For example, $m(3) = 6$, $m(4) = 12$, $m(12) = 30$. Is $m(t) < ct$ for some constant c? In this notation, is it possible to have $x_{i+1} - x_i \leq 2$ for all i? Erdős conjectures that it is not and offers \$10.00 for a solution.

The value of $m(12)$ was incorrect in the second edition. The above value is suggested by the following table calculated by Kevin Brown:

k	denominators of optimum expansion											
3	2	3	6									
4	2	4	6	12								
5	2	4	10	12	15							
6	3	4	6	10	12	15						
7	3	4	9	10	12	15	18					
8	3	5	9	10	12	15	18	20				
9	4	5	8	9	10	15	18	20	24			
10	5	6	8	9	10	12	15	18	20	24		
11	5	6	8	9	10	15	18	20	21	24	28	
12	6	7	8	9	10	14	15	18	20	24	28	30

Dan Hoey confirms the above table and gives the other optimal expansions in this range as

k	denominators of optimum expansion											
8	4	5	6	9	10	15	18	20				
9	4	6	8	9	10	12	15	18	24			
11	5	7	8	9	10	14	15	18	20	24	28	
11	6	7	8	9	10	12	14	15	18	24	28	
12	4	8	9	10	12	15	18	20	21	24	28	30

and says that there can be quite a few duplicates: there are 11 optimal
13-term expansions, 58 optimal 20-term expansions, 37 optimal 23-term
expansions, and 57 optimal 28-term expansions. He also calculated $m(k)$
for $13 \leq 28$ as

k	13	14	15	16	17	18	19	20	21	22	23	24	25	26	27	28
$m(k)$	33	33	35	36	40	42	48	52	52	54	55	55	56	60	63	72

He notices that $m(k)$ is nondecreasing so far, but doesn't see why that
should persist.

Erdős & Graham ask if it is true that any coloring of the integers with
c colors gives a monochromatic solution of

$$\sum \frac{1}{x_i} = 1, \quad x_1 < x_2 < \dots \quad \text{(finite sum)}.$$

This is open even for $c = 2$. If the answer is affirmative, let $f(c)$ be the
smallest integer for which every c-coloring of the integers $1 \leq t \leq f(c)$
contains a monochromatic solution. Determine or estimate $f(c)$.

Erdős also asks that if

$$\frac{1}{x_1} + \frac{1}{x_2} + \dots + \frac{1}{x_k} = 1, \quad x_1 < x_2 < \dots < x_k$$

and k is fixed, what is $\max x_1$? If k varies, what integers can be equal to
x_k, the largest denominator? Not primes, and not several other integers;
do the excluded integers have positive density? Density 1 even? Which
integers can be x_k or x_{k-1} ? Which can be x_k or x_{k-1} or x_{k-2} ? Is
$\liminf \frac{x_k}{x_1} > e$? It is trivial that the limit is $\geq e$. In fact perhaps it is
infinite. If $m(k)$ is $\max(x_k)$ for each k, then Yokota improves a result of
Erdős & Graham by proving that there is an increasing sequence of integers
k for which $m(k)/k \leq (\ln \ln k)^3$. Is there a sequence of k such that $m(k)/k$
is bounded?

Erdős further asks if it is true that for every solution of

$$\frac{1}{x_1} + \frac{1}{x_2} + \dots + \frac{1}{x_k} = 1,$$

$\max(x_{i+1} - x_i) \geq 3$? $\{2,3,6\}$ shows that > 3 is not true but perhaps
this is the only counterexample. Perhaps $\max(x_{i+1} - x_i) \leq c$ has only a

finite number of solutions. If the x_i are the union of r blocks of consecutive integers, the number of solutions is finite and depends only on r. That the sequence $\frac{1}{2} + \frac{1}{3} + \frac{1}{7} + \frac{1}{43} + \cdots$ gives the max x_k as a function of r was proved by Takenouchi; see also Eppstein and Izhboldin & Kurlyandchik.

If $N(n)$ is the set of integers that can be written as a sum of distinct reciprocals of integers $\leq n$, then Yokota shows that every natural number up to

$$\frac{\ln n}{2} \left(1 - \frac{2 \ln \ln n}{\ln n} \right)$$

is in $N(n)$, so that $\#N(n) \geq (\frac{1}{2} + o(1)) \ln n$.

Erdős asked for an estimate for the size of the smallest integer not in $N(n)$, and of the largest in $N(n)$. Croot has since shown that the set of integers n that can be written as the sum of Egyptian fractions each of whose denominators is bounded above by n is $\{1, 2, \ldots, \lfloor H(n) \rfloor\}$, or possibly $\{1, 2, \ldots, \lfloor H(n) \rfloor - 1\}$ if $H(n)$ is only slightly larger than an integer, where $H(n)$ is the partial sum, $\sum_{k \leq n} 1/k$ of the harmonic series.

Given a sequence x_1, x_2, ... of positive density, is there always a finite subset with $\sum 1/x_{i_k} = 1$? If $x_i < ci$ for all i, is there such a finite subset? Erdős again offers \$10.00 for a solution. If $\liminf x_i/i < \infty$, he strongly conjectures that the answer is negative and offers only \$5.00 for a solution.

Denote by $N(t)$ the number of solutions $\{x_1, x_2, \ldots, x_t\}$ of $1 = \sum 1/x_i$ and by $M(t)$ the number of distinct solutions $x_1 \leq x_2 \leq \ldots \leq x_t$. Singmaster calculated

t	1	2	3	4	5	6
$M(t)$	1	1	3	14	147	3462
$N(t)$	1	1	10	215	12231	2025462

and Erdős asked for an asymptotic formula for $M(t)$ or $N(t)$.

Erdős & Joó show that for every $n \geq 1$ there are $2^{\aleph_0}$ numbers q, $1 < q < 2$ such that 1 has exactly $n + 1$ expansions of the form $\sum_{i=1}^{\infty} \epsilon_i q^{-i}$ with $\epsilon_i \in \{0, 1\}$.

Graham has shown that if $n > 77$ we can partition $n = x_1 + x_2 + \ldots + x_t$ into t distinct positive integers so that $\sum 1/x_i = 1$. More generally, that for any positive rational numbers α, β, there is an integer $r(\alpha, \beta)$, which we will take to be the least, such that any integer greater than r can be partitioned into distinct positive integers greater than β, whose reciprocals sum to α. Little is known about $r(\alpha, \beta)$, except that unpublished work of D. H. Lehmer shows that 77 cannot be partitioned in this way, so that $r(1, 1) = 77$.

Graham conjectures that for n sufficiently large (about 10^4 ?) we can similarly partition $n = x_1^2 + x_2^2 + \ldots + x_t^2$ with $\sum 1/x_i = 1$. We can also ask for a decomposition $n = p(x_1) + p(x_2) + \ldots + p(x_t)$ where $p(x)$ is any "reasonable" polynomial; for example $x^2 + x$ is unreasonable since it takes only even values.

In answer to a question of L.-S. Hahn, is there a set of integers, each having an immediate neighbor, the sum of whose reciprocals is an integer, Peter Montgomery gave the examples {1,2,7,8,13,14,39,40,76,77,285,286} and {2,3,4,5,6,7,9,10,17,18,34,35,84,85} whose reciprocals each add to 2.

L.-S. Hahn also asks: if the positive integers are partitioned into a finite number of sets in any way, is there always a set such that *any* positive rational number can be expressed as the sum of the reciprocals of a finite number of distinct members of it? Here it must be possible to choose the set, independent of the rational number. If this is not possible, then given any rational number, can one always choose a set with this property? Now the set can depend on the rational number.

Nagell showed that the sum of the reciprocals of an arithmetic progression is never an integer: see also the paper of Erdős & Niven.

M. H. Ahmadi & M. N. Bleicher, On the conjectures of Erdős and Straus, and Sierpiński on Egyptian fractions, *J. Math. Stat. Sci.*, **7**(1998) 169–185; *MR* **99k**:11049.

P. J. van Albada & J. H. van Lint, Reciprocal bases for the integers, *Amer. Math. Monthly*, **70**(1963) 170–174.

Premchand Anne, Egyptian fractions and the inheritance problem, *Coll. Math. J.*, **29**(1998) 296–300.

Michael Anshel & Dorian Goldfeld, Partitions, Egyptian fractions, and free products of finite abelian groups, *Proc. Amer. Math. Soc.*, **111**(1991) 889–899; *MR* **91h**:11104.

E. J. Barbeau, Expressing one as a sum of odd reciprocals: comments and a bibliography, *Crux Mathematicorum* (=*Eureka* (Ottawa)), **3**(1977) 178–181; and see *Math. Mag.*, **49**(1976) 34.

Laurent Beeckmans, The splitting algorithm for Egyptian fractions, *J. Number Theory*, **43**(1993) 173–185; *MR* **94a**:11043.

Leon Bernstein, Zur Lösung der diophantischen Gleichung $m/n = 1/x + 1/y + 1/z$ insbesondere im Falle $m = 4$, *J. reine angew. Math.*, **211**(1962) 1–10; *MR* **26** #77.

M. N. Bleicher, A new algorithm for the expansion of Egyptian fractions, *J. Number Theory*, **4**(1972) 342–382; *MR* **48** #2052.

M. N. Bleicher and P. Erdős, The number of distinct subsums of $\sum_1^N 1/i$, *Math. Comput.*, **29**(1975) 29–42 (and see *Notices Amer. Math. Soc.*, **20**(1973) A-516).

M. N. Bleicher and P. Erdős, Denominators of Egyptian fractions, *J. Number Theory*, **8**(1976) 157–168; *MR* **53** #7925; II, *Illinois J. Math.*, **20**(1976) 598–613; *MR* **54** #7359.

Lawrence Brenton & Robert R. Bruner, On recursive solutions of a unit fraction equation, *J. Austral. Math. Soc. Ser. A*, **57**(1994) 341–356; *MR* **95i**:11024.

Lawrence Brenton & R. Hill, On the Diophantine equation $1 = \sum(1/n_i) + 1/(\prod n_i)$ and a class of homologically trivial complex surface singularities, *Pacific J. Math.*, **133**(1988) 41–67.

Robert Breusch, A special case of Egyptian fractions, Solution to Advanced Problem 4512, *Amer. Math. Monthly*, **61**(1954) 200–201.

J. L. Brown, Note on complete sequences of integers, *Amer. Math. Monthly*, **68**(1961) 557–560.

Paul J. Campbell, Bibliography of algorithms for Egyptian fractions (preprint), Beloit Coll., Beloit WI 53511, U.S.A.

Cao Zhen-Fu, Liu Rui & Zhang Liang-Rui, On the equation $\sum_{i=1}^{s} 1/x_i + (1/x_1 \cdots x_s) = 1$ and Znam's problem, *J. Number Theory*, **27**(1987); *MR* **89d**:11023.

Robert Cohen, Egyptian fraction expansions, *Math. Mag.*, **46** (1973) 76–80; *MR* **47** #3300.

Ernest S. Croot, On some questions of Erdős and Graham about Egyptian fractions, *Mathematika*, **46**(1999) 359–372; *MR* **2002e**:11044.

Ernest S. Croot, On unit fractions with denominators in short intervals, *Acta Arith.*, **99**(2001) 99–114; *MR* **2002e**:11045.

Ernest S. Croot, On a coloring conjecture about unit fractions, *Ann. of Math.*(2), **157**(2003) 545–556.

Ernest S. Croot, David E. Dobbs, John B. Friedlander, Andrew J. Hetzel & Francesco Pappalardi, Binary Egyptian fractions, *J. Number Theory*, **84**(2000) 63–79; *MR* **2001f**:11052.

D. R. Curtiss, On Kellogg's Diophantine problem, *Amer. Math. Monthly*, **29**(1922) 380–387.

David E. Dobbs & Andrew J. Hetzel, Ahmes expansions of rational numbers of length two, *Internat. J. Math. Ed. Sci. Tech.*, **34**(2003) 742–751.

Christian Elsholtz, Sums of k unit fractions, *Trans. Amer. Math. Soc.*, **353**(2001) 3209–3227; *MR* **2002c**:11030.

David Eppstein, Ten algorithms for Egyptian fractions, *Mathematica in Education and Research*, **4**(1995) 5–15. This article is also available on the web at http://www.ics.uci.edu/~eppstein/numth/egypt/intro.html

P. Erdős, Egy Kürschák-féle elemi számelméleti tétel áltadanositása, *Mat. es Fys. Lapok*, **39**(1932).

P. Erdős, On arithmetical properties of Lambert series, *J. Indian Math. Soc.*, **12**(1948) 63–66.

P. Erdős, The solution in whole numbers of the equation: $1/x_1 + 1/x_2 + \cdots + 1/x_n = a/b$, (Hungarian. Russian and English summaries), *Mat. Lapok*, **1**(1950) 192–210; *MR* **13**, 208.

P. Erdős, On the irrationality of certain series, *Nederl. Akad. Wetensch.* (*Indag. Math.*) **60**(1957) 212–219.

P. Erdős, Sur certaines séries à valeur irrationelle, *Enseignement Math.*, **4**(1958) 93–100.

P. Erdős, *Quelques problèmes de la Théorie des Nombres*, Monographies de l'Enseignement Math. No. 6, Geneva, 1963, problems 72–74.

P. Erdős, Comment on problem E2427, *Amer. Math. Monthly*, **81**(1974) 780–782.

P. Erdős, Some problems and results on the irrationality of the sum of infinite series, *J. Math. Sci.*, **10**(1975) 1–7.

P. Erdős & I. Joó, On the expansion $1 = \sum q^{-n_i}$, *Period. Math. Hungar.*, **23**(1991) 27–30; *MR* **92i**:11030.

P. Erdős & I. Joó, On the number of expansions $1 = \sum q^{-n_i}$, *Ann. Univ. Sci. Budapest. Eötvös Sect. Math.*, **35**(1992) 129–132; *MR* **94a**:11012.

P. Erdős & Ivan Niven, Some properties of partial sums of the harmonic series, *Bull. Amer. Math. Soc.*, **52**(1946) 248–251; *MR* **7**, 413.

Paul Erdős & Sherman Stein, Sums of distinct unit fractions, *Proc. Amer. Math. Soc.*, **14**(1963) 126–131.

P. Erdős & E. G. Straus, On the irrationality of certain Ahmes series, *J. Indian Math. Soc.*, **27**(1968) 129–133.

P. Erdős & E. G. Straus, Some number theoretic results, *Pacific J. Math.*, **36**(1971) 635–646.

P. Erdős & E. G. Straus, Solution of problem E2232, *Amer. Math. Monthly*, **78**(1971) 302–303.

P. Erdős & E. G. Straus, On the irrationality of certain series, *Pacific J. Math.*, **55**(1974) 85–92; *MR* **51** #3069.

P. Erdős & E. G. Straus, Solution to problem 387, *Nieuw Arch. Wisk.*, **23**(1975) 183.

A. Eswarathasan & E. Levine, p-Integral harmonic sums, *Discrete Math.*, **91**(1991) 249–259.

Charles N. Friedman, Sums of divisors and Egyptian fractions, *J. Number Theory*, **44**(1993) 328–339.

S. W. Golomb, An algebraic algorithm for the representation problem of the Ahmes papyrus, *Amer. Math. Monthly* **69**(1962) 785–786.

S. W. Golomb, On the sums of the reciprocals of the Fermat numbers and related irrationalities, *Canad. J. Math.*, **15**(1963) 475–478.

R. L. Graham, A theorem on partitions, *J. Austral. Math. Soc.*, **3**(1963) 435–441; *MR* **29** #64.

R. L. Graham, On finite sums of unit fractions, *Proc. London Math. Soc.* (3), **14**(1964) 193–207; *MR* **28** #3968.

R. L. Graham, On finite sums of reciprocals of distinct nth powers, *Pacific J. Math.*, **14**(1964) 85–92; *MR* **28** #3004.

Thomas R. Hagedorn, A proof of a conjecture on Egyptian fractions, *Amer. Math. Monthly*, **107**(2000) 62–63; *MR* **2000k**:11046.

H. S. Hahn, Old wine in new bottles: Solution to E2327, *Amer. Math. Monthly*, **79**(1972) 1138 (and see Editor's comment).

Liang-Shin Hahn, Problem E2689, *Amer. Math. Monthly*, **85**(1978) 47; Solution, Peter Montgomery & Dean Hickerson, **86**(1979) 224.

Gerd Hofmeister & Peter Stoll, Note on Egyptian fractions, *J. reine angew. Math.*, **362**(1985) 141–145; *MR* **87a**:11025.

O. T. Izhboldin & L. D. Kurlyandchik, Unit fractions, *Proc. St. Petersburg Math. Soc.*, **III**(1995) 193–200; *MR* **99m**:11024.

Allan Wm. Johnson, Letter to the Editor, *Crux Mathematicorum* (= *Eureka* (Ottawa)), **4**(1978) 190.

Ralph W. Jollensten, A note on the Egyptian problem, *Proc. Seventh Southeastern Conf. Combin., Graph Theory & Comput., Congressus Numer.*, **17**(1976), Utilitas Math. Publ. Inc., Winnipeg, 351–364.

O. D. Kellogg, On a diophantine problem, *Amer. Math. Monthly*, **28**(1921) 300–303.

Yoko Kikuchi & Shin-ichi Katayama, Some remarks on Egyptian fractions, *Math. Japon.*, **42**(1995) 127–130; *MR* **96j**:11037.

Chao Ko, Chi Sun & S. J. Chang, On equations $4/n = 1/x + 1/y + 1/z$, *Acta Sci. Natur. Szechuanensis*, **2**(1964) 21–35.

Ilias Kotsireas, The Erdős-Straus conjecture on Egyptian fractions, *Paul Erdős and his mathematics* (*Budapest*, 1999) 140–144, János Bolyai Math. Soc., Budapest 1999; *MR* **2003d**:11052.

Li De-Lang, On the equation $4/n = 1/x + 1/y + 1/z$, *J. Number Theory*, **13**(1981) 485–494; *MR* **83e**:10026.

Dana Mackenzie, Fractions to make an Egyptian scribe blanch, *Science*, **278**:5336(97-10-10) 224.

Greg Martin, Dense Egyptian fractions, *Trans. Amer. Math. Soc.*, **351**(1999) 3641–3657; *MR* **99m**:11035.

Greg Martin, Denser Egyptian fractions, *Acta Arith.*, **95**(2000) 231–260; *MR* **2001m**:11040.

F. Mittelbach, Anzahl- und Dichteuntersuchungen bei Stammbruchdarstellungen von Brüchen, Diplomarbeit, Fachbereich Mathematik, Joh. Gutenberg-Univ., Mainz, 1988.

T. Nagell, *Skr. Norske Vid. Akad. Kristiania I*, 1923, no. 13 (1924) 10–15.

D. J. Newman, Problem 76-5: an arithmetic conjecture, *SIAM Rev.*, **18**(1976) 118.

R. Obláth, Sur l'équation diophantienne $4/n = 1/x_1 + 1/x_2 + 1/x_3$, *Mathesis*, **59**(1950) 308–316; *MR* **12**, 481.

J. C. Owings, Another proof of the Egyptian fraction theorem, *Amer. Math. Monthly*, **75**(1968) 777–778.

G. Palamà, Su di una congettura di Sierpiński relativa alla possibilità in numeri naturali della $5/n = 1/x_1 + 1/x_2 + 1/x_3$, *Boll. Un. Mat. Ital.*(3) **13**(1958) 65–72; *MR* **20** #3821.

G. Palamà, Su di una congettura di Schinzel, *Boll. Un. Mat. Ital.*(3) **14**(1959) 82–94; *MR* **22** #7989.

Jukka Pihko, Remarks on the "greedy odd" Egyptian fraction algorithm, *Fibonacci Quart.*, **39**(2001) 221–227; *MR* **2002h**:11029.

G. J. Rieger, Sums of unit fractions having long continued fractions, *Fibonacci Quart.* **31**(1993) 338–340.

L. A. Rosati, Sull'equazione diofantea $4/n = 1/x_1 + 1/x_2 + 1/x_3$, *Boll. Un. Mat. Ital.*(3), **9**(1954) 59–63; *MR* **15**, 684.

J. W. Sander, On $4/n = 1/x + 1/y + 1/z$ and Rosser's sieve, *Acta Arith.*, **59**(1991) 183–204; *MR* **92j**:11031.

J. W. Sander, On $4/n = 1/x + 1/y + 1/z$ and Iwaniec's half-dimensional sieve, *J. Number Theory*, **46**(1994) 123–136; *MR* **95e**:11044.

J. W. Sander, Egyptian fractions and the Erdős-Straus conjecture, *Nieuw Arch. Wisk.*(4) **15**(1997) 43–50; *MR* **98d**:11039.

Andrzej Schinzel, Sur quelques propriétés des nombres $3/n$ et $4/n$, où n est un nombre impair, *Mathesis*, **65**(1956) 219–222; *MR* **18**, 284.

A. Schinzel, Erdős's work on finite sums of unit fractions, *Paul Erdős and his Mathematics* (*Budapest*, 1999) 629–636, Bolyai Soc. Math. Stud., **11**(2002); *MR*2003k:11055.

Andrzej Schinzel, On sums of three unit fractions with polynomial denominators, *Funct. Approx. Comment. Math.*, **28**(2000) 187–194; *MR* **2002a**:11029.

W. Schwarz, *Einführung in Siebmethoden der analytischen Zahlentheorie*, Bibl. Inst., Mannheim-Wien-Zurich, 1974; *MR* **53** #13147.

Jiří Sedláček, Über die Stammbrüche, *Časopis Pěst. Mat.*, **84**(1959) 188–197; *MR* **23** #A829.

Igor E. Shparlinski, On a question of Erdős and Graham, *Arch. Math. (Basel)*, **78**(2002) 445–448; *MR* **2003i**:11144.

P. Shiu, Egyptian fraction representations of 1 with odd denominators, March 2004 preprint.

W. Sierpiński, Sur les décompositions de nombres rationale en fractions primaires, *Mathesis*, **65**(1956) 16–32; *MR* **17**, 1185.

W. Sierpiński, *On the Decomposition of Rational Numbers into Unit Fractions* (Polish), Pánstwowe Wydawnictwo Naukowe, Warsaw, 1957.

W. Sierpiński, Sur une algorithme pour developper les nombres réels en séries rapidement convergentes, *Bull. Int. Acad. Sci. Cracovie Ser. A Sci. Math.*, **8**(1911) 113–117.

B. M. Stewart, Sums of distinct divisors, *Amer. J. Math.*, **76**(1954) 779–785; *MR* **16**, 336.

B. M. Stewart & W. A. Webb, Sums of fractions with bounded numerators, *Canad. J. Math.*, **18**(1966) 999–1003; *MR* **33** #7297.

R. J. Stroeker & R. Tijdeman, Diophantine equations, *Computational Methods in Number Theory, Part II*, Math. Centrum Tracts **155**, Amsterdam, 1982; *MR* **84i**:10014.

J. J. Sylvester, On a point in the theory of vulgar fractions, *Amer. J. Math.*, **3**(1880) 332–335, 387–388.

Tanzo Takenouchi, On an indeterminate equation, *Proc. Phys.-Math. Soc. Japan*(3), **3**(1921) 78–92.

Gérald Tenenbaum & Hisashi Yokota, Length and denominators of Egyptian fractions III, *J. Number Theory*, **35**(1990) 150–156; *MR* **91g**:11028.

D. G. Terzi, On a conjecture by Erdős-Straus, *BIT*, **11**(1971) 212–216.

R. C. Vaughan, On a problem of Erdős, Straus and Schinzel, *Mathematika*, **17**(1970) 193–198.

H. S. Wilf, Reciprocal bases for the integers, Res. Problem 6, *Bull. Amer. Math. Soc.*, **67**(1961) 456.

Wu Chang-Jiu & Wang Peng-Fei, Remarks on a conjecture of Erdős, *J. Chengdu Univ. Natur. Sci.*, **13**(1994) 9–10, 58.

Koichi Yamamoto, On a conjecture of Erdős, *Mem. Fac. Sci. Kyushū Univ. Ser. A*, **18**(1964) 166–167; *MR* **30** #1968; and see **19**(1965) 37–47; *MR* **31** #2203.

Hisashi Yokota, On number of integers representable as sums of unit fractions, *Canad. Math. Bull.*, **33**(1990) 235–241; *MR* **91a**:11029;

II. *J. Number Theory*, **67**(1997) 162–169; *MR* **98k**:11031; III *J. Number Theory*, **96**(2002) 351–372; *MR* **2003i**:11044.

Hisashi Yokota, On a problem of Erdős and Graham, *J. Number Theory*, **39**(1991) 327–338; *MR* **90d**:11104.

OEIS: A002966-002967, A006585, A028229, A030659, A051882, A069581.

D12 Markoff numbers.

A diophantine equation which has excited a great deal of interest is
$$x^2 + y^2 + z^2 = 3xyz.$$
It obviously has what Cassels has called the singular solutions, $(1,1,1)$
and $(1,1,2)$ (with the usual definition, the variety has only the singular
solution $(0,0,0)$). All solutions can be generated from these since the equation is a quadratic in each of the variables, so one integer solution leads to
a second, and it can be shown that, apart from the singular solutions, all
solutions have distinct values of x, y and z, so that each such solution is
a **neighbor** of just three others (Figure 10). The numbers 1, 2, 5, 13, 29,
34, 89, 169, 194, 233, 433, 610, 985, ... are called **Markoff numbers**. To
avoid trivialities, assume that $0 < x \le y \le z$ (so that the inequalities are
strict if $y \ge 2$). An outstanding problem is whether every Markoff number
z defines a unique integer solution (x,y,z). There are occasional claims to
have proved that the Markoff numbers are unique in this sense, but so far
proofs appear to be fallacious.

If $M(N)$ is the number of triples with $x \le y \le z \le N$, then Zagier has
shown that $M(N) = C(\ln N)^2 + O((\ln N)^{1+\epsilon})$ where $C \approx 0.180717105$, and
calculations lead him to conjecture that the nth Markoff number, m_n, is
$\left(\frac{1}{3} + O(n^{-1/4+\epsilon})\right) A^{\sqrt{n}}$ where $A + e^{1/\sqrt{C}} \approx 10.5101504$. He has no results on
distinctness, but can show that the problem is equivalent to the insolvability
of a certain system of diophantine equations.

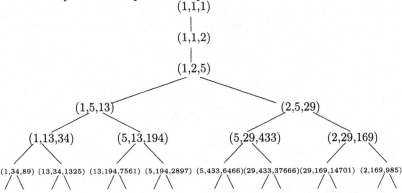

Figure 10. The Tree of Markoff Solutions.

Markoff's equation is a special case of the more general **Hurwitz equation**
$$x_1^2 + x_2^2 + \cdots + x_n^2 = ax_1x_2 \cdots x_n$$
for which there are no integer solutions if $a > n$, and for $a = n$ all integer
solutions can be generated from $(1,1,...,1)$. For any a, $1 \le a \le n$, there
is a finite set of solutions which generates all others. Baragar has shown
that, for any g, there are infinitely many pairs (a,n) so that the equation

requires at least g generators. Let $M(n, N)$ be the number of solutions of the Hurwitz equation with $a = n$ and each $|x_i| \leq N$, then Baragar has also shown that $M(n, N)$ grows like $C(\ln N)^{\alpha(n)+\epsilon}$ for all $\epsilon > 0$ and that $M(n, N) = \Omega((\ln N)^{\alpha(n)-\epsilon})$, where (Zagier) $\alpha(3) = 2$, but $\alpha(4)$ lies between 2.33 and 2.64 (later improved to $2.43 < \alpha(4) < 2.47$). Also that

$$\frac{2\ln n}{\ln 4} \leq \alpha(n) \leq \frac{3\ln n}{\ln 4}.$$

Christian Ballot, Strong arithmetic properties of the integral solutions of $X^3 + DY^3 + D^2Z^3 - 3DXYZ = 1$, where $D + M^3 \pm 1$, $M \in Z^*$, Acta Arith., 89(1999), 259–277.

Arthur Baragar, The Hurwitz equations, Proc. Conf. Markoff Spectrum & Anal. Number Theory, Provo UT, 1991; Lect. Notes Pure Appl. Math., 147 9–16, Dekker, New York, 1993.

Arthur Baragar, Asymptotic growth of Markoff-Hurwitz numbers, Compositio Math., 94(1994) 1–18; MR 95i:11025.

Arthur Baragar, Integral solutions of Markoff-Hurwitz equations, J. Number Theory, 49(1994) 27–44; MR 95g:11066.

Arthur Baragar, On the unicity conjecture for Markoff numbers, Canad. Math. Bull., 39(1996) 3–9; MR 97d:11110.

Arthur Baragar, The exponent for the Markoff-Hurwitz equations, Pacific J. Math., 182(1998) 1–21; MR 99e:11035.

G. V. Belyi, Markov's nubers and quadratic forms. Pontryagin Conference, 8, Algebra (Moscow, 1998), J. Math. Sci. (New York), 106(2001) 3087–3097.

Edward B. Burger, Amanda Folsom, Alexander Pekker, Rungporn Roengpitya & Julia Snyder, On a quantitative refinement of the Lagrange spectrum, Acta Arith., 102(2002) 55–82.

J. O. Button, The uniqueness of the prime Markoff numbers, J. London Math. Soc.(2), 58(1998) 9–17.

J. O. Button, Markoff numbers, principal ideals and continued fraction expansions. J. Number Theory, 87(2001) 77–95; MR 2002a:11074.

J. W. S. Cassels, An Introduction to Diophantine Approximation, Cambridge, 1957, 27–44.

H. Cohn, Approach to Markoff's minimal forms through modular functions, Ann. Math. Princeton(2) 61(1955) 1–12; MR 16, 801e.

T. W. Cusick, The largest gaps in the lower Markoff spectrum, Duke Math. J., 41(1974) 453–463; MR 57 #5902.

T. W. Cusick, On Perrine's generalized Markoff equations, Aequationes Math., 46(1993) 203–211.

L. E. Dickson, Studies in the Theory of Numbers, Chicago Univ. Press, 1930, Chap. VII.

G. Frobenius, Über die Markoffschen Zahlen, S.-B. Preuss. Akad. Wiss. Berlin (1913) 458–487.

Norman P. Herzberg, On a problem of Hurwitz, Pacific J. Math., 50(1974) 485–493; MR 50 #233.

A. Hurwitz, Über eine Aufgabe der unbestimmten Analysis, Arch. Math. Phys., 3(1907) 185–196; Math. Werke, ii, 410–421.

Jeffrey C. Lagarias & Andrew D. Pollington, The continuous Diophantine mapping of Szekeres, *J. Austral. Math. Soc. Ser. A*, **59**(1995) 148–172; *MR* **96f**:11088.

A. Markoff, Sur les formes quadratiques binaires indéfinies, *Math. Ann.*, **15**(1879) 381–409.

Frank Martini, Das Markovspektrum indefiniter quadratische Formen, *Bonner Mathematische Schriften* **290**(1996) 55pp.

Serge Perrine, Sur une généralisation de la théorie de Markoff, *J. Number Theory*, **37**(1991) 211–230; *MR* **92c**:11067.

Serge Perrine, Sur des équations diophantiennes généralisant celle de Markoff, *Ann. Fac. Sci. Toulouse Math*(6) **6**(1997) 127–141; *MR* **99e**:11030.

R. Remak, Über indefinite binäre quadratische Minimalformen, *Math. Ann.*, **92**(1924) 155–182.

R. Remak, Über die geometrische Darstellung der indefiniten binären quadratischen Minimalformen, *Jber. Deutsch Math.-Verein*, **33**(1925) 228–245.

D. Rosen & G. S. Patterson, Some numerical evidence concerning the uniqueness of the Markov numbers, *Math. Comput.*, **25**(1971) 919–921.

Gerhard Rosenberger, The uniqueness of the Markoff numbers, *Math. Comput.*, **30**(1976) 361–365; but see *MR* **53** #280.

Paul Schmutz, Systoles of arithmetic surfaces and the Markoff spectrum, *Math. Ann.*, **305**(1996) 191–203; *MR* **97b**:11090.

Joseph H. Silverman, The Markoff equation $X^2 + Y^2 + Z^2 = aXYZ$ over quadratic imaginary fields, *J. Number Theory*, **35**(1990) 72–104; *MR* **91i**:11028.

L. Ja. Vulakh, The diophantine equation $p^2+2q^2+3r^2 = 6pqr$ and the Markoff spectrum (Russian), *Trudy Moskov. Inst. Radiotehn. Èlektron. i Avtomat. Vyp. 67 Mat.*(1973) 105–112, 152; *MR* **58** #21957.

L. Ya. Vulakh, The Markov spectra for Fuchsian groups, *Trans. Amer. Math. Soc.*, **352**(2000) 4067–4094; *MR* **2000m**:11056.

Don B. Zagier, On the number of Markoff numbers below a given bound *Math. Comput.*, **39**(1982) 709–723; *MR* **83k**:10062.

OEIS: A002559.

D13 The equation $x^x y^y = z^z$.

Erdős asked for solutions of the equation $x^x y^y = z^z$, apart from the trivial ones $y = 1$, $x = z$. Chao Ko found an infinity of solutions, of which the first three are

x	y	z
12^6	6^8	$2^{11}3^7$
224^{14}	112^{16}	$2^{68}7^{15}$
61440^{30}	30720^{32}	$2^{357}15^{31}$

Did he find them all?

Uchiyama showed that if $4xy > z^2$, then there are no nontrivial solutions; if $4xy = z^2$, then Ko's solutions are the only ones; and for each

$Q = xy/z^2 < \frac{1}{4}$, there are at most finitely many solutions, which can be determined effectively. He gave many cases where there are no solutions.

Claude Anderson conjectured that the equation $w^w x^x y^y = z^z$ has no solutions with $1 < w < x < y < z$, but Chao Ko & Sun Qi had earlier found an infinity of counterexamples to a generalization of the conjecture to any number of variables:

$$
\begin{aligned}
x_1 &= k^{k^n(k^{n+1}-2n-k)+2n}(k^n-1)^{2(k^n-1)} \\
x_2 &= k^{k^n(k^{n+1}-2n-k)}(k^n-1)^{2(k^n-1)+2} \\
x_3 &= \cdots = x_k = k^{k^n(k^{n+1}-2n-k)+n}(k^n-1)^{2(k^n-1)+1} \\
z &= k^{k^n(k^{n+1}-2n-k)+n+1}
\end{aligned}
$$

where, for $k \geq 3$, $n > 0$, and, for $k = 2$, $n > 1$. E.g., $w = 3^{12}2^6$, $x = 3^{13}2^5$, $y = 3^{14}2^4$, $z = 3^{14}2^5$. Ajai Choudhry also found a parametric solution for $k = 3$.

Chao Ko, Note on the diophantine equation $x^x y^y = z^z$, *J. Chinese Math. Soc.*, **2**(1940) 205–207; *MR* **2**, 346.

Chao Ko & Sun Qi, On the equation $\prod_{i=1}^k x_i^{x_i}$, *J. Sichuan Univ.*, **2**(1964) 5–9.

W. H. Mills, An unsolved diophantine equation, *Report Inst. Theory of Numbers*, Boulder CO, 1959, 258–268.

Saburô Uchiyama, On the diophantine equation $x^x y^y = z^z$, *Trudy Mat. Inst. Steklov*, **163**(1984) 237–243; *MR* **86i**:11014.

D14 $a_i + b_j$ made squares.

Leo Moser asked for integers a_1, a_2, b_j $(1 \leq j \leq n)$ such that the $2n$ numbers $a_i + b_j$ are all squares. This can be achieved by making $a_2 - a_1$ a sufficiently composite number; for example $a_1 = 0$, $a_2 = 2^{2n+1}$, $b_j = (2^{2n-j} - 2^{j-1})^2$.

John Leech observes that the extension to integers a_1, a_2, a_3, b_j $(1 \leq j \leq n)$ is also solvable for any n. We may take a_1, a_2 to be $(x \pm y)^2$; a_3 can have an arbitrary value $x^2 + \lambda xy + y^2$, which can then be made square by putting $x = u^2 - v^2$, $y = 2uv + \lambda v^2$. Any values of u and v will give triads of squares with differences in this proportion. The problem to find values of u, v so that the scale factor is a rational square reduces to finding rational points on an elliptic curve; arbitrarily many rational values of b_j can then be simultaneously scaled to give integers a_1, a_2, a_3, b_j $(1 \leq j \leq n)$ for any n. A much studied case is $\lambda = 0$, corresponding to sets of rational right-angled triangles of equal area. We can also specialize to fix $a_1 = b_1 = 0$. Provided that a_2, a_3 are squares p^2, q^2 such that q/p has a representation as the product of two *distinct* ratios $(u^2 - v^2)/2uv$, then it is again an elliptic curve problem to find rational squares $b_j = r_j^2$ such that both $p^2 + r_j^2$

and $q^2 + r_j^2$ are squares, and again we can rescale to find integers p, q, r_j $(1 \leq j \leq n)$ for any n such that $p^2 + r_j^2$ and $q^2 + r_j^2$ are integer squares (cf. **D20**). For example, $13/6$ has representations $(u_1, v_1, u_2, v_2) = (9, 4, 5, 1)$ and $(8, 5, 9, 1)$ which yield

$$
\begin{matrix}
0^2 & 351^2 & 650^2 & 1728^2 & 3200^2 \\
720^2 & 801^2 & 970^2 & 1872^2 & 3280^2 \\
1560^2 & 1599^2 & 1690^2 & 2328^2 & 3560^2
\end{matrix}
$$

More generally we seek integers a_i $(1 \leq i \leq m)$, b_j $(1 \leq j \leq n)$. Jean Lagrange has produced the matrix

$$
\begin{bmatrix}
(54, 150, 111)^2 & (56, 150, 79)^2 & (72, 234, 177)^2 & (72, 186, 57)^2 \\
(6, 78, 96)^2 & (16, 48, 56)^2 & (48, 192, 168)^2 & (48, 120, 12)^2 \\
(54, 318, 384)^2 & (56, 312, 376)^2 & (72, 360, 408)^2 & (72, 312, 372)^2 \\
(6, -50, -96)^2 & (16, 0, -56)^2 & (48, 176, 168)^2 & (48, 104, -12)^2
\end{bmatrix}
$$

of squares of quadratic forms, $(a, b, c) = au^2 + buv + cv^2$, which yields an infinity of solutions with $m = n = 4$. For example, $u = 2$, $v = 1$ gives

$$
\begin{bmatrix}
627^2 & 603^2 & 933^2 & 717^2 \\
276^2 & 216^2 & 744^2 & 444^2 \\
1236^2 & 1224^2 & 1416^2 & 1284^2 \\
172^2 & 8^2 & 712^2 & 388^2
\end{bmatrix}
$$

Lagrange, in a letter dated 83-03-13, sends the matrices

$$
\begin{bmatrix}
59^2 & 112^2 & 144^2 & 207^2 & 592^2 & 1351^2 & 4077^2 \\
229^2 & 248^2 & 264^2 & 303^2 & 632^2 & 1369^2 & 4083^2 \\
499^2 & 508^2 & 516^2 & 537^2 & 772^2 & 1439^2 & 4107^2
\end{bmatrix}
$$

and

$$
\begin{bmatrix}
18^2 & 234^2 & 346^2 & 514^2 \\
282^2 & 366^2 & 446^2 & 586^2 \\
477^2 & 531^2 & 589^2 & 701^2 \\
1122^2 & 1146^2 & 1174^2 & 1234^2
\end{bmatrix}
$$

In these examples, the a_i, b_j are not squares. If the a_i, b_j are themselves squares, then they provide configurations relevant to **D20** (which see) where Lagrange & Leech have made considerable progress. Their triad and tetrad of squares a_i^2 $(i = 1, 2, 3)$ and b_j^2 $(j = 1, 2, 3, 4)$ with all $a_i^2 + b_j^2$ squares lead to a 4×5 array

$$
\begin{bmatrix}
0^2 & 7422030^2 & 8947575^2 & 22276800^2 & 44142336^2 \\
9282000^2 & 11184530^2 & 12892425^2 & 24132200^2 & 45107664^2 \\
26822600^2 & 27830530^2 & 28275625^2 & 34867000^2 & 51652664^2 \\
60386040^2 & 60840450^2 & 61045335^2 & 64364040^2 & 74799864^2
\end{bmatrix}
$$

in the present problem.

D15 Numbers whose sums in pairs make squares.

Erdős & Leo Moser (and see earlier references) also asked the analogous question: are there, for every n, n distinct numbers such that the sum of any pair is a square? For $n = 3$ we can take

$$a_1 = \tfrac{1}{2}(q^2 + r^2 - p^2) \quad a_2 = \tfrac{1}{2}(r^2 + p^2 - q^2) \quad a_3 = \tfrac{1}{2}(p^2 + q^2 - r^2)$$

and for $n = 4$ we may augment these by taking s to be any number expressible as the sum of two squares in three distinct ways

$$s = u^2 + p^2 = v^2 + q^2 = w^2 + r^2 \text{ and } a_4 = s - \tfrac{1}{2}(p^2 + q^2 - r^2).$$

Jean Lagrange has given a quite general parametric solution for $n = 5$ and a simplification of it which appears to give a majority of all solutions. He tabulates the first 80 solutions, calculated by J.-L. Nicolas. The smallest is

$$-4878 \quad 4978 \quad 6903 \quad 12978 \quad 31122$$

and the smallest positive solution (at most one number can be negative) is

$$7442 \quad 28658 \quad 148583 \quad 177458 \quad 763442.$$

In a letter dated 72-05-19 he sends the following solution for $n = 6$:

$$-15863902 \quad 17798783 \quad 21126338 \quad 49064546 \quad 82221218 \quad 447422978$$

In fact the problem goes back to T. Baker who found five integers whose sums in pairs were squares, and C. Gill who found five whose sums in threes were squares.

$$1917678, \ 2052219, \ 4152603, \ -1981797, \ 70203$$

which yield the squares of 2850, 1410, 2010, 2022, 2478, 78, 2055, 2505, 375 and 1497. He suggested that it ought to be possible, using his method of trigonometric functions, to find 5 *positive* integers with this property "but the complexity of the formulas do not encourage the attempt, and the object seems unworthy of the effort." With modern equipment Wagon was able to make the effort. His first results involved 48- and 49-digit numbers, but he later discovered the quintuple

$$26072323311568661931, 43744839742282591947, 118132654413675138222,$$
$$186378732807587076747, 519650114814905002347$$

whose sums in threes are squares. What is the smallest solution?

Diophantus, V, 17, gave a method to divide a given number into four parts such that the sum of any three of the parts is a (rational) square. Kausler extended this to n parts, the sum of any $n - 1$ a square.

Lagrange also found sets of n squares of which any $n - 1$ have their sum square. For $n = 3$, 5 and 8, the smallest such are the squares of (44,117,240), (28,64,259,392,680) and (79,112,204,632,896,916,1828,2092).

Martin LaBar asked for a proof or disproof that a 3×3 magic square can be constructed from nine distinct integer squares. This requires that the nine quantities x^2, y^2, z^2, $y^2 + z^2 - x^2$, $z^2 + x^2 - y^2$, $x^2 + y^2 - z^2$, $2x^2 - y^2$, $2x^2 - z^2$, $3x^2 - y^2 - z^2$ be distinct perfect squares; John Robertson showed that this is equivalent to the existence of three points, doubles of those of other points in the group law, whose x-coordinates are in A.P., on a 'congruent number' type (see **D27**) elliptic curve.

Duncan Buell searched for a 'magic hour-glass'

$$
\begin{array}{ccc}
a - b & a + b + c & a - c \\
 & a & \\
a + c & a - b - c & a + b
\end{array}
$$

with all seven entries squares, but found none with $a < 25 \cdot 10^{24}$.

Bremner found a magic square with seven square entries :

$$
\begin{array}{ccc}
373^2 & 289^2 & 565^2 \\
360721 & 425^2 & 23^2 \\
205^2 & 527^2 & 222121
\end{array}
$$

Michael Schweitzer showed that any such square must have entries with at least 9 decimal digits. He gives the following specimens in which only one diagonal fails:

$$
\begin{array}{ccc}
127^2 & 46^2 & 58^2 \\
2^2 & 113^2 & 94^2 \\
74^2 & 82^2 & 97^2
\end{array}
\qquad
\begin{array}{ccc}
188^2 & 194^2 & 118^2 \\
4^2 & 148^2 & 254^2 \\
226^2 & 164^2 & 92^2
\end{array}
\qquad
\begin{array}{ccc}
282^2 & 291^2 & 174^2 \\
6^2 & 222^2 & 381^2 \\
339^2 & 246^2 & 138^2
\end{array}
$$

The magic totals are squares in each case: 147^2, 294^2, 441^2. Does this have to happen?

Andrew Bremner gave the following parametric family of magic squares with seven of the eight sums equal.

$\{\{a, b, c\}, \{d, e, f\}, \{g, h, i\}\}$ is the notation for the square which has the three rows $\{a, b, c\}$, $\{d, e, f\}$ and $\{g, h, i\}$. Only the principal diagonal fails. For it to be truly magic, t must be a root of a computable polynomial of degree 34.

$$
\begin{aligned}
&\{\{t^2(1249 - 2032t^2 + 2824t^4 - 4576t^6 - 392t^8 + 2624t^{10} - 992t^{12} + 128t^{14} + 16t^{16})^2, \\
&\quad 4(-2 + 8t^2 + t^4)^2(-1 - 8t^2 + 2t^4)^2(5 - 2t^2 + 2t^4)^2(-7 + 4t^2 + 2t^4)^2, \\
&\qquad 4t^2(-1 - 8t^2 + 2t^4)^2(5 - 2t^2 + 2t^4)^4(-7 + 4t^2 + 2t^4)^2\}, \\
&\quad \{4t^2(-1 - 8t^2 + 2t^4)^4(5 - 2t^2 + 2t^4)^2(-7 + 4t^2 + 2t^4)^2, \\
&t^2(1201 - 2728t^2 + 1168t^4 + 2384t^6 + 664t^8 - 2272t^{10} + 832t^{12} - 64t^{14} + 16t^{16})^2, \\
&\quad 4(-1 - 8t^2 + 2t^4)^2(5 - 2t^2 + 2t^4)^2(-7 + 4t^2 + 2t^4)^2(-2 - 4t^2 + 7t^4)^2\}, \\
&\quad \{4(-1 - 8t^2 + 2t^4)^2(5 - 2t^2 + 2t^4)^2(-7 + 4t^2 + 2t^4)^2(2 - 2t^2 + 5t^4)^2, \\
&\qquad 4t^2(-1 - 8t^2 + 2t^4)^2(5 - 2t^2 + 2t^4)^2(-7 + 4t^2 + 2t^4)^4, \\
&t^2(-1151 + 3488t^2 + 1240t^4 - 5632t^6 + 3448t^8 - 256t^{10} + 352t^{12} - 256t^{14} + 16t^{16})^2\}\}
\end{aligned}
$$

He also gave the 4×4 magic square of squares:

$$
\begin{array}{cccc}
37^2 & 23^2 & 21^2 & 22^2 \\
1^2 & 18^2 & 47^2 & 17^2 \\
38^2 & 11^2 & 13^2 & 33^2 \\
3^2 & 43^2 & 2^2 & 31^2
\end{array}
$$

It's implicit in the work of Carmichael that there can be no 3×3 magic square with entries which are cubes or are fourth powers.

Magic squares continue to fascinate both amateurs and professionals. The following example, due to David Collison, is 'trimagic' in the sense that it is magic and stays so when you either square or cube the entries.

1160	1189	539	496	672	695	57	10	11	58	631	654	515	558	1123	1152
531	560	675	632	43	66	1179	1132	1133	1180	2	25	651	694	494	523
1155	1089	422	379	831	767	92	45	91	44	790	808	403	360	1118	1126
832	766	99	56	1154	1090	415	368	414	367	1113	1131	80	37	795	803
1106	1135	411	454	716	739	27	74	75	28	757	780	473	430	1143	1172
409	438	717	760	19	42	1115	1162	1163	1116	60	83	779	736	446	475
999	1007	192	235	977	995	164	211	163	210	1018	954	173	216	1036	970
982	990	175	218	994	1012	181	228	180	227	1035	971	156	199	1019	953
183	191	991	1034	195	213	963	1010	962	1009	236	172	972	1015	220	154
200	208	974	1017	178	196	980	1027	979	1026	219	155	955	998	237	171
715	744	20	63	1107	1130	418	465	466	419	1148	1171	82	39	752	781
18	47	1108	1151	410	433	724	771	772	725	451	474	1170	1127	55	84
101	35	1153	1110	423	359	823	776	822	775	382	400	1134	1091	64	72
424	358	830	787	100	36	1146	1099	1145	1098	59	77	811	768	387	395
667	696	46	3	1165	1188	550	503	504	551	1124	1147	22	65	630	659
38	67	1168	1125	536	559	686	639	640	687	495	518	1144	1187	1	30

Has anyone constructed a $5 \times 5 \times 5$ magic cube, or proved its impossibility? Rich Schroeppel notes that the centre cell of a magic 9^5 is always the average cell value, and that a corollary is that there is no magic 9^6.

T. Baker, *The Gentleman's Diary* or *Math. Repository*, London, 1839, 33–35, Question 1385.

Andrew Bremner, On arithmetic progressions on elliptic curves, *Experiment. Math.*, **8**(1999) 409–413; *MR* **2000k**:11068.

Andrew Bremner, On squares of squares, *Acta Arith.*, **88**(1999) 289–297; *MR* **2000f**:11031; II, **99**(2001) 289–308; *MR* **2002d**:11057.

A. Bremner, J. H. Silverman & N. Tzanakis, Integral points in arithmetic progression on $y^2 = x(x^2 - n^2)$, *J. Number Theory*, **80**(2000) 187–208; *MR* **2001i**:11066.

R. D. Carmichael, *Diophantine Analysis*, Dover, 1915, pp. 22 & 70–72.

Watne Chan & Peter Loly, Iterative compounding of square matrices to generare large order magic squares, *Math. Today*, **4**(2002) 113–118.

C. Gill, *Application of the Angular Analysis to Indeterminate Problems of Degree 2*, New York, 1848, p. 60.

John R. Hendricks, David M. Collison's trimagic square, *Math Teacher*, **95** (2002) 406.

Katalin Gyarmati, On a problem of Diophantus, *Acta Arith.*, **97**(2001) 53–65; *MR* **2002d**:11031.

C. F. Kausler, *Mém. Acad. Sci. St. Pétersbourg*,1(1809) 271–282.

Martin LaBar, Problem 270, *Coll. Math. J.*, **15**(1984) 69.

Jean Lagrange, Cinq nombres dont les sommes deux à deux sont des carrés, *Séminaire Delange-Pisot-Poitou* (*Théorie des nombres*) 12e année, **20**(1970-71) 10pp.

Jean Lagrange, Six entiers dont les sommes deux à deux sont des carrés, *Acta Arith.*, **40**(1981) 91–96.

Jean Lagrange, Sets of n squares of which any $n - 1$ have their sum square, *Math. Comput.*, **41**(1983) 675–681.

Jean-Louis Nicolas, 6 nombres dont les sommes deux à deux sont des carrés, *Bull. Soc. Math. France*, Mém. No 49–50 (1977) 141–143; *MR* **58** #482.

John P. Robertson, Magic squares of squares, *Math. Mag.*, **69**(1996) 289–293.

Lee Sallows, The lost theorem, *Math. Intelligencer*, **19**(1997) 51–54.

A. R. Thatcher, A prize problem, *Math. Gaz.*, **61**(1977) 64.

A. R. Thatcher, Five integers which sum in pairs to squares, *Math. Gaz.*, **62**(1978) 25.

Stan Wagon, Quintuples with square triplets, *Math. Comput.*, **64**(1995) 1755–1756; *MR* **95m**:11040.

D16 Triples with the same sum and same product.

The problem to find as many different triples of positive integers as possible with the same sum and the same product has been solved by Schinzel: you can have arbitrarily many. In the interim Stephane Vandemergel found 13 triples each with sum 17116 and product $2^{10}3^35^27^211 \cdot 13 \cdot 19$. It may be of interest to ask for the smallest sums or products with each multiplicity. For example, for 4 triples, J. G. Mauldon finds the smallest common sum to be 118: (14,50,54), (15,40,63), (18,30,70), (21,25,72) and the smallest common product to be 25200: (6,56,75), (7,40,90), (9,28,100), (12,20,105).

Schinzel's solution is derived from points on the elliptic curve $y^2 = x^3 - 9x + 9$ (324C1 in Cremona, rank 1, generator (1,1), inflexion points $(3, \pm 3)$).

Lorraine L. Foster & Gabriel Robins, Solution to Problem E2872, *Amer. Math. Monthly* **89**(1982) 499–500.

J. G. Mauldon, Problem E2872, *Amer. Math. Monthly*, **88**(1981) 148.

Andrzej Schinzel, Triples of positive integers with the same sum and the same product, *Serdica Math. J.*, **22**(1996) 587-588; *MR* **98g**:11033.

D17 Product of blocks of consecutive integers not a power.

Erdős & Selfridge have proved that the product of consecutive integers is never a power, and the binomial coefficient $\binom{n}{k}$ (see **B31**) is never a power for $n \geq 2k \geq 8$. If $k = 2$, then $\binom{n}{k}$ is a square infinitely often, but Tijdeman's methods (see **D9**) will probably show that it is never a nontrivial higher power (for cubes and fourth powers, see Mordell's book), and that $k = 3$ never gives a power, apart from $n = 50$ (see **D3**).

Erdős & Graham ask if the product of two or more disjoint blocks of consecutive integers can be a power. Pomerance has noted that

$$\prod_{i=1}^{4}(a_i - 1)a_i(a_i + 1)$$

is a square if $a_1 = 2^{n-1}$, $a_2 = 2^n$, $a_3 = 2^{2n-1} - 1$, $a_4 = 2^{2n} - 1$, but Erdős & Graham suggest that if $l \geq 4$, then $\prod_{i=1}^{k}\prod_{j=1}^{l}(a_i + j)$ is a square on only a finite number of occasions.

K. R. S. Sastry notes that the product of the blocks $(n-1)n(n+1)$ and $(2n-2)(2n-1)2n$ is a square if $(n+1)(2n-1) = m^2$. This is equivalent to a Bhaskara equation (aka Pell equation) with an infinity of solutions. E.g. $n = 74$ gives

$$(73 \cdot 74 \cdot 75)(146 \cdot 147 \cdot 148) = 73^2 \cdot 74^2 \cdot 210^2$$

Erdős also asks if the product of (more than one) consecutive odd numbers is never a power (higher than the first)? Is the product of 4 consecutive members of an A.P. never a power? Euler showed that it cannot be a square. Fermat had shown that the members cannot be squares individually, while a nonsquare divisor must divide two distinct terms, either (a) 2 divides the first & third or the second & fourth, or (b) 3 divides the first & fourth, or both. (a) alone is impossible mod 8, (b) alone is impossible mod 3, but we could have $6t^2$, u^2, $2v^2$, $3w^2$. But this implies $w^2 + t^2 = v^2$ and $w^2 + 4t^2 = u^2$, which can be disproved by descent – one Pythagorean ratio can't be twice another. For higher powers Leech notes that we can't have three cubes in A.P.

Sastry asks for which k can the product of four consecutive terms of an A.P. be a k-gonal number, where the r-th **k-gonal number** is

$$\tfrac{1}{2}r((k - 2)r - (k - 4)).$$

Not for $k = 4$, as Euler showed, but Sastry finds solutions for all other k except 7, 14 and 37. Are these impossible?

Saradha, Shorey & Tijdeman show that, apart from

$$(k + 1)(k + 2) \cdots (2k) = 2 \cdot 6 \cdots (4k - 2) \quad \text{for} \quad k = 2, 3, 4, \ldots,$$

there are only finitely many A.P.s with given differences of equal lengths ≥ 2 and with equal products.

Bennett, Györy & Hajdu show that the product of consecutive members of a k-term A.P. is not a power for $4 \le k \le 11$, except in cases where there is a common factor, as in $9 \cdot 18 \cdot 27 \cdot 36 = 54^3$.

In a 93-05-07 letter to Ron Graham, Nobuhisa Abe states that $x(x+1)\cdots(x+k) = y^2 - 1$ has the unique solution $(x,y) = (2,71)$ for $k = 5$ and no solutions for $k = 7$ or 11. Ákos Pintér has a method for dealing with the equation when k is odd.

Yismaw Alemu, On the diophantine equation $x(x+1)\cdots(x+n) = y^2 - 1$, *SINET* **18**(1995) 1–22; *MR* **96k**:11034.

Michael A. Bennett, Kálmán Györy & Lajos Hajdu, Powers from products of consectutive terms in arithmetic progression, *J. reine andew. Math.*, submitted June 2003.

B. Brindza, *Publ. Math. Debrecen* **39**(1991) 159–162; *MR* **92j**:11029.

Ajai Choudhry, On arithmetic progressions of equal lengths and equal products of terms, *Acta Arith.*, **82**(1997) 95–97; *MR* **98j**:11004.

M. J. Cohen, The difference between the product of n consecutive integers and the nth power of an integer, *Comput. Math. Appl.*, **39**(2000) 139–157; *MR* **2002b**:11043.

Javier Barja Pérez, J. M. Molinelli & A. Blanco Ferro, Finiteness of the number of solutions of the Diophantine equation $\binom{x}{n} = \binom{y}{2}$, *Bull. Soc. Math. Belg. Sér. A* **45**(1993) 39–43; *MR* **96a**:11027.

H. Darmon & L. Merel, Winding quotients and some variants of Fermat's last theorem, *J. reine angew. Math.*, **490**(1997) 81–100; *MR* **98h**:11076 .

P. Erdős, On consecutive integers, *Nieuw Arch. Wisk.*, **3**(1955) 124–128.

P. Erdős & J. L. Selfridge, The product of consecutive integers is never a power, *Illinois J. Math.*, **19**(1975) 292–301.

P. Filakovszki & L. Hajdu, The resolution of the Diophantine equation $x(x+d)\cdots(x+(k-1)d) = by^2$ for fixed d, *Acta Arith.*, **98**(2001) 151–154; *MR* **2002c**:11028.

Ja. Gabovič, On arithmetic progressions with equal products of terms, (Russian) *Colloq. Math.* **15**(1966) 45–48; *MR* **33** # 2594.

K. Györy, On the diophantine equation $n(n+1)\cdots(n+k-1) = bx^l$, *Acta Arith.*, **83**(1998) 87–92; *MR* **99a**:11036.

K. Györy, Power values of products of consecutive integers and binomial coefficients. *Number Theory and its Applications, (Kyoto 1997)* 145–156, *Dev. Math.*, **2**, Kluwer Acad. Publ., Dordrecht 1999; *MR* **2001a**:11050.

G. Hanrot, N. Saradha & T. N. Shorey, Almost perfect powers in consecutive integers, *Acta Arith.*, **99**(2001) 13–25;*MR* **2002m**:11021.

Jin Yuan, A note on the product of consecutive elements of an arithmetic progression, *Ann. Univ. Sci. Budapest. Sect. Comput.*, **19**(2000) 133–141; *MR* **2003b**:11005.

Koh Young-Mee & Ree Sang-Wook, Divisors of the products of consecutive integers, *Commun. Korean Math. Soc.*, **17**(2002) 541–550.

R. A. MacLeod & I. Barrodale, On equal products of consecutive integers, *Canad. Math. Bull.*, **13**(1970) 255–259; *MR* **42** #193.

R. Marszalek, On the product of consecutive elements of an arithmetic progression, *Monatsh. Math.*, **100**(1985) 215–222; *MR* **87f**:11010.

Richárd Obláth, Über das Produkt fünf aufeinander folgender Zahlen in einer arithmetischen Reihe, *Publ. Math. Debrecen*, **1**(1950) 222–226; *MR* **12** 590e.

Richárd Obláth, Eine Bemerkung über Produkte aufeinander folgender Zahlen, *J. Indian Math. Soc. (N.S.)*, **15**(1951), 135–139 (1952); *MR* **13** 823a.

Csaba Rakaczki, On the Diophantine equation $x(x-1)\cdots(x-(m-1)) = \lambda y(y-1)\cdots(y-(n-1)) + l$, *Acta Arith.*, **110**(2003) 339–360.

N. Saradha, Squares in products with terms in an arithmetic progression. *Acta Arith.* 86 (1998) 27–43; *MR* **99f**:11044.

N. Saradha, On perfect powers in products with terms from arithmetic progressions, *Acta Arith.*, **82**(1997) 147–172; *MR* **99d**:11030.

N. Saradha, On blocks of arithmetic progressions with equal products. *J. Théor. Nombres, Bordeaux*, **9**(1997) 183–199; *MR* **98g**:11012.

N. Saradha & T. N. Shorey, On the ratio of two blocks of consecutive integers, *Proc. Indian Acad. Sci. Math. Sci.*, **100**(1990) 107–132; *MR* **91i**:11033.

N. Saradha & T. N. Shorey, The equations $(x+1)\cdots(x+k)=(y+1)\cdots(y+mk)$ with $m = 3, 4$, *Indag. Math. (N.S.)*, **2**(1991) 489–510; *MR* **93f**:11029.

N. Saradha & T. N. Shorey, On the equation $x(x+d_1)\cdots(x+(k-1)d_1) = y(y+d_2)\cdots(y+(mk-1)d_2)$, *K. G. Ramanathan memorial issue, Proc. Indian Acad. Sci. Math. Sci.*, **104**(1994) 1–12; *MR* **95e**:11039.

N. Saradha & T. N. Shorey, Almost perfect powers in arithmetic progression, *Acta Arith.*, **99**(2001) 363–388;*MR* **2003a**:11029.

N. Saradha & T. N. Shorey, Almost squares in arithmetic progression, *Compositio Math.*, **138**(2003) 73–111.

N. Saradha & T. N. Shorey, Almost squares and factorisations in consecutive integers, *Compositio Math.*, **138**(2003) 113–124.

N. Saradha, T. N. Shorey & R. Tijdeman, On values of a polynomial at arithmetic progressions with equal products, *Acta Arith.*, **72**(1995) 67–76; *MR* **96j**:11041.

N. Saradha, T. N. Shorey & R. Tijdeman, On arithmetic progressions with equal products, *Acta Arith.*, **68**(1994) 89–100; *MR* **96f**:11047.

N. Saradha, T. N. Shorey & R. Tijdeman, On the equation $x(x+1)\cdots(x+k-1) = y(y+d)\cdots(y+(mk-1)d), m = 1, 2$, *Acta Arith.*, **71**(1995) 89–100; *MR* **96d**:11036.

N. Saradha, T. N. Shorey & R. Tijdeman, On arithmetic progressions of equal lengths with equal products, *Math. Proc. Cambridge Philos. Soc.*, **117**(1995) 193–201; *MR* **95j**:11029.

T. N. Shorey & R. Tijdeman, *Exponential Diophantine equations*, Cambridge Tracts in Mathematics, **87**, Cambridge University Press, 1986; *MR* **88h**:11002.

K. Szymiczek, *Elem. Math.*, **27**(1972) 11–12; *MR* **46** #130.

Hideo Wada, On the Diophantine equation $x(x+1)\cdots(x+n)+1 = y^2$ $(17 \le n = \text{odd} \le 27)$, *Proc. Japan Acad. Ser. A Math. Sci.*, **79**(2003) 99–100.

Yuan Ping-Zhi, On a special Diophantine equation $a\binom{x}{n} = by^r + c$, *Publ. Math. Debrecen* **44**(1994) 137–143; *MR* **95j**:11032.

D18 Is there a perfect cuboid? Four squares whose sums in pairs are square. Four squares whose differences are square.

Is there a rational box? Our treatment of this notorious unsolved problem is owed almost entirely to John Leech. Does there exist a **perfect cuboid**, with integer edges x_i, face diagonals y_i and body diagonal z; are there solutions of the simultaneous diophantine equations

(A) $$x_{i+1}^2 + x_{i+2}^2 = y_i^2,$$

(B) $$\sum x_i^2 = z^2 \quad ?$$

($i = 1, 2, 3$; and where necessary, subscripts are reduced modulo 3.)

Martin Gardner asked if any six of x_i, y_i, z could be integers. Here there are three problems: just the body diagonal z irrational; just one edge x_3 irrational; just one face diagonal y_1 irrational.

Problem 1. We require solutions to the three equations (A). Suppose such solutions have **generators** a_i, b_i where

$$x_{i+1} : x_{i+2} : y_i = 2a_i b_i : a_i^2 - b_i^2 : a_i^2 + b_i^2$$

Then we want integer solutions of

(C) $$\prod \frac{a_i^2 - b_i^2}{2a_i b_i} = 1.$$

We can assume that the generator pairs have opposite parity and replace (C) by

(D) $$\frac{a_1^2 - b_1^2}{2a_1 b_1} \cdot \frac{a_2^2 - b_2^2}{2a_2 b_2} = \frac{\alpha^2 - \beta^2}{2\alpha\beta}$$

An example is

(E) $$\frac{6^2 - 5^2}{2 \cdot 6 \cdot 5} \cdot \frac{11^2 - 2^2}{2 \cdot 11 \cdot 2} = \frac{8^2 - 5^2}{2 \cdot 8 \cdot 5}$$

Kraitchik gave $241 + 18 - 2$ cuboids with odd edge less than a million. Lal & Blundon listed all cuboids obtainable from (D) with a_1, b_1; α, $\beta \leq 70$, including the curious pair $(1008, 1100, 1155)$, $(1008, 1100, 12075)$. Leech has deposited a list of all solutions of (D) with two pairs of a_1, b_1; a_2, b_2; α, $\beta \leq 376$.

Reversal of the cyclic order of subscripts in (C) leads to the **derived cuboid**: example (E) gives the least solution $(240, 44, 117)$, known to Euler, and the derived cuboid $(429, 2340, 880)$. Note that $240 \cdot 429 = 44 \cdot 2340 = 117 \cdot 880$.

Several parametric solutions are known: the simplest, also known to Euler, is

(F) $\qquad\qquad \alpha = 2(p^2 - q^2), \qquad a_1 = 4pq, \qquad b_1 = \beta = p^2 + q^2$

For a_1, b_1 fixed, (D) is equivalent to the plane cubic curve

$$\frac{a_1^2 - b_1^2}{2a_1 b_1} = \frac{u^2 - 1}{2u} \cdot \frac{2v}{v^2 - 1}$$

whose rational points are finitely generated, so Mordell tells us that one solution leads to an infinity. But not all rationals a_1/b_1 occur in solutions: $a_1/b_1 = 2$ is impossible, so there is no rational cuboid with a pair of edges in the ratio 3:4.

Problem 2. Just an edge irrational. We want $x_1^2 + x_2^2 = y_3^2$ with $t + x_1^2$, $t + x_2^2$, $t + y_3^2$ all squares. This was proposed by "Mahatma" and readers gave $x_1 = 124$, $x_2 = 957$, $t = 13852800$. Bromhead extended this to a parametric solution. An infinity of solutions is given by

(G) $\qquad (x_1, x_2, y_3) = 2\xi\eta\zeta(\xi, \eta, \zeta), \qquad t = \zeta^8 - 6\xi^2\eta^2\zeta^4 + \xi^4\eta^4$

where (ξ, η, ζ) is a Pythagorean triple. The simplest such is $\xi = 5$, $\eta = 12$; $x_1 = 7800$, $x_2 = 18720$; $t = 211773121$. An earlier solution was given by Flood.

These are not all. We seek solutions of

(H) $\qquad\qquad x_1^2 + x_2^2 = y_3^2, \qquad z^2 = x_1^2 + y_1^2 = x_2^2 + y_2^2$

other than $z = y_3$ $(t = 0)$. Leech found 100 primitive solutions with $z < 10^5$, 46 of which had $t > 0$. The generators for (H) satisfy

(I) $\qquad\qquad \dfrac{\alpha_1^2 + \beta_1^2}{2\alpha_1\beta_1} \cdot \dfrac{2\alpha_2\beta_2}{\alpha_2^2 + \beta_2^2} = \dfrac{a^2 - b^2}{2ab}$

so solutions for fixed x_2/x_1 correspond to rational points on the cubic curve

(J) $\qquad\qquad x_1 v(u^2 + 1) = x_2 u(v^2 + 1)$

The trivial solution $t = 0$ corresponds to several **ordinary points** which generate an infinity of solutions for each ratio x_2/x_1. Solutions form cycles of four

(K) $\qquad \zeta^2 = \xi_1^2 + \eta_1^2 = \xi_2^2 + \eta_2^2 = \xi_3^2 + \eta_3^2 = \xi_4^2 + \eta_4^2, \qquad \xi_1\xi_3 = \xi_2\xi_4,$
$\qquad\qquad \xi_1^2 + \xi_2^2, \ \xi_2^2 + \xi_3^2, \ \xi_3^2 + \xi_4^2, \ \xi_4^2 + \xi_1^2 \qquad$ all squares,

corresponding to two pairs x_2/x_1 which correspond to two points collinear with the point $t = 0$ on (J). Conversely, such a pair of points corresponds to a cycle of four solutions.

Problem 3. Just one face diagonal irrational. There are two closely related problems: find three integers whose sums and differences are all squares; find three squares whose differences are squares. The cuboid form of the problem asks for integers satisfying

(L) $\qquad x_2^2 + y_2^2 = z^2, \qquad x_1^2 + x_3^2 = y_2^2, \qquad x_1^2 + x_2^2 = y_3^2$

whose generators satisfy

(M) $\qquad \dfrac{\alpha_1^2 - \beta_1^2}{2\alpha_1\beta_1} \cdot \dfrac{\alpha_3^2 - \beta_3^2}{2\alpha_3\beta_3} = \dfrac{\alpha_2^2 + \beta_2^2}{2\alpha_2\beta_2}.$

Write $u_i = (\alpha_i^2 - \beta_i^2)^2/4\alpha_i^2\beta_i^2$ and (M) becomes $u_1 u_3 = 1 + u_2$, an equation investigated by many in the contexts of **cycles** and **frieze patterns**. Solutions occur in cycles of five! Leech listed 35 such with $\alpha_1, \beta_1, \alpha_2, \beta_2 \leq 50$ and deposited in UMT a list of all cycles with two pairs $\alpha_i, \beta_i \leq 376$. There is a close connexion with Napier's rules and the construction of rational spherical triangles.

Solutions to (L) are given by $x_2^2 = z^2 - y_2^2 = (p^2 - q^2)(r^2 - s^2)$, $x_3^2 = z^2 - y_3^2 = 4pqrs$, when the products of the numerators and denominators of $(p^2 - q^2)/2pq$ and $(r^2 - s^2)/2rs$ are each squares. Euler made p, q, r, s squares and found differences of fourth powers, e.g., $3^4 - 2^4$, $9^4 - 7^4$, $11^4 - 2^4$, whose products in pairs are squares. The first two give the second smallest solution (117,520,756) of this type, whose cycle includes the smallest (104,153,672), also known to Euler.

We can also express $z^2 = x_2^2 + y_2^2 = x_3^2 + y_3^2$ as the sum of two squares in two different ways: $x_3/x_1 = (\alpha_2^2 - \beta_2^2)/2\alpha_2\beta_2$, $x_2/x_1 = (\alpha_3^2 - \beta_3^2)/2\alpha_3\beta_3$ give $z^2 = 4(\alpha_2^2\alpha_3^2 + \beta_2^2\beta_3^2)(\alpha_2^2\beta_3^2 + \alpha_3^2\beta_2^2)$. Euler made each factor square and found two rational right triangles of equal area $\frac{1}{2}\alpha_2\alpha_3\beta_2\beta_3$. Diophantus solved this with $\beta_2/\alpha_2 = (s+t)/2r$, $\beta_3/\alpha_3 = s/t$, where $r^2 = s^2 + st + t^2$, $s = l^2 - m^2$, $t = m^2 - n^2$. Put $(l, m, n) = (1, 2, -3)$ and we have a cycle containing the third, fourth and fifth smallest these cuboids. Leech found 89 with $z < 10^5$.

For fixed α_1/β_1 a nontrivial solution to (M) corresponds to an ordinary point, which generates an infinity of solutions, on the curve

$$(\alpha_1^2 - \beta_1^2)u(v^2 - 1) = 2\alpha_1\beta_1 v(u^2 + 1).$$

The tangent at the point generates a cycle of special interest.

(M), like (D), but unlike (I), does not have nontrivial solutions for all ratios α/β; e.g., there are none with $\alpha/\beta = 2$ or 3 and again no cuboids with edges in the ratio 3:4. Here we do not have "derived" cuboids.

Two other parametric forms for ratios $(p^2 - q^2)/2pq$ whose product and quotient are squares, are

$$p = 2m^2 \pm n^2, \qquad r = m^2 \pm 2n^2, \qquad q = s = m^2 \mp n^2.$$

Four squares whose sums in pairs are square. A solution of (C) gives three such squares; it may be portrayed as a trivalent **vertex** of a graph, the three edges joining it to **nodes** representing generator pairs for rational triangles. If such a pair occurs in one solution it occurs in infinitely many, so the valence of a *node* is infinite. We seek a subgraph homeomorphic to a tetrahedron K_4 whose vertices give four solutions of (C) and whose edges contain nodes corresponding to generator pairs common to pairs of solutions of (C). Lists of solutions have revealed no such subgraph; indeed, not even a closed circuit! Until a circuit is encountered, we need not distinguish between the pairs a, b and $a \pm b$.

So no example of four such squares is known. Construction of four squares with five square sums of pairs is straightforward.

A. R. Thatcher related the problem to the equation $y^2 = -x^8 + 35x^4 - 25$. The only integer solutions are $(\pm 1, \pm 3)$, but there may be a finite number of other rational solutions. Even if not, this does not preclude solutions to the original problem.

Four squares whose differences are square. This problem extends (M) analogously to the above extension of (C). A *vertex* is now *pentavalent*, adjacent to a cycle of five nodes or generator pairs. The cyclic order of the edges is important, but not the sense of rotation. A solution of (M) corresponds to three *consecutive* edges, and here we must distinguish between α, β and $\alpha \pm \beta$: the corresponding nodes are joined with double edges in Figure 11. The nodes are again of infinite valence. Four squares of the required type would correspond to a subgraph homeomorphic to K_6, with six vertices and 15 nodes, one on each edge. The only cycle so far found is shown in Figure 11 and this will *not* serve as part of such a subgraph. Although a solution is unlikely, there do not appear to be any congruence conditions which forbid it.

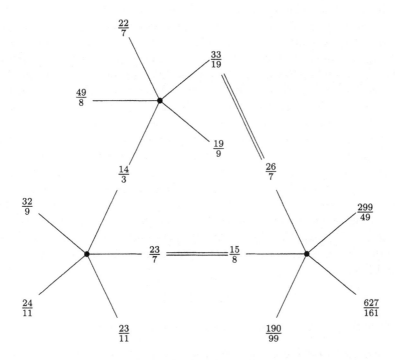

Figure 11. Three Cycles of Five Generator Pairs.

Though there is now no necessary connexion between the generator pairs α, β and $\alpha \pm \beta$, solutions of (M) containing both pairs do occur. Such a solution leads to a sequence of four squares with sums of two or three *consecutive* terms all square. How long can such a sequence be? For more than four we need a sequence of 5-cycles of solutions of (M) each containing adjacent generator pairs α, β; $\bar{\alpha}$, $\bar{\beta}$ where $\alpha \pm \beta$, $\bar{\alpha} \pm \bar{\beta}$ belong one each to the neighboring cycles. Leech found the sequence

$(56,31)(17,6); (23,11)(23,7); (15,8)(26,7); (33,19)(77,19); (48,29)(35,4); (39,31)(13,9)$

where $23, 11 = 17 \pm 6$, etc., which gives a sequence of eight such squares. The squares of the edges of a perfect cuboid would form an infinite (periodic) sequence. Four nonzero squares with differences all square would lead to a sequence with terms three apart in constant ratio: an integer ratio would give an infinite sequence. Leech has since given the longer sequence

$(14,1)(224,37); (261,187)(155,132); (287,23)(23,7);$

$(15,8)(26,7); (33,19)(77,19); (48,29)(35,4); (39,31)(13,9)$

with "both ends surprisingly small" and asks "are they really ends?" He has no proof that a pair (a,b) can occur while $(a+b, a-b)$ does not. He has some other sequences of the same length, but so far none longer. Randall Rathbun was unable to extend this sequence before $(14,1)$ or after $(13,9)$, but he was able to extend the earlier sequence to seven terms by prefixing $(26767,2185)(87,25)$; or $(940,693)(87,25)$; before $(56,31)$.

The perfect rational cuboid. None of the known numerical solutions to problems 1, 2 and 3 gives a perfect cuboid, and many parametric solutions, for example (G), can be shown not to yield one. Spohn used Pocklington's work to show that *one* of the two mutually derived cuboids of (F) is not perfect and E. Z. Chein and Jean Lagrange have each shown that the derived cuboid is never perfect. On the other hand, no known parametric solution is complete so impossibility can't be proved from these alone. A solution of the problem in the previous section need not lead to a perfect cuboid. Korec showed that the least edge of a perfect rational cuboid must exceed 10^6. Extensive searches, mainly by Randall Rathbun and Torbjorn Granlund, have shown that a perfect rational cuboid must have all its edges greater than 333750000. During the search, the results of which have been deposited in UMT, $6800 + 6380 + 6749$ solutions of the three problems were found. Recently Korec has shown that the greatest edge is greater than 10^9. Leech amplifies a result of Horst Bergmann to show that the product of the edges, face diagonals and body diagonal must be divisible by

$$2^8 \times 3^4 \times 5^3 \times 7 \times 11 \times 13 \times 17 \times 19 \times 29 \times 37$$

and Dale Shoults notes that the power of 5 can be raised to 5^4,

Unsolved problems. Do three cycles of solutions of (M) exist whose graph is as in Figure 12? Here we've adopted John Leech's convention of writing the ratio of the edges as a fraction, e.g. $\frac{y_1}{x_1}$, where the same pair of generators belongs to both cycles (as the pair 14, 3 in Figure 11), but writing it as a ratio, e.g. $x_2 : x_3$, where a pair of generators belongs to one cycle, and their sum and difference to the other (as 15, 8 and 23, 7 in Figure 11).

Are there ratios $(p^2 - q^2)/2pq$, $(r^2 + s^2)/2rs$ with product and quotient both of the form $(m^2 - n^2)/2mn$? Is there a nontrivial solution of

$$(a^2c^2 - b^2d^2)(a^2d^2 - b^2c^2) = (a^2b^2 - c^2d^2)^2?$$

Such a solution would lead to a perfect cuboid. Is there a 5-cycle of solutions of (M) with

$$\alpha_1/\beta_1 = (p^2 - q^2)/2pq, \quad \alpha_2/\beta_2 = (r^2 + s^2)/2rs?$$

What circuits, if any, occur in the graph of solutions of (C)? What circuits occur in the graph of solutions of (M)? Are there cycles of solutions of (I) other than those of form (K)? What ratios besides 3/4 can*not* occur as ratios of edges of cuboids in Problem 1? In Problem 3? Is there a parallelepiped with all edges, face diagonals and body diagonals rational? Rathbun has found 41 pairs of primitive cuboids in which two edges of one equal two of the other. Twenty-one of these are pairs of solutions to

Problem 1, with the body diagonal irrational; 13 are pairs of solutions to
Problem 2; three are pairs of solutions to Problem 3, with a face diagonal
irrational. Three are solutions to Problems 1 and 3, while the last is a
solution to Problems 1 and 2.

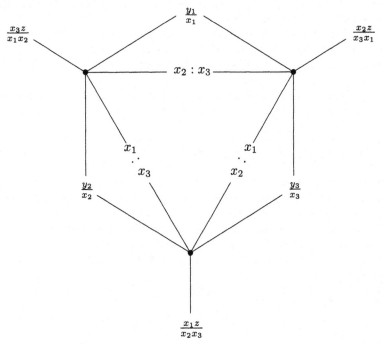

Figure 12. Are there three Cycles of Solutions of (M) like this?

Randall Rathbun has tirelessly pursued the grail of a perfect cuboid. He
has found 4394 cycles of the kind illustrated in Figure 11. He has searched
all integer cuboids with least edge between 44 and $2^{3}2 - 1$. He found no
perfect cuboids among the 17108 body cuboids, 10022 edge cuboids and
16465 face cuboids (total 43595).

Compare end of **D22**.

Andrew Bremner, Pythagorean triangles and a quartic surface, *J. reine
angew. Math.*, **318**(1980) 120–125.

Andrew Bremner, The rational cuboid and a quartic surface, *Rocky Mountain
J. Math.*, **18**(1988) 105–121.

T. Bromhead, On square sums of squares, *Math. Gaz.*, **44**(1960) 219–220;
MR **23** #A1594.

Ezra Brown, $x^4 + dx^2y^2 + y^4$: some cases with only trivial solutions — and
a solution Euler missed, *Glasgow Math. J.*, **31**(1989) 297–307; *MR* **91d**:11026.

W. Burnside, Note on the symmetric group, *Messenger of Math.*, **30**(1900–01)
148–153; *J'buch* **32**, 141–142.

E. Z. Chein, On the derived cuboid of an Eulerian triple, *Canad. Math. Bull.*, **20**(1977) 509–510; *MR* **57** #12375.

W. J. A. Colman, On certain semi-perfect cuboids, *Fibonacci Quart.*, **26** (1988) 54–57 (see also 338–343).

W. J. A. Colman, A perfect cuboid in Gaussian integers, *Fibonacci Quart.*, **32**(1994) 266–268; *MR* **95e**:11038.

J. H. Conway & H. S. M. Coxeter, Triangulated polygons and frieze patterns, *Math. Gaz.*, **57**(1973) 87–94, 175–183 and refs. on pp. 93–94.

H. S. M. Coxeter, Frieze patterns, *Acta Arith.*, **18**(1971) 297–310; *MR* **44** #3980.

L. E. Dickson, *History of the Theory of Numbers*, Vol. 2, Diophantine Analysis, Washington, 1920: ch. 15, ref. 28, p. 448 and cross-refs. to pp. 446–458; ch. 19, refs 1–30, 40–45, pp. 497–502, 505–507.

Marcus Engel, Numerische Lösung eines Quaderproblems, *Wiss. Z. Pädagog. Hochsch. Erfurt/Mühlhausen Math.-Natur. Reihe*, **22**(1986) 78–86; *MR* **87m**: 11021.

P. W. Flood, *Math. Quest. Educ. Times*, **68**(1898), 53.

Martin Gardner, Mathematical Games, *Scientific Amer.*, **223**#1(Jul.1970) 118; correction, #3(Sep.1970)218.

W. Howard Joint, Cycles, Note 1767, *Math. Gaz.*, **28**(1944) 196–197.

Ivan Korec, Nonexistence of a small perfect rational cuboid, I., II., *Acta Math. Univ. Comenian.*, **42/43**(1983) 73–86, **44/45**(1984) 39–48; *MR* **85i**:11004, **86c**:11013.

Ivan Korec, Lower bounds for perfect rational cuboids, *Math. Slovaca*, **42** (1992) 565–582; *MR* **94a**:11196.

Maurice Kraitchik, On certain rational cuboids, *Scripta Math.*, **11**(1945) 317–326; *MR* **8**, 6.

Maurice Kraitchik, *Théorie des nombres*, t.3, Analyse Diophantine et applications aux cuboides rationnels, Paris, 1947.

Maurice Kraitchik, Sur les cuboides rationnels, in *Proc. Internat. Congr. Math.* 1954, Vol. 2, Amsterdam, 33–34.

Jean Lagrange, Sur le dérivé du cuboïde eulerien, *Canad. Math. Bull.*, **22** (1979) 239–241; *MR* **80h**:10022.

Jean Lagrange, Sets of n squares of which any $n-1$ have their sum square, *Math. Comput.*, **41**(1983) 675–681; *MR* **84j**:10012.

Jean Lagrange & John Leech, Pythagorean ratios in arithmetic progression II. Four Pythagorean ratios, *Glasgow Math. J.*, **36**(1994) 45–55; *MR* **99b**:11013.

M. Lal & W. J. Blundon, Solutions of the Diophantine equations $x^2+y^2=l^2$, $y^2+z^2=m^2$, $z^2+x^2=n^2$, *Math. Comput.*, **20**(1966) 144–147; *MR* **32** #4082.

J. Leech, The location of four squares in an arithmetical progression with some applications, in *Computers in Number Theory*, Academic Press, London, 1971, 83–98; *MR* **47** #4913.

J. Leech, The rational cuboid revisited, *Amer. Math. Monthly* **84**(1977) 518–533; corrections (Jean Lagrange) **85**(1978) 473; *MR* **58** #16492.

J. Leech, Five tables related to rational cuboids, *Math. Comput.*, **32**(1978) 657–659.

John Leech, A remark on rational cuboids, *Canad. Math. Bull.*, **24**(1981) 377–378; *MR* **83a**:10022.

John Leech, Four integers whose twelve quotients sum to zero, *Canad. J. Math.*, **38**(1986) 1261–1280; *MR* **88a**:11031; addendum 90-08-18.

J. Leech, Pythagorean ratios in arithmetic progression I. Three Pythagorean ratios, *Glasgow Math. J.*, **35**(1993) 395–407; *MR* **94h**:11002.

R. C. Lyness, Cycles, Note 1581, *Math. Gaz.*, **26**(1942) 62; Note 1847, **29** (1945) 231–233; Note 2952, **45**(1961) 207–209.

"Mahatma", Problem 78, *The A.M.A.* [J. Assist. Masters Assoc. London] **44**(1949) 188; Solutions: J. Hancock, J. Peacock, N. A. Phillips, 225.

Eliakim Hastings Moore, The cross-ratio of $n!$ Cremona-transformations of order $n-3$ in flat space of $n-3$ dimensions, *Amer. J. Math.*, **30**(1900) 279–291; *J'buch* **31**, 655.

L. J. Mordell, On the rational solutions of the indeterminate equations of the third and fourth degrees, *Proc. Cambridge Philos. Soc.*, **21**(1922) 179–192.

H. C. Pocklington, Some diophantine impossibilities, *Proc. Cambridge Philos. Soc.*, **17** (1914) 110–121, esp. p. 116.

W. W. Sawyer, Lyness's periodic sequence, Note 2951, *Math. Gaz.*, **45**(1961) 207.

Waclaw Sierpiński, Pythagorean Triangles, *Scripta Math. Studies*, **9**(1962), Yeshiva University, New York, Chap. 15, pp. 97–107.

W. Sierpiński, *A Selection of Problems in the Theory of Numbers*, Pergamon, Oxford, 1964, p. 112.

W. G. Spohn, On the integral cuboid, *Amer. Math. Monthly* **79**(1972) 57–59; *MR* **46** #7158.

W. G. Spohn, On the derived cuboid, *Canad. Math. Bull.*, **17**(1974) 575–577; *MR* **51** #12693.

D19 Rational distances from the corners of a square.

Is there a point all of whose distances from the corners of the unit square are rational? It was earlier thought that there might not be any nontrivial example (i.e., an example not on the edge of the square) of a point with *three* such rational distances, but John Conway & Mike Guy found an infinity of integer solutions of

$$(s^2 + b^2 - a^2)^2 + (s^2 + b^2 - c^2)^2 = (2bs)^2$$

where a, b, c are the distances of a point from three corners of a square of edge s. There are relations between such solutions as shown in Figure 13.

For the fourth distance d to be an integer we also need $a^2 + c^2 = b^2 + d^2$. In the three-distance problem, one of s, a, b, c is divisible by 3, one by 4, and one by 5. In the four-distance problem, s is a multiple of 4 and a, b, c, d are odd (assuming that there is no common factor). If s is not a multiple of 3 (respectively 5) then two of a, b, c, d are divisible by 3 (resp. 5).

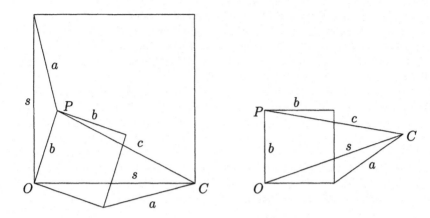

Figure 13. A Solution of the Three-Distance Problem, and its Inverse.

If the problem is generalized to a rational rectangle, then $a^2 + c^2 = b^2 + d^2$ is still required. This is the basis of a Martin Gardner puzzle [Mathematical Games, *Sci. Amer.* **210** #6 (June 1964), Problem 2, p.116]; see also the Dodge reference below. A similar problem with a square with an irrational edge and one irrational distance occurs in Dudeney, Canterbury Puzzles, No. 66, pp. 107–109, 212–213. Gardner sends a copy of correspondence between Leslie J. Upton and J. A. H. Hunter. Hunter's 67-03-21 letter gives an infinity of solutions with the three distances in A.P.: $a = m^2 - 2mn + 2n^2$, $b = m^2 + 2n^2$, $c = m^2 + 2mn + 2n^2$ and $s^2 = 2m^2(m^2 + 4n^2)$ where s is an integer if $m = 2(u^2 + 2uv - v^2)$, $n = u^2 - 2uv - v^2$. For example, there is a point at distances 85, 99 and 113 from three consecutive corners of a square of edge 140. It can be shown that the fourth distance is never rational in such solutions.

John Leech found points solving the three-distance problem which are dense in the plane. These include the Conway-Guy solutions, and the 'inverses' (in the sense of Fig. 13) of the Hunter solutions. But there are other solutions, and in some sense we now know "all" of them. Consider the more general problem of dissecting a rational square into rational triangles. It is known that at least four triangles are needed and there are just four candidate arrangements, the δ-configuration, the κ-configuration, the ν-configuration and the χ-configuration. The first two turn out to be "duals" and solutions are given by the rational points on an infinite family of elliptic curves. The first few hundred have been investigated by Bremner and Guy, who have also dealt similarly with the ν-configurations. The χ-configuration, i.e., the "four distance" problem, remains as an astonishingly hard nut to crack.

There are infinitely many solutions of the corresponding problem of integer distances a, b, c from the corners of an equilateral triangle of edge t. In each of these one of a, b, c, t is divisible by 3, one by 5, one by 7 and one by 8. John Leech has sent us a neat and elementary proof of the fact that the points at rational distances from the vertices of *any* triangle with rational edges are dense in the plane of the triangle. This result was proved earlier by Almering; see reference at **D21**. Arnfried Kemnitz notes that $a = m^2 + n^2$; $b, c = m^2 \pm mn + n^2$ with $m = 2(u^2 - v^2)$, $n = u^2 + 4uv + v^2$ gives $t = 8(u^2 - v^2)(u^2 + uv + v^2)$ and an infinity of solutions in which the points are neither on the edges nor the circumcircle of the triangle. A computer search showed that $(57,65,73,112)$ was the smallest such.

Kevin Buzzard found some parametric solutions to the problem of finding integer-edged triangles whose Fermat-Torricelli point is at an integer distance from each vertex – it should be possible to find "all" such solutions, in the sense of Bremner's approach to equal sums of sixth powers (reference at **D1**). But it is probably a harder question to ask that the triangles also be Heron. Berry writes the earlier displayed equation as

$$2(s^4 + b^4) + a^4 + c^4 = 2(s^2 + b^2)(a^2 + c^2)$$

and notes that this, and the corresponding equation

$$t^4 + a^4 + b^4 + c^4 = t^2a^2 + t^2b^2 + t^2c^2 + b^2c^2 + c^2a^2 + a^2b^2$$

for the equilateral triangle, both represent **Kummer surfaces**, i.e., quartic surfaces with just 16 singular points. They are not isomorphic, but are of the same special type, known as **tetrahedroids**.

There are the following consequences:

- A Kummer surface is not rational: there is no general parametric solution of either problem, in the sense that there are no polynomials (resp. rational functions) giving all integer (resp. rational) solutions.

- One-parameter families of solutions correspond to parametrizable curves on the surface. For example, the 16 conics (which always exist on a Kummer surface) give, in the equilateral triangle problem, points on the edges and circumcircle. The solution given by Arnfried Kemnitz corresponds to the plane section $b + c = 2a$.

- The elliptic curves used by Bremner & Guy to find the delta-lambda configurations form a pencil on the former surface, and since the surfaces are both tetrahedroids, there may be an elliptic pencil on the "equilateral triangle" surface, which allows an analogous attack.

Berry generalizes Almering's result by showing that if the squares on the edges of a triangle are rational and at least one edge is rational, then the set of points at rational distances from all three vertices is dense in the plane of the triangle.

Jerry Bergum asks for what integers n do there exist positive integers x, y with $x \perp y$, x even, and $x^2 + y^2 = b^2$ & $x^2 + (y - nx)^2 = c^2$ both perfect squares. If $n = 2m(2m^2 + 1)$, then $x = 4m(4m^2 + 1)$, $y = mx + 1$ is a solution. There are no solutions if $n = \pm 1$, ± 2, ± 4, ± 11 or $\pm p$ where $p \equiv 3 \bmod 4$ and $p^2 + 4$ is prime, e.g., ± 3, ± 7. Bergum has several infinite families of values of n for which there are solutions, e.g., $n = 8t^2 \pm 4t + 2$ with $t > 0$. There are solutions with $n = \pm 5$, ± 6, ± 8, ± 9, ± 14, ± 19. If $n = 8$, the least x for which there is a y is $x = 2996760 = 2^3 \cdot 3 \cdot 5 \cdot 13 \cdot 17 \cdot 113$ and if $n = 19$ the least x is 2410442371920. One solution is $n = 5$, $x = 120$, $y = 391$ as may be seen from Fig. 15(b)! The connexion between this problem and the original one is that (x, y) are the coordinates of P at distances b and c from the origin O and an adjacent corner of the square of edge $s = nx$ where n is an integer.

Ron Evans notes that the problem may be stated: which integers n occur as the ratios base/height in integer-edged triangles? The sign of n is $\pm$ according as the triangle is acute or obtuse (e.g., $n = -29$, $x = 120$, $y = 119$ is a solution). He also asks the dual problem: find every integer-edged triangle whose base divides its height. Here the height/base ratios 1 and 2 can't occur, but 3 can (e.g., base 4; edges 13, 15; height 12). If a ratio can occur, are there infinitely many primitive triangles for which it occurs? K. R. S. Sastry gives the triangles (3389, 21029, 24360) and (26921, 42041, 68880) in each of which the ratio base/height is 42, and (25,26,3) and (17,113,120) with ratios 1/8 and 8 (the third member of the triple is the base in each case).

J. H. J. Almering, Rational quadrilaterals, *Nederl. Akad. Wetensch. Proc. Ser. A*, **66** = *Indagationes Math.*, **25**(1963) 192–199; II **68** = **27**(1965) 290–304; *MR* **26** #4963, **31** #3375.

T. G. Berry, Points at rational distance from the corners of a unit square, *Ann. Scuola Norm. Sup. Pisa Cl. Sci.*(4), **17**(1990) 505–529; *MR* **92e**:11021.

T. G. Berry, Points at rational distances from the vertices of a triangle, *Acta Arith.*, **62**(1992) 391–398.

T. G. Berry, Triangle distance problems and Kummer surfaces,

Andrew Bremner & Richard K. Guy, The delta-lambda configurations in tiling the square, *J. Number Theory*, **32**(1989) 263–280; *MR* **90g**:11031.

Andrew Bremner & Richard K. Guy, Nu-configurations in tiling the square, *Math. Comput.*, **59**(1992) 195–202, S1–S20; *MR* **93a**:11019.

Charles K. Cook & Gerald E. Bergum, Integer sided triangles whose ratio of altitude to base is an integer, *Applications of Fibonacci numbers, Vol. 5 (St. Andrews, 1992)*, 137–142, Kluwer Acad. Publ., Dordrecht, 1993; *MR* **95c**:11015.

Clayton W. Dodge, Problem 966, *Math. Mag.*, **49**(1976) 43; partial solution **50**(1977) 166–167; comment **59**(1986) 52.

R. B. Eggleton, Tiling the plane with triangles, *Discrete Math.*, **7**(1974) 53–65.

R. B. Eggleton, Where do all the triangles go? *Amer. Math. Monthly*, **82** (1975) 499–501.

Ronald Evans, Problem E2685, *Amer. Math. Monthly*, **84**(1977) 820.

N. J. Fine, On rational triangles, *Amer. Math. Monthly*, **83**(1976) 517–521.

Richard K. Guy, Tiling the square with rational triangles, in R. A. Mollin (ed.) Number Theory & Applications, *Proc. N.A.T.O. Adv. Study Inst., Banff 1988*, Kluwer, Dordrecht, 1989, 45–101.

W. H. Hudson, Kummer's Quartic Surface, reprinted with a foreword by W. Barth, Cambridge Univ. Press, 1990.

Arnfied Kemnitz, Rational quadrangles, Proc. 21st SE Conf. Combin. Graph Theory Comput., Boca Raton 1990, *Congr. Numer.*, **76**(1990) 193–199; *MR* **92k**:11034.

J. G. Mauldon, An impossible triangle, *Amer. Math. Monthly*, **86**(1979) 785–786.

C. Pomerance, On a tiling problem of R. B. Eggleton, *Discrete Math.*, **18** (1977) 63–70.

D20 Six general points at rational distances.

The first edition of this book asked "are there six points in the plane, no three on a line, no four on a circle, all of whose mutual distances are rational?" Leech pointed out that one such configuration is obtained by fitting six copies of a triangle whose edges and medians are all rational. Such triangles were studied by Euler (see **D21**): the simplest has edges 68, 85, 87 and medians the halves of 158, 131, 127 [Fig. 14(a)]. Harborth & Kemnitz have shown that this triangle leads to the minimal configuration of six points, no three collinear, no four concyclic, at integer mutual distances. A related configuration is obtained by inversion in a concentric circle; the six triangles are then similar but no longer congruent. Can any such configuration be extended? Or are there any sets of more than six such points? Kemnitz has exhibited an unsymmetrical set of six points at integer distances, 13 of them distinct, the largest 319 [Fig. 14(b)].

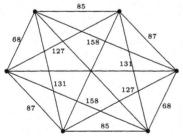
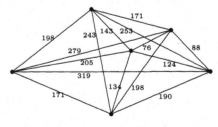
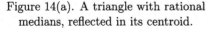

Figure 14(a). A triangle with rational medians, reflected in its centroid.

Figure 14(b). Six points at integer distances, 13 of them distinct.

There are two opposite extreme conjectures: (a) that there is a fixed number c such that any n points in a plane whose mutual distances are rational include at least $n - c$ which are collinear or concyclic, and (b) (ascribed to Besicovitch, but in 1959 he expressed the contrary opinion) that any polygon can be approximated arbitrarily closely by a polygon with all its edges and diagonals rational. If (a) is valid, what is the maximum value of c?

If we have an infinite sequence of points $\{x_i\}$ with all distances rational, can we characterize the set of limit points? It was already known to Euler that they could be dense on a circle. If the sequence is dense in the plane, Ulam conjectured that not all distances could be rational. Does it contain a dense subsequence with all distances irrational?

We can choose a straight line and two points at unit distance from it on a perpendicular line; then points on the first line at distances of the form $(u^2 - v^2)/2uv$ from the intersection form an infinite set of points with rational mutual distances. By inversion in a circle centred at one of the off-line points, we obtain a dense set of points on a circle, together with its centre, with rational mutual distances. This proves Conjecture (b) for *cyclic* polygons. Peeples (who quotes Huff) extended this to a straight line with four points at distances $\pm p$, $\pm q$ from it on a perpendicular line. If q/p has a representation as the product of two *distinct* ratios of the form $(u^2 - v^2)/2uv$, then it has an infinity of them, and there will be an infinity of points on the first line at rational distances from each other and from the four off-line points. Thus c is at least 4 in Conjecture (a). By inversion in a circle centred at one of the four off-line points, we obtain a dense set of points on a circle, together with its centre and a pair of points inverse in the circle, with rational mutual distances. Thus c is at least 3 for the circle in Conjecture (a). Are these the maximum values of c for infinite sets?

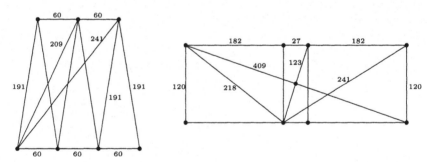

Figure 15 (a) & (b). Leech's Configurations of Points at Rational Distances.

What finite sets surpass these values of c? Leech gave an infinite family of sets of nine points, with no more than four on any line or circle, so $c = 5$ for these sets. They are based on solutions of the simultaneous Diophantine equations

$$x^2 + y^2 = \square, \quad x^2 + z^2 = \square, \quad x^2 + (y+z)^2 = \square, \quad x^2 + (y-z)^2 = \square;$$

most simply $x = 120$, $y = 209$, $z = 182$ [Fig. 15(b)].

Lagrange and Leech have given infinite families of pairs of triads of integers a_1, a_2, a_3, and b_1, b_2, b_3, such that the nine sums $a_i^2 + b_j^2$ are all squares. These lead to sets of 13 points on two perpendicular lines, comprising their intersection, points $(\pm a_i, 0)$ on one line and points $(0, \pm b_j)$ on the other, with integer mutual distances and no more then seven on a line or four on a circle, so $c = 6$ for these sets. The simplest example has $a_i = 952$, 1800, 3536 and $b_j = 960$, 1785, 6630. They also found an example in which one triad is extended to a tetrad; it has

$$a_i = 9282000, \quad 26822600, \quad 60386040,$$
$$b_j = 7422030, \quad 8947575, \quad 22276800, \quad 44142336,$$

but these still give only $c = 6$. Leech has since found three further examples. Can a pair of tetrads be found with all 16 sums $a_i^2 + b_j^2$ square? If so we would obtain a set of 17 points with integer mutual distances and $c = 8$.

Noll and Bell search for configurations with no three points collinear and no four concyclic, but using only lattice points. They call such configurations **N-clusters**. They, and independently William Kalsow & Bryan Rosenburg, found the 6-cluster $(0,0)$, $(132,-720)$, $(546,-272)$, $(960,-720)$, $(1155,540)$, $(546,1120)$. They define the **extent** of an N-cluster to be the radius of the smallest circle, centred at one of the points, which contains them all. They find 91 nonequivalent prime 6-clusters of extent less than 20937, but no 7-clusters.

Harley Flanders sent all cyclic pentagons with integer edges and integer circumradius $R \le 51$. The smallest and largest were

$R = 13$ (1,10,13,22,24)   (1,10,13,23,24)*   (10,13,22,23,24)*
$R = 51$ (6,34,48,90,94)   (6,34,48,90,98)*   (17,48,67,87,90)*

where asterisks denote nonconvex specimens.

Jordan & Peterson find pentagons with integer edges and diagonals, in terms of the Fibonacci numbers u_n:

$$\{u_{n-1}u_n^2, u_{n-2}u_n u_{n+1}, u_n^3, u_n^3 + (-1)^n u_{n-2};$$
$$u_{n+1}u_n^2, u_{n-1}u_n u_{n+2}, u_n^3 + (-1)^n u_{n+2}\}$$

E.g., $n = 3$: $\{4,6,8,7; 12,10,3\}$ — to be interpreted as an integer pentagon with edges 4,6,4,6,4 and opposite diagonals 8,8,7,8,8 and a pentagon with edges 8,8,3,8,8 and diagonals 10,12,12,12,10.

M. Altwegg, Ein Satz über Mengen von Punkten mit ganzzahliger Entfernung, *Elem. Math.*, **7**(1952) 56–58.

D. D. Ang, D. E. Daykin & T. K. Sheng, On Schoenberg's rational polygon problem, *J. Austral. Math. Soc.*, **9**(1969) 337–344; *MR* **39** #6816.

A. S. Besicovitch, Rational polygons, *Mathematika*, **6**(1959) 98; *MR* **22** #1557.

D. E. Daykin, Rational polygons, *Mathematika*, **10**(1963) 125–131; *MR* **30** #63.

D. E. Daykin, Rational triangles and parallelograms, *Math. Mag.*, **38**(1965) 46–47.

H. Harborth, On the problem of P. Erdős concerning points with integral distances, *Annals New York Acad. Sci.*, **175**(1970) 206–207.

H. Harborth, Antwort auf eine Frage von P. Erdős nach fünf Punkten mit ganzzahligen Abständen, *Elem. Math.*, **26**(1971) 112–113.

H. Harborth & A. Kemnitz, Diameters of integral point sets, *Intuitive geometry (Siófok, 1985)*, *Colloq. Math. Soc. János Bolyai*, **48**(1987) 255–266, North-Holland, Amsterdam-New York; *MR* **88k**:52011.

G. B. Huff, Diophantine problems in geometry and elliptic ternary forms, *Duke Math. J.*, **15**(1948) 443–453.

James H. Jordan & Blake E. Peterson, Almost regular integer Fibonacci pentagons, *Rocky Mountain J. Math.*, **23**(1993) 243–247; *MR* **94a**:11023.

A. Kemnitz, Integral drawings of the complete graph K_6, in Bodendiek & Henn, *Topics in Combinatorics and Graph Theory, Essays in honour of Gerhard Ringel*, Physica-Verlag, Heidelberg, 1990, 421–429.

Jean Lagrange, Points du plan dont les distances mutuelles sont rationnelles, *Séminaire de Théorie des Nombres de Bordeaux*, 1982-1983, Exposé no. 27, Talence 1983.

Jean Lagrange, Sur une quadruple équation. *Analytic and elementary number theory* (Marseille, 1983), 107–113, *Publ. Math. Orsay*, 86-1, *Univ. Paris XI, Orsay*, 1986; *MR* **87g**:11036.

Jean Lagrange & John Leech, Two triads of squares, *Math. Comput.*, **46** (1986) 751–758; *MR* **87d**: 11018.

John Leech, Two diophantine birds with one stone, *Bull. London Math. Soc.*, **13**(1981) 561–563; *MR* **82k**:10017.

D. N. Lehmer, Rational triangles, *Annals Math. Ser. 2*, **1**(1899–1900) 97–102.

L. J. Mordell, Rational quadrilaterals, *J. London Math. Soc.*, **35**(1960) 277–282; *MR* **23** #A1593.

L. C. Noll & D. I. Bell, n-clusters for $1 < n < 7$, *Math. Comput.*, **53** (1989) 439–444.

W. D. Peeples, Elliptic curves and rational distance sets, *Proc. Amer. Math. Soc.*, **5**(1954) 29–33; *MR* **15** 645f.

Blake E. Peterson & James H. Jordan, The rational heart of the integer Fibonacci pentagons, *Applications of Fibonacci numbers, Vol. 6 (Pullman WA, 1994)* 381–388; *MR* **97d**:11031.

T. K. Sheng, Rational polygons, *J. Austral. Math. Soc.*, **6**(1966) 452–459; *MR* **35** #137.

T. K. Sheng & D. E. Daykin, On approximating polygons by rational polygons, *Math. Mag.*, **38**(1966) 299–300; *MR* **34** #7463.

D21 Triangles with integer edges, medians and area.

Is there a triangle with integer edges, medians and area? There are, in the literature, incorrect "proofs" of impossibility, but the problem remains open. It may be instructive to would-be solvers to find the fallacies in the arguments of Schubert and of Eggleston, referred to below. For some time it had been suspected that not even two rational medians were possible in a Heron triangle, but discoveries by Randall L. Rathbun, Arnfried Kemnitz and R. H. Buchholz have shown that they can occur. Even more recently, Rathbun has found infinitely many such, and it seems reasonable to conjecture that there are infinitely many infinite families, but this remains undemonstrated, as does the existence or impossibility of three rational medians. Buchholz & Rathbun show that any rational point on the curve $(xy+2)(x-y+1)=3$ with $0 < x, y < 1$ and $2x + y > 1$ corresponds to a triangle with rational sides, rational area and two rational medians.

If we don't require the area to be rational, there are many solutions. Euler gave a parametric solution of degree five,

$$a = 6\lambda^4 + 20\lambda^2 - 18, \quad b, c = \lambda^5 \pm \lambda^4 - 6\lambda^3 \pm 26\lambda^2 + 9\lambda \pm 9$$

with medians $-2\lambda^5 + 20\lambda^3 + 54\lambda$, $\pm\lambda^5 + 3\lambda^4 \pm 26\lambda^3 - 18\lambda^2 \pm 9\lambda + 27$. Recently George Cole has shown that, up to symmetry, there are just two parametric solutions of degree five, Euler's and a new one. These have not so far yielded a non-degenerate triangle with rational area.

A host of problems arise from Pythagorean triangles. Eckert asked if there are two distinct Pythagorean triples whose products are equal, i.e., is there a solution of

$$xy(x^4 - y^4) = zw(z^4 - w^4)$$

in nonzero integers and Prothro asked if one product could be *twice* the other. More generally Leech asks what small integers are the ratios of two such products? Trivially $xy(x^4 - y^4) = 8zw(z^4 - w^4)$ when $x, y = z \pm w$. Since every product is divisible by $3 \cdot 4 \cdot 5$, many integers are possible, from $13 = \frac{5 \cdot 12 \cdot 13}{3 \cdot 4 \cdot 5}$ upwards. More subtly $11 = \frac{21 \cdot 220 \cdot 221}{13 \cdot 84 \cdot 85}$, and imprimitively

$$6 = \frac{6^3 \cdot 24 \cdot 143 \cdot 145}{135 \cdot 352 \cdot 377}.$$

Are these the smallest?

He also notes that one product is *eight* times the other on setting $x, y = z \pm w$.

How many primitive Pythagorean triangles can have the same area? A triple of such, with generators (77,38), (78,55), (138,5) was found by Charles L. Shedd in 1945. In 1986 Rathbun found three more:
(1610,869), (2002,1817), (2622,143);   (2035,266), (3306,61), (3422,55);
and   (2201,1166), (2438,2035), (3565,198).

A fifth triple, (7238,2465), (9077,1122), (10434,731), was found independently on consecutive days by Dan Hoey and Rathbun. Is there an infinity of triples? Are there quadruples? If they are not required to be primitive, there can be arbitrarily many Pythagorean triangles with the same area.

Beiler gives four primitive Pythagorean triangles, each with perimeter 317460. Their generators are (286,269), (330,151), (370,59), (390,17). This inspired Rathbun to find sets of 23 such primitive triangles. For example, those with generators

(254082,248563)	(254930,246043)	(255398,244657)	(257686,237929)
(262922,222823)	(263670,220697)	(274010,192079)	(274170,191647)
(277134,183701)	(283974,165761)	(288002,155443)	(294690,138691)
(295630,136373)	(297330,132203)	(310726,100289)	(311610, 98239)
(323830, 70553)	(330410, 56119)	(332310, 52009)	(333370, 49727)
(333982, 48413)	(348270, 18437)	(351390, 12061)	

each have perimeter 255426093780. There are 11 other, imprimitive, triangles with the same perimeter.

Sastry asks for Pythagorean triangles with a triangular number and a square for legs, and a pentagonal number, $\frac{1}{2}n(3n-1)$, for hypotenuse. Are there any nontrivial examples besides (3,4,5) and (105,100,145)? Does it help to allow pentagonal numbers of negative rank, $\frac{1}{2}n(3n+1)$?

Despairing of solving the problem of the rational box (**D18**), some people have investigated other polyhedra all of whose distances are integers. There are seven topologically different convex hexahedra, for example, and integer examples have been found by Harborth & Kemnitz (see **D20**) and by Peterson & Jordan. Sastry asked for solutions to the rational box problem, but using triangular numbers instead of squares. Charles Ashbacher gave the triangular numbers 66, 105, 105 whose sums of pairs and whose total are all triangular. Alperin finds infinitely many hexahedra with four congruent trapezoidal faces and two congruent rectangular faces, all edges, face- and body-diagonals being integers.

Roger C. Alperin, A quartic surface of integer hexahedra, *Rocky Mountain J. Math.*, **31**(2001) 37–43; *MR* **2002k**:11037.

J. H. J. Almering, Heron problems, thesis, Amsterdam, 1950.

A. S. Anema, Pythagorean triangles with equal perimeters, *Scripta Math.*, **15**(1949) 89.

Raymond A. Beauregard & E. R. Suryanarayan, Arithmetic triangles, *Math. Mag.*, **70**(1997) 105–115; *MR* **98d**:11033.

Albert H. Beiler, *Recreations in the Theory of Numbers – The Queen of Mathematics Entertains*, Dover, 1964, pp.131–132.

Ralph Heiner Buchholz, On triangles with rational altitudes, angle bisectors or medians, PhD thesis, Univ. of Newcastle, Australia, 1989.

Ralph H. Buchholz, Triangles with three rational medians, *J. Number Theory*, **97**(2002) 113–131; *MR* **2003h**:11034.

Ralph Heiner Buchholz & J. A. MacDougall, Heron quadrilaterals with sides in arithmetic or geometric progression, *Bull. Austral. Math. Soc.*, **59**(1999) 263–269; *MR* **2000c**:51019.

Ralph H. Buchholz & Randall L. Rathbun, An infinite set of Heron triangles with two rational medians, *Amer. Math. Monthly*, **104**(1997) 107–115; *MR* **98a**:51015.

Ralph Heiner Buchholz & Randall L. Rathbun, Heron triangles and elliptic curves, *Bull. Austral. Math. Soc.*, **58**(1998) 411–421; *MR* **99h**:11026.

Garikai Campbell & Edray Herber Goins, Heron triangles, Diophantine problems and elliptic curves, preprint, 2003.

George Raymond Cole, Triangles all of whose sides and medians are rational, PhD dissertation, Arizona State University, May 1991.

Ernest J. Eckert, Problem 994, *Crux Mathematicorum*, **10**(1984) 318; comment **12**(1986) 109.

H. G. Eggleston, A proof that there is no triangle the magnitudes of whose sides, area and medians are integers, Note 2204, *Math. Gaz.*, **35**(1951) 114–115.

H. G. Eggleston, Isosceles triangles with integral sides and two integral medians, Note 2347, *Math. Gaz.*, **37**(1953) 208–209.

Albert Fässler, Multiple Pythagorean number triples, *Amer. Math. Monthly* **98**(1991) 505–517; *MR* **92d**:11021.

Martin Gardner, Mathematical Games: Simple proofs of the Pythagorean theorem, and sundry other matters, *Sci. Amer.* **211**#4 (Oct. 1964) 118–126.

Hans Georg Killingbergtrø, Triangles with integral sides and rational angle ratios, *Normat*, **42**(1994) 179–180, 192.

F. L. Miksa, Pythagorean triangles with equal perimeters, *Mathematics*, **24** (1950) 52.

Blake E. Peterson & James H. Jordan, Integer hexahedra equivalent to perfect boxes, *Amer. Math. Monthly*, **102**(1995) 41–45; *MR* **95m**:52026.

E. T. Prothro, *Amer. Math. Monthly*, **95**(1988) 31.

Randall L. Rathbun, Letter to the Editor, *Amer. Math. Monthly*, **99**(1992) 283–284.

K. R. S. Sastry, Problem 1725, *Crux Mathematicorum*, **18**(1992) 75; Problem 1832, **19**(1993) 112.

H. Schubert, *Die Ganzzahligkeit in der algebraischen Geometrie*, Leipzig, 1905, 1–16.

D22 Simplexes with rational contents.

Are there simplexes in any number of dimensions, all of whose contents (lengths, areas, volumes, hypervolumes) are rational? The answer is "yes" in two dimensions; there are infinitely many **Heron triangles** with rational edges and area. An example is a triangle of edges 13, 14, 15 which has area 84. The answer is also "yes" in three dimensions, but can all tetrahedra be approximated arbitrarily closely by such rational ones?

John Leech notes that four copies of an acute-angled Heron triangle will fit together to form such a tetrahedron, provided that the volume is made rational, and this is not difficult. E.g., three pairs of opposite edges of lengths 148, 195, 203. This is the smallest example: he finds the next few triples to be

$$(533, 875, 888), (1183, 1479, 1804), (2175, 2296, 2431), (1825, 2748, 2873),$$

$$(2180, 2639, 3111), (1887, 5215, 5512), (6409, 6625, 8484), (8619, 10136, 11275).$$

He also suggests examining references on p. 224 of Vol II of Dickson's *History* [**A17**]:

Mohammed Aassila, Some results on Heron triangles, *Elem. Math.*, **56**(2001) 143–146.

R. Güntsche, *Sitzungsber. Berlin Math. Gesell.*, **6**(1907) 38–53.

R. Güntsche, *Archiv Math. Phys.*(3), **11**(1907) 371.

E. Haentzschel, *Sitzungsber. Berlin Math. Gesell.*, **12**(1913) 101–108 & **17**(1918) 37–39 (& *cf.* **14**(1915) 371).

O. Schulz, Ueber Tetraeder mit rationalen Masszahlen der Kantenlängen und des Volumen, Halle, 1914, 292 pp.

Dickson appealed for a copy of this last. Did he ever get one? Does anyone know of a copy? Would they be willing to donate it, or offer it for sale, to the Strens Collection?

Leech also notes that this problem is answered positively in three dimensions by solutions to Problem 3 in **D18** (find a box which is rational except for one face diagonal). This problem was published as Problem 930 in *Crux Mathematicorum*, **10**(1984) #3, p. 89, and the solution by the COPS (presumably an acronym for the Carleton (Ottawa) Problem Solvers) is:

Take a tetrahedron with a path of three mutually perpendicular edges, $a = p^2q^2 - r^2s^2$, $b = 2pqrs$, $c = p^2r^2 - q^2s^2$. Then $a^2 + b^2$, $b^2 + c^2$ are squares and $a^2 + b^2 + c^2 = (p^4 + s^4)(q^4 + r^4)$ is a square if

$$p^4 + s^4 = q^4 + r^4.$$

[John Leech notes "but not only if" and gives four casual examples,

$$(1^4 + 2^4)(2^4 + 13^4) = 697^2; \quad (1^4 + 2^4)(38^4 + 43^4) = 9673^2;$$

$$(1^4 + 2^4)(314^4 + 863^4) = 1275643^2; \quad (1^4 + 3^4)(9^4 + 437^4) = 1729298^2,$$

which imply further ones of type $(2^4 + 13^4)(38^4 + 43^4)$.]

This equation was solved by Euler. The solution mentioned in **D9** is

$$p, q = x^7 + x^5y^2 - 2x^3y^4 \pm 3x^2y^5 + xy^6$$
$$r, s = x^6y \pm 3x^5y^2 - 2x^4y^3 + x^2y^5 + y^7$$

"but this is not in any sense complete".

Buchholz found that the only rational tetrahedron with edge lengths ≤ 156 was the second of those listed in the next paragraph, with face-areas 1800, 1890, 2016, 1170. He also showed that a *regular d*-dimensional simplex with rational edge has rational d-dimensional volume just if d is of shape $4k(k+1)$ or $2k^2 - 1$.

Randall Rathbun has found 614 tetrahedra with integer edges and rational surface area and volume. The first ten are

area	volume			edges			
6384	8064	160	153	25	39	56	120
6876	18144	117	84	51	52	53	80
17220	35280	225	200	65	119	156	87
48384	206976	318	221	203	42	175	221
54600	611520	203	195	148	203	195	148
64584	170016	595	429	208	116	276	325
64584	200928	595	507	116	208	275	176
79200	399168	468	340	232	65	225	297
83160	1034880	319	318	175	175	210	221

Dove & Sumner relax the condition that the faces of a tetrahedron have rational area and find two tetrahedra with volume 3 and pairs of opposite edges (32,76) (33,70) (35,44) and (21,58) (32,76) (47,56). They ask if there are infinitely many tetrahedra with integer edges and the same integer volume. Is there such a tetrahedron with volume any given multiple of 3? They have examples from 3 to 99, except for 87.

The tetrahedron with a pair of opposite edges 896 and 990 and the other four edges each 1073, two face areas 436800 and 471240, and volume 62092800, is mentioned by Sierpiński and by Leitzmann.

Mohammed Aassila shows the existence of infinitely many pairs of Heron triangles with common area and perimeter. Ronald van Luijk has outlined (02-10-21 email) a proof that there are arbitrarily large sets of Heron triangles with common perimeter and area. Randall Rathbun has given a set of eight with perimeter 128700 and area 594594000:

$$\{53262, 52338, 23100\} \quad \{55440, 48950, 24310\}$$
$$\{55530, 48750, 24420\} \quad \{55770, 48180, 24750\} \quad \{57200, 42790, 28710\}$$
$$\{57310, 42075, 29315\} \quad \{57420, 41250, 30030\} \quad \{57750, 36630, 34320\}.$$

Mohammed Aassila, Some results on Heron triangles, *Elem. Math.*, **56**(2001) 143–146; *MR* **2002j**:11023.

Ralph Heiner Buchholz, Perfect pyramids, *Bull. Austral. Math. Soc.*, **45**(1991) 353–368.

Kevin L. Dove & John L. Sumner, Tetrahedra with integer edges and integer volume, *Math. Mag.*, **65**(1992) 104–111.

K. È. Kalyamanova, Rational tetrahedra (Russian), *Izv. Vyssh. Uchebn. Zaved. Mat.*, **1990** 73–75; *MR* **92b**:11014.

W. Lietzmann, Der pythagoreisch Lehrsatz, Leipzig, 1965, p. 91 [not in 1930 edition].

Florian Luca, Fermat primes and Heron triangles with prime power sides, *Amer. Math. Monthly*, **110**(2003) 46–49.

W. Sierpiński, *Pythagorean Triangles*, New York, 1962, p. 107.

D23 Some quartic equations.

See also **D3, D6, D9, D10**.

Another of many unsolved diophantine equations is

$$(x^2 - 1)(y^2 - 1) = (z^2 - 1)^2$$

though Schinzel & Sierpiński have found all solutions for which $x - y = 2z$. Cao Zhen-Fu has shown that the only solutions satisfying $x - y = lz$ for other values of $l \leq 30$ are $|x| = |y|$ or $|z| = 1$, and Wang Yan Bin that these are the only solutions with $x - y = z^2 + 1$. Szymiczek considered the equation and excluded trivial solutions by assuming that $x < z < y$. He noted the nontrivial solution (3,17,7) and Sierpiński's observation that the problem of finding three triangular numbers in geometric progression is equivalent to finding odd nontrivial solutions of the equation. But this had been solved by Gérardin and was solved again by Szymiczek: if the nth triangular number, t_n, is a square, m^2, then so is $t_{3n+4m+1}$ and t_n, t_{n+2m}, $t_{3n+4m+1}$ form a G.P.

Among sporadic solutions (not of the special kind considered by Schinzel & Sierpiński) Szymiczek notes (4,11,31), (2,13,97), (155,2729,48049) of which the third yields a triple of triangular numbers, t_{77}, t_{1364}, t_{24024}, in G.P., but not of the above form. Szymiczek conjectures that the answer to Sierpiński's question: are there four triangular numbers in G.P.? is "no!"

Ronald van Luijk observes that the equation represents a singular K3 surface with an elliptic fibration of positive rank, so that the rational points are dense. [For experts, a correspondent writes that 'it can certainly be approached with the relatively standard technique of computing the Neron-Severi group, in order to find parametrizable curves of small degree.']

Another K3 surface, enquired about by Henri Cohen, has equation $(x^2 - y^2)(1 - x^2y^2) = z^2$. Daniel Coray, in an 02-10-25 email message, gives some of the infinitely many parametric solutions.

Kashihara has shown that all solutions of

$$(x^2 - 1)(y^2 - 1) = (z^2 - 1)$$

can be derived from the trivial solutions $(n, 1, 1)$ and $(1, n, 1)$.

For the equation $x^2 - 1 = y^2(z^2 - 1)$, Mignotte has shown that if z is large, then the greatest prime factor of y is at least $c \ln \ln y$.

Ron Graham has observed that the diophantine equations

$$2x^2(x^2 - 1) = 3(y^2 - 1) \quad \text{and} \quad (2x - 1)^2 = 2^n - 7$$

each have the solutions $x = 0, 1, 2, 3, 6$ and 91. Is this merely an example of the Law of Small Numbers? Evidently so! Stroeker & de Weger find that Graham's equation has one pair and five quadruples of solutions. They note that it is unfair to count the solutions $x = 0$ and $x = 1$ of the Ramanujan-Nagell equation as separate, while 'forgetting' the solutions $x = -1, -2, -5, -90$.

Ljunggren showed that $x^4 - Dy^2 = 1$ has at most two solutions; Le Mao-Hua showed that it has at most one if $\ln D > 64$, and Cipu that it has at most one if $D \neq 1785$.

Venkatesh showed that $x^4 + 5x^2y^2 + y^4 = z^2$ has no solutions with $xy \neq 0$.

Baragar has shown that the equation

$$x(x + 1)y(y + 1) = z(z + 1),$$

studied by Katayama, is equivalent to a Markoff type equation (see **D12**)

$$x^2 + y^2 + z^2 = 2xyz + 5$$

and has counted the number of solutions of size less than N.

Not a quartic, but probably suggested by the opening equation in this section: Michael Bennett (W. No. Theory Problem 97:08) asked if there are any solutions of

$$\frac{x^2 - 1}{y^2 - 1} = (z^2 - 1)^2$$

with $x, y, z > 1$ other than $(x, y, z) = (m(4m^2 - 3), m, 2m)$ with $m \geq 2$ and $(8m^4 + 16m^3 + 8m^2 - 1, 2m^2 + 2m, 2m + 1)$ with $m \geq 1$. He has shown that there are at most three solutions of two simultaneous Pell equations and conjectures that there are at most two.

Mohammed Al-Kadhi & Omar Kihel, Some remarks on the Diophantine equation $(x^2 - 1)(y^2 - 1) = (z^2 - 1)^2$, *Proc. Japan Acad. Ser. A Math. Sci.*, **77**(2001) 155–156; *MR* **2002i**:11029.

Michael A. Bennett, On the number of solutions of simultaneous Pell equations, *J. reine angew. Math.*, **498**(1998) 173–199.

Michael A. Bennett, Solving families of simultaneous Pell equations, *J. Number Theory*, **67**(1997) 246–251; *MR* **98j**:11015.

Michael A. Bennett & P. Gary Walsh, The Diophantine equation $b^2X^4 - dY^2 = 1$, *Proc. Amer. Math. Soc.*, **127**(1999) 3481–3491; *MR* **2000b**:11025.

Michael A. Bennett & P. Gary Walsh, Simultaneous quadratic equations with few or no solutions, *Indag. Math. (N.S.)* **11**(2000) 1–12; *MR* **2002k**:11040.

Andrew Bremner & John William Jones, On the equation $x^4 + mx^2y^2 + y^4 = z^2$, *J. Number Theory*, **50**(1995) 286–298; *MR* **95k**:11032.

Richard T. Bumby, The Diophantine equation $3x^2 - 2y^2 = 1$, *Math. Scand.*, **21**(1967) 144–148.

Cao Zhen-Fu, A study of some Diophantine equations, *J. Harbin Inst. Tech.*, (1988) 1–7.

Cao Zhen-Fu, A generalization of the Schinzel-Sierpiński system of equations (Chinese; English summary), *J. Harbin Inst. Tech.*, **23**(1991) 9–14; *MR* **93b**:11026.

Chen Jian-Hua & Paul Voutier, Complete solution of the Diophantine equation $X^2 + 1 = dY^4$ and a related family of quartic Thue equations, *J. Number Theory*, **62**(1997) 71–99; *MR* **97m**:11039.

Ajai Choudhry, On a quartic diophantine equation, *Math. Student*, **70**(2001) 181–183.

Mihai Cipu, Diophantine equations with at most one positive solution, *Manuscripta Math.*, **93**(1997) 349–356; *MR* **98j**:11017.

J. H. E. Cohn, On the Diophantine equation $z^2 = x^4 + Dx^2y^2 + y^4$, *Glasgow Math. J.*, **36**(1994) 283–285; *MR* **95k**:11035.

J. H. E. Cohn, Twelve Diophantine equations, *Arch. Math. (Basel)*, **65**(1995) 130–133; *MR* **96d**:11031. [the equations $x^2 - kxy^2 \pm y^4 = \pm c$, $c = 1, 2, 4$.]

J. H. E. Cohn, The Diophantine equation $x^4 + 1 = Dy^2$, *Math. Comput.*, **66**(1997) 1347–1351; *MR* **98e**:11032.

J. H. E. Cohn, The Diophantine equation $x^4 - Dy^2 = 1$, II, *Acta Arith.*, **78**(1997) 401–403; *MR* **98e**:11033.

J. H. E. Cohn, The Diophantine system $x^2 - 6y^2 = -5$, $x = 2z^2 - 1$, *Math. Scand.*, **82**(1998) 161–164; *MR* **99h**:11024.

Péter Filakovszky, The Diophantine equation $x^4 + kx^2y^2 + y^4 = z^2$, *Math. Student*, **64**(1995) 105–108.

Kenji Kashihara, The Diophantine equation $x^2 - 1 = (y^2 - 1)(z^2 - 1)$ (Japanese; English summary), *Res. Rep. Anan College Tech. No.* **26**(1990) 119–130; *MR* **91d**:11025.

Kenji Kashihara, On the Diophantine equation $x(x + 1) = y(y + 1)z^2$, *Proc. Japan Acad. Ser. A Math. Sci.*, **72**(1996) 91; *MR* **97b**:11039.

Shin-ichi Katayama & Kenji Kashihara, On the structure of the integer solutions of $z^2 = (x^2 - 1)(y^2 - 1) - a$, *J. Math. Tokushima Univ.*, **24**(1990) 1–11; *MR* **93c**:11013.

Le Mao-Hua, On the Diophantine equation $D_1x^4 - D_2y^2 = 1$, *Acta Arith.*, **76**(1996) 1–9; *MR* **97b**:11040.

W. Ljunggren, Über die Gleichung $x^4 - Dy^2 = 1$, *Arch. Math. Naturvid.*, **45**(1942) 61–70; *MR* **7**, 471.

W. Ljunggren, Sätze über unbestimmte Gleichungen, *Skr. Norske Vid. Akad. Oslo*, (1942) No. 9; *MR* **6**, 169e.

W. Ljunggren, Zur Theorie der Gleichung $x^2 + 1 = Dy^4$, *Avh. Norske Vid. Akad. Oslo*, (1942) No. 5; *MR* **8**, 6f.

W. Ljunggren, On the Diophantine Equation $Ax^4 - By^2 = C$, $(C = 1, 4)$, *Math. Scand.*, **21**(1967) 149–158; *MR* **39** #6820.

Florian Luca, A generalization of the Schinzel-Sierpiński system of equations, *Bull. Math. Soc. Sci. Math. Roumanie (N.S.)*, **41(89)**(1998) 181–195; *MR* **2002m**:11020,

Florian Luca & P. Gary Walsh, A generalization of a theorem of Cohn on the equation $x^3 - Ny^2 = \pm1$, *Rocky Mountain J. Math.*, **31**(2001) 503–509; *MR* **2002g**:11027.

Maurice Mignotte, A note on the equation $x^2 - 1 = y^2(z^2 - 1)$, *C. R. Math. Rep. Acad. Sci. Canada*, **13**(1991) 157–160; *MR* **92j**:11026.

Mu Shan-Zhi, The Diophantine equation $x^4 - 4x^2y^2 + y^4 = 526$, *Natur. Sci. J. Harbin Normal Univ.*, **13**(1997) 12–15.

Paulo Ribenboim, An algorithm to determine the points with integral coordinates in certain elliptic curves, *J. Number Theory*, **74**(1999) 19–38; *MR* **99j**:11028.

A. Schinzel & W. Sierpiński, Sur l'equation diophantienne $(x^2 - 1)(y^2 - 1) = [((y - x)/2)^2 - 1]^2$, *Elem. Math.*, **18**(1963) 132–133; *MR* **29** #1180.

Roel J. Stroeker & Benjamin M. M. de Weger, On a quartic Diophantine equation, *Proc. Edinburgh Math. Soc.*(2), **39**(1996) 97–114.

K. Szymiczek, *Elem. Math.*, **18**(1963) 132–133.

K. Szymiczek, The equation $(x^2 - 1)(y^2 - 1) = (z^2 - 1)^2$, *Eureka*, **35**(1972) 21–25.

Gheorghe Udrea, The Diophantine equations $x^2 - k = 2 \cdot T_n((b^2 \pm 2)/2)$, *Portugal. Math.*, **55**(1998) 255–259.

T. Venkatesh, On integer solutions of $x^4 + 5x^2y^2 + y^4 = z^2$, *Indian J. Pure Appl. Math.*, **26**(1995) 837–839; *MR* **96d**:11035.

Peter Gary Walsh, A note on Ljunggren's theorem about the Diophantine equation $aX^2 - bY^4 = 1$, *C.R. Math. Acad. Sci. Soc. R. Can.*, **20**(1998) 113–118; *MR* **99j**:11029.

Peter Gary Walsh, The Diophantine equation $X^2 - db^2Y^4 = 1$, *Acta. Arith.*, **87**(1998) 179–188; *MR* **2000e**:11031.

P. G. Walsh, An improved method for solving the family of Thue equations $X^4 - 2rX^2Y^2 - sY^4 = 1$, *Number Theory for the Millenium III* (*Urbana IL*, 2000) 375–383, AKPeters, Natick MA, 2002; *MR* **2003k**:11052.

Wang Yan-Bin, On the Diophantine equation $(x^2 - 1)(y^2 - 1) = (z^2 - 1)^2$ (Chinese, English summary), *Heilongjiang Daxue Ziran Kexue Xuebao*, **1989** no. 4 84–85; *MR* **91e**:11028.

H. C. Williams, Note on a diophantine equation, *Elem. Math.*, **25**(1970) 123–125; *MR* **43** # 3207.

Wu Hua-Ming & Le Mao-Hua, A note on the Diophantine equation $(x^2 - 1)(y^2 - 1) = (z^2 - 1)^2$, *Colloq. Math.*, **71**(1996) 133–136.

Xu Xue-Wen, The Diophantine equation $y(y + 1)(y + 2)(y + 3) = p^{2k}x(x + 1)(x + 2)(x + 3)$, *J. Central China Normal Univ. Natur. Sci.*, **31**(1997) 257–259; *MR* **98m**:11018.

Zhong Mei & Deng Mou-Jie, The Diophantine equation $3y(y + 1)(y + 2)(y + 3) = 4x(x + 1)(x + 2)(x + 3)(x + 4)$, *Natur. Sci. J. Harbin Normal Univ.*, **12**(1996) 30–34.

D24 Sum equals product.

For the case $k = 3$ with sum and product 1 there is an extensive literature. See *MR* **99a**:11034 for a pocket survey.

For $k > 2$ the equation $a_1a_2 \cdots a_k = a_1 + a_2 + \cdots + a_k$ has the solution $a_1 = 2$, $a_2 = k$, $a_3 = a_4 = \cdots = a_k = 1$. Schinzel showed that there is no other solution in positive integers for $k = 6$ or $k = 24$. Misiurewicz has shown that $k = 2, 3, 4, 6, 24, 114$ (misprinted as 144 in *Elem. Math.* and in

the first edition of this book), 174 and 444 are the only $k < 1000$ for which
there is exactly one solution. The search has been extended by Singmaster,
Bennett & Dunn to $k \leq 1440000$. They let $N(k)$ be the number of different
'sum = product' sequences of size k, and conjecture that $N(k) > 1$ for all
$k > 444$. They find that $N(k) = 2$ for 49 values of k up to 120000, the
largest being 6174 and 6324, and conjecture that $N(k) > 2$ for $k > 6324$.
They also find that $N(k) = 3$ for 78 values of k in the same range, the
largest being 7220 and 11874, and conjecture that $N(k) > 3$ for $k > 11874$;
also that $N(k) \to \infty$.

This problem seems first to have been asked by Trost, arising from the
solution of $a_1 a_2 \cdots a_k = a_1 + a_2 + \cdots + a_k = 1$ in rationals. For $k = 3$ this
is due to Sierpiński; for $k > 3$ to Schinzel.

On 96-08-19, a month before he died, Erdős wrote:

Do you know anything about the equation

$$x_1 \cdot x_2 \cdots \cdots x_n = n(x_1 + x_2 + \cdots + x_n),$$

$x_1 \leq x_2 \leq \ldots \leq x_n$ where the x_i are positive integers? Denote
the number of solutions in n by $f(n)$; $f(n) > n^\epsilon$ for some ϵ
seems to be true? n is a **champion** if $f(n) > f(m)$ for all
$m < n$, n is an **antichampion** if $f(m) > f(n)$ for all $m > n$.
The antichampions are perhaps always primes, but this would
be the Law of Small Numbers.

Let $a_1 < a_2 < \ldots < a_k \leq x$ be a sequence of integers $\leq x$ no
one of which divides any other. $\max k = \lfloor (x+1)/2 \rfloor$ is of course
well known. Let $A = \{a_1 < a_2 < \ldots < a_k \leq x\}$; denote by
$f(A; n)$ the number of $a_i \leq n$. How large can $\sum_{n=1}^{x} f(A; n)$ be?
Is it true that the maximum is $(x^2)/6 + O(x)$? The maximum
is given perhaps by the integers $x/3 \leq a_i < 2x/3$.

M. L. Brown, On the diophantine equation $\sum X_i = \prod X_i$, *Math. Comput.*,
42(1984) 239–240; *MR* **85d**:11030.

Michael W. Ecker, When does a sum of positive integers equal their product?
Math. Mag., **75**(2002) 41–47.

M. Z. Garaev, On the Diophantine equation $(x + y + z)^3 = nxyz$, *Vestnik
Moskov. Univ. Ser. I Mat. Mekh.*, **2001** 66–67, 72; transl. *Moscow Univ. Math.
Bull.*, **56**(2001) 46; *MR* **2002e**:11032.

M. Misiurewicz, Ungelöste Probleme, *Elem. Math.*, **21**(1966) 90.

A. Schinzel, Triples of positive integers with the same sum and the same
product, *Serdica Math. L.*, **22**(1996) 587–588; *MR* **98g**:11033.

David Singmaster, Untitled Brain Twister, *The Daily Telegraph* weekend section, 88-04-09 & 16.

David Singmaster, Mike Bennett & Andrew Dunn, Sum = product sequences,
(July 1988 preprint).

E. P. Starke & others, Solution to Problem E2262 [1970, 1008] & editorial comment, *Amer. Math. Monthly* **78**(1971) 1021–1022.

E. Trost, Ungelöste Probleme, Nr. **14**, *Elem. Math.*, **11**(1956) 134–135; & **21**(1966) 90.

C. Viola, On the diophantine equations $\prod_0^k x_i - \sum_0^k x_i = n$ and $\sum_0^k \frac{1}{x_i} = \frac{a}{n}$, *Acta Arith.*, **22**(1972/3) 339–352; MR **48** #234.

OEIS: A033178-033179.

D25 Equations involving factorial n.

Are the only solutions of $n! + 1 = x^2$ given by $n = 4$, 5 and 7? Overholt has related this problem to a conjecture of Szpiro. Erdős & Obláth dealt with the equation $n! = x^p \pm y^p$ with $x \perp y$ and $p > 2$. For the case $p = 2$ with the plus sign, see Leech's remark at **D2**; and for the minus sign, split $n!$ into two even factors: $4! = 5^2 - 1^2 = 7^2 - 5^2$; $5! = 11^2 - 1^2 = 13^2 - 7^2 = 17^2 - 13^2 = 31^2 - 29^2$. The number of solutions is $\frac{1}{2}d(n!/4)$.

Simmons notes that $n! = (m-1)m(m+1)$ for $(m,n) = (2,3)$, (3,4), (5,5) and (9,6) and asks if there are other solutions. More generally he asks if there are any other solutions of $n! + x = x^k$. This is a variation on the question of asking for $n!$ to be the product of k consecutive integers in a nontrivial way ($k \neq n+1-j!$). Compare **B23**.

Berend & Osgood have shown that the set of n for which the equation $P(x) = n!$ has an integer solution x has density zero if $P(x)$ is a polynomial of degree ≥ 2 with integer coefficients.

Yu & Liu show that the only solutions of $(p-1)! + a^{p-1} = p^k$ are $(p,a,k) = (3,1,1)$, (5,1,2), (3,5,3).

Zoltán Sasvári mentions the equation

$$\frac{(n+1)(n+2)\cdots(n+k)}{(n-1)(n-2)\cdots(n-k)} = m$$

When $m = 2$ the only solution is $(n,k) = (3,1)$ and for $m = 3$, $(n,k) = (2,1)$. Solutions always occur for $k = n-1$ and $k = n-2$, namely $m = $ the binomial coefficient $\binom{2n-1}{n}$ and the Catalan number $\frac{1}{n}\binom{2n-2}{n-1}$. Selfridge notes that whenever $k = n-3$ gives a solution, then so does $k = n-4$ (e.g., $n = 7$, 16, 21, 29, 43, 46, 67, 78, 89, 92, 105, 111, 127, 141, 154, 157, 171, 188, 191, ...), but not necessarily vice versa (e.g., $n = (4)$, 13, 19, 85, 100, 121, 199, ...). Sasvári found the sporadic solution $n = 101$, $k = 95$, and Marc Paulhus found several others. Here are some other examples where the entries, 8, 7, ... are values of i where $k = n - i$.

$n =$ 211			6	5	4	3
552			6		4	3
990			6	5	4	3
991	8	7		5	4	
1378					5	4
2480			6	5	4	3
2481	8	7	6	5	4	3

Daniel Berend & Charles F. Osgood, On the equation $P(x) = n!$ and a question of Erdős, *J. Number Theory*, **42**(1992) 189–193; *MR* **93e**:11016.

B. Brindza & P. Erdős, On some Diophantine problems involving powers and factorials, *J. Austral. Math. Soc. Ser. A*, **51**(1991) 1–7; *MR* **92i**:11036.

H. Brocard, Question 166, *Nouv. Corresp. Math.*, **2**(1876) 287; Qu. 1532, *Nouv. Ann. Math.*(3) **4**(1885) 391.

Andrzej Bogdan Dąbrowski, On the Diophantine equation $x! + A = y^2$, *Nieuw Arch. Wisk.*,(4) **14**(1996) 321–324; *MR* **97h**:11028.

P. Erdős & R. Obláth, Über diophantische Gleichungen der Form $n! = x^p \pm y^p$ und $n! \pm m! = x^p$, *Acta Szeged*, **8**(1937) 241–255.

Hansraj Gupta, *Math. Student*, **3**(1935) 71.

Le Mao-Hua, On the Diophantine equation $x^{p-1} + (p-1)! = p^n$, *Publ. Math. Debrecen*, **48**(1996) 145–149; *MR* **96m**:11021.

M. Kraitchik, Recherches sur la Théorie des Nombres, tome 1, Gauthier-Villars, Paris, 1924, 38–41.

Florian Luca, The Diophantine equation $P(x) = n!$ and a result of M. Overholt, *Glas. Mat. Ser. III*, **37(57)**(2002) 269–273; *MR* **2003i**:11045.

Florian Luca, On the equation $1!^k + 2!^k + \cdots n!^k = x^2$, *Period. Math. Hungar.*, **44**)(2002) 219–224; *MR* **2003d**:11048.

Marius Overholt, The Diophantine equation $n! + 1 = m^2$, *Bull. London Math. Soc.*, **25**(1993) 104; *MR* **93m**:11026.

Eli Passow, On the solutions of $n! = m!r!$, *Far East J. Math. Sci.*, **6**(1998) 381– 389; *MR* **99i**:11025.

Richard M. Pollack & Harold N. Shapiro, The next to last case of a factorial diophantine equation, *Comm. Pure Appl. Math.*, **26**(1973) 313–325; *MR* **50** #12915.

Gustavus J. Simmons, A factorial conjecture, *J. Recreational Math.*, **1**(1968) 38.

Yu Kun-Rui & Liu De-Hua, A complete resolution of a problem of Erdős & Graham, *Rocky Mountain J. Math.*, **26**(1996) 1235–1244; *MR* **97k**:11049.

OEIS: A000290, A000292, A002378, A005563, A007531, A013468.

D26 Fibonacci numbers of various shapes.

Stark asks which Fibonacci numbers (see **A3**) are half the difference or sum of two cubes. This is related to the problem of finding all complex quadratic fields of class number 2. Examples: $1 = \frac{1}{2}(1^3 + 1^3)$, $8 = \frac{1}{2}(2^3 + 2^3)$,

$13 = \frac{1}{2}(3^3 - 1^3)$. Antoniadis has related all such fields to solutions of certain diophantine equations, and solved them all but two, which were later settled by de Weger.

Cohn showed that the only square Fibonacci numbers are 0, 1 and 144, and Luo Ming confirmed Vern Hoggatt's conjecture that the only triangular ones, i.e. of the form $\frac{1}{2}m(m+1)$ are 0, 1, 3, 21 and 55, and later that the only such Lucas numbers are 1, 3 and 5778. He also shows that the only pentagonal numbers in the two sequences are respectively 1, 5 and 2, 1, 7.

Grossman & Luca show that the largest Fibonacci and Lucas numbers which can be expressed as the sum or difference of two factorials are $144 = 5! + 4!$ and $18 = 4! - 3!$ hGidxfactorial

B. M. M. de Weger has proved that the largest Fibonacci number which is of the form $y^2 - y - 1$ is $u_{11} = 55 = 8^2 - 8 - 1$. His proof is more than 10 pages long and uses the Thue equations. Elementary proof wanted. This reported by Flammenkamp who asks if the only members of the Lucas sequence $v_0 = 2$, $v_1 = 1$ and $v_{n+2} = v_{n+1} + v_n$ which are differences of two consecutive cubes are $v_1 = 1^3 - 0^3$, $v_4 = 2^3 - 1^3$ and $v_{17} = 35^3 - 34^3$.

On 2003-11-25, Yann Bugeaud emailed that he, M. Mignotte & S. Siksek have shown that 0, 1, 8, 144 are the only perfect powers in the Fibonacci sequence. One might also allow -1 and -8.

U. Alfred, On square Lucas numbers, *Fibonacci Quart.*, **2**(1964) 11–12.

Jannis A. Antoniadis, Über die Kennzeichnung zweiklassiger imaginär-quadratischer Zahlkörper durch Lösungen diophantischer Gleichungen, *J. reine angew. Math.*, **339**(1983) 27–81; *MR* **85g**:11098.

S. A. Burr, On the occurrence of squares in Lucas sequences, Abstract 63T-302, *Amer. Math. Soc. Notices*, **10**(1963) 11–12.

J. H. E. Cohn, On square Fibonacci numbers, *J. London Math. Soc.*, **39**(1964) 537–540; *MR* **29** #1166.

J. H. E. Cohn, Square Fibonacci numbers, etc., *Fibonacci Quart.*, **2**(1964) 109–113.

J. H. E. Cohn, Lucas and Fibonacci numbers and some diophantine equations, *Proc. Glasgow Math. Assoc.*, **7**(1965) 24–28; *MR* **31** #2202.

J. H. E. Cohn, Eight Diophantine equations, *Proc. London Math. Soc.*(3), **16**(1966) 153–166; addendum **17**(1967) 381; *MR* **32** #7492; **34** #5748.

J. H. E. Cohn, Five Diophantine equations, *Math. Scand.*, **21**(1967) 61–70; **39** #2702.

J. H. E. Cohn, Squares in some recurrent sequences, *Pacific J. Math.*, **41**(1972) 631–646; *MR* **47** #4914.

J. H. E. Cohn, Perfect Pell powers, *Glasgow Math. J.*, **38**(1996) 19–20; *MR* **97b**:11047.

R. Finkelstein, On Fibonacci numbers which are one more than a square, *J. reine angew. Math.*, **262/263**(1973) 171–178; *MR* **48** #3858.

R. Finkelstein, On Lucas numbers which are one more than a square, *Fibonacci Quart.*, **13**(1975) 340–342; *MR* **54** #10126.

Clemens Fuchs & Robert F. Tichy, Perfect powers in linear recurring sequences, *Acta Arith.*, **107**(2003) 9–25.

George Grossman & Florian Luca, Sums of factorials in binary recurrence series, *J. Number Theory*, **93**(2002) 87–107; *MR* **2002m**:11011.

D. G. Gryte, R. A. Kingsley & H. C. Williams, On certain forms of Fibonacci numbers, *Proc. 2nd Louisiana Conf. Combin. Graph Theory & Comput.*, *Congr. Numer.*, **3**(1971) 339–344.

V. E. Hoggatt, Problem 3, *WA St. Univ. Conf. Number Theory*, 1971, p. 225.

Jiang Ren Yi, The Lucas numbers of the form $n^2 - n - 1$, *J. Zhijiang Univ. Sci. Ed.*, **30**(2003) 121–124; *MR* **2004b**:11017.

B. Krishna Gandhi& M. Janga Reddy, Triangular numbers in the generalized associated Pell sequence, *Indian J. Pure Appl. Math.*, **34**(2003) 1237–1248.

J. C. Lagarias & D. P. Weisser, Fibonacci and Lucas cubes, *Fibonacci Quart.*, **19**(1981) 39–43; *MR* **82d**:10023.

H. London & R. Finkelstein, On Fibonacci and Lucas numbers which are perfect powers, *Fibonacci Quart.*, **7**(1969) 476–481, 487; Corr. **8**(1970) 248; *MR* **41** #144.

Florian Luca, Palindromes in Lucas sequences, *Monatsh. Math.*, **138**(2003) 209–223.

Florian Luca & P. G. Walsh, Squares in Lehmer sequences and some Diophantine applications, *Acta Arith.*, **100**(2000) 47–62; *MR* **2002h**:11024.

Luo Ming, On triangular Fibonacci numbers, *Fibonacci Quart.*, **27**(1989) 98–108; *MR* **90f**:11013.

Luo Ming, On triangular Lucas numbers, *Applications of Fibonacci numbers*, Vol. 4 (1990), 231–240, Kluwer, Dordrecht, 1991; *MR* **93i**:11016.

Luo Ming, Pentagonal numbers in the Lucas sequence, *Portugal. Math.*, **53**(1996) 325–329; *MR* **97e**:11-24; **97j**:11008.

Wayne L. McDaniel, The g.c.d. in Lucas sequences and Lehmer number sequences, *Fibonacci Quart.*, **29**(1991) 24–29.

Wayne L. McDaniel, Pronic Fibonacci numbers, *Fibonacci Quart.*, **36**(1998) 56–59.

Wayne L. McDaniel, Pronic Lucas numbers, *Fibonacci Quart.*, **36**(1998) 60–62.

Wayne L. McDaniel, Perfect squares in the sequence 3, 5, 7, 11, ..., *Portugal. Math.*, **55**(1998) 349–353; *MR* **99j**:11018.

Wayne L. McDaniel & Paulo Ribenboim, Square-classes in Lucas sequences having odd parameters, *J. Number Theory*, **73**(1998) 14–27; *MR* **99j**:11017.

A. Pethö, Perfect powers in second order linear recurrences, *J. Number Theory*, **15**(1982) 5–13; *MR* **84f**:10024.

A. Pethö, Perfect powers in second order recurrences, in: *Topics in Classical Number Theory (Budapest 1981)*, *Colloq. Math. Soc. János Bolyai*, **34**, North-Holland, Amsterdam, 1984, 1217–1227; *MR* **86i**:11009.

A. Pethö, Full cubes in the Fibonacci sequence, *Publ. Math. Debrecen*, **30**(1983) 117–127; *MR* **85j**:11027.

V. Siva Rama Prasad & B. Srinivasa Rao, Pentagonal numbers in the Pell sequence and Diophantine equations $2x^2 = y^2(3y - 1)^2 \pm 2$, *Fibonacci Quart.*, **40**(2002) 233–241; *MR* **2003j**:11028.

Paulo Ribenboim, The terms Cx^h ($h \geq 3$) in Lucas sequences: an algorithm and applications to Diophantine equations, *Acta Arith.*, **106**(2003) 105–114.

Paulo Ribenboim & Wayne L. McDaniel, Square classes of Lucas sequences, *Portugal. Math.*, **48**(1991) 469–473; *MR* **92k**:11015.

Paulo Ribenboim & Wayne L. McDaniel, The square terms in Lucas sequences, *J. Number Theory*, **58**(1996) 104–123; *MR* **99c**:11021; addendum **61**(1996) 420.

Paulo Ribenboim & Wayne L. McDaniel, The square classes in Lucas sequences with odd parameters, *C.R. Math. Rep. Acad. Sci. Canada*, **18**(1996) 223–227; *MR* **97h**:11011.

Paulo Ribenboim & Wayne L. McDaniel, Squares in Lucas sequences having an even first parameter, *Colloq. Math.*, **78**(1998) 29–34; *MR* **99j**:11019.

Neville Robbins, Fibonacci and Lucas numbers of the forms $w^2 - 1$, $w^3 \pm 1$, *Fibonacci Quart.*, **19**(1981) 369–373.

N. Robbins, On Pell numbers of the form PX^2, where P is a prime, *Fibonacci Quart.*, **22**(1984) 340–348.

T. N. Shorey & C. L. Stewart, Pure powers in recurrence sequences and some related Diophantine equations, *J. Number Theory*, **27**(1987) 324–352; *MR* **89a**:11024.

B. Srinivasa Rao, Heptagonal numbers in the Lucas sequence and Diophantine equations $x^2(5x - 3)^2 = 20y^2 \pm 16$, *Fibonacci Quart.*, **40**(2002) 319–322; *MR* **2003d**:11024; *MR* **2003d**:11024.

H. M. Stark, Problem 23, *Summer Institute on Number Theory*, Stony Brook, 1969.

Ray Steiner, On Fibonacci numbers which are one more than a square, *J. reine angew. Math.*, **262/263**(1973) 171–182.

Ray Steiner, On triangular Fibonacci numbers, *Utilitas Math.*, **9**(1976) 319–327.

László Szalay, On the resolution of the equations $U_n = \binom{x}{3}$ and $V_n = \binom{x}{3}$, *Fibonacci Quart.*, **40**(2002) 9–12; *MR* **2002j**:11008.

M. H. Tallman, Problem H23, *Fibonacci Quart.*, **1**(1963) 47.

Paul M. Voutier, Primitive divisors of Lucas and Lehmer sequences III, *Math. Proc. Cambridge Philos. Soc.*, **123**(1998) 407–419.

Charles R. Wall, On triangular Fibonacci numbers, *Fibonacci Quart.*, **23** (1985) 77–79.

B. M. M. de Weger, A diophantine equation of Antoniadis, *Number Theory and Applications* (Banff, 1988), 547–553, *NATO Adv. Sci. Inst. Ser. C: Math. Phys. Sci.*, **265**, Kluwer, 1989; *MR* **92f**:11048.

B. M. M. de Weger, A hyperelliptic Diophantine equation related to imaginary quadratic number fields with class number 2, *J. reine angew. Math.*, **427**(1992) 137–156; *MR* **93d**:11034.

B. M. M. de Weger, A curious property of the eleventh Fibonacci number, *Rocky Mountain J. Math.*, **25**(1995) 977–994; *MR* **96k**:11017.

H. C. Williams, On Fibonacci numbers of the form $k^2 + 1$, *Fibonacci Quart.*, **13**(1975) 213–214; *MR* **52** #250.

O. Wyler, Squares in the Fibonacci series, *Amer. Math. Monthly*, **71**(1964) 220–222.

Minoru Yabuta, Perfect squares in the Lucas numbers, *Fibonacci Quart.*, **40**(2002) 460–466; *MR* **2003m**:11025.

Zhou Chi-Zhong, A general conclusion on Lucas numbers of the form px^2 where p is a prime, *Fibonacci Quart.*, **37**(1999) 39–45; *MR* **99m**:11014.

D27 Congruent numbers.

Congruent numbers are perhaps confusingly named; they are related to Pythagorean triangles and have an ancient history. Several examples (5, 6, 14, the seventeen entries CA in Table 7, and ten more greater than 1000) are in an Arab manuscript more than a thousand years ago. But it is only for the last twenty years, since the work of Tunnell, that we have a reasonably complete understanding of them. They are those integers a for which

$$x^2 + ay^2 = z^2 \quad \text{and} \quad x^2 - ay^2 = t^2$$

have simultaneous integer solutions. Part of the fascination is the often inordinate size of the smallest solutions. For example, $a = 101$ is a congruent number and Bastien gave the smallest solution:

$$x = 20\,1524246294\,9760001961 \qquad y = 1\,1817143185\,2779451900$$

$$z = 23\,3914843530\,6225006961 \qquad t = 16\,2812437072\,7269996961$$

and it spite of improved computing techniques and machines, it may still be some time before some other of the more recalcitrant examples are discovered. Some other large specimens, found by J. A. H. Hunter, M. R. Buckley and K. Gallyas, are given in the first edition of this book.

Congruent numbers are equivalently defined as those a for which there are solutions of the diophantine equation

$$x^4 - a^2y^4 = u^2$$

Dickson's *History* gives many early references, including Leonardo of Pisa (Fibonacci); Genocchi; and Gérardin, who gave 7, 22, 41, 69, 77, the twenty Arabic examples and the forty-three entries CG in Table 7. We need consider only squarefree values of a; of the 608 such that are less than 1000, 361 are congruent and 247 are noncongruent. It has long been conjectured that squarefree numbers are congruent if they are $\equiv 5$, 6 or 7 mod 8. Since the work of Tunnell, this is now known to be true. He has also shown that if n is odd and congruent, then the number of triples satisfying $2x^2 + y^2 + 8z^2 = n$ is twice the number satisfying $2x^2 + y^2 + 32z^2 = n$. Is the converse true?

The entries C5 or C7, and C6 in Table 7 are for primes $\equiv 5$ or 7 mod 8, and the doubles of primes $\equiv 3$ mod 8. Bastien observed that the following are noncongruent: primes $\equiv 3$ mod 8; products of two such primes; the doubles of primes $\equiv 5$ mod 8; the doubles of the products of two such primes; and the doubles of primes $\equiv 9 \bmod 16$; these are the respective entries N3, N9, NX, NL and N2 in Table 7. He gave some other noncongruent numbers (entries NB, though $a = 1$ is due to Fermat, and many others were known earlier, e.g. to Genocchi) and stated that a was noncongruent if it was a prime $\equiv 1 \bmod 8$ with $a = b^2 + c^2$ and $b + c$ a nonresidue (see **F5**) of a. This accounts for several of the entries N1.

Note that the entries 1, 3, 5 and 7 serve as a table of primes in these residue classes mod 8.

Entries C& and N& are from Alter, Curtz & Kubota, and CJ and NJ from Jean Lagrange's thesis.

Table 7. Congruent (C) and Noncongruent (N) Numbers less than 1000. The entry for $a = 40c+r$ is in column   c   and row   r.

r \ c	0	1	2	3	4	5	6	7	8	9	10	11	12	13	14	15	16	17	18	19	20	21	22	23	24	r
1	NB	C1	□	□	CG	N9	N1	N1	N9	□	N1	□	NJ	N1	CG	N1	N1	N9	C&	C1	□	□	N1	N9	□	1
2	NB	NB	N2	NX	□	NX	□	NT	NJ	NX	NJ	CG	NT	□	N2	CG	NJ	NJ	□	NJ	NT	NX	□	NX	NL	2
3	N3	N3	N3	N&	N3	NJ	□	N3	C&	□	NJ	N3	NJ	N3	N3	□	N3	N3	CJ	NJ	NJ	NJ	N3	NJ	□	3
4	□	□	□	□	□	□	□	□	□	□	□	□	□	□	□	□	□	□	□	□	□	□	□	□	□	4
5	C5	□	CG	□	CG	CG	□	C&	□	CJ	□	C&	CJ	□	C&	□	C&	CJ	□	□	C&	□	CJ	□	CL	5
6	C6	C6	C6	□	C6	C6	C&	CA	C6	C&	CJ	C6	□	C6	C6	CJ	CG	□	□	C6	CG	□	C6	C6	CG	6
7	C7	C7	CG	C7	C7	□	CJ	C&	CJ	C7	CJ	CJ	C7	C&	□	C7	C7	CJ	C7	CJ	CJ	□	C7	□	C7	7
8	□	□	□	□	□	□	□	□	□	□	□	□	□	□	□	□	□	□	□	□	□	□	□	□	□	8
9	□	N1	N9	□	N9	N9	□	NJ	□	N1	N1	N9	□	N1	C&	N9	C&	□	N1	N1	N9	C&	N1	NJ		9
10	NX	□	□	NL	N&	CA	□	NL	CA	NL	CG	□	□	NL	NJ	NL	□	NJ	NJ	NJ	□	□	CG	NJ	NJ	10
11	N3	NB	NB	N3	□	N3	N3	CG	N3	CG	NJ	NJ	N3	□	N3	NJ	CG	N3	C&	NJ	N3	NJ	□	□	N3	11
12	□	□	□	□	□	□	□	□	□	□	□	□	□	□	□	□	□	□	□	□	□	□	□	□	□	12
13	C5	C5	CG	CJ	C5	CJ	CJ	C5	□	C5	CJ	CJ	CJ	C&	CL	C5	C5	□	C5	C5	C&	C5	CL	CL	CJ	13
14	C6	□	C6	C6	CG	C6	C6	□	C6	CJ	□	C6	CJ	CJ	C&	C6	CJ	C6	C6	□	C&	CJ	CJ	C6	C&	14
15	CA	CG	CG	□	C&	CG	CJ	CJ	□	CJ	C&	□	C&	□	C&	CJ	CJ	□	□	CJ	□	CJ	C&	□		15
16	□	□	□	□	□	□	□	□	□	□	□	□	□	□	□	□	□	□	□	□	□	□	□	□	□	16
17	N1	N9	N1	C1	N9	NJ	C1	□	N1	NJ	N9	C1	NJ	N9	N1	N1	□	NJ	N9	C&	N9	N1	NT	N1	N1	17
18	□	NX	□	CG	N2	NX	NJ	NX	□	□	NJ	NX	NJ	NX	□	NJ	C&	NX	□	NX	N2	NJ	NT	NT	NJ	18
19	N3	N3	□	N3	N3	C&	NJ	CG	NJ	N3	N3	□	N3	□	NJ	N3	N3	NJ	N3	NJ	□	N3	NJ	NT	NJ	19
20	□	□	□	□	□	□	□	□	□	□	□	□	□	□	□	□	□	□	□	□	□	□	□	□	□	20
21	CA	C5	C5	C&	C5	CA	□	CJ	CJ	CJ	C5	C5	CJ	C5	CJ	□	C5	C5	CG	CJ	C5	CJ	C&	C5	□	21
22	C6	C6	CG	C6	C&	CJ	C6	C6	□	C6	C6	CG	C6	C6	C&	C6	□	C6	CJ	CJ	C6	CJ	CJ	C6	C6	22
23	C7	□	C7	C&	CJ	C7	C7	CJ	□	C7	□	C7	C7	CL	CG	CL	C&	CJ	C7	□	C7	C7	C&	C&	C7	23
24	□	□	□	□	□	□	□	□	□	□	□	□	□	□	□	□	□	□	□	□	□	□	□	□	□	24
25	□	CA	NJ	CG	NJ	□	CG	NJ	NJ	NJ	□	CG	C&	NJ	□	□	NJ	NJ	NJ	NJ	□	NJ	C&	□	C&	25
26	NX	NB	NX	N2	N&	C&	NJ	□	NX	C&	C&	N2	NJ	CA	NX	N2	□	NT	NX	NJ	NJ	C&	NJ	NJ	NJ	26
27	□	N3	N3	□	N&	N3	NJ	N3	□	NJ	N3	□	N3	N3	NT	NJ	NJ	□	N3	N3	□	N3	N3	C&		27
28	□	□	□	□	□	□	□	□	□	□	□	□	□	□	□	□	□	□	□	□	□	□	□	□	□	28
29	C5	CG	C5	C5	□	C5	C5	CJ	C5	C5	CA	CJ	C5	□	CJ	C&	C&	C5	CJ	CJ	C5	CJ	□	C&	CJ	29
30	CA	CA	CA	□	CA	CJ	□	CA	CJ	CG	CG	□	CJ	□	C&	C&	□	CJ	C&	C&	□	□	□	□		30
31	C7	C7	CG	C7	C7	CA	C7	C7	□	C&	C7	CJ	C&	CJ	C7	C&	□	C7	C&	CJ	CJ	C7	C&	C7		31
32	□	□	□	□	□	□	□	□	□	□	□	□	□	□	□	□	□	□	□	□	□	□	□	□	□	32
33	N9	N1	N1	□	N1	N1	NJ	C1	C1	N9	N1	N9	□	NJ	N1	N9	N1	NJ	N9	C&	□	□	N9	N1	N9	33
34	CA	NX	N&	CA	C&	□	N2	NX	NJ	NX	CG	NJ	C&	NX	□	NX	C&	NJ	NL	NX	NJ	NJ	N2	□	NJ	34
35	NB	□	N&	NJ	NJ	NJ	□	NJ	C&	NJ	□	NJ	NJ	NJ	NJ	□	NJ	NJ	NJ	NJ	□	C&	NJ	C&	NJ	35
36	□	□	□	□	□	□	□	□	□	□	□	□	□	□	□	□	□	□	□	□	□	□	□	□	□	36
37	C5	CG	□	C5	C5	C5	CJ	C5	C5	CJ	CL	□	C5	C5	C5	CL	□	C5	CL	C5	C5	□	C5	CL	C5	37
38	C6	CG	C6	C6	□	CJ	C6	C&	C6	C6	CG	C6	C&	□	CJ	CJ	CJ	C6	C6	CG	C6	C6	□	C6	C6	38
39	CG	C7	C&	C&	C7	C7	□	C&	C7	C&	C7	C7	CJ	CJ	C7	□	CJ	C7	CG	C&	C7	C&	C7	C&	□	39
40	□	□	□	□	□	□	□	□	□	□	□	□	□	□	□	□	□	□	□	□	□	□	□	□	□	40
r \ c	0	1	2	3	4	5	6	7	8	9	10	11	12	13	14	15	16	17	18	19	20	21	22	23	24	r

Ronald Alter, The congruent number problem, *Amer. Math. Monthly* **87** (1980) 43–45.

R. Alter & T. B. Curtz, A note on congruent numbers, *Math. Comput.*, **28**(1974) 303–305; *MR* **49** #2527 (not #2504 as in *MR* indexes); correction **30**(1976) 198; *MR* **52** #13629.

R. Alter, T. B. Curtz & K. K. Kubota, Remarks and results on congruent numbers, *Proc. 3rd S.E. Conf. Combin. Graph Theory Comput.*, *Congr. Numer.* **6** (1972) 27–35; *MR* **50** #2047.

L. Bastien, Nombres congruents, *Intermédiaire Math.*, **22**(1915) 231–232.

B. J. Birch, Diophantine analysis and modular functions, *Proc. Bombay Colloq. Alg. Geom.*, 1968.

J. W. S. Cassels, Diophantine equations with special reference to elliptic curves, *J. London Math. Soc.*, **41**(1966) 193–291.

L. E. Dickson, *History of the Theory of Numbers*, Vol. 2, Diophantine Analysis, Washington, 1920, 459–472.

Feng Ke-Qin, Non-congruent numbers, odd graphs and the Birch–Swinnerton-Dyer conjecture, *Acta Arith.*, **75**(1996) 71–83; *MR* **97e**:11077.

A. Genocchi, Note analitiche sopra Tre Sritti, *Annali di Sci. Mat. e Fis.*, **6**(1855) 273–317.

A. Gérardin, Nombres congruents, *Intermédiaire Math.*, **22**(1915) 52–53.

H. J. Godwin, A note on congruent numbers, *Math. Comput.*, **32** (1978) 293–295; **33** (1979) 847; *MR* **58** #495; **80c**:10018.

D. R. Heath-Brown, The size of Selmer groups for the congruent number problem, *Invent. Math.*, **111**(1993) 171–195; II **118**(1994) 331–370; *MR* **93j**:11038; **95h**:11064.

Boris Iskra, Non-congruent numbers with arbitrarily many prime factors congruent to 3 modulo 8, *Proc. Acad. Japan Ser. A Math. Sci.*, **72**(1996) 168–169; *MR* **97h**:11003.

Jean Lagrange, Thèse d'Etat de l'Université de Reims, 1976.

Jean Lagrange, Construction d'une table de nombres congruents, *Bull. Soc. Math. France Mém.* No. 49–50 (1977) 125–130; *MR* **58** #5498.

Franz Lemmermeyer, Some families of non-congruent numbers, *Acta Arith.*, **110**(2003) 15–36.

Paul Monsky, Mock Heegner points and congruent numbers, *Math. Z.*, **204** (1990) 45–68; *MR* **91e**:11059.

Paul Monsky, Three constructions of rational points on $Y^2 = X^3 \pm NX$, *Math. Z.*, **209**(1992) 445–462; *MR* **93d**:11058.

L. J. Mordell, *Diophantine Equations*, Academic Press, London, 1969, 71–72.

Fidel Ronquillo Nemenzo, All congruent numbers less than 40000, *Proc. Japan Acad. Ser. A Math. Sci.*, **74**(1998) 29–31; *MR* **99e**:11070.

Kazunari Noda & Hideo Wada, All congruent numbers less than 10000, *Proc. Japan Acad. Ser. A Math. Sci.*, **69**(1993) 175–178; *MR* **94e**:11064.

S. Roberts, Note on a problem of Fibonacci's, *Proc. London Math. Soc.*, **11**(1879–80) 35–44.

P. Serf, Congruent numbers and elliptic curves, in *Computational Number Theory* (Proc. Conf. Number Theory, Debrecen, 1989), de Gruyter, 1991, 227–238; *MR* **93g**:11068.

N. M. Stephens, Congruence properties of congruent numbers, *Bull. London Math. Soc.*, **7**(1975) 182–184; *MR* **52** #260.

Jerrold B. Tunnell, A classical diophantine problem and modular forms of weight 3/2, *Invent. Math.* **72**(1983) 323–334; *MR* **85d**:11046.

Hideo Wada & Mayako Taira, Computations of the rank of elliptic curve $y^2 = x^3 - \eta^2 x$, *Proc. Japan Acad. Ser. A Math. Sci.*, **70**(1994) 154–157; *MR* **95g**:11052.

Shin-ichi Yoshida, Some variants on the congruent number problem, *Kyushu J. Math.*, **55**(2001) 387–404.

Zhao Chun-Lai, A criterion for elliptic curves with (second) lowest 2-power in $L(1)$, *Math. Proc. Cambridge Philos. Soc.*, **121**(1997) 385–400; **131**(2001) 385–404; II bf134(2003) 407–420; *MR* **97m**:11088; **2003c**:11073.

OEIS: A003273, A006991, A072068-072071.

D28 A reciprocal diophantine equation.

Mordell asked for the integer solutions of

$$\frac{1}{w} + \frac{1}{x} + \frac{1}{y} + \frac{1}{z} + \frac{1}{wxyz} = 0.$$

Several papers have appeared, giving parametric families of solutions. For example, Takahiro Nagashima sends solutions to Mordell's equation: (w, x, y, z) = $(5, 3, 2, -1)$, $(-7, -3, -2, 1)$, $(31, -5, -3, 2)$, $(1366, -15, 7, -13)$, $(n+1, -n, -1, 1)$ and more generally $w = xyz + 1$ with

$$x = -2\epsilon h^3 - \delta \epsilon h^2(n-3) + \epsilon h(n - 1 - 2\delta\epsilon) + 1,$$

$$y = 2\delta\epsilon h^2 + \epsilon h(n-3) - \epsilon\delta(n-1) + 1,$$

$$z = -2\delta\epsilon h^2 - \epsilon h(n-1) - 1,$$

where ϵ, $\delta = \pm 1$ independently, but there seems to be no guarantee that these four two-parameter families give all solutions.

Zhang has shown how to obtain all solutions below a given bound, while Clellie Oursler and Judith Longyear have sent extensive analyses which each give a procedure for finding all solutions. That of Longyear extends to the equation $\sum(1/x_i) + \prod(1/x_i) = 0$ with $n(\geq 3)$ variables x_i in place of Mordell's $n = 4$.

Lawrence Brenton & Daniel S. Drucker, On the number of solutions of $\sum_{j=1}^{s}(1/x_j) + 1/(x_1 \cdots x_s) = 1$, *J. Number Theory*, **44**(1993) 25–29.

Cao Zhen-Fu, Mordell's problem on unit fractions. (Chinese. English summary) *J. Math.* (Wuhan) **7**(1987) 239–244; *MR* **90a**:11032.

Sadao Saito, A diophantine equation proposed by Mordell (Japanese. English summary), *Res. Rep. Miyagi Nat. College Tech.* No. 25 (1988) 101–106; II, *No. 26* (1990) 159–160; *MR* **91c**:11016-7.

Chan Wah-Keung, Solutions of a Mordell Diophantine equation, *J. Ramanujan Math. Soc.*, **6**(1991) 129-140; *MR* **93d**:11033.

Wen Zhang-Zeng, Investigation of the integer solutions of the Diophantine equation $1/w + 1/x + 1/y + 1/z + 1/wxyz = 0$ (Chinese) *J. Chengdu Univ. Natur. Sci.*, **5**(1986) 89–91; *MR* **89c**:11051.

Wu Wei[6], The indefinite equation $\sum_{i=1}^{k} 1/x_i - 1/x_1 \cdots x_k = 1$, *Sichuan Daxue Xuebao*, **34**(1997) 1–5; *MR* **98h**:11038.

Zhang Ming-Zhi, On the diophantine equation $\frac{1}{x} + \frac{1}{y} + \frac{1}{z} + \frac{1}{w} + \frac{1}{xyzw} = 0$, *Acta Math. Sinica (N.S.)*, **1**(1985) 221–224; *MR* **88a**:11033.

D29 Diophantine m-tuples.

A set of positive integers $\{a_1, \ldots, a_m\}$ is called a Diophantine m-tuple with property $D(n)$ if $a_i a_j + n$ is a perfect square for all $1 \le i < j \le m$. Fermat found the property $D(1)$ Diophantine 4-tuple $\{1, 3, 8, 120\}$. Baker & Davenport proved that the Diophantine triple $\{1, 3, 8\}$ can be extended in only one way to a Diophantine 4-tuple, so that it can't be extended to a Diophantine 5-tuple. Dujella & Pethö generalize this by showing that the pair $\{1, 3\}$ can be extended to infinitely many Diophantine 4-tuples, but not to a Diophantine 5-tuple. In fact a Diophantine 5-tuple with property $D(1)$ is not known. Dujella has shown that there cannot exist a 9-tuple with property $D(1)$). A Diophantine 5-tuple with property $D(256)$ and a 6-tuple with property $D(2985984)$ are known, namely $\{1, 33, 105, 320, 18240\}$ (Dujella) and $\{99, 315, 9920, 32768, 44460, 19534284\}$ (Gibbs).

Dujella, Fuchs & Clemens extend the problem from integers to polynomials. The set $\{a_1, \ldots, a_m\}$ forms a polynomial $D(n)$-m-tuple if, for all $1 \le i < j \le m$, $a_i a_j + n$ is the square of a polynomial with integer coefficients. They show that if n is a linear polynomial, then a polynomial $D(n)$-m-tuple has at most 26 elements. The example $\{x, 16x + 8, 25x + 14, 36x + 20\}$ with $n = 16x + 9$ has 4 elements.

Bugeaud & Dujella have extended the problem to higher powers.

A. Baker & H. Davenport, The equations $3x^2 - 2 = y^2$ and $8x^2 - 7 = z^2$, *Quart. J. Math. Oxford Ser.*(2), **20**(1969) 129–137; *MR* **40** #1333.

Yann Bugeaud & Andrej Dujella, On a problem of Diophantus for higher powers, *Math. Proc. Cambridge Philos. Soc.*, **135**(2003) 1–10; *MR* **2004b**:11035.

Andrej Dujella, Some estimates of the number of Diophantine quadruples, *Publ. Math. Debrecen*, **53**(1998) 177–189; *MR* **99k**:11052.

Andrej Dujella, A note on Diophantine quintuples, *Algebraic number theory and Diophantine analysis* (*Graz*, 1998) 123–127, de Gruyter, Berlin, 2000; *MR* **2001f**:11044.

Andrej Dujella, Diophantine m-tuples and elliptic curves, 21st Journées Arithmétiques (Rome, 2001), *J. Théor. Nombres Bordeaux*, **13**(2001) 111–124; *MR* **2002e**:11069.

Andrej Dujella, An absolute bound for the size of Diophantine m-tuples, *J. Number Theory*, **89**(2001) 126–150; *MR* **2002g**:11026.

Andrej Dujella, On the size of Diophantine m-tuples, *Math. Proc. Cambridge Philos. Soc.*, **132**(2002) 23–33; *MR* **2002g**:11036.

Andrej Dujella, Clemens Fuchs & Robert Tichy, Diophantine m-tuples for linear polynomials, *Period. Math. Hungar.*, **45**(2002) 21–33; *MR* **2003j**:11026.

Andrej Dujella & Attila Pethö, A generalization of a theorem of Baker and Davenport, *Quart. J. Math. Oxford Ser.*(2), **49**(1998) 291–306; *MR* **99g**:11035.

P. Gibbs, http://arXiv.org/abs/math.NT/9902081, Some rational Diophantine sextuples. (preprint)

M. J. Jacobson & H. C. Williams, Modular arithmetic on elements of small norm in quadratic fields, *Des. Codes Cryptogr.*, **27**(2002) 93–110; *MR* **2003g**: 11124.

E. Sequences of Integers

Here we are mainly, but not entirely, concerned with infinite sequences; there is some overlap with sections **C** and **A**. An excellent text and source of problems is

H. Halberstam & K. F. Roth, *Sequences*, 2nd edition, Springer, New York, 1982.

Other references are

P. Erdős, A. Sárközi & E. Szemerédi, On divisibility properties of sequences of integers, in *Number Theory, Colloq. Math. Soc. János Bolyai*, **2**, North-Holland, 1970, 35–49; *MR* **43** #4790.

H. Ostmann, *Additive Zahlentheorie* I, II, Springer-Verlag, Heidelberg, 1956.

Carl Pomerance & András Sárközi, Combinatorial Number Theory, in R. Graham, M. Grötschel & L. Lovász (editors) *Handbook of Combinatorics*, North-Holland, Amsterdam, 1993.

N. J. A. Sloane, An on-line version of the encyclopedia of integer sequences, *Electron. J. Combin.*, **1**(1994), Feature 1, approx. 5 pp. (electronic); *MR* **95b**:05001.

N. J. A. Sloane, The on-line encyclopedia of integer sequences, *Notices Amer. Math. Soc.*, **50**(2003) 912–915.

N. J. A. Sloane & Simon Plouffe, *The Encyclopedia of Integer Sequences*, Academic Press, San Diego, 1995; *MR* **96a**:11001.

A. Stöhr, Gelöste und ungelöste Fragen über Basen der natürlichen Zahlenreihe I, II, *J. reine ungew. Math.*, **194**(1955) 40–65, 111–140; *MR* **17**, 713.

Paul Turán (editor), *Number Theory and Analysis; a collection of papers in honor of Edmund Landau* (1877–1938), Plenum Press, New York, 1969, contains several papers, by Erdős and others, on sequences of integers.

We will denote by $\mathcal{A} = \{a_i\}$, $i = 1, 2, \ldots$ a possibly infinite strictly increasing sequence of nonnegative integers. The number of a_i which do not exceed x is denoted by $A(x)$. By the **density** of a sequence we will mean $\lim A(x)/x$, if it exists.

E1 A thin sequence with all numbers equal to a member plus a prime.

Erdős offers \$50.00 for a solution of the problem: does there exist a sequence thin enough that $A(x) < c \ln x$, but with every sufficiently large integer expressible in

311

the form $p + a_i$ where p is a prime?

For the analogous problem with squares in place of primes, Leo Moser showed that $A(x) > (1 + c)\sqrt{x}$ for some $c > 0$, while Erdős showed that there was a sequence with $A(x) < c\sqrt{x}$. Moser's best value for c was 0.06, but this has been improved to 0.147 by Abbott, to 0.245 by Balasubramanian & Soundarajan, and to 0.273 by Cilleruelo. For the rth powers, Cilleruelo obtains

$$A(x) > \frac{x^{1 - \frac{1}{r}}}{\Gamma(2 - \frac{1}{r})\Gamma(1 + \frac{1}{r})}$$

For the problem with powers of two in place of primes, Ruzsa obtained the analog of Erdős's result, but it is not known if there is a constant $c > 0$ such that every sequence A for which every positive integer is representable in the form $a + 2^k$ has $A(x) > (1 + c)\log_2 x$.

H. L. Abbott, On the additive completion of sets of integers, *J. Number Theory*, **17**(1983) 135–143; *MR* **85b**:11011.

R. Balasubramanian & K. Soundarajan, On the additive completion of squares, II, *J. Number Theory*, **40**(1992) 127–129; *MR* **92m**:11020.

Javier Cilleruelo, The additive completion of kth powers, *J. Number Theory*, **44**(1993) 237–243; *MR* **94f**:11093.

P. Erdős, Problems and results in additive number theory, *Colloque sur la Théorie des Nombres, Bruxelles*, 1955, 127–137, Masson, Paris, 1956.

Leo Moser, On the additive completion of sets of integers, *Proc. Symp. Pure Math.*, **8**(1965) Amer. Math. Soc., Providence RI, 175–180; *MR* **31** #150.

I. Ruzsa, On a problem of P. Erdős, *Canad. Math. Bull.*, **15**(1972) 309–310; *MR* **46** #3479.

E2 Density of a sequence with l.c.m. of each pair less than x.

What is the maximum value of $A(x)$ if the least common multiple $[a_i, a_j]$ of each pair of members of the sequence is at most x? It is known that

$$(9x/8)^{1/2} \leq \max A(x) \leq (4x)^{1/2}.$$

The lower bound is obtained by taking all the numbers from 1 up to $\sqrt{x/2}$ and then the even numbers up to $\sqrt{2x}$.

And how many numbers less than x can we find with the greatest common divisor of any pair $< t$ for a given t? If $t < n^{\frac{1}{2}+\epsilon}$, the number $\sim \pi(n)$, while if $t = n^{\frac{1}{2}+c}$, it is $\sim (1 + c')\pi(n)$.

Erdős also asks for bounds on $B(x)$, the smallest number so that any subset of $[1, x]$ of cardinality $B(x)$ always contains three members which have pairwise the same least common multiple. Perhaps $B(x) = o(x)$. Again, let $C(x)$ be the corresponding smallest cardinality, so that there are always three numbers with pairwise the same greatest common divisor. No doubt

$$¿ \quad e^{c_1 (\ln x)^{1/2}} < C(x) < e^{c_2 (\ln x)^{1/2}} \quad ?$$

but the best that Erdős has proved is $C(x) < x^{3/4}$.

Given a sequence A, $a_1 < a_2 < \ldots$, Erdős & Szemerédi denote by $F(A, x, k)$ the number of i for which the l.c.m. $[a_{i+1}, a_{i+2}, \ldots, a_{i+k}] < x$, and ask if it is true that for every $\epsilon > 0$ there is a k for which $F(A, x, k) < x^\epsilon$? They proved that $F(A, x, 3) < c_1 x^{1/3} \ln x$ for every A, and that there is an A for which $F(A, x, 3) > c_2 x^{1/3} \ln x$ for infinitely many x, but they don't know if there is an A for which this is true for *all* x.

Graham, Spencer & Witsenhausen ask how dense can a sequence of integers be so that $\{n, 2n, 3n\}$ never occur?

There is a good bibliography and many unsolved problems in this area in the paper of Erdős, Sárközy & Szemerédi. See also the references at **B24**.

P. Erdős, Problem, *Mat. Lapok* **2**(1951) 233.

P. Erdős & A. Sárközy, On the divisibility properties of sequences of integers, *Proc. London Math. Soc.* (3), **21**(1970) 97–100; *MR* **42** #222.

P. Erdős, A. Sárközy & E. Szemerédi, On divisibility properties of sequences of integers, in Number Theory *Colloq. János Bolyai Math. Soc., Debrecen 1968*, North-Holland, Amsterdam (1970) 35–49; *MR* **43** #4790.

P. Erdős & E. Szemerédi, Remarks on a problem of the *American Mathematical Monthly*, *Mat. Lapok*, **28**(1980) 121–124; *MR* **82c**:10066.

R. L. Graham, J. H. Spencer & H. S. Witsenhausen, On extremal density theorems for linear forms, in H. Zassenhaus (ed), *Number Theory and Algebra*, Academic Press, New York, 1977, 103–109; *MR* **58** #569.

E3 Density of integers with two comparable divisors.

Is it true that the density of those integers

6, 12, 15, 18, 20, 24, 28, 30, 35, 36, 40, 42, 45, 48, 54, 56, 60, 63, 66, 70, 72, ...

which have two divisors d_1, d_2 such that $d_1 < d_2 < 2d_1$, is one? Erdős has shown that the density exists. There is a connexion with covering congruences (**F13**). Since the first edition this has been solved affirmatively by Maier & Tenenbaum.

Let $\mathcal{A}$ be a strictly increasing sequence of integers exceeding 1 and let $\mathcal{M}(\mathcal{A}) := \{ma : a \in \mathcal{A}, m \geq 1\}$ denote its set of multiples. Call $\mathcal{A}$ a **Behrend sequence** if $\mathcal{M}(\mathcal{A})$ has asymptotic density 1. Erdős notes that a central problem is to find general criteria to decide if a given sequence $\mathcal{A}$ is Behrend.

H. Davenport & P. Erdős, On sequences of positive integers, *Acta Arith.*, **2**(1937) 147–151; *J. Indian Math. Soc.*, **15**(1951) 19–24; *MR* **13**,326c.

P. Erdős, On the density of some sequences of integers, *Bull. Amer. Math. Soc.*, **54**(1948) 685–692; *MR* **10**, 105.

P. Erdős, R. R. Hall & G. Tenenbaum, On the densities of sets of multiples, *J. reine angew. Math.*, **454**(1994) 119–141; *MR* **95k**:11115.

R. R. Hall & G. Tenenbaum, On Behrend sequences, *Math. Proc. Cambridge Philos. Soc.*, **112**(1992) 467–482; *MR* **93h**:11026.

Helmut Maier & G. Tenenbaum, On the set of divisors of an integer, *Invent. Math.*, **76**(1984) 121–128; *MR* **86b**:11057.

I. Ruzsa & G. Tenenbaum, A note on Behrend sequences, *Acta Math. Hungar.*, **72**(1996) 327–337; *MR* **98e**:11107.

G. Tenenbaum, Uniform distribution on divisors and Behrend sequences, *Enseign. Math.*(2), **42**(1996) 153–197; *MR* **97e**:11111.

G. Tenenbaum, On block Behrend sequences, *Math. Proc. Cambridge Philos. Soc.*, **120**(1996) 355–367; *MR* **97b**:11008.

OEIS: A005279, A010814.

E4 Sequence with no member dividing the product of r others.

If no member of the sequence $\{a_i\}$ divides the product of $r \geq 2$ other terms, Erdős shows that

$$\pi(x) + c_1 x^{2/(r+1)} (\ln x)^{-2} < A(x) < \pi(x) + c_2 x^{2/(r+1)} (\ln x)^{-2}$$

where $\pi(x)$ is the number of primes $\leq x$ If, however, we suppose that the products of any number, not greater than r, of the a_i are distinct, what is max $A(x)$? For $r \geq 3$, Erdős shows

$$\max A(x) < \pi(x) + O(x^{2/3+\epsilon}).$$

If $r = 1$, so that no term divides any other, the sequence is called **primitive**. Zhang has shown that for a primitive sequence whose members each contain at most four prime factors,

$$\sum_{a_i \leq n} \frac{1}{a_i \ln a_i} \leq \sum_{p \leq n} \frac{1}{p \ln p}$$

(and hence less than 1.64) for $n > 1$, the sums being taken over all members of the sequence up to n and all primes up to n.

A month before Erdős died he emailed:

> Let $a_1 < a_2 < ... < a_k \leq x$ be a sequence of integers $\leq x$ no one of which divides any other. max $k = [(x+1)/2]$ is of course well known. Let $A = \{a_1 < a_2 < \cdots < a_k \leq x\}$, denote by $f(A; n)$ the number of $a_i \leq n$. How large can $\sum_{n=1}^{x} f(A; n)$ be? Is it true that the maximum is $(x^2)/6 + O(x)$? The maximum is given perhaps by the integers $x/3 \leq a_i < 2x/3$.

P. Erdős, On sequences of integers no one of which divides the product of two others and on some related problems, *Inst. Math. Mec. Tomsk*, **2**(1938) 74–82.

P. Erdős, Extremal problems in number theory V (Hungarian), *Mat. Lapok*, **17**(1966) 135–155.

P. Erdős, On some applications of graph theory to number theory, *Publ. Ramanujan Inst.*, **1**(1969) 131–136.

P. Erdős & Zhang Zhen-Xiang, Upper bound of $\sum 1/(a_i \log a_i)$ for primitive sequences, *Proc. Amer. Math. Soc.*, **117**(1993) 891–895.

Zhang Zhen-Xiang, On a conjecture of Erdős on the sum $\sum_{p \leq n} 1/(p \log p)$, *J. Number Theory*, **39**(1991) 14–17; *MR* **92f**:11131.

Zhang Zhen-Xiang, On a problem of Erdős concerning primitive sequences, *Math. Comput.*, **60**(1993) 827–834; *MR* **93k**:11120.

E5 Sequence with members divisible by at least one of a given set.

Let $D(x)$ be the number of numbers not greater than x which are divisible by at least one a_i where $a_1 < a_2 < \cdots < a_k \leq n$ is a finite sequence. Is $D(x)/x < 2D(n)/n$ for all $x > n$? The number 2 cannot be reduced: for example, $n = 2a_1 - 1$, $x = 2a_1 < a_2$. In the other direction it is known that for each $\epsilon > 0$ there is a sequence which does *not* satisfy the inequality $D(x)/x > \epsilon D(n)/n$.

A. S. Besicovitch, On the density of certain sequences, *Math. Ann.*, **110**(1934) 335–341.

P. Erdős, Note on sequences of integers no one of which is divisible by any other, *J. London Math. Soc.*, **10**(1935) 126–128.

E6 Sequence with sums of pairs not members of a given sequence.

Let $n_1 < n_2 < \cdots$ be a sequence of integers such that $n_{i+1}/n_i \to 1$ as $i \to \infty$, and the $\{n_i\}$ are distributed uniformly mod d for every d; i.e., the number $N(c, d; x)$ of the $n_i \leq x$ with $n_i \equiv c \bmod d$ is such that

$$N(c, d; x)/N(1, 1; x) \to 1/d \quad \text{as} \quad x \to \infty$$

for each c, $0 \leq c < d$, and all d. If $a_1 < a_2 < \cdots$ is an infinite sequence for which $a_j + a_k \neq n_i$ for any i, j, k then Erdős asks: is it true that the density of the a_j is less than $\frac{1}{2}$?

A. Khalfalah, S. Lodha & E. Szemerédi, Tight bound for the density of sequence of integers the sum of no two of which is a perfect square, *Discrete Math.*, **256**(2002) 243–255; *MR* **2003e**:11007.

E7 A series and a sequence involving primes.

If p_n is the n th prime, Erdős asks if $\sum(-1)^n n/p_n$ converges. He notes that the series $\sum(-1)^n (n \ln n)/p_n$ diverges.

He also asks if, given three distinct primes and $a_1 < a_2 < a_3 < \cdots$ are all the products of their powers arranged in increasing order, it is true infinitely often that a_i and a_{i+1} are both prime powers. And what if we use k primes or even infinitely many in place of three? Meyer & Tijdeman have asked a similar question for two finite sets S and T of primes with $a_1 < a_2 < a_3 < \cdots$ formed from $S \cup T$. Are there infinitely many i for which a_i is a product of powers of primes from S, while a_{i+1} is a product of powers of primes from T?

E8 Sequence with no sum of a pair a square.

Paul Erdős & David Silverman consider k integers $1 \le a_1 < a_2 < \cdots < a_k \le n$ such that no sum $a_i + a_j$ is a square. Is it true that $k < n(1+\epsilon)/3$, or even that $k < n/3 + O(1)$? The integers $\equiv 1 \bmod 3$ show that if this is true, then it is best possible. They suggest that the same question could be asked for other sequences instead of the squares.

Erdős & Graham added to their book at the proof stage that J. P. Massias has discovered that the sum of any two integers
$$\equiv 1, 5, 9, 13, 14, 17, 21, 25, 26, 29, 30 \bmod 32$$
is never a square mod 32, so k can be chosen to be at least $11n/32$. This is best possible for the modular version of the problem since Lagarias, Odlyzko & Shearer have shown that if $S \subseteq \mathbb{Z}_n$ and $S + S$ contains no square of $\mathbb{Z}_n$, then $|S| \le 11n/32$.

J. C. Lagarias, A. M. Odlyzko & J. B. Shearer, On the density of sequences of integers the sum of no two of which is a square, I. Arithmetic progressions, *J. Combin. Theory Ser. A*, **33**(1982) 167–185; II. General sequences, **34**(1983) 123–139; *MR* **85d**:11015ab.

E9 Partitioning the integers into classes with numerous sums of pairs.

The conjecture of K. F. Roth, that there exists an absolute constant c so that for every k there is an $n_0 = n_0(k)$ with the following property: For $n > n_0$, partition the integers not exceeding n into k classes $\{a_i^{(j)}\}$, $(1 \le j \le k)$; then the number of distinct integers not exceeding n which can be written in the form $a_{i_1}^{(j)} + a_{i_2}^{(j)}$ for some j is greater than cn, has been confirmed by Erdős, Sárközy & Sós.

They also investigate the corresponding problem with products in place of sums, where the problem for $k = 2$ remains open.

Noga Alon, M. Nathanson & I. Z. Ruzsa, Adding distinct congruence classes modulo a prime, *Amer. Math. Monthly*, **102**(1995) 250–255; *MR* **95k**:11009.

Y. Bilu, Addition of sets of integers of positive density, *J. Number Theory*, **64**(1997) 233–275; *MR* **98e**:11013.

P. Erdős & A. Sárközy, On a conjecture of Roth and some related problems, II, in R. A. Mollin (ed.) *Number Theory*, Proc. 1st Conf. Canad. Number Theory Assoc., Banff 1988, de Gruyter, 1990, 125–138.

P. Erdős, A. Sárközy & V. T. Sós, On a conjecture of Roth and some related problems, I, *Colloq. Math. Soc. Janos Bolyai* (1992) –

Andrew Granville & Friedrich Roesler, The set of differences of a given set, *Amer. Math. Monthly*, **106**(1999) 338–344.

I. Z. Ruzsa, Sets of sums and differences, *Sém. Théorie Nombres*, (*Paris* 1982/1983) 267–273.

I. Z. Ruzsa, On the number of sums and differences, *Acta Math. Hungar.*, **59**(1992) 439–447; *MR* **93h**:11025.

I. Z. Ruzsa, Generalized arithmetical progressions and sumsets, *Acta Math. Hungar.*, **65**(1994) 379–388; *MR* **95k**:11011.

E10 Theorem of van der Waerden. Szemerédi's theorem. Partitioning the integers into classes; at least one contains an A.P.

The well-known theorem of van der Waerden states that for every l there is a number $n(h, l)$ such that if the integers not exceeding $n(h, l)$ are partitioned into h classes, then at least one class contains an arithmetic progression (A.P.) containing $l + 1$ terms. More generally, given $l_0, l_1, \ldots, l_{h-1}$, there is always a class V_i ($0 \le i \le h - 1$) containing an A.P. of $l_i + 1$ terms. Denote by $W(h, l)$, or more generally $W(h; l_0, l_1, \ldots, l_{h-1})$, the least such $n(h, l)$.

Chvátal computed $W(2; 2, 2) = 9$, $W(2; 2, 3) = 18$, $W(2; 2, 4) = 22$, $W(2; 2, 5) = 32$ and $W(2; 2, 6) = 46$ and Beeler & O'Neil give $W(2; 2, 7) = 58$, $W(2; 2, 8) = 77$ and $W(2; 2, 9) = 97$. The values $W(2; 3, 3) = 35$ and $W(2; 3, 4) = 55$ were found by Chvátal and $W(2; 3, 5) = 73$ by Beeler & O'Neil. Stevens & Shantaram found $W(2; 4, 4) = 178$; Chvátal found $W(3; 2, 2, 2) = 27$ and Brown $W(3; 2, 2, 3) = 51$. Beeler & O'Neil also found $W(4; 2, 2, 2, 2) = 76$.

Most proofs of van der Waerden's theorem give poor estimates for $W(h, l)$. Erdős & Rado showed that $W(h, l) > (2lh^l)^{\frac{1}{2}}$ and Moser, Schmidt, and Berlekamp successively improved this to

$$W(h,l) > lh^{c\ln h} \quad \text{and} \quad W(h,l) > h^{l+1-c\sqrt{(l+1)\ln(l+1)}}$$

Moser's bound has been improved for $l \geq 5$ by Abbott & Liu to
$$W(h,l) > h^{c_s(\ln h)^s}$$
where s is defined by $2^s \leq l < 2^{s+1}$, and Everts has shown that $W(h,l) > lh^l/4(l+1)^2$, a result which is sometimes better than Berlekamp's. For $h = 2$ Szabó has recently shown that $W(2,l) > 2^l/l^\varepsilon$. All upper bounds were 'ackermanic' in size, until Shelah's proof reduced them to 'wowser' — for an explanation of these terms see the book by Graham, Rothschild & Spencer.

A closely related function, with $l + 1 = k$, is the now famous $r_k(n)$, introduced long years ago by Erdős & Turán: the least r such that the sequence $1 \leq a_1 < a_2 < \cdots < a_r \leq n$ of r numbers not exceeding n must contain a k-term A.P. The best bounds when $k = 3$ are due to Behrend, Roth, and Moser:
$$n\exp(-c_1\sqrt{\ln n}) < r_3(n) < c_2 n/\ln\ln n$$
and for larger k Rankin showed that
$$r_k(n) > n^{1-c_s/(\ln n)^{s/(s+1)}}$$
where s, much as before, is defined by $2^s < k \leq 2^{s+1}$.

A big breakthrough was Szemerédi's proof that $r_k(n) = o(n)$ for all k, but neither his proof, nor those of Furstenberg and of Katznelson & Ornstein (see Thouvenot) give estimates for $r_k(n)$. Erdős conjectures that

$$¿ \qquad r_k(n) = o(n(\ln n)^{-t}) \quad \text{for every } t \qquad ?$$

This would imply that for every k there are k primes in A.P. See **A5** for a potentially remunerative conjecture of Erdős, which, if true, would imply Szemerédi's theorem.

Gowers and others have shown that $r_3(n) < cn/(\ln\ln n)^{2/3}$ and that $r_k(n) < n/(\ln\ln\ln n)^{c_k}$.

Another closely related problem was considered by Leo Moser, who wrote the integers in base three, $n = \sum a_i 3^i$ ($a_i = 0$, 1 or 2) and examined the mapping of n into lattice points $(a_1, a_2, a_3, \ldots)$ of infinite-dimensional Euclidean space. He called integers **collinear** if their images are collinear; e.g., $35 \to (2,2,0,1,0,\ldots)$, $41 \to (2,1,1,1,0,\ldots)$ and $47 \to (2,0,2,1,0,\ldots)$ are collinear. He conjectured that every sequence of integers with no three collinear has density zero. If integers are collinear, they are in A.P., but not necessarily conversely (e.g., $16 \to (1,2,1,0,0,\ldots)$, $24 \to (0,2,2,0,0,\ldots)$ and $32 \to (2,1,0,1,0,\ldots)$ are not collinear) so truth of the conjecture would imply Roth's theorem that $r_3(n) = o(n)$.

If $f_3(n)$ is the largest number of lattice points with no three in line in the n-dimensional cube with three points in each edge, then Moser showed that $f_3(n) > c3^n/\sqrt{n}$. It is easy to see that $f_3(n)/3^n$ tends to a limit; is it zero? Chvátal improved the constant in Moser's result to $3/\sqrt{\pi}$ and found the values $f_3(1) = 2$, $f_3(2) = 6$, $f_3(3) = 16$. It is known that $f_3(4) \geq 43$.

More generally, if the n-dimensional cube has k points in each edge, Moser asked for an estimate of $f_k(n)$, the maximum number of lattice points with no k collinear. It is a theorem of Hales & Jewett, with applications to n-dimensional k-in-a-row (tic-tac-toe), that for sufficiently large n, any partition of the k^n lattice points into h classes has a class with k points in line. This implies van der Waerden's theorem on letting the point $(a_0, a_1, \ldots, a_{n-1})$, $(0 \leq a_i \leq k - 1)$ correspond to the base k expansion of the integer $\sum a_i k^i$. It is not known whether, for every c and sufficiently large n, it is possible to choose $ck^n/\sqrt{n}$ points without including k in line. It *is* known for *some* c. Inequality (4) in Riddell's second paper quoted below implies that

$$f_k(n) > k^{n+1}/(2\pi e^3(k-1)n)^{\frac{1}{2}}$$

so that one can choose a "line-free" set of $ck^n/\sqrt{n}$ points for some c. In the other direction he obtains $f_3(n) \leq 16 \cdot 3^{n-3}$. He acknowledges Leo Moser's inspiration in obtaining these results.

If you use the **greedy algorithm** to construct sequences not containing an A.P. you don't get a very dense sequence, but you do get some interesting ones. Odlyzko & Stanley construct the sequence $S(m)$ of positive integers with $a_0 = 0$, $a_1 = m$ and each subsequent a_{n+1} is the least number greater than a_n so that $a_0, a_1, \ldots, a_{n+1}$ does not contain a three-term A.P. For example

$S(1)$: 0, 1, 3, 4, 9, 10, 12, 13, 27, 28, 30, 31, 36, 37, 39, 40, 81, 82, 84, 85, 90, 91, 93, 94, 108, 109, 111, 112, 117, 118, 120, ...

$S(4)$: 0, 4, 5, 7, 11, 12, 16, 23, 26, 31, 33, 37, 38, 44, 49, 56, 73, 78, 80, 85, 95, 99, 106, 124, 128, 131, 136, 143, ...

If m is a power of three, or twice a power of three, then the members of the sequence are fairly easy to describe (write $S(1)$ in base 3), but for other values the sequences behave quite erratically. Their rates of growth seem to be similar, but this has yet to be proved.

The "simplest" such sequence containing no four-term A.P. is

0, 1, 2, 4, 5, 7, 8, 9, 14, 15, 16, 18, 25, 26, 28, 29, 30, 33, 36, 48, 49, 50, 52, 53, 55, 56, 57, 62, ...

Is there a simple description of this? How fast does it grow?

If we define the **span** of a set S to be $\max S - \min S$, what is the smallest span $\mathrm{sp}(k, n)$ of a set of n integers containing no k-term A.P.? The following values have been corrected and extended by Rainer Rosenthal:

$$n = 3\ 4\ 5\ 6\ \ \ 7\ \ \ 8\ \ \ 9\ \ 10\ 11\ 12\ 13\ 14\ 15\ 16\ 17 \ldots$$
$$\mathrm{sp}(3,n) = 3\ 4\ 8\ 10\ 12\ 13\ 19\ 23\ 25\ 29\ 31\ 35\ 39\ 40\ 50 \ldots$$
$$\mathrm{sp}(4,n) = \ \ \ \ 4\ 5\ 7\ \ \ 8\ \ \ 9\ \ 12\ 14\ 16\ 18\ 20\ 22\ 24\ 26\ 27 \ldots$$

Abbott notes that it follows from Szemerédi's theorem that for each

$k \geq 3$, the sequence $\{\text{sp}(k, n + 1) - \text{sp}(k, n)\}$ is unbounded, and asks if it contains a bounded subsequence.

A paper of Alfred Brauer, with a magnificent early bibliography, which is relevant to sections **E10** to **E14**, is referred to in **F6**.

H. L. Abbott & D. Hanson, Lower bounds for certain types of van der Waerden numbers, *J. Combin. Theory*, **12**(1972) 143–146.

H. L. Abbott & A. C. Liu, On partitioning integers into progression free sets, *J. Combin. Theory*, **13**(1972) 432–436; *MR* **46** #3325.

H. L. Abbott, A. C. Liu & J. Riddell, On sets of integers not containing arithmetic progressions of prescribed length, *J. Austral. Math. Soc.*, **18**(1974) 188–193; *MR* **57** #12441.

Noga Alon & Ayal Zaks, Progressions in sequences of nearly consecutive integers, *J. Combin. Theory Ser. A*, **84**(1998) 99–109; *MR* **99g**:11018.

Michael D. Beeler & Patrick E. O'Neil, Some new van der Waerden numbers, *Discrete Math.*, **28**(1979) 135–146; *MR* **81d**:05003.

F. A. Behrend, On sets of integers which contain no three terms in arithmetical progression, *Proc. Nat. Acad. Sci. USA* **32**(1946) 331–332; *MR* **8**, 317.

V. Bergelson & A. Leibman, Polynomial extensions of van der Waerden's and Szemerédi's theorems, *J. Amer. Math. Soc.*, **9**(1996) 725–753; *MR* **99j**:11013.

E. R. Berlekamp, A construction for partitions which avoid long arithmetic progressions, *Canad. Math. Bull.*, **11**(1968) 409–414; *MR* **38** #1066.

E. R. Berlekamp, On sets of ternary vectors whose only linear dependencies involve an odd number of vectors, *Canad. Math. Bull.*, **13**(1970) 363–366; *MR* **43** #3277.

J. Bourgain, On arithmetic progressions in sums of sets of integers, *A tribute to Paul Erdős*, 105–109, Cambridge Univ. Press, 1990; *MR* **92e**:11011.

Thomas C. Brown, Some new Van der Waerden numbers, Abstract 74T-A113, *Notices Amer. Math. Soc.*, **21**(1974) A-432.

T. C. Brown, Behrend's theorem for sequences containing no k-element progression of a certain type, *J. Combin. Theory Ser. A*, **18**(1975) 352–356; *MR* **51** #5543.

Tom C. Brown & Peter Shiue Jau-Shyong, On the history of van der Waerden's theorem on arithmetic progressions, *Tamkang J. Math.*, **32**(2001) 335–341; *MR* **2002h**:11012.

Tom C. Brown, Ronald L. Graham & Bruce M. Landman, On the set of common differences in van der Waerden's theorem on arithmetic progressions, *Canad. Math. Bull.*, **42**(1999) 25–36; *MR* **2000f**:11010.

Ashok K. Chandra, On the solution of Moser's problem in four dimensions, *Canad. Math. Bull.*, **16**(1973) 507–511; *MR* **50** #1944.

V. Chvátal, Some unknown van der Waerden numbers, in *Combinatorial Structures and their Applications*, Gordon and Breach, New York, 1970, 31–33; *MR* **42** #1793.

Thierry Coquand, A constructive topological proof of van der Waerden's theorem, *J. Pure Appl. Algebra* **105**(1995) 251–259; *MR* **97b**:03080.

J. A. Davis, R. C. Entringer, R. L. Graham & G. J. Simmons, On permutations containing no long arithmetic progressions, *Acta Arith.*, **34**(1977/78) 81–90; *MR* **58** #10705.

W. Deuber, On van der Waerden's theorem on arithmetic progressions, *J. Combin. Theory Ser. A*, **32**(1982) 115–118; *MR* **83e**:10082.

P. Erdős, Some recent advances and current problems in number theory, in *Lectures on Modern Mathematics*, Wiley, New York, **3**(1965) 196–244.

P. Erdős & R. Rado, Combinatorial theorems on classifications of subsets of a given set, *Proc. London Math. Soc.*(3), **2**(1952) 417–439; *MR* **16**, 445.

P. Erdős & J. Spencer, *Probabilistic Methods in Combinatorics*, Academic Press, 1974, 37–39.

P. Erdős & P. Turán, On some sequences of integers, *J. London Math. Soc.*, **11**(1936) 261–264.

F. Everts, PhD thesis, Univ. of Colorado, 1977.

H. Furstenberg, Ergodic behaviour of diagonal measures and a theorem of Szemerédi on arithmetic progressions, *J. Analyse Math.*, **31**(1977) 204–256; *MR* **58** #16583.

H. Furstenberg & E. Glasner, Subset dynamics and van der Waerden's theorem, *Topological dynamics and applications* (*Minneapolis MN*, 1995) 197–203, *Contemp. Math.*, **215** Amer. Math. Soc., 1998; *MR* **99d**11010.

Joseph L. Gerver & L. Thomas Ramsey, Sets of integers with no long arithmetic progressions generated by the greedy algorithm, *Math. Comput.*, **33**(1979) 1353–1359; *MR* **80k**:10053.

Joseph Gerver, James Propp & Jamie Simpson, Greedily partitioning the natural numbers into sets free of arithmetic progressions, *Proc. Amer. Math. Soc.*, **102**(1988) 765–772; *MR* **89f**:11026.

W. T. Gowers, A new proof of Szemerédi's theorem for arithmetic progressions of length four, *Geom. Funct. Anal.*, **8**(1998) 529–551; *MR* **2000d**:11019.

W. T. Gowers, A new proof of Szemerédi's theorem, *Geom. Funct. Anal.*, **11**(2001) 465–588; *MR* **2002k**:11014; erratum **11**(2001) 869.

W. T. Gowers, Arithmetic progressions in sparse sets, *Current developments in mathematics*, 2000, 149–196, Int. Press, Somerville MA, 2001; *MR* **2003a**:11011.

R. L. Graham & B. L. Rothschild, A survey of finite Ramsey theorems, *Proc. 2nd Louisiana Conf. Combin., Graph Theory, Comput., Congr. Numer.*, **3**(1971) 21–40; *MR* **47** #4805.

R. L. Graham & B. L. Rothschild, A short proof of van der Waerden's theorem on arithmetic progressions, *Proc. Amer. Math. Soc.*, **42**(1974) 385–386; *MR* **48** #8257.

Ronald L. Graham, Bruce L. Rothschild & Joel H. Spencer, *Ramsey Theory*, 2nd edition, Wiley-Interscience, 1990.

G. Hajós, Über einfache und mehrfache Bedeckungen des n-dimensionalen Raumes mit einem Würfelgitter. *Math. Z.*, **47**(1942) 427–467; *MR* **3**, 302b.

A. W. Hales & R. I. Jewett, Regularity and positional games, *Trans. Amer. Math. Soc.*, **106**(1963) 222–229; *MR* **26** #1265.

A. Y. Khinchin, *Three Pearls of Number Theory*, Graylock Press, Rochester NY, 1952, 11-17.

Bruce M. Landman, An upper bound for van der Waerden-like numbers using k colors, *Graphs Combin.*, **9**(1993) 177–184; *MR* **94d**:11006.

Bruce M. Landman & Raymond N. Greenwell, Some new bounds and values for van der Waerden-like numbers, *Graphs Combin.*, **6**(1990) 287–291; *MR* **91k**:11023.

G. Mills, A quintessential proof of van der Waerden's theorem on arithmetic progressions, *Discrete Math.*, **47**(1983) 117–120; *MR* **84k**:05014.

L. Moser, On non-averaging sets of integers, *Canad. J. Math.*, **5**(1953) 245–252; *MR* **14**, 726d, 1278.

Leo Moser, Notes on number theory II. On a theorem of van der Waerden, *Canad. Math. Bull.*, **3**(1960) 23–25; *MR* **22** #5619.

L. Moser, Problem 21, *Proc. Number Theory Conf.*, Univ. of Colorado, Boulder, 1963, 79.

L. Moser, Problem 170, *Canad. Math. Bull.*, **13**(1970) 268.

Irene Mulvey, Recurrent ideas in number theory: the multiple Birkhoff recurrence theorem used to prove van der Waerden's theorem, *Math. Mag.*, **70**(1997) 358–361.

A. M. Odlyzko & R. P. Stanley, Some curious sequences constructed with the greedy algorithm, Bell Labs. internal memo, 1978.

Carl Pomerance, Collinear subsets of lattice-point sequences – an analog of Szemerédi's theorem, *J. Combin. Theory Ser. A*, **28**(1980) 140–149; *MR* **81m**: 10104.

Jim Propp, What are the laws of greed?, *Amer. Math. Monthly* **96**(1989) 334–336.

John R. Rabung, On applications of van der Waerden's theorem, *Math. Mag.*, **48**(1975) 142–148.

John R. Rabung, Some progression-free partitions constructed using Folkman's method, *Canad. Math. Bull.*, **22**(1979) 87–91; *MR* **51** #3092.

R. Rado, Note on combinatorial analysis, *Proc. London Math. Soc.*, **48**(1945) 122–160; *MR* **5**, 87a.

R. A. Rankin, Sets of integers containing not more than a given number of terms in arithmetical progression, *Proc. Roy. Soc. Edinburgh Sect. A*, **65** (1960/61) 332–334; *MR* **26** #95.

J. Riddell, On sets of numbers containing no l terms in arithmetic progression, *Nieuw Arch. Wisk.* (3), **17**(1969) 204–209; *MR* **41** #1678.

J. Riddell, A lattice point problem related to sets containing no l-term arithmetic progression, *Canad. Math. Bull.*, **14**(1971) 535–538; *MR* **48** #265.

K. F. Roth, Sur quelques ensembles d'entiers, *C.R. Acad. Sci. Paris*, **234**(1952) 388–390; *MR* **13**, 724d.

K. F. Roth, On certain sets of integers, *J. London Math. Soc.*, **28**(1953) 104–109; *MR* **14**, 536; (& see **29**(1954) 20–26; *MR* **15**, 288f); *J. Number Theory*, **2**(1970) 125–142; *MR* **41** #6810; *Period. Math. Hungar.*, **2**(1972) 301–326; *MR* **51** #5546.

R. Salem & D. C. Spencer, On sets of integers which contain no three terms in arithmetical progession, *Proc. Nat. Acad. Sci.*, **28**(1942) 561–563; *MR* **4**, 131.

R. Salem & D. C. Spencer, On sets which do not contain a given number in arithmetical progession, *Nieuw Arch. Wisk.* (2), **23**(1950) 133–143; *MR* **11**, 417e.

H. Salié, Zur Verteilung natürlicher Zahlen auf elementfremde Klassen, *Ber. Verh. Sächs. Akad. Wiss. Leipzig*, 4(1954) 2–26; *MR* **15**, 934e.

Wolfgang M. Schmidt, Two combinatorial theorems on arithmetic progressions, *Duke Math. J.*, **29**(1962) 129–140; *MR* **25** #1125.

S. Shelah, Primitive recursive bounds for van der Waerden numbers, *J. Amer. Math. Soc.*, **1**(1988) 683–697; *MR* **89a**:05017.

G. J. Simmons & H. L. Abbott, How many 3-term arithmetic progressions can there be if there are no longer ones? *Amer. Math. Monthly* **84**(1977) 633–635; *MR* **57** #3056.

R. S. Stevens & R. Shantaram, Computer generated van der Waerden partitions, *Math. Comput.*, **32**(1978) 635–636; *MR* **58** #10714.

Zoltán István Szabó, An application of Lovász' local lemma—a new lower bound for the van der Waerden number, *Random Structures Algorithms*, **1**(1990) 343–360; *MR* **92c**:11011.

E. Szemerédi, On sets of integers containing no four terms in arithmetic progression, *Acta Math. Acad. Sci. Hungar.*, **20**(1969) 89–104; *MR* **39** #6861.

E. Szemerédi, On sets of integers containing no k elements in arithmetic progression, *Acta Arith.*, **27**(1975) 199–245; *MR* **51** #5547.

J. P. Thouvenot, La démonstration de Furstenberg du théorème de Szemerédi sur les progressions arithmétiques, *Lect. Notes in Math.*, Springer Berlin, **710**(1979) 221–232; *MR* **81c**:10072.

V. H. Vu, On a question of Gowers, *Ann. Comb.*, **6**(2002) 229–233; *MR* **2003k**:11013.

B. L. van der Waerden, Beweis einer Baudet'schen Vermutung, *Nieuw Arch. Wisk.* (2), **15**(1927) 212–216.

B. L. van der Waerden, How the proof of Baudet's conjecture was found, in *Studies in Pure Mathematics*, Academic Press, London, 1971, 251–260; *MR* **42** #5764 (no review).

E. Witt, Ein kombinatorische Satz der Elementargeometrie, *Math. Nachr.*, **6**(1952) 261–262; *MR* **13**, 767d.

OEIS A003278, A005048, A005487, A005823, A005836, A005839, A032924, A054591.

E11 Schur's problem. Partitioning integers into sum-free classes.

Schur proved that if the integers less than $n!e$ are partitioned into n classes in any way, then $x + y = z$ can be solved in integers within one class. Let $s(n)$ be the largest integer such that there exists a partition of the integers $[1, s(n)]$ into n classes with no solutions in any class. Abbott & Moser obtained the lower bound $s(n) > (89)^{n/4 - c \ln n}$ for some c and all sufficiently large n and Abbott & Hanson obtained $s(n) > c(89)^{n/4}$, improving Schur's own estimate of $s(n) \geq (3^n + 1)/2$. This last result is in fact sharp for $n = 1$, 2 and 3, but it is too low for larger values of n. The value $s(4) = 44$ was computed by Baumert: for example, the first 44 numbers may be split into four sum-free classes

$$\{1,3,5,15,17,19,26,28,40,42,44\}, \qquad \{2,7,8,18,21,24,27,33,37,38,43\},$$
$$\{4,6,13,20,22,23,25,30,32,39,41\}, \qquad \{9,10,11,12,14,16,29,31,34,35,36\}.$$

Later Fredricksen showed that $s(5) \geq 157$ (see **E12** for his example) and this improves the lower bound for all subsequent Schur numbers: $s(n) \geq c(315)^{n/5}$ $(n > 5)$.

Robert Irving has slightly improved Schur's upper bound from $\lfloor n!e \rfloor$ to $\lfloor n!(e - \frac{1}{24}) \rfloor$. This result also appears in O'Sullivan's Ph.D. thesis (see **E28**). Eugene Levine says that this seems to be the best that can be deduced from Jon Folkman's result that the Ramsey number $R(3,3,3,3) \leq 65$. Also, Schinzel notes that the result ascribed to Irving was attributed by the latter to Earl Glen Whitehead.

If s and y are required to be distinct, then Irving showed that the corresponding bound is $\lfloor \frac{1}{2}(2n + 1)e \cdot n! \rfloor + 2$, which Bornsztein slightly improves to $\lfloor n!ne \rfloor + 1$.

Denote by $v = \sigma(m, n)$ the least integer v such that any partition of $\{1, 2, \ldots, v\}$ into n subsets has a part containing $a_1, \ldots, a_m$ (not necessarily distinct) which satisfy $a_1 + \ldots + a_{m-1} = a_m$, i.e., $s(n) = \sigma(3, n)$. Beutelspacher & Brestovansky note that $\sigma(m, 1) = m - 1$ and $\sigma(2, n) = 1$ and prove that $\sigma(m, 2) = m^2 - m - 1$. They exhibit 3-sumfree 6- and 7-partitions that show that $\sigma(3, 6) \geq 476$ & $\sigma(3, 7) \geq 1430$. Hence $\sigma(3, n) \geq \frac{1}{2}(2859 \cdot 3^{n-7} + 1)$ for $n \geq 7$. They also define and investigate Schur numbers of arithmetic progressions.

E. & G. Szekeres and Schönheim have considered what Bill Sands calls an un-Schur problem. Call a partition of the integers $[1, n]$ into three classes **admissible** if there is *no* solution to $x + y = z$ with x, y, z in *distinct* classes. There is no admissible partition with the size of each class $> \frac{1}{4}n$.

Is it true that if the integers are split into r classes, then some class contains three distinct integers x, y, z satisfying $\frac{1}{x} + \frac{1}{y} = \frac{1}{z}$? T. C. Brown has verified this for $r = 2$.

Harvey L. Abbott, PhD thesis, Univ. of Alberta, 1965.

H. L. Abbott & D. Hanson, A problem of Schur and its generalizations, *Acta Arith.*, **20**(1972) 175–187; *MR* **47** #8475.

H. L. Abbott & L. Moser, Sum-free sets of integers, *Acta Arith.*, **11**(1966) 393–396; *MR* **34** #69.

L. D. Baumert, Sum-free sets, *Jet Propulsion Lab. Res. Summary, No. 36-10,* **1**(1961) 16–18.

Albrecht Beutelspacher & Walter Brestovansky, Generalized Schur numbers, in Combinatorial Theory, *Springer Lecture Notes in Math.*, **969**(1982) 30–38; *MR* **84f**:10019.

Pierre Bornsztein, On an extension of a theorem of Schur, *Acta Arith.*, **101** (2002) 395–399; *MR* **2002k**:11028.

S. L. G. Choi, The largest sum-free subsequence from a sequence of n numbers, *Proc. Amer. Math. Soc.*, **39**(1973) 42–44; *MR* **47** #1771.

S. L. G. Choi, J. Komlós & E. Szemerédi, On sum-free subsequences, *Trans. Amer. Math. Soc.*, **212**(1975) 307–313; *MR* **51** #12769.

Paul Erdős, Some problems and results in number theory, in *Number Theory and Combinatorics* (Japan, 1984) World Sci. Publishing, Singapore, 1985, 65–87;

MR **87g**:11003.

H. Fredricksen, Five sum-free sets, *Proc. 6th SE Conf. Graph Theory, Combin. & Comput., Congressus Numerantium* **14** Utilitas Math., 1975, 309–314; *MR* **52** #13715.

R. W. Irving, An extension of Schur's theorem on sum-free partitions, *Acta Arith.*, **25**(1973/74) 55–64; *MR* **49** #2418.

J. Komlós, M. Sulyok & E. Szemerédi, Linear problems in combinatorial number theory, *Acta Math. Acad. Sci. Hungar.*, **26**(1975) 113–121; *MR* **51** #342.

L. Mirsky, The combinatorics of arbitrary partitions, *Bull. Inst. Math. Appl.*, **11**(1975) 6–9; *MR* **56** #148.

J. Schönheim, On partitions of the positive integers with no x, y, z belonging to distinct classes satisfying $x + y = z$, *Number Theory, Proc. 1st Conf. Canad. Number Theory Assoc., Banff* 1988, de Gruyter, 1990, 515–528; *MR* **92d**:11018.

I. Schur, Über die Kongruenz $x^m + y^m \equiv z^m$ mod p, *Jahresb. Deutsche Math.-Verein.*, **25**(1916) 114–117.

Esther & George Szekeres, Adding numbers, *James Cook Math. Notes*, **4** no. 35(1984) 4073-4075.

W. D. Wallis, A. P. Street & J. S. Wallis, *Combinatorics: Room Squares, Sum-free Sets, Hadamard Matrices*, Springer-Verlag, 1972.

Earl Glen Whitehead, The Ramsey number $N(3, 3, 3, 3; 2)$, *Discrete Math.*, **4**(1973) 389–396; *MR* **47** #3229.

Š. Znám, Generalisation of a number-theoretic result, *Mat.-Fyz. Časopis*, **16**(1966) 357–361; *MR* **35** #78.

Š. Znám, On k-thin sets and n-extensive graphs, *Math. Časopis*, **17**(1967) 297–307; *MR* **38** #1022.

OEIS: A030126, A045652.

E12 The modular version of Schur's problem.

A similar problem to Schur's was considered by Abbott & Wang. Let $t(n)$ be the largest integer m so that there is a partition of the integers from 1 to m into n classes, with no solution to the congruence

$$x + y \equiv z \bmod (m + 1)$$

in any class. Clearly $t(n) \leq s(n)$, where $s(n)$ is as in Schur's problem (**E11**), but for $n = 1$, 2 or 3, we have equality, $t(1) = s(1) = 1$, $t(2) = s(2) = 4$, $t(3) = s(3) = 13$. Indeed, the only three partitions of [1,13] into three sets satisfying the sum-free condition,

$$\{1, 4, 10, 13\} \quad \{2, 3, 11, 12\} \quad \{5, 6, 8, 9\}$$

(with 7 in any of the three sets) all satisfy the seemingly more restrictive congruence-free condition, modulo 14, while Baumert's example (**E11**) shows only one failure: $33 + 33 \equiv 21$ mod 45 in the second set. In fact Baumert found 112 ways of partitioning [1,44] into four sum-free sets, and some of these are sum-free mod 45, so $t(4) = 44$. An example is

$$\{\pm 1, \pm 3, \pm 5, 15, \pm 17, \pm 19\}, \quad \{\pm 2, \pm 7, \pm 8, \pm 18, \pm 21\}$$
$$\{\pm 4, \pm 6, \pm 13, \pm 20, \pm 22, 30\}, \quad \{\pm 9, \pm 10, \pm 11, \pm 12, \pm 14, \pm 16\}.$$

Abbott & Wang obtained the inequality

$$f(n_1 + n_2) \geq 2f(n_1)f(n_2)$$

which holds for $f(n) = s(n) - \frac{1}{2}$ and leads to the same lower bound that Schur obtained for his problem, $t(n) \geq (3^n + 1)/2$. Indeed, they obtain evidence that $t(n) = s(n)$. Moreover, the example of Fredricksen

$$\pm\{1, 4, 10, 16, 21, 23, 28, 34, 40, 43, 45, 48, 54, 60\},$$
$$\pm\{2, 3, 8, 9, 14, 19, 20, 24, 25, 30, 31, 37, 42, 47, 52, 65, 70\},$$
$$\pm\{5, 11, 12, 13, 15, 29, 32, 33, 35, 36, 39, 53, 55, 56, 57, 59, 77, 79\},$$
$$\pm\{6, 7, 17, 18, 22, 26, 27, 38, 41, 46, 50, 51, 75\},$$
$$\pm\{44, 49, 58, 61, 62, 63, 64, 66, 67, 68, 69, 71, 72, 73, 74, 76, 78\},$$

which shows that $s(5) \geq 157$, is also sum-free mod 158 so that $t(5) \geq 157$ and $t(n) > c(315)^{n/5}$ as well.

Erdős lets $f(n)$ be the smallest integer for which the integers less than n can be partitioned into $f(n)$ classes so that n is not the sum of distinct members of the same class. For example, $f(11) = 2$, because of the partition $\{1, 3, 4, 5, 9\}$, $\{2, 6, 7, 8, 10\}$, but $f(12) = 3$. Erdős can prove $f(n) < n^{1/3}/\ln n$ but is unable to show that $f(n) > n^{1/3-\epsilon}$.

Alon & Kleitman call a subset A of a commutative group **sum-free** if no sum of two elements of A is in A, i.e., $(A + A) \cap A$ is empty, and they show that every set of n nonzero elements of such a group contains a sum-free subset of cardinality $> \frac{2}{7}n$. That $\frac{2}{7}$ is best possible follows from a result of Rhemtulla & Street, though it can be improved for particular groups. They also show that any set of n nonzero integers contains a sum-free subset of cardinality $> \frac{1}{3}n$, where $\frac{1}{3}$ cannot be replaced by $\frac{12}{29}$. Füredi notes that the set $\{1,2,3,4,5,6,8,9,10,18\}$ shows that it cannot be replaced by $\frac{2}{5}$: is $\frac{1}{3}$ best possible?

See also **C9**.

H. L. Abbott & E. T. H. Wang, Sum-free sets of integers, *Proc. Amer. Math. Soc.*, **67**(1977) 11-16; *MR* **58** #5571.

Noga Alon, Independent sets in regular graphs and sum-free subsets of finite groups, *Israel J. Math.*, **73**(1991) 247–256; *MR* **92k**:11024.

Noga Alon & Daniel J. Kleitman, Sum-free subsets, *A tribute to Paul Erdős*, Cambridge Univ. Press, Cambridge, 1990, 13–26; *MR* **92f**:11020.

Neil J. Calkin, On the number of sum-free sets, *Bull. London Math. Soc.*, **22**(1990) 141–144; *MR* **91b**:11015.

H. Fredricksen, Schur numbers and the Ramsey number $N(3, 3, \ldots, 3; 2)$, *J. Combin. Theory Sec. A*, **27**(1979), 376–377; *MR* **81d**:05001.

Vsevolod F. Lev & Tomasz Schoen, Cameron-Erdős modulo a prime, *Finite Fields Appl.*, **8**(2002) 108–119; *MR* **2002k**:11029.

Vsevolod F. Lev, Tomasz Łuczak & Tomasz Schoen, Sum-free sets in abelian groups, *Israel J. Math.*, **125**(2001) 347–367; *MR* **2002i**:11023.

A. H. Rhemtulla & Anne Penfold Street, Maximum sum-free sets in elementary Abelian p-groups, *Canad. Math. Bull.*, **14**(1971) 73–80; *MR* **45** #2017; and see *Bull. Austral. Math. Soc.*, **2**(1970) 289–297; *MR* **41** #8519.

OEIS: A030126, A045652.

E13 Partitioning into strongly sum-free classes.

Turán has shown that if the integers $[m, 5m + 3]$ are partitioned into two classes in any way, then in at least one of them the equation $x + y = z$ is solvable with $x \neq y$, and that this is not true for the integers $[m, 5m + 2]$. The uniqueness of the partition of $[m, 5m + 2]$ into two sum-free sets has been demonstrated by Znám.

Turán also considered the problem where x, y are not necessarily distinct. Define $s(m, n)$ as the least integer s such that however the interval $[m, m + s]$ is partitioned into n classes, one of them contains a solution of $x + y = z$. His result corresponding to the first problem is $s(m, 2) = 4m$. Clearly $s(1, n) = s(n) - 1$, where $s(n)$ is as in **E11**, and Irving's result implies that $s(m, n) \leq m \lfloor n!(e - \frac{1}{24}) - 1 \rfloor$. Abbott & Znám (see **E11**) independently noted that $s(m, n) \geq 3s(m, n-1) + m$ so that $s(m, n) \geq m(3^n - 1)/2$.

Abbott & Hanson call a class **strongly sum-free** if it contains no solution to either of the equations $x + y = z$ or $x + y + 1 = z$. They show that if $r(n)$ is the least r such that however $[1, r]$ is partitioned into n classes, one of them contains such a solution, then
$$r(m + n) \geq 2r(n)s(m) - r(n) - s(m) + 1.$$
They used this to improve the lower bound for $s(m, n)$; their method, with Fredericksen's example, now gives $s(m, n) > cm(315)^{n/5}$.

Š. Znám, Megjegyzések Turán Pál egy publikálatlan ereményéhez, *Mat. Lapok*, **14** (1963) 307–310.

E14 Rado's generalizations of van der Waerden's and Schur's problems.

Rado has considered a number of generalizations of van der Waerden's and Schur's problems. For example he shows that for any natural numbers a, b, c, there is a number u so that however the numbers $[1, u]$ are partitioned into two classes, there is a solution of $ax + by = cz$ in at least one of the classes. He gives a value for u, but, as in Schur's original problem, this is not best possible. For example, for $2x + y = 5z$, the theorem gives $u = 20$, whereas it is true even for $u = 15$, though not for any smaller value of u: neither of the sets
$$\{1, 4, 5, 6, 9, 11, 14\} \qquad \{2, 3, 7, 8, 10, 12, 13\}$$
contains a solution of $2x + y = 5z$. If we are allowed *three* sets, then 45 is the least value for u, since the following three sets contain all the

numbers [1,44], even with 6, 7, 8 and 9 duplicated.

$$\{1,4,5,6,9,11,14,16,19,20,21,24,26,29,31,34,36,39,41,44\},$$
$$\{2,3,7,8,10,12,13,15,17,18,22,23,27,28,32,33,37,38,42,43\},$$
$$\{6,7,8,9,25,30,35,40\}$$

Rado called the equation $\sum a_i x_i = 0$, where the a_i are nonzero integers, **n-fold regular** if there is a number $u(n)$, which we can assume to be minimal, such that however the interval $[1, u(n)]$ is partitioned into n classes, at least one class contains a solution to the equation. He called it **regular** if it was n-fold regular for all n, and showed that an equation was regular just if $\sum a_j = 0$ for some subset of the a_i. For example, if $a_1 = a_2 = 1$ and $a_3 = -1$, we have Schur's original problem with $u(n) = s(n)$. alié and Abbott considered the problem of finding lower bounds for $u(n)$; see **E10** and **E11** for references.

The example with $a_1 = 2$, $a_2 = 1$, $a_3 = -5$ is *not* regular, since, although we have seen that it is both 2-fold and 3-fold regular, it is not 4-fold regular. For, put every number $5^k l$, where $5 \nmid l$, into just one of four classes, according as k is even or odd, and l is $\equiv \pm 1$ or ± 2 mod 5. It can be verified that none of these four classes contains a solution of $2x + y = 5z$.

Rado asked if there exist, for every k, equations which are k-regular, but not $(k + 1)$-regular.

For the equations $2x_1 + x_2 = 2x_3$ and $x_1 + x_2 + x_3 = 2x_4$, Salié, Abbott, and Abbott & Hanson obtained successively better lower bounds, culminating in $u(n) > c(12)^{n/3}$ and $c(10)^{n/3}$ respectively.

Vera Sós asks for the maximum size of subset of $[1, n]$ such that Rado's equation has no solution in the subset. For example if $a_1 = a_2 = 1$ and $a_3 = -2$, the answer is in the interval $[n \exp(-\sqrt{\ln n}), n/(\ln n)^\alpha]$. If $a_1 = a_2 = 1$ and $a_3 = a_4 = -1$, we have a Sidon set (compare **C9**) and the answer is $\approx \sqrt{n}$. If $a_1 = a_2 = 1$ and $a_3 = -1$, the answer is $n/2$. It is known more generally that the answer is $o(n)$ just if $x_1 = x_2 = \ldots = 1$ is a solution of Rado's equation. Can the answer ever be comparable to n_α with $\frac{1}{2} < \alpha < 1$?

Compare problems **E10–14** with **C14–16**.

Fan R. K. Chung & John L. Goldwasser, Integer sets containing no solution to $x+y = 3z$, *The Mathematics of Paul Erdős, I*, Springer, Berlin, 1997, 218–227; *MR* **97h**:11016.

Walter Deuber, Partitionen und lineare Gleichungssysteme, *Math. Z.*, **133** (1973) 109–123; *MR* **48** #3753.

R. Rado, Studien zur Kombinatorik, *Math. Z.*, **36**(1933) 424–480.

E. R. Williams, M.Sc. thesis, Memorial University, 1967.

E15 A recursion of Göbel.

F. Göbel has remarked that the recursion $x_0 = 1$,

$$x_n = (1 + x_0^2 + x_1^2 + \ldots + x_{n-1}^2)/n \qquad n = 1, 2, \ldots$$

[or, for $n > 0$, $(n+1)x_{n+1} = x_n(x_n + n)$] yields integers

$$x_1 = 2, 3, 5, 10, 28, 154, 3520, 1551880, 267593772160, \ldots$$

for a long time, but Hendrik Lenstra found that x_{43} was not an integer!

The corresponding sequence with cubes in place of squares holds out as far as x_{89}. Henry Ibstedt has made extensive calculations, for various powers k and various initial values a_0. The table shows the rank of the first noninteger member of the sequence

k	2	3	4	5	6	7	8	9	10	11
$x_1 = 2$	43	89	97	214	19	239	37	79	83	239
$x_1 = 3$	7	89	17	43	83	191	7	127	31	389
$x_1 = 4$	17	89	23	139	13	359	23	158	41	239
$x_1 = 5$	34	89	97	107	19	419	37	79	83	137
$x_1 = 6$	17	31	149	269	13	127	23	103	71	239
$x_1 = 7$	17	151	13	107	37	127	37	103	83	239
$x_1 = 8$	51	79	13	214	13	239	17	163	71	239
$x_1 = 9$	17	89	83	139	37	191	23	103	23	169
$x_1 = 10$	7	79	23	251	347	239	7	163	41	239
$x_1 = 11$	34	601	13	107	19	478	37	79	31	389

Raphael Robinson has observed that, in contrast to Göbel's sequence, the recurrence

$$x_n x_{n-k} = a x_{n-p} x_{n-k+p} + b x_{n-q} x_{n-k+q} + c x_{n-r} x_{n-k+r}$$

appears to generate integers from the starting values $x_0 = x_1 = \ldots = x_k = 1$ for any integers $a \geq 0$, $b \geq 0$, $c \geq 0$, $p \geq 1$, $q \geq 1$, $r \geq 1$, k such that $p + q + r = k$.

David Boyd has studied Göbel sequences p-adically, generalizing to the recurrences $x_n = f(n, x_{n-1})$ where $f(n, x) = x + g(x)/n$ with $g(x)$ a polynomial with integer coefficients. For example, $g(x) = x(x-1)$ above. He has an example with $g(x)$ a degree 7 polynomial, and (only!) the first 35530 terms integers.

Jim Propp reports that combinatorial interpretations have been found for the Somos-4 and Somos-5 sequences 1,1,1,1,2,3,7,23,59,314,..., and 1,1,1,1,1,2,3,5,11,37,83,..., given respectively by

$$a(n)a(n-4) = a(n-1)a(n-3) + a(n-2)^2, \quad \text{and}$$
$$b(n)b(n-5) = b(n-1)b(n-4) + b(n-2)b(n-3),$$

and described in David Gale's "Tracking the Automatic Ant".

David Boyd, Some sequences of nonintegers, *J. Integer Sequences* (submitted)

David Gale, Mathematical Entertainments, *Math. Intelligencer*, **13**(1991) No. 1, 40–43.

David Gale, *Tracking the Automatic Ant*, Springer, 1998, Chap. 1. Simple sequences with puzzling properties.

Henry Ibstedt, Some sequences of large integers, *Fibonacci Quart.*, **28**(1990) 200–203; *MR* **91h**:11011.

Janice L. Malouf, An integer sequence from a rational recursion, *Discrete Math.*, **110**(1992) 257-261; *MR* **93m**:11011.

Raphael M. Robinson, Periodicity of Somos sequences, *Proc. Amer. Math. Soc.*, **116**(1992) 613–619; *MR* **93a**:11012.

Michael Somos, Problem 1470, *Crux Mathematicorum*, **15**(1989) 208.

E16 The $3x + 1$ problem.

When he was a student, L. Collatz asked if the sequence defined by $a_{n+1} = a_n/2$ (a_n even), $a_{n+1} = 3a_n + 1$ (a_n odd) is tree-like in structure, apart from the cycle 4, 2, 1, 4, ... (Figure 16) in the sense that, starting from any integer a_1, there is a value of n for which $a_n = 1$.

Erdős has said that "Mathematics may not be ready for such problems." Do not attempt it without first reading Jeff Lagarias's 1985 article or visiting Eric Roosendaal's "On the $3x + 1$ problem" page.

In September 2003 a new record replaced that of the previous March. If $a_1 = 25587533\,6134000063$, then a maximum of

$$4830\,8572251691\,7423129398\,7863972468$$

is attained, but this is still a long way from infinity.

Shalom Eliahou showed that if the minimum of a cycle is $> 2^{40}$ then its length is $17087915b + kc$ for some positive integer b, nonnegative integer c and $k = 301994$ or 85137581. Halbeisen & Hungerbühler sharpen Eliahou's criterion, and deduce from a computation by Oliviera e Silva, that the $3x+1$ conjecture is true for integers below $3 \cdot 2^{50}$, and that any nontrivial cycle has period length at least 102225496. In September,2003 this lower bound is at least 630000000.

Krasikov & Lagarias show by computer-aided proof that at least $x^{0.84}$ of the integers $< x$ contain 1 in their forward orbit under the $3x + 1$ map.

Daniel Bernstein shows that the conjecture is equivalent to the following: For an increasing sequence of nonnegative integers $0 \le d_0 < d_1 < \ldots$ define two elements in the ring $\mathbb{Z}_2$ of 2-adic integers by $Q = 2^{d_0} + 2^{d_1} + \ldots$, $N = \frac{-1}{3}2^{d_0} + \frac{-1}{9}2^{d_1} + \ldots$. Then the function $\Phi(Q) = N$ is a bijection of $\mathbb{Z}_2$ with itself, and he conjectures that the set of positive integers is contained in $\Phi((\frac{1}{3})\mathbb{Z})$, where $\mathbb{Z}$ is the ordinary integers.

Crandall conjectured that for any odd $q > 3$, there is an m whose orbit in the '$qx + 1$ problem' does not contain 1. E.g., for $q = 5$ there is 13, 33,

83, 13 and for $q = 181$ there is 27, 611, 27. Franco & Pomerance show that the conjecture is true for almost all q in the sense of asymptotic density.

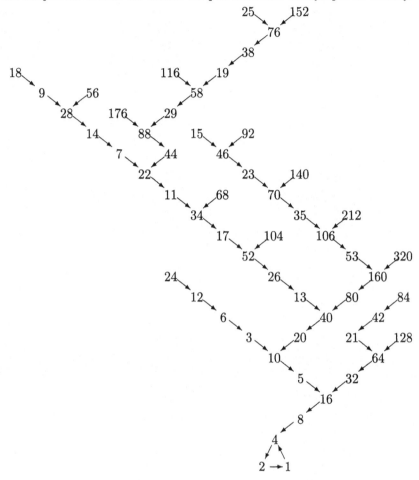

Figure 16. Is the Collatz Sequence Tree-like?

If we iterate the function $T(n) = n/2$ (n even), $(3n+1)/2$ (n odd), then we can define the stopping time, $s(n)$, as the least number k of iterations that give $T^k(n) < n$, and the maximum excursion, $t(n)$, as the maximum value of $T^k(n)$ for $k > 0$. Are $s(n)$ and $t(n)$ always finite? A distributed computing project at http://personal.computrain.nl/eric/wondrous has verified the $3x + 1$ conjecture for $n \leq 2.52 \times 10^{17}$. The lower bound for the size of a hypothetical new loop is 630 million.

There is evidence that $t(n) < n^2 f(n)$ where $f(n)$ is either constant or very slowly increasing. The highest value found for $t(n)/n^2$ is 7.527 for $n = 3716509988199$. For only 7 of the 76 record-holders is the value greater than one.

If $3a_n + 1$ is replaced by $3a_n - 1$ (or if we allow negative integers) then it seems likely that any sequence concludes with one of the cycles $\{1,2\}$, $\{5,14,7,20,10\}$ or

$\{17,50,25,74,37,110,55,164,82,41,122,61,182,91,272,136,68,34\}$.

This is true for all $a_1 \leq 10^8$.

David Kay and others define the sequence more generally by $a_{n+1} = a_n/p$ if $p|n$ and $a_{n+1} = a_n q + r$ if $p \nmid a_n$ and asks if there are numbers p, q, r for which the problem can be settled. For $(p,q,r) = (2,5,1)$ or $(2,7,1)$ it seems plausible that any sequence will increase rapidly, if erratically, but it seems to be just as hard to prove anything as it is for the original problem. The literature is enormous; would-be solvers are urged to study carefully the writings of Lagarias; see his *Amer. Math. Monthly* article and the 50-page annotated bibliography, http://arXiv.org/abs/math/0309224

Define $f(n)$ to be the largest odd divisor of $3n + 1$. Zimian asked if

$$\prod_{i=1}^{m} n_i = \prod_{i=1}^{m} f(n_i)$$

holds for any (multi)set $\{n_i\}$ of integers $n_i > 1$. Erdős found that

$$65 \cdot 7 \cdot 7 \cdot 11 \cdot 11 \cdot 17 \cdot 17 \cdot 13 = 49 \cdot 11 \cdot 11 \cdot 17 \cdot 17 \cdot 13 \cdot 13 \cdot 5.$$

Call an integer n **self-contained** if n divides $f^k(n)$ for some $k \geq 1$. If this happens and if the Collatz sequence $n^* = f^k(n)/n$ reaches 1, then the set

$$\{n, f(n), \dots, f^{k-1}(n), n^*, f(n^*), \dots, 1\}$$

is a set such as the above. A computer search for $n \leq 10^4$ yielded five self-contained integers: 31, 83, 293, 347 and 671.

These mappings are not one-to-one; you can't retrace the history of a sequence, since there is often no unique inverse.

Farkas proves that a modified algorithm, in which $3n + 1)/2$ is replaced by $(n + 1)/2$ when $n \equiv 1 \pmod 4$ and in which powers of 3 are divided out, always leads to a cycle, e.g.

55, 83, 125, 63, 7, 11, 17, 9, 1, 1, 1, ...

He believes that it also does, even if one is not allowed to divide out the powers of 3.

J.-P. Allouche, Sur la conjecture de "Syracuse-Kakutani-Collatz," *Séminaire de Théorie des Nombres*, 1978/79, Exp. No. 9, Talence, 1979; *MR* **81g**:10014.

Ştefan Andrei, Manfred Kudlek & Radu Ştefan Niculescu, Some results on the Collatz problem, *Acta Inform.*, **37**(2000) 145–160; *MR* **2002c**:11022.

David Applegate & Jeffrey C. Lagarias, Density bounds for the $3x+1$ problem, I. Tree-search method, II. Krasikov inequalities, *Math. Comput.*, **65**(1995) 411–426, 427–438; *MR* **95c**:11024–5.

David Applegate & Jeffrey C. Lagarias, The distribution of $3x + 1$ trees, *Experiment. Math.*, **4**(1995) 193–209; *MR* **97e**:11033.

David Applegate & Jeffery C. Lagarias, Lower bounds for the total stopping time of $3x + 1$ iterates, *Math. Comput.*, **72**(2003) 1035–1049; *MR* **2004a**:11016.

Enzo Barone, A heuristic probabilistic argument for the Collatz sequence, *Ital. J. Pure Appl. Math.*, **4**(1998) 151–153; *MR* **2000d**:11033.

Michael Beeler, William Gosper & Rich Schroeppel, Hakmem, Memo 239, Artificial Intelligence Laboratory, M.I.T., 1972, p. 64.

Lothar Berg & Günter Meinardus, Functional equations connected with the Collatz problem, *Results Math.*, **25**(1994) 1–12; *MR* **95d**:11025.

Lothar Berg & Günter Meinardus, The $3n + 1$ Collatz problem and functional equations, *Rostock. Math. Kolloq.*, **48**(1995) 11–18; *MR* **99e**:11034.

Daniel J. Bernstein, A noniterative 2-adic statement of the $3N + 1$ conjecture, *Proc. Amer. Math. Soc.*, **121**(1994) 405–408; *MR* **94h**:11108.

Daniel J. Bernstein & Jeffrey C. Lagarias, The $3x + 1$ conjugacy map, *Canad. J. Math.*, **48**(1996) 1154–1169; *MR* **98a**:11027.

David Boyd, Which rationals are ratios of Pisot sequences? *Canad. Math. Bull.*, **28**(1985) 343–349; *MR* **86j**:11078.

Stefano Brocco, A note on Mignosi's generalization of the $(3X + 1)$-problem, *J. Number Theory*, **52**(1995) 173–178; *MR* **96d**:11025.

S. Burckel, Functional equations associated with congruential functions, *Theoret. Comput. Sci.*, **123**(1994) 397–407; *MR* **94m**:11147.

Marc Chamberland, A continuous extension of the $3x + 1$ problem to the real kine, *Dynam. Contin. Discrete Impuls. Systems*, **2**(1996) 495–509; *MR* **97m**:11028.

Busiso P. Chisala, Cycles in Collatz sequences, *Publ. Math. Debrecen*, **45**(1994) 35–39; *MR* **95h**:11019.

Dean Clark, Second order difference equations related to the Collatz $3n + 1$ conjecture, *J. Differ. Equations Appl.*, **1**(1995) 73–85; *MR* **96e**:11031.

Dean Clark & James T. Lewis, A Collatz-type difference equation, *Proc. 26th SE Internat. Conf. Combin., Graph Theory, Comput.; Congr. Numer.*, **111**(1995) 129–135; *MR* **98b**:11008.

R. E. Crandall, On the "$3x + 1$" problem, *Math. Comput.*, **32**(1978) 1281–1292; *MR* **58** #494.

J. L. Davidson, Some comments on an iteration problem, Proc. 6th Manitoba Conf. Numerical Math., 1976, *Congressus Numerantium*, **18**(1977) 155–159.

Jeffrey P. Dumont & Clifford A. Reiter, Real dynamics of a 3-power extension of the $3x + 1$ function, *Dyn. Contin. Discrete Impuls. Syst. Ser. A Math. Anal.*, **10**(2003) 875–893.

S. Eliahou, The $3x+1$ problem: new lower bounds on nontrivial cycle lengths, *Discrete Math.*, **118**(1993) 45–56; *MR* **94h**:11017.

C. J. Everett, Iteration of the number-theoretic function $f(2n) = n$, $f(2n + 1) = 3n + 2$, *Advances in Math.*, **25**(1977) 42–45; *MR* **56** #15552.

Hershel M. Farkas, Variants of the $3N + 1$ conjecture, *Proceedings on Geometry, Groups, Dynamics and Spectral Theory. In memory of Robert Brooks.*

Edited Michael Entov, Yehuda Pinchover & Michah Sageev. (submitted)

P. Filipponi, On the $3n + 1$ problem: something old, something new, *Rend. Mat. Appl.*(7) **11**(1991) 85–103; *MR* **92i**:11031.

Zachary Franco & Carl Pomerance, On a conjecture of Crandall concerning the $qx + 1$ problem, *Math. Comput.*, **64**(1995) 1333–1336; *MR* **95j**:11019.

Gao Guo-Gang, On consecutive numbers of the same height in the Collatz problem, *Discrete Math.*, **112**(1993) 261–267; *MR* **94i**:11018.

L. E. Garner, On the Collatz $3n + 1$ algorithm, *Proc. Amer. Math. Soc.*, **82**(1981) 19–22; *MR* **82j**:10090.

Lynn E. Garner, On heights in the Collatz $3n + 1$ problem, *Discrete Math.*, **55**(1985) 57–64; *MR* **86j**:11005.

David Gluck & Brian D. Taylor, A new statistic for the $3x + 1$ problem, *Proc. Amer. Math. Soc.*, **130**(2002) 1293–1301; *MR* **2002k**:11031.

Lorenz Halbeisen & Norbert Hungerbühler, Optimal bounds for the length of rational Collatz cycles, *Acta Arith.*, **78**(1997) 227–239; *MR* **98g**:11025.

E. Heppner, Eine Bemerkung zum Hasse-Syracuse-Algorithmus, *Arch. Math. (Basel)*, **31**(1977/79) 317–320; *MR* **80d**:10007.

I. N. Herstein & I. Kaplansky, *Matters Mathematical*, 2nd ed., Chelsea, 1978, pp. 44–45.

David C. Kay, *Pi Mu Epsilon J.*, **5**(1972) 338.

I. Korec, The $3x + 1$ problem, generalized Pascal triangles and cellular automata, *Math. Slovaca*, **42**(1992) 547–563; *MR* **94g**:11019.

Ivan Korec, A density estimate for the $3x+1$ problem, *Math. Slovaca*, **44**(1994) 85–89; *MR* **95h**:11022.

I. Korec & Š. Znám, A note on the $3x + 1$ problem, *Amer. Math. Monthly* **94**(1987) 771–772; *MR* **90g**:11023.

I. Krasikov, How many numbers satisfy the $3x + 1$ conjecture? *Internat. J. Math. Math. Sci.*, **12**(1989) 791–796; *MR* **90k**:11013.

Ilia Krasikov & Jeffrey C. Lagarias, Bounds for the $3x + 1$ problem using difference inequalities; *Acta Arith.*, **109**(2003) 237–258.

James R. Kuttler, On the $3x + 1$ problem, *Adv. in Appl. Math.*, **15**(1994) 183–185.

Jeffrey C. Lagarias, The $3x + 1$ problem and its generalizations, *Amer. Math. Monthly*, **92**(1985) 3–23; *MR* **86i**:11043.

Jeffrey C. Lagarias, The set of rational cycles for the $3x + 1$ problem, *Acta Arith.*, **56**(1990) 33–53, *MR* **91i**:11024.

J. C. Lagarias, H. A. Porta & K. B. Stolarsky, Asymmetric tent map expansions I: eventually periodic points, *J. London Math. Soc.*, **47**(1993) 542–556; *MR* **94h**:58139; II: Purely periodic points, *Illinois J. Math.*, **38**(1994) 574–588; *MR* **96b**:58093.

Jeffrey C. Lagarias & A. Weiss, The $3x + 1$ problem: two stochastic models, *Ann. Appl. Probab.*, **2**(1992) 229–261; *MR* **92k**:60159.

K. R. Matthews & A. M. Watts, A generalization of Hasse's generalization of the Syracuse algorithm, *Acta Arith.*, **43**(1984) 167–175; *MR* **85i**:11068.

K. R. Matthews & A. M. Watts, A Markov approach to the generalized Syracuse algorithm, *Acta Arith.*, **45**(1985) 29–42; *MR* **87c**:11071.

Filippo Mignosi, On a generalization of the $3x+1$ problem, *J. Number Theory*, **55**(1995) 28–45; *MR* **99m**:11016.

Tomoaki Mimuro, On certain simple cycles of the Collatz conjecture, *SUT J. Math.*, **37**(2001) 79–89; *MR* **2002j**:11018 .

Herbert Möller, Über Hasses Verallgemeinerung der Syracuse-Algorithmus (Kakutani's problem), *Acta Arith.*, **34**(1978) 219–226; *MR* **57** #16246.

Kenneth G. Monks, $3x + 1$ minus the $+$, *Discrete Math. Theor. Comput. Sci.*, **5**(2002) 47–53; *MR* **2003f**:11030.

Helmut Müller, Das "$3n + 1$"-Problem, *Mitt. Math. Ges. Hamburg*, **12**(1991) 231–251; *MR* **93c**:11053.

Helmut A. Müller, Über eine Klasse 2-adischer Funktionen im Zusammenhang mit dem "$3x + 1$"-Problem, *Abh. Math. Sem. Univ. Hamburg*, **64**(1994) 293–302; *MR* **95e**:11032.

Tomás Oliviera e Silva, Maximum excursion and stopping time record-holders for the $3x+1$ problem: computational results, *Math. Comput.*, **68**(1999) 371–384; *MR* **2000g**:11015.

Qiu Wei-Xing, Observations on the "$3x + 1$" problem, *J. Shanghai Univ. Nat. Sci.*, **3**(1997) 462–464.

Daniel A. Rawsthorne, Imitation of an iteration, *Math. Mag.*, **58**(1985) 172–176; *MR* **86i**:40001.

J. L. Rouet & M. R. Feix, A generalization of the Collatz problem. Building cycles and a stochastic approach, *J. Statist. Phys.*, **107**(2002) 1283–1298; *MR* **2003i**:11035.

J. W. Sander, On the $(3N + 1)$-conjecture, *Acta Arith.*, **55**(1990) 241–248; *MR* **91m**:11052.

J. Shallit, The "$3x + 1$" problem and finite automata, *Bull. Europ. Assoc. Theor. Comput. Sci.*, **46**(1991) 182–185.

J. Shallit & D. Wilson, The "$3x + 1$" problem and finite automata, *Bull. Europ. Assoc. Theor. Comput. Sci.*, **46**(1991) 182–185.

Ray P. Steiner, On the "$Qx + 1$ problem", Q odd, *Fibonacci Quart.*, **19**(1981) 285–288; II, 293–296; *MR* **84m**:10007ab.

Kenneth B. Stolarsky, A prelude to the $3x + 1$ problem, *J. Differ. Equations Appl.*, **4**(1998) 451–461; *MR* **99k**:11037.

Riho Terras, A stopping time problem on the positive integers, *Acta Arith.*, **30**(1976) 241–252; *MR* **58** #27879 (and see **35**(1979) 100–102; *MR* **80h**:10066).

Ilan Vardi, Computational Recreations in *Mathematica*©, Addison-Wesley, Redwood City CA, 1991, Chap. 7.

G. Venturini, Iterates of number-theoretic functions with periodic rational coefficients (generalization of the $3x + 1$ problem), *Stud. Appl. Math.* **86**(1992) 185–218; *MR* **93b**:11102.

G. Venturini, On a generalization of the $3x + 1$ problem, *Adv. in Appl. Math.*, **19**(1997) 295–305; *MR* **98j**:11013.

Stan Wagon, The Collatz problem, *Math. Intelligencer*, **7**(1985) 72–76; *MR* **86d**:11103.

Günther Wirsching, An improved estimate concerning $3n+1$ predecessor sets, *Acta Arith.*, **63**(1993) 205–210; *MR* **94e**:11018.

Günther Wirsching, A Markov chain underlying the backward Syracuse algorithm, *Rev. Roumaine Math. Pures Appl.*, **39**(1994) 915–926; *MR* **96d**:11027.

Günther Wirsching, On the combinatorial structure of $3n + 1$ predecessor sets, *Discrete Math.*, **148**(1996) 265–286; *MR* **97b**:11029.

Günther J. Wirsching, *The dynamical system generated by the* $3x+1$ *function*, Lecture Notes in Mathematics, **1681**, Springer-Verlag, 1998; *MR* **99g**:11027.

Günther J. Wirsching, A functional differential equation and $3n+1$ dynamics, *Topics in functional differential and difference equations* (*Lisbon*, 1999), 369–378, *Fields Inst. Commun.*, **29**, AMS 2001; *MR* **2002b**:11035.

Günter J. Wirsching, Über das $3x + 1$ Problem, *Elem. Math.*, **55**(2000) 142–155; *MR* **2002h**:11022.

Wu Jia-Bang, The trend of changes of the proportion of consecutive numbers of the same height in the Collatz problem, *J. Huazhong Univ. Sci. Tech.*, **20**(1992) 171–174; *MR* **94b**:11024.

Wu Jia-Bang, On the consecutive positive integers of the same height in the Collatz problem, *Math. Appl.*, **6**(1993) 150–153.

Masaji Yamada, A convergence proof about an integral sequence, *Fibonacci Quart.*, **18**(1980) 231–242; see *MR* **82d**:10026 for errors.

Zhou Chuan-Zhong, Some discussion on the $3x + 1$ problem, *J. South China Normal Univ. Natur. Sci. Ed.*, **1995** 103–105; *MR* **97h**:11021.

Zhou Chuan-Zhong, Some recurrence relations connected with the $3x + 1$ conjecture, *J. South China Normal Univ. Natur. Sci. Ed.*, **1997** 7–8.

OEIS: A001281, A003079, A003124, A005184, A006370, A006460, A006513, A006577, A006666-006667, A008873-008884, A008894-008903, A008908, A037082, A066756, A066773, A068529-068530.

E17 Permutation sequences.

The situation is different, though no more clear, in the case of **permutation sequences**. A simple example, probably the inverse of Collatz's original problem (see **E16** and Lagarias's article cited there), is

$$a_{n+1} = 3a_n/2 \quad (a_n \text{ even}), \qquad a_{n+1} = \lfloor (3a_n + 1)/4 \rfloor \quad (a_n \text{ odd}),$$

or, perhaps more perspicuously,

$$2m \to 3m \qquad 4m - 1 \to 3m - 1 \qquad 4m + 1 \to 3m + 1$$

from which it is clear that the inverse operation works just as well. So the resulting structure consists only of disjoint cycles and doubly infinite chains. It is not known whether there is a finite or infinite number of each of these, nor even whether an infinite chain exists. It is conjectured that the only cycles are $\{1\}$, $\{2,3\}$, $\{4,6,9,7,5\}$ and

$$\{44, 66, 99, 74, 111, 83, 62, 93, 70, 105, 79, 59\}.$$

Mike Guy, with the help of TITAN, showed that any other cycles have period greater than 320. What is the status of the sequence containing the number 8?

$$\ldots, 97, 73, 55, 41, 31, 23, 17, 13, 10, 15, 11, \mathbf{8},$$
$$12, 18, 27, 20, 30, 45, 34, 51, 38, 57, 43, 32, 48, 72, \ldots$$

Do the numbers 8, 14, 40, 64, 80, 82, 104, 136, 172, 184, 188, 242, 256, 274, 280, 296, 352, 368, 382, 386, 424, 472, 496, 526 530, 608, 622, 638, 640, 652, 670, 688, 692, 712, 716, 752, 760, 782, 784, 800, 814, 824, 832, 860, 878, 904, 910, 932, 964, 980, ... each belong to a separate sequence?

There are some intriguing paradoxes: as you go "forward" you multiply by 3/2 if the current number is even, and by about 3/4 if it's odd — and get an erratic "pseudo-GP" of common ratio $3/\sqrt{8} \approx 1.060660172$. On the other hand, as you go "backward" you multiply by 2/3 if the current number is a multiple of 3, and by about 4/3 otherwise — a "pseudo-GP" of common ratio $32^{1/3}/3 \approx 1.058267368$. These two numbers should be reciprocal! We have a sort of discrete analog of an everywhere non-differentiable function. The "derivative" on the right is positive; that on the left is negative. Note that, when going "forwards" each successor of an even number is a multiple of 3 — half the numbers are multiples of three!

J. H. Conway, Unpredictable iterations, in *Proc. Number Theory Conf.*, Boulder CO, 1972, 49–52; *MR* **52** #13717.

David Gale, Mathematical Entertainments, *Math. Intelligencer*, **13**(1991) No. 3, 53–55.

G. Venturini, Iterates of number theoretic functions with periodic rational coefficients (generalization of the $3x + 1$ problem), *Stud. Appl. Math.*, **86**(1992) 185–218; *MR* **93b**:11102.

OEIS: A006368-006369.

E18 Mahler's Z-numbers.

Mahler considered the following problem: given any real number α, let r_n be the fractional part of $\alpha(3/2)^n$. Do there exist **Z-numbers**, for which $0 \leq r_n < \frac{1}{2}$ for all n? Probably not. Mahler shows that there is at most one between each pair of consecutive integers, and that, for x large enough, at most $x^{0.7}$ less than x. Flatto has improved on Mahler's results, but the problem remains unsolved.

A similar question is: is there a rational number r/s ($s \neq 1$) such that $\lfloor (r/s)^n \rfloor$ is odd for all n? Tijdeman proved that for every odd integer $r > 3$, there are real numbers α such that the fractional part of $\alpha(r/2)^n$ is in $[0, \frac{1}{2})$ for all n.

Littlewood once remarked that it was not known that the fractional part of e^n did not tend to 0 as $n \to \infty$.

Leopold Flatto, Z-numbers and β-transformations, *Symbolic Dynamics and its Applications, Contemporary Math.*, **135**, Amer. Math. Soc., 1992, 181–201; *MR* **94c**:11065.

K. Mahler, An unsolved problem on the powers of 3/2, *J. Austral. Math. Soc.*, **8**(1968) 313–321; *MR* **37** #2694.

R. Tijdeman, Note on Mahler's $\frac{3}{2}$-problem, *Kongel. Norske Vidensk. Selsk. Skr.*, **16**(1972) 1–4.

E19 Are the integer parts of the powers of a fraction infinitely often prime?

Forman & Shapiro have proved that infinitely many integers of the form $\lfloor (4/3)^n \rfloor$ and also of the form $\lfloor (3/2)^n \rfloor$ are composite. A. L. Whiteman conjectures that these two sequences also each contain infinitely many primes. The method appears not to work for other rationals.

W. Forman & H. N. Shapiro, An arithmetic property of certain rational powers, *Comm. Pure Appl. Math.*, **20**(1967) 561–573; *MR* **35** #2852.

OEIS: A043037-043038, A046037-046038, A067904-067905, A070758-070759, A070761-070762.

E20 Davenport-Schinzel sequences.

Form sequences from an alphabet $[1, n]$ of n letters such that there are no immediate repetitions $\ldots aa \ldots$ and no alternating subsequences
$$\ldots a \ldots b \ldots a \ldots b \ldots$$
of length greater than d. Denote by $N_d(n)$ the maximal length of any such sequence; then a sequence of this length is a **Davenport-Schinzel sequence**. The problem is to determine all D-S sequences, and in particular to find $N_d(n)$. We need only consider **normal** sequences in which the first appearance of an integer of the alphabet comes after the first appearance of every smaller one.

The sequences 12131323, 12121213131313232323, and
$$1\,2\,1\,3\,1\,4\,1 \ldots 1\,\overline{n-1}\,1\,\overline{n-1}\,\overline{n-2} \ldots 3\,2\,n\,2\,n\,3\,n \ldots n\,\overline{n-1}\,n$$
show that $N_4(3) \geq 8$, $N_8(3) \geq 20$, and $N_4(n) \geq 5n - 8$. Davenport & Schinzel showed that $N_1(n) = 1$, $N_2(n) = n$, $N_3(n) = 2n - 1$; that $N_4(n) = O(n \ln n / \ln \ln n)$, $\lim N_4(n)/n \geq 8$ and, with J. H. Conway, that $N_4(lm + 1) \geq 6lm - m - 5l + 2$, so that $N_4(n) = 5n - 8$ ($4 \leq n \leq 10$). Z. Kolba showed that $N_4(2m) \geq 11m - 13$ and Mills obtained the values of $N_4(n)$ for $n \leq 21$. For example, the sequence

$$abacadaeafafedcbgbhbhgcicigdjdjgekekgkjihflflhliljlkl$$

(which, Günter Rote notes, was misprinted in the first edition) is part of the proof that $N_4(12) = 53$.

Roselle & Stanton fixed n rather than d and obtained $N_d(2) = d$, $N_d(3) = 2\lfloor 3d/2 \rfloor - 4$ $(d > 3)$, $N_d(4) = 2\lfloor 3d/2 \rfloor + 3d - 13$ $(d > 4)$ and $N_d(5) = 4\lfloor 3d/2 \rfloor + 4d - 27$ $(d > 5)$, though Peterkin observed that this last parenthesis should be $(d > 6)$ since $N_6(5) = 34$. Roselle & Stanton also showed that normal D-S sequences of length $N_{2d+1}(5)$ are unique and that there are just two of length $N_{2d+1}(4)$ and $N_{2d}(5)$. Peterkin exhibited the 56 D-S sequences of length $N_5(6) = 29$ and showed that $N_5(n) \geq 7n - 13$ $(n > 5)$ and $N_6(n) \geq 13n - 32$ $(n > 5)$.

Rennie & Dobson gave an upper bound for $N_d(n)$ in the form

$$(nd - 3n - 2d + 7)N_d(n) \leq n(d - 3)N_d(n - 1) + 2n - d + 2 \qquad (d > 3)$$

thus generalizing the result of Roselle & Stanton for $d = 4$.

Table 8. Values of $N_d(n)$.

d \ n	1	2	3	4	5	6	7	8	9	10	11	12	13	14	15	16	17	18	19	20	21
1	1	1	1	1	1	1	1	1	1	1	1	1	1	1	1	1	1	1	1	1	1
2	1	2	3	4	5	6	7	8	9	10	11	12	13	14	15	16	17	18	19	20	21
3	1	3	5	7	9	11	13	15	17	19	21	23	25	27	29	31	33	35	37	39	41
4	1	4	8	12	17	22	27	32	37	42	47	53	58	64	69	75	81	86	92	98	104
5	1	5	10	16	22	29															
6	1	6	14	23	34																
7	1	7	16	28	41																
8	1	8	20	35	53																
9	1	9	22	40	61																
10	1	10	26	47	73																

Szemerédi showed that $N_d(n) < c_d n \log^* n$, where $\log^* n$ is a slow-growing function, the least number of iterations of the exponential function needed to exceed n. More recent work, mainly by Sharir, has shown that the order of $N_d(n)$ is $\Theta(n\alpha(n))$, where $\alpha(n)$ is the incredibly slow-growing inverse of the Akermann function. See the paper of Agarwal, Sharir & Shor for precise details.

P. K. Agarwal, M. Sharir & P. Shor, Sharp upper and lower bounds on the length of general Davenport-Schinzel sequences, *J. Combin. Theory Ser. A*, **52**(1989) 228–274; *MR* **90m**:11034.

H. Davenport & A. Schinzel, A combinatorial problem connected with differential equations, *Amer. J. Math.*, **87**(1965) 684–694; II, *Acta Arith.*, **17**(1970/71) 363–372; *MR* **32** #7426; **44** #2619.

Annette J. Dobson & Shiela Oates Macdonald, Lower bounds for the lengths of Davenport-Schinzel sequences, *Utilitas Math.*, **6**(1974) 251–257; *MR* **50** #9781.

Z. Füredi & P. Hajnal, Davenport-Schinzel theory of matrices, *Discrete Math.*, **103**(1992) 233–251; *MR* **93g**:05026.

D. Gardy & D. Gouyou-Beauchamps, Enumerating Davenport-Schinzel sequences, *RAIRO Inform. Théor. Appl.* **26**(1992) 387–402; *MR* **93k**:05011; see also *MR* **95h**:05008.

S. Hart & M. Sharir, Nonlinearity of Davenport-Schinzel sequences and of generalized path compression schemes, *Combinatorica*, **6**(1986) 151–177; *MR* **88f**:05004.

Martin Klazar, Generalized Davenport-Schinzel sequences: results, problems and applications, *Integers*, **2**(2002) A11, 39pp.; *MR* **2003e**:11028.

P. Komjáth, A simplified construction of nonlinear Davenport-Schinzel sequences, *J. Combin. Theory Ser. A*, **49**(1988) 262–267.

W. H. Mills, Some Davenport-Schinzel sequences, *Congress. Numer.* **9**, Proc. 3rd Manitoba Conf. Numer. Math., 1973, pp. 307–313; *MR* **50** #135.

W. H. Mills, On Davenport-Schinzel sequences, *Utilitas. Math.* **9**(1976) 87–112; *MR* **53** #10703.

R. C. Mullin & R. G. Stanton, A map-theoretic approach to Davenport-Schinzel sequences, *Pacific J. Math.*, **40**(1972) 167–172; *MR* **46** #1745.

C. R. Peterkin, Some results on Davenport-Schinzel sequences, *Congress. Numer.* **9**, Proc. 3rd Manitoba Conf. Numer. Math., 1973, pp. 337–344; *MR* **50** #136.

B. C. Rennie & Annette J. Dobson, Upper bounds for the lengths of Davenport-Schinzel sequences, *Utilitas Math.*, **8**(1975) 181–185; *MR* **52** #13624.

D. P. Roselle, An algorithmic approach to Davenport-Schinzel sequences, *Utilitas Math.*, **6**(1974) 91–93; *MR* **50** #9780.

D. P. Roselle & R. G. Stanton, Results on Davenport-Schinzel sequences, *Congress. Numer.* **1** Proc. Louisiana Conf. Combin. Graph Theory, Comput., (1970) 249–267; *MR* **43** #68.

D. P. Roselle & R. G. Stanton, Some properties of Davenport-Schinzel sequences, *Acta Arith.*, **17**(1970/71) 355–362; *MR* **44** #1641.

M. Sharir, Almost linear upper bounds on the length of generalized Davenport-Schinzel sequences, *Combinatorica*, **7**(1987) 131–143; *MR* **89f**:11037.

Micha Sharir, Improved lower bounds on the length of Davenport-Schinzel sequences, *Combinatorica*, **8**(1988) 117–124; *MR* **89j**:11024.

Micha Sharir & Pankaj K. Agarwal, *Davenport-Schinzel Sequences and their Geometric Applications*, Cambridge Univ. Press, 1995; *MR* **96e**:11034.

R. G. Stanton & D. P. Roselle, A result on Davenport-Schinzel sequences, *Combinatorial Theory and its Applications*, Proc. Colloq., Balatonfüred, 1969, North-Holland, 1970, pp. 1023–1027; *MR* **46** #3324.

R. G. Stanton & P. H. Dirksen, Davenport-Schinzel sequences, *Ars Combin.*, **1**(1976) 43–51; *MR* **53** #13106.

E. Szemerédi, On a problem of Davenport and Schinzel, *Acta Arith.*, **25** (1973/74) 213–224; *MR* **49** #244.

OEIS: A002004, A005004-005006, A005280-005281.

E21 Thue-Morse sequences.

Thue showed that there are infinite sequences on 3 symbols which contain no two identically equal consecutive segments, and sequences on 2 symbols which contain no three identically equal consecutive segments, and many others have rediscovered these results.

If, instead of identically equal segments, we ask to avoid consecutive segments which are *permutations* of one another, Justin constructed a sequence on 2 symbols without five consecutive segments which are permutations of each other, and Pleasants constructed a sequence on 5 symbols without two such consecutive segments. Dekking has solved the (2,4) and (3,3) problems, and the (4,2) problem has been solved by Keränen.

Fraenkel & Simpson show that if $g(k)$ is the length of a longest binary word containing at most k distinct squares (two identical adjacent substrings), then $g(3) = \infty$.

Let t be the infinite fixed point, starting with 1, of the morphism μ : $0 \to 01$, $1 \to 10$. An infinite word over $\{0,1\}$ is said to be overlap-free if it contains no factor of the form $axaxa$, where $a \in \{0,1\}$ and $x \in \{0,1\}^*$. Allouche, Currie & Shallit prove that the lexicographically least infinite overlap-free binary word beginning with any specified prefix, if it exists, has a suffix which is a suffix of t. In particular, the lexicographically least infinite overlap-free binary word is 001001 t.

Theoret. Comput. Sci., **218**(1999), No.1 is devoted to words. Thanks to Shallit for amplification of the following bibliography.

S. Arshon, Démonstration de l'éxistence des suites asymétriques infinies (Russian. French summary), *Mat. Sb.*, **2**(44)(1937) 769–779.

Jean-Paul Allouche, André Arnold, Jean Berstel, Srečko Brlek, William Jokusch, Simon Plouffe & Bruce E. Sagan, A relative of the Thue-Morse sequence, *Discrete Math.*, **139**(1995) 455–461; *MR* **96c**:11029.

Jean-Paul Allouche, James Currie & Jeffrey Shallit, Extremal infinite overlap-free binary words, *Electron. J. Combin.*, **5**(1) R27; *MR* **99d**:68201. [can be viewed at http://www.combinatorics.org].

Jean-Paul Allouche, J. Peyrière, Wen Zhi-Xiong & Wen Zhi-Ying, Hankel determinants of the Thue-Morse sequence, *Ann. Inst. Fourier (Grenoble)*, **48**(1998) 1–27; *MR* **99a**:11024.

Jean-Paul Allouche & Jeffrey Shallit, The ubiquitous Prouhet-Thue-Morse sequence, *Sequences and their applications (Singapore, 1998)*, 1–16, *Springer Ser. Discrete Math. Theor. Comput. Sci.*, 1999; *MR* **2002e**:11025.

M. Baake, V. Elser & U. Grimm, The entropy of square-free words, *Combinatorics and physics (Marseilles 1995)*, *Math. Comput. Modelling*, **26**(1997) 13–26; *MR* **99d**:68202.

K. A. Baker, G. F. McNulty & W. Taylor, Growth problems for avoidable words, *Theor. Comput. Sci.*, **69**(1989) 319–345.

D. A. Bean, G. Ehrenfeucht & G. McNulty, Avoidable patterns in strings of symbols, *Pacific J. Math.*, **85**(1979) 261–294.

J. Berstel, Sur la construction de mots sans carré, *Séminaire de Théorie des Nombres*, **1978–1979** 18.01–18.15; *MR* **82a**:68156.

J. Berstel, Some recent results on squarefree words, *STACS '84*, *Springer Lect. Notes Comput. Sci.*, **166**(1984) 14–25; *MR* **86e**:68056.

J. Berstel, Langford strings are squarefree, *Bull. Europ. Assoc. Theor. Comput. Sci.*, **37**(1989) 127–129.

F. J. Brandenburg, Uniformly growing k-th power-free homomorphisms, *Theor. Comput. Sci.*, **23**(1983) 69–82.

C. H. Braunholtz, Solution to Problem 5030 [1962,439], *Amer. Math. Monthly*, **70**(1963) 675–676.

J. Cassaigne, Counting overlap-free binary words, *STACS '93, Springer Lect. Notes Comput. Sci.*, **665**(1993) 216–225; *MR* **94j**:68152.

J. Cassaigne, Unavoidable binary patterns, *Acta Inform.*, **30**(1993) 385–395; *MR* **94m**:68155.

R. Cori & M. R. Formisano, Partially abelian squarefree words, *RAIRO Inform. Théor.*, **24**(1990) 509–520; *MR* **92e**:68154.

Max Crochemore, Sharp characterizations of squarefree morphisms, *Theor. Comput. Sci.*, **18**(1982) 221–226; *MR* **84a**:20070.

Max Crochemore, Recherche linéaire d'un carré dans un mot, *C. R. Acad. Sci. Paris Sér. I Math.*, **296**(1983) 781–784; *MR* **85a**:68058.

Max Crochemore, Michel Le Rest & Phillippe Wender, An optimal test on finite unavoidable sets of words, *Inform. Process. Lett.*, **16**(1983) 179–180; *MR* **85a**:68088.

M. Crochemore & W. Rytter, Squares, cubes, and time-space efficient string searching. *Algorithmica*, **13**(1995) 405–425; *MR* **96d**:68080.

L. J. Cummings, On the construction of Thue sequences, *Proc. 9th Southeast Conf. Combin. Comput., Congr. Numer.*, **21**(1978) 235–242; *MR* **80j**:05013.

L. J. Cummings, Overlapping substrings and Thue's problem, *Proc. 3rd Carib. Conf. Combin. Comput.*, 1981, 99–109; *MR* **84b**:68096.

Richard A. Dean, A sequence without repeats on x, x^{-1}, y, y^{-1}, *Amer. Math. Monthly*, **72**(1965) 383–385; *MR* **31** #3500.

F. Dejean, Sur un théorème de Thue, *J. Combin. Theory Ser. A*, **13**(1972) 90–99; *MR* **46** #119.

F. M. Dekking, On repetitions of blocks in binary sequences, *J. Combin. Theory Ser. A*, **20**(1976) 292–299; *MR* **55** #2739.

F. M. Dekking, Strongly non-repetitive sequences and progression-free sets, *J. Combin. Theory Ser. A*, **27**(1979) 181–185; *MR* **81b**:05027.

Gergely Dombi, Additive properties of certain sets, —it Acta Arith., **103**(2002) 137–146; *MR* **2003b**:11006.

T. Downarowicz & Y. Lacroix, Merit factors and Morse sequences, *Theor. Comput. Sci.*, **209**(1998) 377–387; *MR* **99i**:11015.

Michael Drmota & Mariusz Skalba, Rarified sums of the Thue-Morse sequence, *Trans. Amer. Math. Soc.*, **352**(2000) 609–642; *MR* **2000c**:11038.

A. Ehrenfeucht & G. Rozenberg, Repetitions in homomorphisms and languages, *ICALP '82, Springer Lect. Notes Comput. Sci.*, **140**(1982) 192–196; *MR* **83m**:68135.

R. C. Entringer, D. E. Jackson & J. A. Schatz, On non-repetitive sequences, *J. Combin. Theory Ser. A*, **16**(1974) 159–164; *MR* **48** #10860.

P. Erdős, Some unsolved problems, *Magyar Tud. Akad. Mat. Kutató Int. Közl.*, **6**(1961) 221–254, esp. p. 240.

A. A. Evdokimov, Strongly asymmetric sequences generated by a finite number of symbols, *Dokl. Akad. Nauk SSSR*, **179**(1968) 1268–1271; *Soviet Math. Dokl.*, **7**(1968) 536–539; *MR* **38** #3156.

Earl Dennet Fife, Binary sequences which contain no BBb, PhD thesis, Wesleyan Univ., Middletown CT, 1976.

Nathan J. Fine & Herbert S. Wilf, Uniqueness theorem for periodic sequences, *Proc. Amer. Math. Soc.*, **16**(1965) 109–114; *MR* **30** #5124.

Aviezri S. Fraenkel & R. Jamie Simpson, How many squares must a binary sequence contain? *Electron. J. Combin.*, **2**(1995) Res. Paper 2; *MR* **96b**:11023.

Jean-Pierre Gazeau & Jacek Miękisz, A symmetry group of a Thue-Morse quasicrystal, *J. Phys. A*, **31**(1998) L435–L440; *MR* **99j**:11023.

D. Hawkins & W. E. Mientka, On sequences which contain no repetitions, *Math. Student*, **24**(1956) 185–187; *MR* **19**, 241.

G. A. Hedlund, Remarks on the work of Axel Thue on sequences, *Nordisk Mat. Tidskr.*, **15**(1967) 148–150; *MR* **37** #4454.

G. A. Hedlund & W. H. Gottschalk, A characterization of the Morse minimal set, *Proc. Amer. Math. Soc.*, **15**(1964) 70–74; *MR* **28** #1609.

J. Justin, Généralisation du théorème de van de Waerden sur les semi-groupes répétitifs, *J. Combin. Theory Ser. A*, **12**(1972) 357–367; *MR* **46** #5485.

J. Justin, Semi-groupes répétitifs, *Sém. IRIA, Log. Automat.*, 1971, 101–105, 108; *Zbl.* 274.20092.

J. Justin, Characterization of the repetitive commutative semigroups, *J. Algebra*, **21**(1972) 87–90; *MR* **46** #277; **12**(1972) 357–367; *MR* **46** #5485.

J. Karhumäki, On cube-free ω-words generated by binary morphisms, *Discr. Appl. Math.*, **5**(1983) 279–297; *MR* **84j**:03081.

V. Keränen, On k-repetition free words generated by length uniform morphisms over a binary alphabet, *ICALP '85, Springer Lect. Notes Comput. Sci.*, **194**(1985) 338–347; *MR* **87d**:68060.

V. Keränen, On the k-freeness of morphisms on free monoids, *STACS '87, Springer Lect. Notes Comput. Sci.*, **247**(1987) 180–188; *MR* **88g**:68061.

V. Keränen, Abelian squares are avoidable on 4 letters, *Automata, Languages and Programming, Springer Lect. Notes Comput. Sci.*, **623**(1992) 41–52; *MR* **94j**:68244.

A. J. Kfoury, Square-free and overlap-free words, *Mathematical problems in computation theory*, Banach Centre Publications, **21**(1968) 285–297; Polish Scientific Publishers, Warsaw.

A. J. Kfoury, A linear-time algorithm to decide whether a binary word contains an overlap, *RAIRO Inform. Théor.*, **22**(1988) 135–145; *MR* **89h**:68056.

Y. Kobayashi, Repetition-free words, *Theor. Comput. Sci.*, **44**(1986) 175–197; *MR* **88a**:20084.

M. Leconte, Kth power-free codes, *Automata on Infinite Words, Springer Lect. Notes Comput. Sci.*, **192**(1984) 172–187; *MR* **87e**:20107.

John Leech, A problem on strings of beads, *Math. Gaz.*, **41**(1957) 277–278.

W. F. Lunnon & P. A. B. Pleasants, Characterization of two-distance sequences, *J. Austral. Math. Soc. Ser. A*, **53**(1992) 198–218; *MR* **93h**:11027.

M. G. Main, An infinite square-free co-CFL, *Inform. Process. Lett.*, **20**(1985) 105–107.

M. G. Main, W. Bucher & D. Haussler, Applications of an infinite square-free co-CFL, *Theor. Comput. Sci.*, **49**(1987) 113–119.

M. G. Main & R. J. Lorentz, Linear time recognition of squarefree strings, in A. Apostolico & Z. Galil (editors), *Combinatorial Algorithms on Words*, Springer-Verlag, 1985, 271–278.

M. G. Main & R. J. Lorentz, An $O(n \log n)$ algorithm for finding all repetitions in a string, *J. Algorithms*, **5**(1984) 422–432; *MR* **85h**:68019.

Christian Mauduit & András Sárközy, On finite pseudorandom binary sequences II. The Champernowne, Rudin-Shapiro and Thue-Morse sequences, a further construction, *J. Number Theory*, **73**(1998) 256–276; *MR* **99m**:11084.

F. Mignosi & G. Pirello, Repetitions in the Fibonacci infinite word, *RAIRO Inform. Théor.*, **26**(1992) 199–204.

Marston Morse, A solution of the problem of infinite play in chess, *Bull. Amer. Math. Soc.*, **44**(1938) 632.

Marston Morse & Gustav A. Hedlund, Unending chess, symbolic dynamics and a problem in semigroups, *Duke Math. J.*, **11**(1944) 202; *MR* **5**, 202e.

J.-J. Pansiot, A propos d'une conjecture de F. Dejean sur les répétitions dans les mots, *Discr. Appl. Math.*, **7**(1984) 297–311; *MR* **85g**:05010, 68036.

P. A. B. Pleasants, Non-repetitive sequences, *Proc. Cambridge Philos. Soc.*, **68**(1970) 267–274; *MR* **42** #85.

H. Prodinger, Nonrepetitive sequences and Gray code, *Discrete Math.*, **43** (1983) 113–116; *MR* **84d**:05026.

Helmut Prodinger & Friedrich J. Urbanek, Infinite 0-1 sequences without long adjacent identical blocks, *Discrete Math.*, **28**(1979) 277–289; *MR* **84d**:05026.

E. Prouhet, Mémoire sur quelques relations entre less puissances des nombres, *C. R. Acad. Sci. Paris*, **33**(1851) 225.

A. Restivo & S. Salemi, Overlap free words on two symbols, *Automata on Infinite Words, Springer Lect. Notes Comput. Sci.*, **192**(1984) 198–206; *MR* **87c**:20101.

A. Restivo & S. Salemi, Some decision results on nonrepetitive words, *Combinatorial Algorithms on Words*, Springer-Verlag, 1985, 289–295; *MR* **87g**:68041.

H. E. Robbins, On a class of recurrent sequences, *Bull. Amer. Math. Soc.*, **43**(1937) 413–417.

P. Roth, Every binary pattern of length six is avoidable on the two-letter alphabet, *Acta Inform.*, **29**(1992) 95–107; *MR* **93c**:68084.

U. Schmidt, Avoidable patterns on 2 letters, *STACS '87, Springer Lect. Notes Comput. Sci.*, **247**(1987) 189–197; *MR* **88h**:68046; *Theor. Comput. Sci.*, **63**(1989) 1–17; *MR* **90d**:68043.

P. Séébold, Morphismes itérés, mot de Morse, et mot de Fibonacci, *C. R. Acad. Sci. Paris*, **295**(1982) 439–441; *MR* **83m**:20096.

P. Séébold, Overlap-free sequences, *Automata on Infinite Words, Springer Lect. Notes Comput. Sci.*, **192**(1984) 207–215; *MR* **87f**:20082.

P. Séébold, Complément à l'étude des suites de Thue-Morse généralisées, *RAIRO Inform. Théor.*, **20**(1986) 157–181; *MR* **88i**:68051.

Jeffrey Shallit, On the maximum number of distinct factors in a binary string, *Graphs Combin.*, **9**(1993) 197–200; *MR* **94b**:05015.

R. Shelton, Aperiodic words on three symbols, *J. reine angew. Math.*, **321** (1981) 195–209; II **327**(1981) 1–11; *MR* **82m**:05004ab.

R. O. Shelton & R. P. Soni, Aperiodic words on three symbols III, *J. reine angew. Math.*, **330**(1982) 44–52; *MR* **82m**:05004c.

H. J. Shyr, A strongly primitive word of arbitrary length and its application, *Internat. J. Comput. Math.*, **6**(1977) 165–170; *MR* **57** #4671.

A. Thue, Über unendliche Zeichenreihen, *Norske Vid. Selsk. Skr. I Mat.-Nat. Kl. Christiana*, 1906, No. 7, 1–22.

A. Thue, Über die gegensetige Lage gleicher Teile gewisser Zeichenreihen, *Norske Vid. Selsk. Skr. I Mat.-Nat. Kl. Christiana*, 1912, No. 1, 1–67.

Robert Tijdeman, On the minimum complexity of infinite words, *Indag. Math.* (*N.S.*), **10**(1999) 123–129; *MR* **2000d**:11032.

Yao Jia-Yan, Généralisations de la suite de Thue-Morse, *Ann. Sci. Math. Québec*, **21**(1997) 177–189; *MR* **98k**:11108.

M. A. Zaks, A. S. Pikovskiĭ & Jürgen Kurths, On the correlation dimension of the spectral measure for the Thue-Morse sequence, *J. Statist. Phys.*, **88**(1997) 1387–1392; *MR* **98k**:11021; and see *J. Phys. A*, **32**(1999) 1523–1530; *MR* **2000c**:37007.

E22 Cycles and sequences containing all permutations as subsequences.

Hansraj Gupta asked, for $n \geq 2$, to find the least positive integer $m = m(n)$ for which a cycle $a_1, a_2, \ldots, a_m$ of positive integers, each $\leq n$, exists such that any given permutation of the first n natural numbers appears as a subsequence (not necessarily consecutive) of

$$a_j, a_{j+1}, \ldots, a_1, a_2, \ldots, a_{j-1}$$

for at least one j, $1 \leq j \leq m$. For example, for $n = 5$, such a cycle is 1, 2, 3, 4, 5, 4, 3, 2, 1, 5, 4, 5, so that $m(5) \leq 12$. He conjectures that $m(n) \leq \lfloor n^2/2 \rfloor$.

Motzkin & Straus used the **ruler function** (exponent of the highest power of 2 which divides k), e.g., $n = 5$, $1 \leq k \leq 31$,

$$1,2,1,3,1,2,1,4,1,2,1,3,1,2,1,5,1,2,1,3,1,2,1,4,1,2,1,3,1,2,1,$$

but this doesn't make use of the cyclic options and gives only $m(n) \leq 2^n - 1$.

E23 Covering the integers with A.P.s.

If S is the union of n arithmetic progressions, each with common difference $\geq k$, where $k \leq n$, Crittenden & Vanden Eynden conjecture that S contains all positive integers whenever it contains those $\leq k2^{n-k+1}$. If this is true, it's best possible. They have proved it for $k = 1$ and 2, and Simpson has proved it for $k = 3$.

R. B. Crittenden & C. L. Vanden Eynden, Any n arithmetic progressions covering the first 2^n integers, *Proc. Amer. Math. Soc.*, **24**(1970) 475–481; *MR* **41** #3365.

R. B. Crittenden & C. L. Vanden Eynden, The union of arithmetic progressions with differences not less than k, *Amer. Math. Monthly* **79**(1972) 630.

R. J. Simpson, On a conjecture of Crittenden and Vanden Eynden concerning coverings by arithmetic progressions, *J. Austral. Math. Soc. Ser. A*, **63**(1997) 396–420; *MR* **98k**:11005.

E24 Irrationality sequences.

Erdős & Straus called a sequence of positive integers $\{a_n\}$ an **irrationality sequence** if $\sum 1/a_n b_n$ is irrational for all integer sequences $\{b_n\}$. What are the irrationality sequences? Find some interesting ones. If $\limsup(\log_2 \ln a_n)/n > 1$, where the log is to base 2, then $\{a_n\}$ is an irrationality sequence. $\{n!\}$ is not an irrationality sequence, because $\sum 1/n!(n+2) = \frac{1}{2}$. Erdős has shown that $\{2^{2^n}\}$ is an irrationality sequence. The sequence 2, 3, 7, 43, 1807, ..., where $a_{n+1} = a_n^2 - a_n + 1$, is *not* an irrationality sequence, since we may take $b_n = 1$ and the sum of the reciprocals is 1. We asked about the sequence of alternate terms, 2, 7, 1807, ... and Golomb showed that the growth rate of 2, 3, 7, 43, 1807, ... is like θ^{2^n} where $\theta \approx 1.5979102$ so that alternate terms are asymptotic to $\theta^{2^{2n}}$ and the given criterion shows that they form an irrationality sequence.

Analogously, Hančl defines a sequence $\{a_n\}$ of positive reals to be a **transcendental sequence** if $\sum(a_n c_n)^{-1}$ is transcendental for every sequence $\{c_n\}$ of positive integers and shows that $\{a_n/b_n\}$ is transcendental if $\{a_n\}$, $\{b_n\}$ are sequences of positive integers satisfying $\ln \ln a_n > u^n$ and $\ln \ln b_n < v^n$ for all large n, with u, v real numbers $u > v > 3$.

C. Badea, *Glasgow Math. J.*, **29**(1987) 221–228; *MR* **88i**:11044

P. Erdős, On the irrationality of certain series, *Nederl. Akad. Wetensch. Proc. Ser. A = Indagationes Math.*, **19**(1957) 212–219; *MR* **19**, 252.

P. Erdős, On the irrationality of certain series, *Math. Student*, **36**(1968) 222–226 (1969); *MR* **41** #6787.

P. Erdős, Some problems and results on the irrationality of the sum of infinite series, *J. Math. Sci.*, **10**(1975) 1–7.

P. Erdős & E. G. Straus, On the irrationality of certain Ahmes series, *J. Indian Math. Soc.*, **27**(1963) 129–133 (1969); *MR* **41** #6787.

S. Golomb, On certain non-linear recurring sequences, *Amer. Math. Monthly*, **70**(1963) 403–405; *MR* **26** #6112.

I. J. Good, *Fibonacci Quart.*, **12**(1974) 346; *MR* **50** #4465.

Jaroslav Hančl, Transcendental sequences, *Math. Slovaca*, **46**(1996) 177–179; *MR* **97k**:11107.

Wayne L. McDaniel, The irrationality of certain series whose terms are reciprocals of Lucas sequence terms, *Fibonacci Quart.*, **32**(1994) 346–351; *MR* **95d**:11018.

Zhu Yao-Chen, Irrationality of certain series, *Acta Math. Sinica*, **40**(1997) 857–860; *MR* **99c**:11090.

E25 Golomb's self-histogramming sequence.

The sequence

$$1, 2, 2, 3, 3, 4, 4, 4, 5, 5, 5, 6, 6, 6, 6, 7, 7, 7, 7, 8, 8, 8, 8, 9, 9, 9, 9, 9, \ldots$$

defined by $f(1) = 1$ and $f(n)$ as the number of occurrences of n in a nondecreasing sequence of integers was attributed to David Silverman in the first edition. It was given as a problem by Golomb, and solved by him, by van Lint, and by Marcus & Fine: the asymptotic expression for the n-th term is indeed $\tau^{2-\tau} n^{\tau-1}$ where τ is the golden ratio $(1+\sqrt{5})/2$. The error term $E(n)$ has been investigated by Ilan Vardi, who conjectured that

$$E(n) = \Omega_\pm \left(\frac{n^{\tau-1}}{\ln n} \right)$$

where $E(n) = \Omega_\pm(g(n))$ means that there are constants c_1, c_2 such that $E(n) > c_1 g(n)$ and $E(n) < -c_2 g(n)$ are each true for infinitely n. This has been proved by Jean-Luc Rémy, and he and Pétermann have shown that

$$E(n) = \frac{n^{\tau-1}}{\ln n} h \left(\frac{\ln \ln n}{\ln \tau} \right) + o \left(\frac{n^{\tau-1}}{\ln n} \right)$$

where h is a function satisfying $h(x) = -h(x+1)$ and h not identically 0.

A vaguely similar sequence has been proposed by Lionel Levine, consisting of the last member of each row in the array

```
1   1
1   2
1   1   2
1   1   2   3
1   1   1   2   2   3   4
1   1   1   1   2   2   2   3   3   4   4   5   6   7
...
```

in which each row is obtained from the previous one by reading it from *right to left*, e.g., read the penultimate line as 'four ones, three twos, two threes, two fours, one five, one six, one seven'. Sloane knows fifteen members of the sequence,

1, 2, 2, 3, 4, 7, 14, 42, 213, 2837, 175450, 139759600, 6837625106787, 266437144916648607844, 5080094713794888214442619865035440

and wonders if the 20th member will ever be calculated.

A **counting sequence** is defined as a sequence of sequences $\{S_i\}_{i=0}^\infty$ of positive integers. The sequence S_{i+1} is obtained from S_i by counting the number m_k of times an integer k occurs in S_i and writing down in S_{i+1} the pair (m_k, k) in increasing order of k, for all k for which $m_k > 0$. See the Sauerberg & Shu paper.

Marshall Hall proved the existence of a sequence such that every positive integer occurs uniquely as the difference of two members of the sequence. For example,

$$1, 2, 4, 8, 16, 21, 42, 51, 102, 112, 224, 235, 470, 486, 972, 990, 1980, 2001, \ldots$$

defined by $a_1 = 1$, $a_2 = 2$, $a_{2n+1} = 2a_{2n}$, $a_{2n+2} = a_{2n+1} + r_n$, where r_n is the least natural number which cannot be represented in the form $a_j - a_i$ with $1 \leq i < j \leq 2n + 1$. There are other possible sequences, but the problem of finding the ones with smallest asymptotic growth remains open.

V. Bronstein & A. S. Fraenkel, On a curious property of counting sequences, *Amer. Math. Monthly*, **101**(1994) 560–563; *MR* **95b**:05014.

J. Browkin, Solution of a certain problem of A. Schinzel (Polish), *Prace Mat.*, **3**(1959) 205–207; *MR* **23** #A3703.

Solomon W. Golomb, Influence of data processing on the design and communication of experiments, *Radio Science*, **68D**(1964) 1021–1024.

Solomon W. Golomb, Problem 5407, *Amer. Math. Monthly*, **73**(1966) 674.

R. L. Graham, Problem E1910, *Amer. Math. Monthly*, **73**(1966) 775; remark by C. B. A. Peck, **75**(1968) 80–81.

M. Hall, Cyclic projective planes, *Duke Math. J.*, **4**(1947) 1079–1090; *MR* **9**, 370b.

Steven Kahan, A curious sequence, *Math. Mag.*, **48**(1975) 290–293; *MR* **53** #5464.

Hervé Lehning, Computer-aided or analytic proof? *Coll. Math. J.*, **21**(1990) 228–239.

Daniel Marcus & N. J. Fine, Solutions to Problem 5407, *Amer. Math. Monthly*, **74** (1967) 740–743.

Michael D. McKay & Michael S. Waterman, Self-descriptive strings, *Math. Gaz.*, **66**(1982) 1–4; *MR* **84h**:05013.

Y.-F. S. Pétermann, On Golomb's self-describing sequence, I, *J. Number Theory*, **53**(1995) 13–24; *MR* **97g**:11022; II, *Arch. Math. (Basel)* **67**(1996) 473–477; *MR* **98a**:11029.

Y.-F. S. Pétermann & Jean-Luc Rémy, Golomb's self-described sequence and functional-differential equations, *Illinois J. Math.*, **42**(1998) 420–440; *MR* **99k**:11035.

Jean-Luc Rémy, Sur la suite autodécrite de Golomb, *J. Number Theory*, **66**(1997) 1–28; *MR* **98i**:11077.

Lee Sallows & Victor L. Eijkhout, Co-descriptive strings, *Math. Gaz.*, **70**(1986) 1–10.

Jim Sauerberg & Shu Ling-Hsueh, The long and the short on counting sequences, *Amer. Math. Monthly*, **104**(1997) 306–317; *MR* **98i**:11013.

W. Sierpiński, *Elementary Theory of Numbers*, 2nd English edition (A. Schinzel) PWN, Warsaw, 1987, chap. 12,4 p. 444.

Ilan Vardi, The error term in Golomb's sequence, *J. Number Theory*, **40**(1992) 1–11; *MR* **93d**:11103.

OEIS: A000002, A001462-001463, A001856, A054540.

E26 Epstein's Put-or-Take-a-Square game.

Richard Epstein's Put-or-Take-a-Square game is played with one heap of beans. Two players play alternately. A move is to add or subtract the largest perfect square number of beans that is in the heap. That is, the two players alternately name nonnegative integers a_n, where

$$a_{n+1} = a_n \pm \lfloor \sqrt{a_n} \rfloor^2,$$

the winner being the first to name zero. This is a loopy game and many numbers lead to a draw. For example, from 2 the next player will not take 1, allowing his opponent to win, so she does to 3. Now to add 1 is a bad move, so her opponent goes back to 2. Similarly 6 leads to a draw with best play: 6, 10, 19!, 35, 60, 109!, 209!, 13!, 22!, 6, ..., where ! means a good move, not factorial!

For example, 405 is a bad move after 209, since the next player can go to 5 which is a $\mathcal{P}$-**position** (previous player winning). Similarly, from 60, it is bad to go to 11, an $\mathcal{N}$-**position**, one in which the next player can win (by going to 20).

Do either of the sequences of $\mathcal{P}$-positions

    0, 5, 20, 29, 45, 80, 101, 116, 135, 145, 165, 173, 236, 257, 397,
    404, 445, 477, 540, 565, 580, 629, 666, 836, 845, 885, 909, 944,
    949, 954, 975, 1125, 1177, ...

or of $\mathcal{N}$-positions

    1, 4, 9, 11, 14, 16, 21, 25, 30, 36, 41, 44, 49, 52, 54, 64, 69, 71,
    81, 84, 86, 92, 100, 105, 120, 121, 126, 136, 141, 144, 149, 164,
    169, 174, 189, 196, 201, 208, 216, 225, 230, 245, 252, 254, 256,
    261, ...

have positive density?

E. R. Berlekamp, J. H. Conway & R. K. Guy, *Winning Ways for your Mathematical Plays*, Academic Press, London, 1982, Chapter 15.

OEIS: A005240-005241, A014586-014589, A019509.

E27 Max and mex sequences.

In his master's thesis, Roger Eggleton discussed **max sequences**, in which a given finite sequence a_0, a_1, ..., a_n is extended by defining $a_{n+1} = \max_i(a_i + a_{n-i})$. One of the main results is that the first differences are ultimately periodic. For example, starting from 1, 4, 3, 2 we get 7, 8, 11, 12, 15, 16, ... with differences 3, −1, −1, 5, 1, 3, 1, 3, 1, ... What happens to **mex sequences**, where the **mex** of a set of nonnegative integers is the minimum excluded number, or least nonnegative integer which does not appear in the set? Now the sequence 1, 4, 3, 2 continues

0, 0, 0, 0, 0, 5, 1, 1, 1, 1, 1, 6, 2, 2, 0, 0, 0, 0, 0, 5, 1, 1, 1, 1, 1, 6, ...

Are such sequences ultimately periodic?

A. S. Fraenkel is reminded of the sequence $a_i = \lfloor i\alpha \rfloor$, where α is any real number, which satisfies the inequalities

$$\max(a_{n-i} + a_i) \le a_n \le 1 + \min(a_{n-i} + a_i), \qquad 1 \le i < n, \qquad n = 2, 3, \ldots .$$

E.g., $\alpha = \frac{1}{2}(1 + \sqrt{5})$ generates the sequence 1, 3, 4, 6, 8, 9, 11, 12, 14, 16, 17, 19, 21, ... which complements that for $b_i = \lfloor i\beta \rfloor$, where $1/\alpha + 1/\beta = 1$. These **Beatty sequences** combine to form **Wythoff pairs**.

Motivation for this problem comes from the analysis of **octal games** using the Sprague-Grundy theory, where ordinary addition is replaced by **nim addition**, i.e., addition in base 2 without carry, or XOR. The behavior of such sequences remains a considerable mystery, clarification of which would lead to results about nim-like games.

S. Beatty, Problem 3173, *Amer. Math. Monthly* **33** (1926) 159 (and see J. Lambek & L. Moser, **61** (1954) 454.

E. R. Berlekamp, J. H. Conway & R. K. Guy, *Winning Ways for your Mathematical Plays*, Academic Press, London, 1982, Chapter 4.

Michael Boshernitzan & Aviezri S. Fraenkel, Nonhomogeneous spectra of numbers, *Discrete Math.*, **34**(1981) 325–327; *MR* **82d**: 10077.

Michael Boshernitzan & Aviezri S. Fraenkel, A linear algorithm for nonhomogeneous spectra of numbers, *J. Algorithms*, **5**(1984) 187–198; *MR* **85j**:11183.

R. B. Eggleton, Generalized integers, M.A. thesis, Univ. of Melbourne, 1969.

Ronald L. Graham, Lin Chio-Shih & Lin Shen, Spectra of numbers, *Math. Mag.*, **51**(1978) 174–176; *MR* **58** #10808.

OEIS: A067016-067018.

E28 B$_2$-sequences. Mian-Chowla sequences.

Call an infinite sequence $1 \le a_1 < a_2 < \ldots$ an A-**sequence** if no a_i is the sum of distinct members of the sequence other than a_i. Erdős proved that for every A-sequence, $\sum 1/a_i < 103$, and Levine and O'Sullivan improved this to 4. They also gave an A-sequence whose sum of reciprocals is > 2.035. Abbott, and later Zhang, have given the example

$$\{1, 2, 4, 8, 1 + 24k, 35950 + 24t : 1 \le k \le 55, 0 \le t \le 44\}$$

which improves this to 2.0648. A further block of terms will push this past 2.0649, but not as far as 2.065.

If $1 \leq a_1 < a_2 < \ldots$ is a B_2-**sequence** (compare **C9**), i.e., a sequence where all the sums of pairs $a_i + a_j$ are different, what is the maximum of $\sum 1/a_i$? There are two problems, according as $i = j$ is permitted or not, but Erdős was unable to solve either of them.

The most obvious B_2-sequence is that obtained by the greedy algorithm (compare **E10**). Each term is the least integer greater than earlier terms which does not violate the distinctness of sums condition; $i = j$ is permitted:

$$1, 2, 4, 8, 13, 21, 31, 45, 66, 81, 97, 123, 148, 182, \ldots.$$

Mian & Chowla used this to show the existence of a B_2-sequence with $a_k \ll k^3$. If M is the maximum of $\sum 1/a_i$ over all B_2-sequences and S^* is the sum of the reciprocals of the Mian–Chowla sequence, then $M \geq S^* > 2.156$. But Levine observes that if $t_n = n(n+1)/2$, then $M \leq \sum 1/(t_n + 1) < 2.374$, and asks if $M = S^*$? Zhang disproves this by showing that $S^* < 2.1596$ and $M > 2.1597$. The latter result is obtained by replacing the next term, 204, in the Mian-Chowla sequence, by 229, and then continuing with the greedy algorithm.

Let $a_1 < a_2 < \cdots$ be an infinite sequence of integers for which all the triple sums $a_i + a_j + a_k$ are distinct. Erdős offers \$500.00 for a proof or disproof of an old conjecture of his, that $\lim a_n/n^3 = \infty$.

Let $K(m) = \max(m - \sqrt{a_m})$ over all m-element B_2-sequences. Zhang Zhen-Xiang finds lower bounds for $K(p)$ when p is prime; e.g., $K(829) > 10.279$. The steady growth of these lower bounds suggests that the Erdős-Turán conjecture $K(m) = O(1)$ may be false.

Compare **C9, C14, E10, E32**

H. L. Abbott, On sum-free sequences, *Acta Arith.*, **48**(1987) 93–96; *MR* **88g**:11007.

P. Erdős, Problems and results in combinatorial analysis and combinatorial number theory, *Graph Theory, Combinatorics and Applications*, Vol. 1 (Kalamazoo MI, 1988) 397–406, Wiley, New York, 1991.

Jia Xing-De, On B_{2k}-sequences, *J. Number Theory*, **48**(1994) 183–196; *MR* **95c**:11027.

Eugene Levine, An extremal result for sum-free sequences, *J. Number Theory*, **12**(1980) 251–257; *MR* **82d**:10078.

Eugene Levine & Joseph O'Sullivan, An upper estimate for the reciprocal sum of a sum-free sequence, *Acta Arith.*, **34**(1977) 9–24; *MR* **57** #5900.

Abdul Majid Mian & S. D. Chowla, On the B_2-sequences of Sidon, *Proc. Nat. Acad. Sci. India Sect. A*, **14**(1944) 3–4; *MR* **7**, 243.

J. O'Sullivan, On reciprocal sums of sum-free sequences, PhD thesis, Adelphi University, 1973.

Zhang Zhen-Xiang, A B_2-sequence with larger reciprocal sum, *Math. Comput.*, **60**(1993) 835–839; *MR* **93m**:11012.

Zhang Zhen-Xiang, Finding finite B_2-sequences with larger $m - a_m^{1/2}$, *Math. Comput.*, **61**(1993); *MR* **94i**:11109.

OEIS: A005282, A025582, A051788.

E29 Sequence with sums and products all in one of two classes.

Partition the integers into two classes. Is it true that there is always a sequence $\{a_i\}$ so that all the sums $\sum \epsilon_i a_i$ and all the products $\prod a_i^{\epsilon_i}$ where the ϵ_i are 0 or 1 with all but a finite number zero, are in the same class? Hindman answered this question of Erdős negatively.

Is there a sequence $a_1 < a_2 < \cdots$ so that all the sums $a_i + a_j$ and products $a_i a_j$ are in the same class? Graham proved that if we partition the integers [1,252] into two classes, there are four distinct numbers x, y, $x + y$ and xy all in the same class. Moreover, 252 is best possible. Hindman proved that if we partition the integers [2,990] into two classes, then one class always contains four distinct numbers x, y, $x + y$ and xy. No corresponding result is known for the integers ≥ 3.

Hindman also proved that if we partition the integers into two classes, there is always an infinite sequence $\{a_i\}$ so that all the sums $a_i + a_j$ ($i = j$ permitted) are in the same class. On the other hand he found a decomposition into three classes so that no such infinite sequence exists.

J. Baumgartner, A short proof of Hindman's theorem, *J. Combin. Theory Ser. A*, **17**(1974) 384–386; *MR* **50** #6873.

Neil Hindman, Finite sums with sequences within cells of a partition of n, *J. Combin. Theory Ser. A*, **17**(1974) 1–11; *MR* **50** #2067.

Neil Hindman, Partitions and sums and products of integers, *Trans. Amer. Math. Soc.*, **247**(1979) 227–245; *MR* **80b**:10022.

Neil Hindman, Partitions and sums and products — two counterexamples, *J. Combin. Theory Ser. A*, **29**(1980) 113–120; *MR* **82b**:05020.

E30 MacMahon's prime numbers of measurement.

MacMahon's "prime numbers of measurement,"

$$1, 2, 4, 5, 8, 10, 14, 15, 16, 21, 22, 25, 26, 28, 33, 34, 35, 36, 38, 40, 42, \ldots$$

are generated by excluding all the sums of two or more consecutive earlier members of the sequence.

If m_n is the nth member of the sequence, and M_n is the sum of the first n members, then George Andrews conjectures that

$$\text{¿} \quad m_n \sim n(\ln n)/\ln\ln n \quad ? \quad \text{and} \quad \text{¿} \quad M_n \sim n^2(\ln n)/\ln(\ln n)^2 \quad ?$$

and poses the following, presumably easier, problems: prove $\lim n^{-\Delta} m_n = 0$ for some $\Delta < 2$; prove $\lim m_n/n = \infty$; prove $m_n < p_n$ for every n, where p_n is the nth prime.

Jeff Lagarias suggests excluding only the sums of *two* or *three* consecutive earlier members, and asked if the resulting sequence

$$1, 2, 4, 5, 8, 10, 12, 14, 15, 16, 19, 20, 21, 24, 25, 27, 28, 32, 33, 34,$$
$$37, 38, 40, 42, 43, 44, 46, 47, 48, 51, 53, 54, 56, 57, 58, 59, 61, \ldots$$

has density $\frac{3}{5}$. Don Coppersmith has a better heuristic, suggesting that the answer is 'no.'

More generally, if $1 \le a_1 < a_2 < \cdots < a_k \le n$ is a sequence in which no a is the sum of consecutive earlier members, then Pomerance found that $\max k \ge \lfloor \frac{n+3}{2} \rfloor$ and Róbert Freud later showed that $\max k \ge \frac{19}{36} n$. They notice, with Erdős, that $\max k \le \frac{2}{3} n$, even if we only forbid sums of *two* consecutive earlier members. Coppersmith & Phillips have since shown that $\max k \ge \frac{13}{24} n - O(1)$ and they lower the upper bound to

$$\max k \le \left(\frac{2}{3} - \epsilon\right) n + O(\ln n) \quad \text{with} \quad \epsilon = \frac{1}{896}.$$

Erdős asks if the lower density of the sequence is zero; perhaps

$$¿ \quad \frac{1}{\ln x} \sum_{a_i < x} \frac{1}{a_i} \to 0 \quad ?$$

G. E. Andrews, MacMahon's prime numbers of measurement, *Amer. Math. Monthly*, **82**(1975) 922–923.

Don Coppersmith & Steven Phillips, On a question of Erdős on subsequence sums, *SIAM J.Discrete Math.*, **9**(1996) 173–177; *MR* **97d**:05273.

Róbert Freud, *James Cook Math. Notes*, Jan. 1993.

R. L. Graham, Problem 1910, *Amer. Math. Monthly*, **73**(1966) 775; solution **75**(1968) 80–81.

Jeff Lagarias, Problem 17, W. Coast Number Theory Conf., Asilomar, 1975.

P. A. MacMahon, The prime numbers of measurement on a scale, *Proc. Cambridge Philos. Soc.*, **21**(1923) 651–654.

Štefan Porubský, On MacMahon's segmented numbers and related sequences, *Nieuw Arch. Wisk.*(3) **25**(1977) 403–408; *MR* **58** #5575.

OEIS: A002048-002049, A005242.

E31 Three sequences of Hofstadter.

Doug Hofstadter has defined three intriguing sequences.

(a) $a_1 = a_2 = 1$ and $a_n = a_{n-a_{n-1}} + a_{n-a_{n-2}}$ for $n \geq 3$. What is the
     general behavior of this sequence?

$$1, 1, 2, 3, 3, 4, 5, 5, 6, 6, 6, 8, 8, 8, 10, 9, 10, 11, 11,$$
$$12, 12, 12, 12, 16, 14, 14, 16, 16, 16, 16, 20, 17, 17, \ldots$$

Are there infinitely many integers 7, 13, 15, 18, ... that get
missed out?

(b) $b_1 = 1$, $b_2 = 2$ and for $n \geq 3$, b_n is the least integer greater than
     b_{n-1} which can be expressed as the sum of two or more *consecutive*
     terms of the sequence, so it goes

$$1, 2, 3, 5, 6, 8, 10, 11, 14, 16, 17, 18, 19, 21, 22, 24, 25, 29,$$
$$30, 32, 33, 34, 35, 37, 40, 41, 43, 45, 46, 47, 49, 51, \ldots$$

This is a sort of dual of MacMahon's prime numbers of measurement (**E30**). How does the sequence grow?

(c) $c_1 = 2$, $c_2 = 3$, and when $c_1, \ldots, c_n$ are defined, form all possible
     expressions

$c_i c_j - 1$ $(1 \leq i < j \leq n)$ and append them to the sequence:

$$2, 3, 5, 9, 14, 17, 26, 27, 33, 41, 44, 50, 51, 53, 65, 69, 77,$$
$$80, 81, 84, 87, 98, 99, 101, 105, 122, 125, 129, \ldots$$

Does the result include almost all of the integers?

A similar sequence to the first of these three was given by Conway:

$$1, 1, 2, 2, 3, 4, 4, 4, 5, 6, 7, 7, 8, 8, 8, 8, 9, \ldots$$

defined, for $n \geq 3$, by

$$a(n) = a(a(n-1)) + a(n - a(n-1)).$$

The difficult questions were answered in an entertaining paper by Mallows.
Several identities have been obtained by Zeitlin. A variation is to define

$$b(n) = b(b(n-1)) + b(n - 1 - b(n-1)),$$

but this is related to the previous sequence by $b(n-1) = n - a(n)$. But if
we write

$$c(n) = c(c(n-2)) + c(n - c(n-2))$$

then the increments become very irregular, and it is not even clear that
$c(n)/n$ tends to a limit.

Higham & Tanny observe that the variant defined by

$$e(n) = e(n - 1 - e(n-1)) + e(n - 2 - e(n-2))$$

with $k \geq 4$ initial values $\lfloor i/2 \rfloor$, $0 \leq i < k$ is well behaved.

R. B. J. T. Allenby & Rachael C. Smith, Some sequences resembling Hofstadter's, *J. Korean Math. Soc.*, **40**(2003) 921–932.

Ed Barbeau & Stephen Tanny, On a strange recursion of Golomb, *Electron.
J. Combin.*, **3**(1996) RP8; *MR* **96k**:11024.

W. A. Beyer, R. G. Schrandt & S. M. Ulam, Computer studies of some
history-dependent random processes, LA-4246, Los Alamos Nat. Lab., 1969.

J. H. Conway, The weird and wonderful chemistry of audioactive decay, in *Open Problems in Communication and Computation* (T. M. Cover & B. Gopinath, eds.) pp. 173-188, Springer-Verlag, New York, 1987; see also *Eureka*, **46**(1986) 5–16.

J. H. Conway, Some crazy sequences, videotaped talk at A.T. & T. Bell Laboratories, 88-07-15.

Peter J. Downey & Ralph E. Griswold, On a family of nested recurrences, *Fibonacci Quart.*, 22(1984) 310–317; *MR* **86e**:11013.

P. Erdős & R. L. Graham, Old and New Problems and Results in Combinatorial Number Theory, *Monographie de L'Enseignement Mathématique, Genève*, **28**(1980) 83–84.

Yoshihiro Hamaya, On the asymptotic behavior of a delay Fibonacci equation, *Dynamic Systems and applications*, 3(*Atlanta GA*, 1999, 273–277, Dynamic, Atlanta GA, 2001; *MR* **2002i**:11013.

Jeff Higham & Stephen Tanny, A tamely chaotic meta-Fibonacci sequence, 23rd Manitoba Conf. Numer. Math. & Comput. (Winnipeg, 1993), *Congr. Numer.*, **99**(1994) 67–94; *MR* **95h**:11013.

Jeff Higham & Stephen Tanny, More well-behaved meta-Fibonacci sequences, 24th Southeastern Conf. Combin., Graph Theory & Comput. (Boca Raton FL 1993), *Congr. Numer.*, **98**(1993) 3–17; *MR* **95a**:11021.

Douglas R. Hofstadter, *Gödel, Escher, Bach*, Vintage Books, New York, 1980, p. 137.

Mark Kac, A history-dependent random sequence defined by Ulam, *Adv. in Appl. Math.*, **10**(1989) 270–277; *MR* **91c**:11042.

Péter Kiss & Béla Zay, On a generalization of a recursive sequence, *Fibonacci Quart.*, **30**(1992) 103–109; *MR* **90e**:11022.

Tal Kubo & Ravi Vakil, On Conway's recursive sequence, *Discrete Math.*, **152**(1996) 225–252; *MR* **99c**:11015.

Colin L. Mallows, Conway's challenge sequence, *Amer. Math. Monthly*, **98** (1991) 5–20; *MR* **92e**:39007.

David Newman, Problem E3274, *Amer. Math. Monthly*, **95**(1988) 555.

Klaus Pinn, Order and chaos in Hofstadter's $Q(n)$ sequence, *Complexity*, 4(1999) 41–46; *MR* **99k**:11036.

Stephen M. Tanny, A well-behaved cousin of the Hofstadter sequence, *Discrete Math.*, **105**(1992) 227–239; *MR* **93i**:11029.

David Zeitlin, Explicit solutions and identities for Conway's iterated sequence, *Abstracts Amer. Math. Soc.*, **12**(1991).

OEIS: A0004001, A005185, A005206, A005229, A005243-005244, A005374-005375, A005378-005379, A048973.

E32 B_2-sequences from the greedy algorithm.

An old problem of Dickson is still unsolved. Given a set of k integers, $a_1 < a_2 < \ldots < a_k$, define a_{n+1} for $n \geq k$ as the least integer greater than a_n which is *not* of the form $a_i + a_j$, $i, j \leq n$. Except for the prescribed section

at the beginning of the sequence, these are sum-free sequences formed by
the greedy algorithm (compare **C9, C14, E10, E28**).

Is the sequence of differences $a_{n+1} - a_n$ ultimately periodic?

Such sequences may take a long time before the periodicity appears.
For example, even for $k = 2$, if we take $a_1 = 1$, $a_2 = 6$, the sequence is

1,6,8,10,13,15,17,22,24,29,31,33,36,38,40,45,47,52,54,56,59,61,63,68,...

and one can be forgiven for not immediately recognizing the pattern. Try
starting with the set $\{1,4,9,16,25\}$; after 82 irregular differences, it settles
down to a period of length 224.

Queneau (see reference at **C4**) considered the similar problem with
$i < j \le n$ in place of $i, j \le n$. Such **0-additive sequences** have also
been conjectured to have ultimately periodic differences. Steven Finch has
calculated $1\frac{1}{2}$ million terms of the sequence whose first 6 terms are given
as $\{3, 4, 6, 9, 10, 17\}$ without detecting any sign of ultimate periodicity of
the differences.

Selmer tells me that Dickson's problem is the particular case $h = 2$ of
Stöhr sequences : let $a_1 = 1$ and define a_{n+1} for $n \ge k$ to be the least
integer greater than a_n which can *not* be written as the sum of at most
h addends among $a_1, a_2, \ldots, a_n$. Compare the h-bases of **C12**. In the
great majority of cases, the sequence of differences $a_{n+1} - a_n$ turns out to
be ultimately periodic, but there are a few of the examined cases where
periodicity has not been established.

Calkin & Erdős show that certain natural aperiodic sum-free sets are
incomplete in the sense that there are infinitely many n not in S which are
not the sum of two elements in S. They note that the reference to Dickson
does not contain the problem ascribed to him.

Compare **C14, E12**.

Neil J. Calkin & Paul Erdős, On a class of aperiodic sum-free sets, *Math.
Proc. Cambridge Philos. Soc.*, **120**(1996) 1–5; *MR* **97b**:11030.

Neil J. Calkin, On the number of sum-free sets, *Bull. London Math. Soc.*,
22(1990) 141–144; *MR* **91b**:11015.

Neil J. Calkin & Steven R. Finch, Conditions on periodicity for sum-free sets,
Experiment. Math., **5**(1996) 131–137; *MR* **98c**:11010.

Neil J. Calkin & Andrew Granville, On the number of co-prime-free sets,
Number theory (New York, 1991–1995), 9–18, Springer, New York, 1996; *MR*
97j:11006.

Neil J. Calkin & Angela C. Taylor, Counting sets of integers, no k of which
sum to another, *J. Number Theory*, **57**(1996) 323–327; *MR* **97d**:11015.

Neil J. Calkin & Jan McDonald Thomson, Counting generalized sum-free
sets, *J. Number Theory*, **68**(1998) 151–159; *MR* **98m**:11010.

Peter J. Cameron, Portrait of a typical sum-free set, Surveys in Combinatorics
1987, *London Math. Soc. Lecture Notes*, **123**(1987) Cambridge Univ. Press, 13–
42; *MR* **88k**:05138.

P. J. Cameron & P. Erdős, On the number of sets of integers with various properties, *Number theory (Banff, 1988)* 61–79, de Gruyter, Berlin, 1990; *MR* **92g**:11010.

L. E. Dickson, The converse of Waring's problem, *Bull. Amer. Math. Soc.*, **40**(1934) 711–714.

Steven R. Finch, Are 0-additive sequences always regular? *Amer. Math. Monthly*, **99**(1992) 671–673.

Ernst S. Selmer, On Stöhr's recurrent h-bases for N, *Kgl. Norske Vid. Selsk. Skrifter*, **3**(1986) 1–15.

Ernst S. Selmer & Svein Mossige, Stöhr sequences in the postage stamp problem, No. **32**(Dec. 1984) Dept. Pure Math., Univ. Bergen, ISSN 0332-5407.

E33 Sequences containing no monotone A.P.s.

Erdős & Graham say that a sequence $\{a_i\}$ has a **monotone** A.P. of length k if there are subscripts $i_1 < i_2 < \cdots < i_k$ such that the subsequence a_{i_j}, $1 \le j \le k$ is either an increasing or a decreasing A.P. If $M(n)$ is the number of permutations of $[1, n]$ which have no monotone 3-term A.P., then Davis et al have shown that

$$M(n) \ge 2^{n-1} \qquad M(2n-1) \le (n!)^2 \qquad M(2n) \le (n+1)(n!)^2$$

They ask if $M(n)^{1/n}$ is bounded.

Davis et al have also shown that any permutation of (all) the positive integers must contain an increasing 3-term A.P., but there are permutations with no monotone 5-term A.P. It is not known whether a monotone 4-term A.P. must always occur.

If the positive integers are arranged as a doubly-infinite sequence then a monotone 3-term A.P. must still occur, but it's possible to prevent the occurrence of 4-term ones.

If *all* the integers are to be permuted then Tom Odda has shown that no 7-term A.P. need occur in the singly-infinite case, but little else is known.

J. A. Davis, R. C. Entringer, R. L. Graham & G. J. Simmons, On permutations containing no long arithmetic progressions, *Acta Arith.*, **34**(1977) 81–90; *MR* **58** #10705.

Tom Odda, Solution to Problem E2440, *Amer. Math. Monthly* **82**(1975) 74.

E34 Happy numbers.

Reg. Allenby's daughter came home from school in Britain with the concept of **happy numbers**. The problem may have originated in Russia. If you iterate the process of summing the squares of the decimal digits of a number, then it's easy to see that you either reach the cycle

$$4 \to 16 \to 37 \to 58 \to 89 \to 145 \to 42 \to 20 \to 4$$

or arrive at 1. In the latter case you started from a happy number. The first hundred happy numbers are

$$\begin{array}{cccccccccccccccccccc}
1 & 7 & 10 & 13 & 19 & 23 & 28 & 31 & 32 & 44 & 49 & 68 & 70 & 79 & 82 & 86 & 91 & 94 & 97 & 100 \\
\end{array}$$

> 1   7   10   13   19   23   28   31   32   44   49   68   70   79   82   86   91   94   97   100
> 103 109 129 130 133 139 167 176 188 190 192 193 203 208 219 226 230 236 239 262
> 263 280 291 293 301 302 310 313 319 320 326 329 331 338 356 362 365 367 368 376
> 379 383 386 391 392 397 404 409 440 446 464 469 478 487 490 496 536 556 563 565
> 566 608 617 622 623 632 635 637 638 644 649 653 655 656 665 671 673 680 683 694

It seems that about $1/7$ of all numbers are happy, but what bounds on the density can be proved? How many consecutive happy numbers can you have? Can there be arbitrarily many? The first pair is 31, 32; the first triple is 1880, 1881, 1882 and an example of five consecutive happy numbers is 44488, 44489, 44490, 44491, 44492. Jud McCranie finds that the first quadruple of happy numbers ends with 7842, and that the first quintuple is the one above. He found no sextuples below 10^{12}. However, Allenby tells me that Hendrik Lenstra can show that there are arbitrarily long sequences of consecutive happy numbers.

How large can the gaps be in the sequence of happy numbers? We can define the **height** of a happy number to be the number of iterations needed to reach 1. For example, the least happy numbers of

$$\begin{array}{ccccccccc}
\text{height} & 0 & 1 & 2 & 3 & 4 & 5 & 6 \\
\text{are} & 1 & 10 & 13 & 23 & 19 & 7 & 356.
\end{array}$$

McCranie verifies that 78999 is the least happy number of height 7, that that of height 8 is the 977-digit number 37889999999...999, and that that of height 9 has about 10^{977} digits. Warut Roonguthai confirms these results and gives the last as

$$78889 \cdot 10^{(421 \cdot 10^{973} - 34)/9} - 1,$$

a number of $(421 \cdot 10^{973} + 11)/9$ decimal digits.

How big is the least happy number of height h?

If we replace squares by cubes, then the situation is dominated, at least in base 10, by the fact that perfect cubes are congruent to 0 or ± 1 mod 9. The density of the corresponding numbers (1, 10, 100, 112, 121, 211, 778, ...) may be zero. Numbers $\equiv 0$ mod 3 converge to 153 and numbers $\equiv 2$ mod 3 converge to 371 or 407, so that interest is confined to numbers $\equiv 1$ mod 3. These converge to 370, or to one of the two 3-cycles (55, 250, 133) or (160, 217, 352), or to one of the two 2-cycles (919, 1459) or (136, 244), or, just occasionally, to 1. What proportion goes to each?

Somjit Datta considers fourth and fifth powers. For fourth powers he notes that about $8/9$ of the numbers $< 10^4$ converge to the 7-cycle (1138, 4179, 9219, 13139, 6725, 4339, 4514), about 1 in 23 to the 2-cycle (2178, 6514) and the rest, apart from powers of 10, converge to 8208.

For powers of five he found 12 nontrivial cycles: with cycle lengths

1	1	1	1	2	4	6	10	10	12	22	28

and least member

4150	4151	54748	93084	58618	10933	8299	8294	9044	39020	9045	244

The appearance of (4150,4151) and (9044,9045) is intriguing. What of higher powers? And different bases? Some results are given by Grundman & Teeple.

Alan F. Beardon, Sums of squares of digits, *Math. Gaz.*, **82**(1998) 379–388.

Henry Ernest Dudeney, *536 Puzzles & Curious Problems* (edited Martin Gardner), Scribner's, New York, 1967, Problem 143, pp. 43, 258–259.

Esam El-Sedy & Samir Siksek, On happy numbers, *Rocky Mountain J. Math.*, **30**(2000) 565–570; *MR* **2002c**:11011.

Tony Gardiner, *Discovering Mathematics*, Oxford University Press, pp. 18, 21.

H. G. Grundman & E. A. Teeple, Generalized happy numbers, *Fibonacci Quart.*, **39**(2001) 462–466; *MR* **2002h**:11010.

Joseph S. Madachy, *Mathematics on Vacation*, Scribner's, New York, 1966, pp. 163–165.

OEIS: A001273, A007770, A018785, A031177, A035497, A035502–035504, A046519.

E35 The Kimberling shuffle.

Clark Kimberling considered the array:

```
[1]  2    3    4    5    6    7    8   9  10  11  12  13  14  15  16  17  18  19
 2  [3]   4    5    6    7    8    9  10  11  12  13  14  15  16  17  18  19  20
 4   2   [5]   6    7    8    9   10  11  12  13  14  15  16  17  18  19  20  21
 6   2    7   [4]   8    9   10   11  12  13  14  15  16  17  18  19  20  21  22
 8   7    9    2  [10]   6   11   12  13  14  15  16  17  18  19  20  21  22  23
 6   2   11    9   12   [7]  13    8  14  15  16  17  18  19  20  21  22  23  24
13  12    8    9   14   11  [15]   2  16   6  17  18  19  20  21  22  23  24  25
 2  11   16   14    6    9   17   [8] 18  12  19  13  20  21  22  23  24  25  26
18  17   12    9   19    6   13   14 [20] 16  21  11  22   2  23  24  25  26  27
...  ..  ..   ..   ..   ..   ..   ..  ..  ..  ..  ..  ..  ..  ..  ..  ..  ..  ..
```

in which each row is obtained from the previous by boxing (and expelling) the main diagonal element, and then reading the first number after the box, the first before the box, the second after the box, the second before the box, and so on until all the initial numbers are read off, and then continuing with all the remaining numbers (still in numerical order). Is every number eventually expelled?

The numbers	1	2	3	4	5	6	7	8	9	10
are expelled on rows	1	25	2	4	3	22	6	8	10	5

and the numbers	11	12	13	14	15	16	17	18	19	20
are expelled on rows	32	83	44	14	7	66	169	11	49595	9

and the numbers	40	68	106	147
are expelled on rows	93167	181393	270186	8765242

and the numbers	242	322	502	669
are expelled on rows	16509502	38293016	118850522	653494691

In the alternate reading from the right and left of the box, one could instead start from the left first, leading to the array

$\boxed{1}$	2	3	4	5	6	7	8	9	10	11	12	13	14	15	16	17	18	19
2	$\boxed{3}$	4	5	6	7	8	9	10	11	12	13	14	15	16	17	18	19	20
2	4	$\boxed{5}$	6	7	8	9	10	11	12	13	14	15	16	17	18	19	20	21
4	6	2	$\boxed{7}$	8	9	10	11	12	13	14	15	16	17	18	19	20	21	22
2	8	6	9	$\boxed{4}$	10	11	12	13	14	15	16	17	18	19	20	21	22	23
9	10	6	11	8	$\boxed{12}$	2	13	14	15	16	17	18	19	20	21	22	23	24
8	2	11	13	6	14	$\boxed{10}$	15	9	16	17	18	19	20	21	22	23	24	25
14	15	6	9	13	16	11	$\boxed{17}$	2	18	8	19	20	21	22	23	24	25	26
11	2	16	18	13	8	9	19	$\boxed{6}$	20	15	21	14	22	23	24	25	26	27

. . . .

which displays a similar chaotic behavior.

In each array there's a small amount of pattern: some knight's move arithmetic progressions of common difference 3. For example, in the original array, for each $y \geq 0$, the number $n + 3y$ is in position $2y + 1$ on row $x + y$, where, for each $t \geq 0$,

$$n = 9 \cdot 2^t - 3t - 7, \quad 12 \cdot 2^t - 3t - 8, \quad \text{resp.} \quad 15 \cdot 2^t - 3t - 9$$
when $\quad x = 3 \cdot 2^t - 1, \quad 4 \cdot 2^t - 1, \quad \text{resp.} \quad 5 \cdot 2^t - 1$

so that $n + 3y$ is expelled on row $2x - 1 = 2y + 1$. I.e.,

$$n = \quad 18 \cdot 2^t - 3t - 13, \; 24 \cdot 2^t - 3t - 14, \; \text{resp.} \; 30 \cdot 2^t - 3t - 15$$
is expelled on row $6 \cdot 2^t - 3, \quad 8 \cdot 2^t - 3, \quad \text{resp.} \quad 10 \cdot 2^t - 3$.

for $t = -1, 0, 1, \ldots$.

Clark Kimberling, Problem 1615, *Crux Mathematicorm*, **17**#2(Feb 1991) 44.

OEIS: A006852, A007063, A035486, A035505, A038807.

E36 Klarner-Rado sequences.

The sequence

1, 2, 4, 5, 8, 9, 10, 14, 15, 16, 17, 18, 20, 26, 27, 28, 29, 30, 32, 33, 34, 36, 40, 44, 47, 50, 51, 52, 53, 54, 56, 57, 58, 60, 62, 63, 64, 66, 68, 72, 80, 83, 86, 87, 88, 89, 92, 93, 94, 98, 99, 100, 101, 102, 104, 105, 106, 108, 110, 111, 112, 114, 116, 120, 122, 123, 124, 126, 128, 132, 134, 136, ...

is the thinnest which contains 1, and whenever it contains x, also contains $2x$, $3x + 2$ and $6x + 3$. Does it have positive density?

Several questions of this type were asked in a paper

David A. Klarner & Richard Rado, Arithmetic properties of certain recursively defined sets, *Pacific J. Math.*, **53**(1974) 445–463; In the review, *MR* **50** #9784, it was stated that a subsequent paper "Sets generated by a linear operation", same J., to appear, settles many of the conjectures stated in this paper. Did it ever appear? See also

David A. Klarner & Richard Rado, Linear combinations of sets of consecutive integers, *Amer. Math. Monthly*, **80**(1973) 985–989; *MR* **48** #8378.

David A. Klarner & Karel Post, Some fascinating integer sequences, *Discrete Math.*, **106/107**(1992) 303–309; *MR* **93i**:11031.

E37 Mousetrap.

Cayley introduced a permutation problem he called **Mousetrap** which is loosely based on the card game Treize. Suppose that the numbers $1, 2, \ldots, n$ are written on cards, one to a card. After shuffling (permuting) the cards, start counting the deck from the top card down. If the number on the card does not equal the count, transfer the card to the bottom of the deck and continue counting. If the two are equal then set the card aside and start counting again from 1. The game is **won** if all the cards have been set aside, but lost if the count reaches $n + 1$. Cayley proposed two questions.

1. For each n find all the winning permutations of $1, 2, \ldots, n$.

2. For each n find the number of permutations that eliminate precisely i cards for each i, $1 \le i \le n$.

A third question arose during our investigations. Consider a permutation for which every number is set aside. The list of numbers in the order that they were set aside is another permutation. Any permutation obtained in this way we call a **reformed** permutation.

3. Characterize the reformed permutations.

The permutation 4213 is a winning permutation which gives rise to the permutation 2134; this in turn gives the reformed permutation 3214 which is not a winning permutation.

4. For a given n, what is the longest sequence of reformed permutations?

5. Are there sequences of arbitrary length? Are there any cycles other

than
$$1 \to 1 \to 1 \to 1 \ldots \quad \text{and} \quad 12 \to 12 \to 12 \to 12 \ldots ?$$

Modular Mousetrap. We can play Mousetrap, but instead of counting n, $n+1$, ..., we can start again, ..., n, 1, 2, ... Now at least as many cards get set aside. In fact if n is prime, then either the initial deck is a derangement, or all cards get set aside, so every sequence cycles or terminates in a derangement. The identity permutation $123 \ldots n$ will always form a 1-cycle and now there are also examples of nontrivial cycles.

6. Are there k-cycles for every k ? What is the least value of n which yields a k-cycle?

A. Cayley, A Problem in Permutations, *Quart. Math. J.*, **I**(1857), 79.

A. Cayley, On the Game of Mousetrap, *Quart. J. Pure Appl. Math.*, **XV** (1877), 8-10.

A. Cayley, A Problem on Arrangements, *Proc. Roy. Soc. Edinburgh*, **9**(1878) 338–342.

A. Cayley, Note on Mr. Muir's Solution of a Problem of Arrangement, *Proc. Roy. Soc. Edinburgh*, **9**(1878) 388–391.

Richard K. Guy & Richard J. Nowakowski, Mousetrap, Proc. Erdős 80 Kesthely Combin. Conf., 1993; see also *Amer. Math. Monthly*, **101**(1994) 1007–1010.

T. Muir, On Professor Tait's Problem of Arrangement, *Proc. Royal Soc. Edinburgh*, **9**(1878) 382–387.

T. Muir, Additional Note on a Problem of Arrangement, *Proc. Royal Soc. Edinburgh*, **11**(1882) 187–190.

Adolf Steen, Some Formulae Respecting the Game of Mousetrap, *Quart. J. Pure Appl. Math.*, **XV**(1878), 230-241.

Peter Guthrie Tait, *Scientific Papers*, vol. 1, Cambridge, 1898, 287.

OEIS: A000142, A000166, A002467-002469, A007709, A007711-007712, A028305-028306, A055459, A068106.

E38 Odd sequences

Call a sequence of n zeros and ones, $\{a_1, \ldots, a_n\}$, **odd** if each of the n sums $\sum_{i=1}^{n-k} a_i a_{i+k}$, $(k = 0, 1, \ldots, n-1)$ is odd. For example, 1101 is odd. It was thought that there were no odd sequences if $n \geq 5$, but Peter Alles showed that there are infinitely many: if **o** is an odd sequence of length n and **x** and **z** are sequences of $n-1$ and $3n-2$ zeros, then **oxozo** and **ozoxo** are odd sequences of length $7n - 3$. For

$n =$	1	4	12	16	24	25	36	37	40	45
he finds	1	2	2	8	2	4	2	16	2	16

odd sequences and no others with $n \leq 50$. For example, 101011100011 and its reversal. He asks if n is always congruent to 0 or 1 modulo 4, and if, when there *are* odd sequences, they are always a power of two in number. Adam Sikora reported that he has answered both of these questions in the affirmative, but he was anticipated by Inglis & Wiseman, who show further that there is an odd sequence of length $n > 1$ just if the order of 2 is odd in the multiplicative group of integers modulo $2n - 1$. However, all these people were anticipated by MacWilliams & Odlyzko in 1977.

Pieter Moree has written:

> ... an odd sequence of length n exists if and only if there exists a $[2n, n]$ extended binary cyclic code, the celebrated $[24, 12, 8]$ Golay code arising in that way for $n = 12$. Alles's results and questions have previously been established by MacWilliams and Odlyzko in 1977, furthermore they gave a simple characterization of the n for which odd sequences exist:
>
> **Theorem.** A sequence of length n is odd if and only if $2n - 1$ is in P, where P is a set having the following properties:
>
> (i) An integer r belongs to P if and only if all the prime factors of r belong to P.
>
> (ii) If p is a prime, then $p \in P$ if and only if p divides $2^{2s+1} - 1$ for some s; in other words if and only if $\exp_p(2)$ is odd, where $\exp_p(a)$ is the smallest integer m such that $a^m \equiv 1 \bmod p$. $\square$
>
> This implies that if $p \equiv -1 \bmod 8$, then $p \in P$, and if $p \equiv \pm 3 \bmod 8$, then $p \notin P$. Moreover, from their proof it is clear that if there exists odd sequences of length n, they can be constructed easily, and that their number is a power of 2. This solves Question (ii) posed by Alles.
>
> The last part of the theorem leaves open the behaviour of the primes $p \equiv 1 \bmod 8$. Sometimes they turn out to be in P and sometimes not. Let $\pi_P(x)$ denote the number of primes in P not exceeding x. MacWilliams and Odlyzko stated that they could prove that $\pi_P(x) \sim (7/24)\pi(x)$ as $x \to \infty$, where $\pi(x)$ denotes the number of primes not exceeding x. Theorem 2 of Wiertelak implies the stronger result:
>
> $$\pi_P(x) = \frac{7}{24}\pi(x) + O\left(\frac{x(\ln \ln x)^4}{(lnx)^3}\right).$$

From the theorem it follows that if an odd sequence of length n exists, then $n \equiv 0 \bmod 4$ or $n \equiv 1 \bmod 4$. This answers Question (i) of Alles. From the theorem and the fact that $2^3 \equiv 1 \bmod 7$, it follows that if $2n - 1 \in P$, then $7(2n - 1) \in P$. Thus if there is an odd sequence of length n, then there also must be an odd sequence of length $7n - 3$. This observation was used by Alles to prove that there are infinitely many odd sequences. Indeed, the above display combined with Theorem

4 of my preprint (referred to below) allows one to make this much more precise. Let $N(x)$ be the number of odd sequences of length not exceeding x. Then, as $x \to \infty$,

$$N(x) = c\frac{x}{(\ln x)^{17/24}}\left(1 + O\left(\frac{(\ln \ln x)^5}{\ln x}\right)\right),$$

where c is a positive constant.

Moree also adds:

> If K is a field then the minimal number, $s(K)$, if it exists, of elements of K such that $-1 = \alpha_1^2 + \cdots + \alpha_s^2$ is called the Stufe (level) of K. For $m \geq 3$ let $K_m := Q(e^{2\pi im})$. Hilbert has proved that the Stufe of such a field is either 1, 2 or 4. From MacWilliams & Odlyzko and the results of Fein, Gordon & Smith it follows that for $m \geq 2$, K_{2m-1} has Stufe 4 if and only if there exists an odd sequence of length m.

Moree & Solé conjecture that if $S(n)$ is the number of odd sequences of length n, then, for each $e \geq 1$, there are infinitely many n for which $S(n) = 2^e$. Under the Generalised Riemann Hypothesis, they prove that this is true if $2e + 1$ is not a prime $\equiv 5 \bmod 8$. If $N_e(x)$ is the number of $n \leq x$ for which $S(n) = 2^e$, and $r(e) = \max \omega_1(2n - 1)$ taken over such n, where ω_1 is the number of prime divisors occurring only to the first power, then they also conjecture that if $r(e) \geq 1$, then there is a constant c_e such that

$$N_e(x) \sim c_e\frac{x}{\ln x}(\ln \ln x)^{r(e)-1}$$

as $x \to \infty$.

Peter Alles, On a conjecture of J. Pelikán, *J. Combin. Theory Ser. A*, **60** (1992) 312–313; *MR* **93i**:11028.

B. Fein, B. Gordon & J. H. Smith, On the representation of -1 as a sum of two squares in an algebraic number field, *J. Number Theory*, **3**(1971) 310-315; *MR* **47** #8481.

Nicholas F. J. Inglis & Julian D. A. Wiseman, Very odd sequences, *J. Combin. Theory Ser. A*, **71**(1995) 85–96; *MR* **96c**:11004.

F. J. MacWilliams & A. M. Odlyzko, Pelikan's conjecture and cyclotomic cosets, *J. Combin. Theory Ser. A*, **22**(1977) 110-114; *MR* **54** #12724.

Pieter Moree, On the divisors of $a^k + b^k$, *Acta Arith.*, **80**(1997) 197–212; *MR* **98e**:11105.

J. Pelikán, Problem, in *Infinite and Finite Sets, Vol. III* (*Keszthely*,1973), 1549, *Colloq. Math. Soc. János Bolyai*, **10**, North-Holland, Amsterdam, 1975.

Adam Sikora, On odd sequences, preprint, 95-04-11.

K. Wiertelak, On the density of some sets of primes. IV, *Acta Arith.*, **43**(1984) 177-190; *MR* **95f**:11069.

OEIS: A036259, A053006.

F. None of the Above

The first few problems in this miscellaneous section are about **lattice points**, whose Euclidean coordinates are integers. Most of them are two-dimensional problems, but some can be formulated in higher dimensions as well. Some interesting books are

J. W. S. Cassels, *Introduction to the Geometry of Numbers*, Springer-Verlag, New York, 1972.

L. Fejes Tóth, *Lagerungen in der Ebene, auf der Kugel und in Raum*, Springer-Verlag, Berlin, 1953.

J. Hammer, *Unsolved Problems Concerning Lattice Points*, Pitman, 1977.

O.-H. Keller, *Geometrie der Zahlen*, Enzyklopedia der Math.Wissenschaften **12**, B. G. Teubner, Leipzig, 1954.

C. G. Lekkerkerker, *Geometry of Numbers*, Bibliotheca Mathematica **8**, Walters-Noordhoff, Groningen; North-Holland, Amsterdam, 1969.

Lin Ke-Pao & Stephen S.-T. Yau, Analysis for a sharp polynomial upper estimate of the number of positive integral points in a 4-dimensional tetrahedron, *J. reine angew. Math.*, **547**(2002) 191–205; *MR* **2003a**:11128.

C. A. Rogers, *Packing and Covering*, Cambridge Univ. Press, 1964.

F1 Gauß's lattice point problem.

A very difficult unsolved problem is **Gausss problem@Gauß's problem**. How many lattice points are there inside the circle with centre at the origin and radius r? If the answer is $\pi r^2 + h(r)$, then Hardy & Landau showed that $h(r)$ is *not* $o(r^{1/2}(\ln r)^{1/4})$. It is conjectured that $h(r) = O(r^{1/2+\epsilon})$. Iwaniec & Mozzochi have shown that $h(r) = O(r^{7/11+\epsilon})$, and the best that was known was $h(r) = O(r^{46/73+\epsilon})$, by Huxley. But see his recent article in *Number Theory for the Millenium* where the exponent is improved to $131/208$; and the corresponding exponent for the divisor problem is $131/416$. Here, Dirichlet's divisor problem is to estimate the error, $\Delta(x)$, in the formula

$$\sum_{n=1}^{x} d(n) = x \ln x + (2\gamma - 1)x + \Delta(x)$$

for the sum of $d(n)$, the divisor function, i.e., the lattice point problem for the rectangular hyperbola.

One can ask analogous questions in three dimensions for the sphere and regular tetrahedron. For the rectangular tetrahedron of **F22** see the paper of Lehmer and also that of Xu & Yau, but their suggested counterexample to Overhagen's upper bound for arbitrary convex bodies is incorrect.

The corresponding problems with cubes or higher powers in place of squares has also attracted much attention. See the literature quoted below.

V. Bentkus & Friedrich Götze, On the number of lattice points in a large ellipsoid, *Dokl. Akad. Nauk*, **343**(1995) 439–440; *MR* **96i**:11111.

Pavel M. Bleher & Freeman J. Dyson, Mean square limit for lattice points in a sphere, *Acta Arith.*, **68**(1994) 383–393; *MR* **96a**:11104.

Pavel M. Bleher, Cheng Zhe-Ming, Freeman J. Dyson & Joel L. Lebowitz, Distribution of the error term for the number of lattice points inside a shifted circle, *Comm. Math. Phys.*, **154**(1993) 433–469; *MR* **94g**:11081.

Chen Jing-Run, The lattice points in a circle, *Sci. Sinica*, **12**(1963) 633–649; *MR* **27** #4799.

Chen Yong-Gao & Cheng Lin-Feng, Visibility of lattice points, *Acta Arith.*, **107**(2003) 203–207.

Javier Cilleruelo, The distribution of the lattice points on circles, *J. Number Theory*, **43**(1993) 198–202; *MR* **94c**:11097.

Javier Cilleruelo, Lattice points on circles, *J. Austral. Math. Soc.*, **72**(2002) 217–222; *MR* bf2002k:11160.

K. S. Gangadharan, Two classical lattice point problems, *Proc. Cambridge Philos. Soc.*, **57**(1961) 699–721; *MR* **24** #A92.

Andrew Granville, The lattice points of an n-dimensional tetrahedron, *Aequationes Math.*, **41**(1991) 234–241; *MR* **92b**:11070.

James Lee Hafner, New omega theorems for two classical lattice point problems, *Invent. Math.*, **63**(1981) 181–186; *MR* **82e**:10076.

Martin Huxley, Exponential sums and lattice points. II, *Proc. London Math. Soc.*, **66**(1993) 279–301; *MR* **94b**:11100.

M. N. Huxley, *Area, lattice points, and exponential sums*, London Mathematical Society Monographs. New Series, **13**. The Clarendon Press, Oxford, 1996; *MR* **97g**:11088.

M. N. Huxley, Integer points, exponential sums and the Riemann zeta function, *Number theory for the millennium*, II (Urbana, IL, 2000) 275–290, A K Peters, Natick, MA, 2002; *MR* **2004d**:11098.

A. E. Ingham, On two classical lattice point problems, *Proc. Cambridge Philos. Soc.*, **36**(1940) 131–138; *MR* **2**, 149f.

Aleksandar Ivić, Large values of the error term in divisor problems and the mean square of the zeta-function, *Invent. Math.*, **71**(1983) 513–520; *MR* **84i**:10046.

H. Iwaniec & C. J. Mozzochi, On the divisor and circle problems, *J. Number Theory*, **29**(1988) 60–93; *MR* **89g**:11091.

I. Kátai, The number of lattice points in a circle (Russian), *Ann. Univ. Sci. Budapest. Eötvös Sect. Math.*, **8**(1965) 39–60; *MR* **33** #5587.

Ekkehard Krätzel, *Lattice points*, Mathematics and its Applications (East European Series), **33**. Kluwer, 1988; *MR* **90e**:11144.

E. Krätzel, A sum formula related to ellipsoids with applications to lattice point theory, *Abh. Math. Sem. Univ. Hamburg*, **71**(2001) 143–159.

Manfred Kühleitner & Werner Georg Nowak, On sums of two kth powers: a mean-square asymptotics over short intervals, *Acta Arith.*, **99**(2001) 191–203; *MR* **2002f**:11135.

M. Kühleitner & W. G. Nowak, The average number of solutions of the Diophantine equation $U^2 + V^2 = W^3$ and related arithmetic functions, http://arxiv.org/abs/math/0307221

M. Kühleitner, W. G. Nowak, J. Schoissengeier & T. D. Wooley, On sums of two cubes: an Ω_+-estimate for the error term, *Acta Arith.*, **85**(1998) 179–195; *MR* **99g**:11119.

D. H. Lehmer, The lattice points of an n-dimensional tetrahedron, *Duke Math. J.*, **7**(1940) 341–353.

J. van de Lune & E. Wattel, Systematic computations on Gauss' lattice point problem (in commemoration of Johannes Gualtherus van der Corput, 1890–1975), *Summer course* 1990: *number theory and its applications*, 25–56, *CWI Syllabi*, 27, *Math. Centrum, Amsterdam*, 1990; *MR* **94h**:11093.

Wolfgang Müller, Lattice points in bodies with algebraic boundary, *Acta Arith.*, **108**(2003) 9–24.

Werner Georg Nowak, Sums of two kth powers: an omega estimate for the error term, *Arch. Math. (Basel)*, **68**(1997) 27–35; *MR* **97i**:11101; see also *Analysis*, **16**(1996) 297–304; *MR* **97g**:11116.

T. Overhagen, Zur Gitterpunktanzahl konvexer Körper im 3-dimensionalen euklidischen Raum, *Math. Ann.*, **216**(1975) 217–224; *MR* **57** #281.

K. Soundararajan, Omega results for the divisor and circle problems, *Int. Math. Res. Not.*, **2003** 1987–1998.

Xu Yijing & Stephen Yau S.-T., A sharp estimate of the number of integral points in a tetrahedron, *J. reine angew. Math.*, **423**(1992) 199–219; *MR* **93d**:11067; A sharp estimate of the number of integral points in a 4-dimensional tetrahedra, **473**(1996) 1–23; *MR* **97d**:11151.

OEIS: A007882.

F2 Lattice points with distinct distances.

What is the largest number k, of lattice points (x,y), $1 \leq x, y \leq n$, which can be chosen so that their $\binom{k}{2}$ mutual distances are all distinct? It is easy to see that $k \leq n$. This bound can be attained for $n \leq 7$, for example for $n = 7$ with the points (1,1), (1,2), (2,3), (3,7), (4,1), (6,6) and (7,7), but not for any larger value of n. Erdős & Guy showed that

$$n^{2/3-\epsilon} < k < cn/(\ln n)^{1/4}$$

and they conjecture that

$$\text{¿} \qquad k < cn^{2/3}(\ln n)^{1/6} \qquad ?$$

One can also ask for "saturated" configurations, containing a *minimum* number of points which determine distinct distances, but such that *no* lattice point may be added without duplicating a distance. Erdős observes that this needs at least $n^{2/3-\epsilon}$ lattice points. In one dimension he cannot improve on $O(n^{1/3})$ and suspects that $O(n^{1/2+\epsilon})$ is best possible.

P. Erdős & R. K. Guy, Distinct distances between lattice points, *Elem. Math.*, **25**(1970) 121–123; *MR* **43** #7406.

F3 Lattice points, no four on a circle.

Erdős & Purdy ask how many of the n^2 lattice points (x, y), $1 \leq x, y \leq n$ can you choose with no four of them on a circle. It is easy to show $n^{2/3-\epsilon}$, but more should be possible.

What is the smallest t so that you can choose t of the lattice points so that the $\binom{t}{2}$ lines that they determine contain all the n^2 lattice points? It is not hard to show $t > cn^{2/3}$ and Noga Alon has since obtained the bounds

$$cn^{d(d-1)/(2d-1)} \leq t(n, d) \leq Cn^{d(d-1)/(2d-1)} \ln n$$

for the problem in any number, d, of dimensions.

Noga Alon, Economical coverings of sets of lattice points, *Geom. Funct. Anal*, **1**(1991) 224–230; *MR* **92g**:52017.

F4 The no-three-in-line problem.

The no-three-in-line problem. Can $2n$ lattice points (x, y) with $(1 \leq x, y \leq n)$ be selected with no three in a straight line? This has been achieved for $2 \leq n \leq 32$ and for several larger even values of n. Guy & Kelly make four conjectures.

1. There are no configurations with the symmetry of a rectangle which do not have the full symmetry of the square.

2. The only configurations having the full symmetry of the square are those in Figure 17. The $n = 10$ configuration was first found by Acland-Hood. This conjecture has been verified by Flammenkamp for $n \leq 60$.

3. For large enough n, the answer to the initial question is "no," i.e., there are only finitely many solutions to the problem. The total number of configurations, not counting reflexions and rotations, is, for

n	2	3	4	5	6	7	8	9	10	11	12	13
#	1	1	4	5	11	22	57	51	156	158	566	499

and the configurations with specific symmetries have been enumerated for larger values of n.

4. For large n we may select at most $(c+\epsilon)n$ lattice points with no three in line, where $3c^3 = 2\pi^2$, i.e. $c \approx 1.85$. As recently as March, 2004, Gabor Ellmann notes an error in the original paper, so that the result can be impoved to: $3c^2 = \pi^2$, $c \approx 1.813799$.

In the opposite direction, Erdős showed that if n is prime, it is possible to choose n points with no three in line, and Hall et al have shown that for n large, $(\frac{3}{2} - \epsilon)n$ such points can be found.

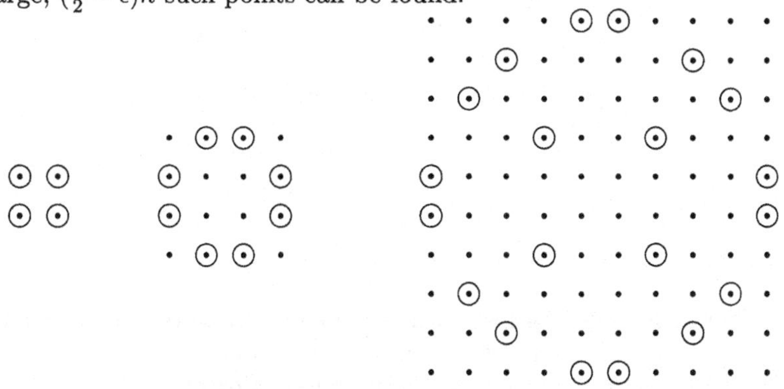

Figure 17. $2n$ Lattice Points, No Three in Line, $n = 2, 4, 10$.

Let S_n be the set of n^2 lattice points with cordinates in $[1, n]$. Two points (x, y), (u, v) are mutually visible if the segment joining them contains no other lattice point. Let $f(n)$ be the minimum cardinality of a subset $X_n \subset S_n$ such that every point of S_n is visible from at least one point of X_n. Abbott showed that for n sufficiently large,

$$\frac{\ln n}{2 \ln \ln n} < f(n) < 4 \ln n$$

and Adhikari & Subramanian improve the lower bound to

$$\frac{c \ln n \ln \ln \ln n}{\ln \ln n}$$

Buchholz has examined the no-3-in-line problem on an equilateral triangular lattice of side n. Here is a breakdown into symmetry classes of the numbers of solutions.

$n =$	2	3	4	5	6	7	8	9	10	11	12	13	14	15	16	17
# counters	3	4	6	7	8	10	11	12	13	15	16	17	19	20	22	23
D3 - equil	1		1					1								
C3 -1/3 turn								3	2	2				1		
C2 - flip		2		1	4		2	6	15	3	3	7				
C1 - asymm				1	11	1	3	68	954	15	274	2889	5	199	1	4
total	1	2	1	2	15	1	5	78	971	20	277	2896	5	200	1	4

The unique order 16 solution is

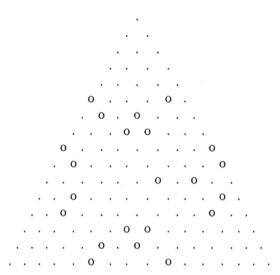

T. Thiele modifies Erdős's construction to show that one can find $(\frac{1}{4} - \epsilon)n$ points with no 3 in line and no 4 on a circle.

The no-three-in-line problem is a discrete analog of an old problem of Heilbronn. Place n (≥ 3) points in a disk (or square, or equilateral triangle) of unit area so as to maximize the smallest area of a triangle fromed by three of the points. If we denote this maximum area by $\Delta(n)$, then Heilbronn originally conjectured that $\Delta(n) < c/n^2$, but Komlós, Pintz & Szemerédi disproved this by showing that $\Delta(n) > (\ln n)/n^2$. Roth showed that $\Delta(n) \ll 1/n(\ln \ln n)^{1/2}$; Schmidt improved this to $\Delta(n) \ll 1/n(\ln n)^{1/2}$ and Roth subsequently made the further improvement $\Delta(n) \ll 1/n^{\mu - \epsilon}$, first with $\mu = 2 - 2/\sqrt{5} > 1.1055$ and later with $\mu = (17 - \sqrt{65})/8 > 1.1172$.

Given $3n$ points in the unit square, $n \geq 2$, they determine n triangles in many ways. Choose them so as to minimize the sum of the areas, and let $a^*(n)$ be the maximum value of this minimum sum, taken over all configurations of $3n$ points. Then Odlyzko & Stolarsky show that $n^{-1/2} \ll a^*(n) \ll n^{-1/24}$. If the n triangles are required to be *area disjoint* it is not even clear that the sum of their areas tends to zero.

Harvey L. Abbott, Some results in combinatorial geometry, *Discrete Math.*, **9**(1974) 199–204; *MR* **50** #1936.

Acland-Hood, *Bull. Malayan Math. Soc.*, **0**(1952-53) E11–12.

Michael A. Adena, Derek A. Holton & Patrick A. Kelly, Some thoughts on the no-three-in-line problem, Proc. 2nd Austral. Conf. Combin. Math., *Springer Lecture Notes*, **403**(1974) 6–17; *MR* **50** #1890.

Sukumar Das Adhikari & Ramachandran Balasubramanian, On a question regarding visibility of lattice points, *Mathematika*, **43**(1996) 155–158; *MR* **97k**:11105.

David Brent Anderson, Update on the no-three-in-line problem, *J. Combin. Theory Ser. A*, **27**(1979) 365–366.

W. W. Rouse Ball & H. S. M. Coxeter, *Mathematical Recreations & Essays*, 12th edition, University of Toronto, 1974, p. 189.

Francesc Comellas & J. Luis A. Yebra, New lower bounds for Heilbronn numbers, *Electron. J. Combin.*, **9**(2002) R6, 10 pp.; *MR* **2002k**:05019.

C. E. Corzatt, Some extremal problems of number theory and geometry, PhD dissertation, Univ. of Illinois, Urbana, 1976.

D. Craggs & R. Hughes-Jones, On the no-three-in-line problem, *J. Combin. Theory Ser. A*, **20**(1976) 363–364; *MR* **53** #10590.

Hallard T. Croft, Kenneth J. Falconer & Richard K. Guy, Unsolved Problems in Geometry, Springer-Verlag, New York, 1991, §**E5**.

H. E. Dudeney, *The Tribune*, 1906-11-07.

H. E. Dudeney, *Amusements in Mathematics*, Nelson, Edinburgh, 1917, pp. 94, 222.

P. Erdős, *Math. Scand.*, **10**(1962) 163–170; *MR* **26** #3651.

Achim Flammenkamp, Progress in the no-three-in-line problem, *J. Combin. Theory Ser. A*, **60**(1992) 305–311; *MR* **939**:05040.

Martin Gardner, Mathematical Games, *Sci. Amer.*, **226** #5 (May 1972) 113–114; **235** #4 (Oct 1976) 133–134; **236** #3 (Mar 1977) 139–140.

Michael Goldberg, Maximizing the smallest triangle made by N points in a square, *Math. Mag.*, **45**(1972) 135–144.

R. Goldstein, K. W. Heuer & D. Winter, Partition of S into n triples; solution to Problem 6316, *Amer. Math. Monthly*, **89**(1982) 705–706.

Richard K. Guy, *Bull. Malayan Math. Soc.*, **0**(1952-53) E22.

Richard K. Guy & Patrick A. Kelly, The no-three-in-line problem, *Canad. Math. Bull.*, **11**(1968) 527–531.

R. R. Hall, T. H. Jackson, A. Sudbery & K. Wild, Some advances in the no-three-in-line problem, *J. Combin. Theory Ser. A*, **18**(1975) 336–341.

Heiko Harborth, Philipp Oertel & Thomas Prellberg, No-three-in-line for seventeen and nineteen, *Discrete Math.*, **73**(1989) 89–90; *MR* **90f**:05041.

P. A. Kelly, The use of the computer in game theory, M.Sc. thesis, Univ. of Calgary, 1967.

Torliev Kløve, On the no-three-in-line problem II, III, *J. Combin. Theory Ser. A*, **24**(1978) 126–127; **26**(1979) 82–83; *MR* **57** #2962; **80d**:05020.

J. Komlós, J. Pintz & E. Szemerédi, A lower bound for Heilbronn's problem, *J. London Math. Soc.*(2) **25**(1982) 13–24; *MR* **83i**:10042.

Andrew M. Odlyzko, J. Pintz & Kenneth B. Stolarsky, Partitions of planar sets into small triangles, *Discrete Math.*, **57**(1985) 89–97; *MR* **87e**:52007.

Carl Pomerance, Collinear subsets of lattice point sequences – an analog of Szemerédi's theorem, *J. Combin. Theory Ser. A*, **28**(1980) 140–149.

K. F. Roth, On a problem of Heilbronn, *J. London Math. Soc.*, **25**(1951) 198–204, esp. p. 204; II, III, *Proc. London Math. Soc.*, **25**(1972) 193–212; 543–549.

K. F. Roth, Developments in Heilbronn's triangle problem, *Advances in Math.*, **22**(1976) 364–385; *MR* **55** #2771.

Wolfgang M. Schmidt, On a problem of Heilbronn, *J. London Math. Soc.*, **4**(1971/72) 545–550; *MR* **45** #7612.

Torsten Thiele, The no-four-on-circle problem, *J. Combin. Theory Ser. A*, **71**(1995) 332–334; *MR* **96e**:52042.

Robert C. Vaughan, *Proc. Edinburgh Math. Soc.*(2), **20**(1976/77) 329–331; *MR* **56** #11937.

OEIS: A000755, A000769, A037185-037189, A047840.

F5 Quadratic residues. Schur's conjecture.

The **quadratic residues** of a prime p are the nonzero numbers r for which the congruence $r \equiv x^2 \bmod p$ has solutions. There are $\frac{1}{2}(p-1)$ of them in the interval $[1, p-1]$ and they are symmetrically distributed if p is of shape $4k + 1$. If $p = 4k - 1$, there are more quadratic residues in the interval $[1, 2k-1]$ than in $[2k, 4k-2]$, but all known proofs use Dirichlet's class-number formula. Is there an elementary proof?

For the first few values of d it is easy to remember which primes have d as a quadratic residue:

$$
\begin{array}{llll}
d = -1 & p = 4k+1 & & \\
d = -2 & p = 8k+1, 3 & d = 2 & p = 8k \pm 1 \\
d = -3 & p = 6k+1 & d = 3 & p = 12k \pm 1 \\
d = -5 & p = 20k+1, 3, 7, 9 & d = 5 & p = 10k \pm 1 \\
d = -6 & p = 24k+1, 5, 7, 11 & d = 6 & p = 24k \pm 1, 5
\end{array}
$$

However, it's just an example of the Law of Small Numbers that in these small cases the residues are just those in the first half or the end quarters of the period, according to the sign of d.

That there are equally many residues as nonresidues in the interval $[1, 2, \ldots, 2k]$ for primes $8k + 3$ and in the interval $[2k, 2k+1, \ldots, 4k-1]$ for primes $8k - 1$ was proved by Lebesgue.

The **Legendre symbol**, $\left(\frac{a}{p}\right)$, is often used to indicate the quadratic character of a number a, $a \perp p$, relative to the prime p. Its value is ± 1 according as a is, or is not, a quadratic residue of p. For example, $\left(\frac{-1}{p}\right) = \pm 1$ according as $p = 4k \pm 1$. The important properties of this symbol are that $\left(\frac{a}{p}\right) = \left(\frac{c}{p}\right)$ if $a \equiv c \bmod p$; and Gauß's famous **quadratic reciprocity law**, that for odd primes p and q, $\left(\frac{p}{q}\right) = \left(\frac{q}{p}\right)$ unless p and q are both $\equiv -1 \bmod 4$, in which case $\left(\frac{p}{q}\right) = -\left(\frac{q}{p}\right)$. These can be used for making quick verifications of the quadratic character of quite large numbers. For example,

$$
\left(\frac{173}{211}\right) = \left(\frac{211}{173}\right) = \left(\frac{38}{173}\right) = \left(\frac{2}{173}\right)\left(\frac{19}{173}\right) = -\left(\frac{19}{173}\right) = -\left(\frac{173}{19}\right) = -\left(\frac{2}{19}\right) = +1
$$

in fact $173 \equiv 54^2 \bmod 211$.

Robin Chapman has a number of conjectures which concern the distribution of the quadratic residues. For example, if $p = 2k + 1$ is a prime > 3 and $\equiv 3 \pmod 4$ is the matrix whose (i,j)-th entry is $\left(\frac{i+j}{p}\right)$ singular? i.e., is its determinant zero? For example, $p = 11$, $k = 5$,

$$\begin{vmatrix} -1 & 1 & 1 & 1 & -1 \\ 1 & 1 & 1 & -1 & -1 \\ 1 & 1 & -1 & -1 & -1 \\ 1 & -1 & -1 & -1 & 1 \\ -1 & -1 & -1 & 1 & -1 \end{vmatrix} = 0$$

A useful generalization of the Legendre symbol is the **Jacobi symbol** $\left(\frac{a}{b}\right)$ which is defined for $a \perp b$ and b any positive odd number, by the product

$$\prod \left(\frac{a}{p_i}\right)$$

of Legendre symbols, where $b = \prod p_i$ is the prime factorization of b, *with repetitions counted*. It has similar properties to the Legendre symbol, but note that, if b is not prime, then $\left(\frac{a}{b}\right) = +1$ does not necessarily imply that a is a quadratic residue of b.

If R (respectively N) is the maximum number of consecutive quadratic residues (respectively nonresidues) modulo an odd prime p, then A. Brauer showed that for $p \equiv 3 \bmod 4$, $R = N < \sqrt{p}$. On the other hand, if $p = 13$, then $N = 4 > \sqrt{13}$, since 5, 6, 7, 8 are all nonresidues of 13. Schur conjectured that $N < \sqrt{p}$ if p is large enough. Hudson proved Schur's conjecture; moreover, he believes that $p = 13$ is the only exception; this was recently proved by Hummel. Calculations by Jud McCranie suggest that $N = O(p^{1/4})$.

Fletcher, Lindgren & Pomerance say that two odd primes p, q form a **symmetric pair** if the numbers of lattice points above and below the main diagonal of the rectangle $(0,0)$, $(p/2,0)$, $(p/2,q/2)$, $(0,q/2)$ are equal, and call a prime **symmetric** if it belongs to a symmetric pair and **asymmetric** otherwise. They show that twin primes form symmetric pairs. They note that about 5/6 of the first 100000 odd primes are symmetric, but, asymptotically, almost all primes are asymmetric.

Ribet asks for a discussion of the function $S(p)/p$, where $S(p)$ is the sum of the quadratic residues, mod p.

A. Brauer, Über die Verteilung der Potenzreste, *Math. Z.*, **35**(1932) 39–50; *Zbl.* **3**, 339.

Mihai Caragiu, Counting the maximal sequences of consecutive quadratic residues modulo p, *Rev. Roumaine Math. Pures Appl.*, **38**(1993) 745–749; *MR* **95c**:11003.

Robin Chapman, Steinitz classes of unimodular lattices, preprint, 2003-02-07.

Robin Chapman, Determinants of Legendre matrices, preprint, 2003-04-22.

Peter Fletcher, William Lindgren & Carl Pomerance, Symmetric and asymmetric primes, *J. Number Theory*, **58**(1996) 89–99.

H. Davenport, *The Higher Arithmetic*, Fifth edition, Cambridge University Press, 1982, pp.74–77.

Richard H. Hudson, On sequences of quadratic nonresidues, *J. Number Theory*, **3**(1971) 178–181; *MR* **43** #150.

Richard H. Hudson, On a conjecture of Issai Schur, *J. reine angew. Math.*, **289**(1977) 215–220; *MR* **58** #16481.

Patrick Hummel, On consecutive quadratic non-residues: a conjecture of Issai Schur (preprint, 2003).

V. A.Lebesgue, Démonstration de quelques théorèmes relatifs aux résidus et aux non-résidus quadratiques, *J. Math. Pures Appl.*, **7**(1842) 137–159.

Kenneth A. Ribet, Modular forms and Diophantine questions, *Challenges for the 21st Century (Singapore* 2000 162–182, World Sci. Publishing, River Edge NJ 2001; *MR* **2002i**:11030.

F6 Patterns of quadratic residues.

What patterns of quadratic residues are sure to occur? It is easy to see that a pair of neighboring ones always do, since at least one of 2, 5 and 10 is a residue, so that (1,2), (4,5) or (9,10) is such a pair. In the same way at least one of (1,3), (2,4) or (4,6) is a pair of residues differing by 2; (1,4) is a pair differing by 3; (1,5), (4,8), (6,10) or (12,16) is a pair differing by 4; and so on.

Suppose that each of r, $r + a$, $r + b$ is a quadratic residue modulo p. Emma Lehmer asks: for which pairs (a, b) will such a triplet occur for *all* sufficiently large p? Denote by $\Omega(a, b)$ the least number such that a triplet is assured with $r \leq \Omega(a, b)$ for all $p > p(a, b)$, and write $\Omega(a, b) = \infty$ if there is no such finite number. For example, Emma Lehmer showed that $\Omega(1, 2) = \infty$, and more generally that $\Omega(a, b) = \infty$ if $(a, b) \equiv (1, 2)$ mod 3; or if $(a, b) \equiv (1, 3)$, $(2, 3)$ or $(2, 4)$ mod 5; or if $(a, b) \equiv (1, 5)$, $(2, 3)$ or $(4, 6)$ mod 7. Is $\Omega(a, b)$ finite in all other cases? Emma Lehmer conjectures that it is finite if a and b are squares. Of course, $\Omega(a, b) = 1$ if a, b are each one less than a square. As an example, let us see why $\Omega(5, 23) = 16$. If the triplets $(1, 6, 24)$ and $(4, 9, 27)$ are not all residues then 6 and 3 are not, and 2 must be a residue. If the triplets $(2, 7, 25)$ and $(13, 18, 36)$ are not all residues, then 7 and 13 must be nonresidues. Under these circumstances, $(r, r + 5, r + 23)$ are not all residues for $1 \leq r \leq 15$, but when $r = 16$, $(16, 21, 39)$ are residues.

Table 9, with nine entries corrected by Chris Thompson, contains what are believed to be the (minimum) values of $\Omega(a, b)$. They provide good evidence for the conjectured finiteness in all cases except those already noted. Can an upper bound be obtained in terms of a and b?

Table 9. Values of $\Omega(a,b)$ for $a < b \leq 25$.

a	$b=4$	5	6	7	8	9	10	11	12	13	14
1	45	∞	24	38	∞	84	26	∞	∞	∞	∞
2	∞	25	20	∞	∞	∞	∞	70	30	∞	∞
3	174	39	∞	∞	1	∞	55	∞	∞	36	105
4		∞	∞	∞	∞	91	36	∞	∞	∞	∞
5			49	∞	∞	121	∞	25	4	∞	28
6				57	∞	33	28	∞	24	∞	42
7					∞	∞	75	∞	74	∞	∞
8						66	∞	∞	∞	∞	26
9							∞	54	∞	55	66
10								∞	60	85	∞
11									28	∞	119
12										∞	∞
13											∞

a	$b=15$	16	17	18	19	20	21	22	23	24	25
1	77	35	∞	∞	∞	∞	15	35	∞	21	69
2	54	∞	∞	∞	∞	25	98	∞	∞	∞	∞
3	1	∞	∞	18	36	95	∞	∞	∞	1	51
4	126	60	∞	38	168	∞	90	∞	∞	60	77
5	∞	∞	64	110	∞	100	4	∞	16	64	∞
6	60	36	38	∞	62	78	60	78	∞	45	∞
7	27	9	∞	∞	∞	∞	70	42	∞	∞	45
8	1	∞	∞	77	∞	48	∞	∞	42	1	∞
9	57	66	∞	36	27	16	72	∞	21	∞	119
10	55	∞	∞	32	102	∞	77	26	∞	28	56
11	49	∞	39	∞	∞	∞	64	∞	∞	25	∞
12	∞	65	98	∞	∞	36	4	∞	∞	∞	90
13	42	∞	∞	∞	36	∞	∞	∞	∞	36	∞
14	49	∞	∞	42	∞	52	56	∞	64	81	∞
15		66	27	69	∞	49	25	99	110	1	105
16			∞	∞	102	∞	169	95	∞	∞	56
17				∞	∞	76	64	∞	∞	∞	∞
18					50	∞	∞	∞	62	192	144
19						∞	33	∞	∞	36	96
20							74	∞	40	25	∞
21								93	∞	70	100
22									∞	∞	98
23										∞	∞
24											63

What about patterns of four residues, $r, r+a, r+b, r+c$? Of course these won't necessarily occur if any of the four subpatterns of three residues aren't forced to do so. We need examine only $(a,b,c) = (2,5,6)$, $(1,6,7)$, $(1,4,9)$, $(5,6,9)$, $(1,6,10)$, $(1,7,10)$, ... where $\Omega(a,b)$, $\Omega(a,c)$, $\Omega(b,c)$ and $\Omega(b-a, c-a)$ are each known to be finite. Some corresponding values of $\Omega(a,b,c)$ are $\Omega(1,4,9) = 357$ (Peter Montgomery corrects an error in the first edition), $\Omega(1,4,15) = 675$, and of course $\Omega(3,8,15) = 1$.

But although $\Omega(1,6) = 24$, $\Omega(1,7) = 38$, $\Omega(5,6) = 49$ and $\Omega(6,7) = 57$, it appears that $\Omega(1,6,7) = \infty$. In fact the pattern r, $r+a$, $r+b$, $r+c$, $r+d$, with $(a,b,c,d) = (1,6,7,10)$ is such that, for each of the five subpatterns of four, $\Omega(1,6,7) = \Omega(1,6,10) = \Omega(1,7,10) = \Omega(5,6,9) = \Omega(6,7,10) = \infty$.

It is customary to define k-**th power residues** as numbers r for which $x^k \equiv r \bmod p$ has a solution, only with respect to those primes for which k divides $p-1$. In the same way that we remarked that every prime > 10 has a pair of consecutive quadratic residues not exceeding the pair $(9, 10)$, Hildebrand has shown that for each k there is a fixed bound, $\Lambda(k, 2)$ so that every sufficiently large prime has a pair of consecutive k-th power residues below this bound. There is no such bound for 3 consecutive quadratic or quartic, etc., residues. The argument consists in forcing primes of the form $3k+1$ to be residues and those of the form $3k+2$ to be nonresidues. Similarly by making 2 a residue and as many odd primes as necessary nonresidues, there is no such bound below which four consecutive k-th power residues must appear for any k. This leaves open only the question of three consecutive k-th powers for k odd. The case $k = 3$ was solved by the Lehmers, Mills & Selfridge.

Alfred Brauer, Combinatorial methods in the distribution of kth power residues, in *Probability and Statistics*, **4**, University of North Carolina, Chapel Hill, 1969, 14–37.

Adolf Hildebrand, On consecutive k-th power residues, *Monatsh. Math.*, **102** (1986) 103–114; *MR* **88a**:11089.

Adolf Hildebrand, On consecutive k-th power residues II, *Michigan Math. J.*, **38** (1991) 241–253; *MR* **92d**:11097.

D. H. Lehmer & Emma Lehmer, On runs of residues, *Proc. Amer. Math. Soc.*, **13**(1962) 102–106; *MR* **25** #2035.

D. H. Lehmer, Emma Lehmer & W. H. Mills, Pairs of consecutive power-residues, *Canad. J. Math.*, **15**(1963) 172–177; *MR***26** #3660.

D. H. Lehmer, Emma Lehmer, W. H. Mills & J. L. Selfridge, Machine proof of a theorem on cubic residues, *Math. Comput.*, **16**(1962) 407–415; *MR* **28** #5578.

René Peralta, On the distribution of quadratic residues andnonresidues modulo a prime number, *Math. Comput.*, **58**(1992) 433–440; *MR* **93c**:11115.

F7 A cubic analog of a Bhaskara equation.

Hugh Williams observes that if $p \equiv 3 \bmod 4$, then the Bhaskara equation $x^2 - py^2 = 2$ has integer solutions just if the congruence $w^2 \equiv 2 \bmod p$ does, i.e. just if $\left(\frac{2}{p}\right) = 1$, and asks for a cubic analog in the cases where $p \not\equiv \pm 1 \bmod 9$: $x^3 + py^3 + p^2 z^3 - 3pxyz = 3$ is solvable just if $w^3 \equiv 3 \bmod p$ is. Barrucand & Cohn have shown that this is true for $p \equiv 2$ or $5 \bmod 9$. What about $p \equiv 4$ or $7 \bmod 9$? This is a special case of a more general conjecture of Barrucand. If true, it would be useful in abbreviating the calculations needed to find the fundamental unit (regulator) of the cubic field $\mathbb{Q}\left(\sqrt[3]{p}\right)$. There was a rumor that Harold Stark had settled this, but he denies it, so the problem is still open.

P.-A. Barrucand & Harvey Cohn, A rational genus, class number divisibility and unit theory for pure cubic fields, *J. Number Theory*, **2**(1970) 7–21; *MR* **40** #2643.

H. C. Williams, Improving the speed of calculating the regulator of certain pure cubic fields, *Math. Comput.*, **35**(1980) 1423–1434; *MR* **82a**:12003.

F8 Quadratic residues whose differences are quadratic residues.

Gary Ebert asks us to find the largest collection of quadratic residues $r_i \bmod p^n$, given $p^n \equiv 1 \bmod 4$, such that $r_i - r_j$ is a quadratic residue for all pairs (i, j). Fabrykowski shows that the maximum cardinality of such a set is bounded above by $p^{1/2}$ for all such primes, and bounded below by $(1 - \epsilon) \ln p / (2 \ln 2)$ for all sufficiently large primes. He later removed the ϵ for all primes $p \geq 29$ (and $\equiv 1 \bmod 4$).

J. Fabrykowski, On maximal residue difference sets modulo p, *Canad. Math. Bull.*, **36**(1993) 144–146; *MR* **94b**:11006.

J. Fabrykowski, On quadratic residues and nonresidues in difference sets modulo m, *Proc. Amer. Math. Soc.*, **122**(1994) 325–331; *MR* **95a**:11003.

F9 Primitive roots

A **primitive root**, g, of a prime p is a number such that the residue classes of g, g^2, ..., $g^{p-1} = 1$ are all distinct. For example, 5 is a primitive root of 23 because

$$5, \ 5^2 \equiv 2, \ 5^3 \equiv 10, \quad 4, \ -3, \quad 8, \ -6, \ -7, \quad 11, \quad 9, \ -1,$$
$$-5, \qquad -2, \qquad -10, \ -4, \quad 3, \ -8, \quad 6, \quad 7, \ -11, \ -9, \quad 1$$

all belong to different residue classes mod 23.

There is a famous conjecture of Artin, that for each integer $g \neq -1$, g not a square, there are infinitely many primes p with g as a primitive root. Hooley proved this assuming the extended Riemann hypothesis, and Gupta & Murty proved it unconditionally for infinitely many g. Heath-Brown has proved the remarkable theorem that, but for at most two exceptional primes p_1, p_2 the following is true: For each prime p there are infinitely many primes q with p a primitive root of q. For example, there are infinitely many primes q with either 2 or 3 or 5 as a primitive root.

Erdős asks: if p is large enough, is there always a prime $q < p$ so that q is a primitive root of p?

Given a prime $p > 3$, Brizolis asks if there is always a primitive root g of p and x $(0 < x < p)$ such that $x \equiv g^x \bmod p$. If so, can g also be chosen so that $0 < g < p$ and $g \perp (p - 1)$? Zhang Wen-Peng answers this for sufficiently large p.

On 2002-08-29 Carl Pomerance wrote:

... the Brizolis problem that you mention has now been solved by Mariana Campbell and myself. We show that for every prime $p > 3$ there is a primitive root g in the interval $[1, p - 1]$ that is relatively prime to $p - 1$. A consequence is that there is a primitive root g and an integer x in the interval $[1, p - 1]$ such that $g^x \equiv x \pmod{p}$.

We also plan to solve the Vegh problem you mention, [see next paragraph but one] in fact a stronger form of it: For every prime $p > 61$ and for every pair of nonzero residues a, b (mod p), there are primitive roots g, h such that $ag + h = b$ (mod p). So, Vegh is the case $a = -1$. The case $a = 1$ is a conjecture of Golomb. The general case has been considered by Stephen Cohen.

Pieter Moree notes that Brizolis was asking for a 1-cycle of the discrete logarithm, and that one can also ask for 2-cycles, i.e., pairs (g, h) such that $g^h = a \bmod p$ and $g^a = h \bmod p$ with $0 < a < p$. His work with Joshua Holden is listed below.

Vegh asks whether, for all primes $p > 61$, every integer can be expressed as the difference of two primitive roots of p. W. Narkiewicz notes that there is an affirmative answer for $p > 10^{19}$, so that this can, in theory, be answered by computer.

If p and $q = 4p^2 + 1$ are both primes, Gloria Gagola asks if 3 is a primitive root of q for all $p > 3$; is $p = 193$ the only odd prime for which 2 is not a primitive root of q; is $p = 653$ the only prime for which 5 is neither a quadratic residue nor a primitive root of q; and is there a number, perhaps a function of p (such as $2p - 1$ for large p), which is always a primitive root of q? Jud McCranie verifies that 2 is a primitive root of q for all $p < 2.5 \cdot 10^8$ except $p = 193$.

The idea of primitive root can be extended to composite moduli by letting a primitive root for a modulus n be an integer whose multiplicative order modulo n is maximal. This maximal order is the Carmichael function, $\lambda(n)$. Clearly $\lambda(n) \le \phi(n)$, but for composite n can be dramatically smaller. E.g., 3 and 7 are primitive roots of 10 in this sense, and the primitive roots of 105 are $\pm\{2, 17, 23, 32, 37, 38, 47, 52\}$, each of order 12. Note that the maximal order is not necessarily $\phi(n)$. For properties of such primitive roots and the difference in behavior of the analog of Artin's conjecture, see the papers of Li & Pomerance.

Golomb has conjectured that the finite field GF(q) of characteristic $q \ge 2$ has two primitive roots, α, β (not necessarily distinct) with $\alpha + \beta = 1$, and if $q \ge 3$. with $\alpha + \beta = -1$, and, for sufficiently large q, every nonzero element of GF(q) can be represented as $\alpha + \beta$. Zhang has studied some similar problems. See also papers of Chang, Cohen & Mullen, Gao, Le, Moreno and Truong & Reed.

Mariana Campbell, MSc thesis, Univ. of California, Berkeley, 2003.

Chang Yan-Xun & Kang Qing-De, A representation of nonzero elements in finite field, *Sci. China Ser. A*, **34**(1991) 641–649; *MR* **92m**:11140.

Cristian Cobeli & Alexandru Zaharescu, An exponential congruence with solutions in primitive roots, *Rev. Roumaine Math. Pures Appl.*, **44**(1999) 15–22; *MR* **2002d**:11005,

C. Cobeli, M. Vâjâitu & A. Zaharescu, A class of algebraic-exponential congruences modulo p, *Acta Math. Univ. Comenian.* (*N.S.*), **71**(2002) 113–117; *MR* **2003m**:11208.

Stephen D. Cohen, & Gary L. Mullen, Primitive elements in finite fields and Costas arrays, *Appl. Algebra Engrg. Comm. Comput.*, **2**(1991) 45–53; *MR* **94i**:11100.

Anton Dumitriu, Congruences du premier degré, *Rev. Roumaine Math. Pures Appl.*, **10**(1965) 1201–1234; *MR* **34** #126.

Paul Erdős, Carl Pomerance & Eric Schmutz, Carmichael's lambda function, *Acta Arith.*, **58**(1991) 363–385; *MR* **92g**:11093.

John B. Friedlander, Carl Pomerance & Igor E. Shparlinski, Period of the power generator and small values of Carmichael's function, *Math. Comput.*, **70**(2001) 1591–1605; *MR* **2002g**:11112.

Gao Wei Dong, On the Golomb conjecture (Chinese), *Dongbei Shida Xuebao*, **1988** 11–14; *MR* **90j**:11139.

Solomon W. Golomb, Algebraic constructions for Costas arrays, *J. Combin. Theory Ser. A*, **37**(1984) 13–21; *MR* **85f**:05031.

Rajiv Gupta & Maruti Ram Murty, A remark on Artin's conjecture, *Invent. Math.*, **78**(1984) 127–130; *MR* **86d**:11003.

D. R. Heath-Brown, Artin's conjecture for primitive roots, *Quart. J. Math. Oxford Ser.*(2) **37**(1986) 27–38; *MR* **88a**:11004.

J. Holden & P. Moree, Some heuristics and results for small cycles of the discrete logarithm, arXiv:math.NT/0401013.

J. Holden & P. Moree, New conjecturs and results for small cycles of the discrete logarithm, arXiv:math.NT/0305305, to appear as: Primes and Misdemeanours, in Lectures in Honour of the Sixtieth Birthday of Prof. H. C. Williams.

C. Hooley, On Artin's conjecture, *J. reine ungew. Math.*, **225**(1967) 209–220; *MR* **34** #7445.

Le Mao Hua, On Golomb's conjecture, *J. Combin. Theory Ser. A*, **54**(1990) 304–308; *MR* **91j**:11107.

Hendrik W. Lenstra, On Artin's conjecture and Euclid's algorithm in global fields, *Inventiones Math.*, **42**(1977) 201–224; *MR* **58** #576.

Li Shu-Guang & Carl Pomerance, On generalizing Artin's conjecture on primitive roots to composite moduli, *J. reine angew. Math.*, **556**(2003) 205–224; *MR* **2004c**:11177.

Li Shu-Guang & Carl Pomerance, Primitive roots: a survey, *Number theoretic methods* (*Iizuka*, 2001) 219–231, *Dev. Math.*, **8**, Kluwer Acad. Publ., Dordrecht, 2002; *MR* **2004a**:11106.

Greg Martin & Carl Pomerance, The normal order of iterates of the Carmichael λ-function, (2003 preprint).

Oscar Moreno, On the existence of a primitive quadratic of trace 1 over $GF(p^m)$, *J. Combin. Theory Ser. A*, **51**(1989) 104–110; *MR* **90b**:11133.

Leo Murata, On the magnitude of the least primitive root, *Astérisque No.* **198-200**(1991) 253–257; *MR* **93b**:11127.

Maruti Ram Murty & Seshadri Srinivasan, Some remarks on Artin's conjecture, *Canad. Math. Bull.*, **30**(1987) 80–85; *MR* **88e**:11094.

Francesco Pappalardi, Filip Saidak & Igor E. Shparlinski, Square-free values of the Carmichael function, *J. Number Theory*, **103**(2003) 122–131.

A. Paszkiewicz & A. Schinzel, On the least prime primitive root modulo a prime, *Math. Comput.*, **71**(2002) 1307–1321; *MR* **2003d**:11006.

Michael Szalay, On the distribution of the primitive roots of a prime, *J. Number Theory*, **7**(1975) 184–188; *MR* **51** #5524.

T. K. Truong & I. S. Reed, Golomb's conjecture is true for certain classes of finite fields, *Kexue Tongbao* (English Ed.) **28**(1983) 1310–1312; *MR* **85k**:11067.

Emanuel Vegh, Pairs of consecutive primitive roots modulo a prime, *Proc. Amer. Soc.*, **19**(1968) 1169–1170; *MR* **37** #6240.

Emanuel Vegh, Primitive roots modulo a prime as consecutive terms of an arithmetic progression, *J. reine angew. Math.*,**235**(1969) 185–188; II, **244**(1970) 185–188; III, **256**(1972) 130–137; *MR* **39** #4086; **42** #1755; **46** #7137.

Emanuel Vegh, A note on the distribution of the primitive roots of a prime, *J. Number Theory*, **3**(1971) 13–18; *MR* **44** #2694.

Samuel S. Wagstaff, Pseudoprimes and a generalization of Artin's conjecture, *Acta Arith.*, **41**(1982) 141–150; *MR* **83m**:10004.

Zhang Wen-Peng, On a problem of Brizolis (Chinese), *Pure Appl. Math.*, **11**(1995), *suppl.*, 1–3; *MR* **98d**:11099.

Zhang Wen-Peng, On a problem related to Golomb's conjectures, *J. Syst. Sci. Complex.*, **16**(2003) 13–18; *MR* **2004c**:11179.

F10 Residues of powers of two.

Graham asked about the residue of 2^n mod n. There are no solutions of $2^n \equiv 1 \bmod n$ with $n > 1$. $2^n \equiv 2 \bmod n$ whenever n is a pseudoprime base 2 (see **A12**) or is a prime. In the early sixties the Lehmers found the smallest solution, $n = 4700063497 = 19 \cdot 47 \cdot 5263229$, of $2^n \equiv 3 \bmod n$. Peter Montgomery found the second solution to be

    63130707451134435989380140059866138830623361447484274774099906755.

Of course, n has to be composite, and it is not divisible by 2 or 3. In fact, Mąkowski (see reference at **B5**) notes that if $\left(\frac{2}{p}\right)$ and $\left(\frac{3}{p}\right)$ are of opposite sign, i.e. if $p = 24k \pm 7$ or ± 11, then n is not divisible by p.

Rotkiewicz (compare **A12**) notes that if m satisfies $2^m \equiv 3 \bmod m$, then $n = 2^m - 1$ is a solution of $2^{n-2} \equiv 1 \bmod n$.

Benkoski asked if there was a solution to $2^n \equiv 4 \pmod{n}$ which didn't end in 7 when written in decimal notation. Zhang Ming-Zhi gave the solutions where $n \equiv 1$ or $3 \pmod{1)0}$ and asked if there were any with $n \equiv 9 \pmod{1)0}$.

Victor Meally reports that $2^n \equiv -1 \pmod{n}$ for $n = 3^k$ and $2^n \equiv -2 \pmod{n}$ for $n = 2, 6, 66, 946, \ldots$. Schinzel notes that the existence of infinitely many n such that $2^n \equiv -2 \pmod{n}$ is proved in the remark to Exercise 4 on p. 235 of Sierpiński's *Elementary Theory of Numbers*, 2nd English Edition, 1987.

The Lehmers found solutions of $2^n \equiv k \pmod{n}$ for all $|k| < 100$, except, of course, $k = 1$. Do solutions exist for all other k?

Zhang Ming-Zhi, A note on the congruence $2^{n-2} \equiv 1 \bmod n$ (Chinese. English summary), *Sichuan Daxue Xuebao*, **27** (1990) 130–131; MR **92b**: 11003 (where the wrong Benkoski reference appears to be given).

OEIS: A015910, A036236-036237.

F11 Distribution of residues of factorials.

What is the distribution of $1!, 2!, 3!, \ldots, (p-1)!, p!$ modulo p? About p/e of the residue classes are not represented. Here are the missing ones for the first few values of p:

$p = 2$ or 3,   none. $p = 5$, $\{-2\}$. $p = 7$, $\{-2, -3\}$.
$p = 11$, $\{-2, \pm 3, \pm 4\}$. $p = 13$, $\{-3, 4, -5\}$.
$p = 17$, $\{4, 5, -6, -7, -8\}$. $p = 19$, $\{3, -5, -6, \pm 7, \pm 8\}$.
$p = 23$, $\{-3, -4, -6, -7, -8, 10\}$.
$p = 29$, $\{-2, -4, 7, -8, -9, -10, -11, -12, 13, -14\}$.
$p = 31$, $\{\pm 3, 4, 8, \pm 10, 11, 12, 13, 14\}$.
$p = 37$, $\{3, 4, \pm 5, -9, 10, 11, -14, \pm 15, -18\}$.

Until we reach the last two entries we might be tempted to conjecture that there were always at least as many negative entries as positive ones. Are there infinitely many examples of each case? The value $p = 23$ is remarkable in that the only duplicates are ± 1.

In answer to an Erdős question, Rokowska & Schinzel show that if the residues of $2!, 3!, \ldots, (p-1)!$ are all distinct, then the missing residue must be that of $-\frac{p-1}{2}!$, that $p \equiv 5 \bmod 8$, and that there are no such p with $5 < p \le 1000$.

B. Rokowska & A. Schinzel, Sur une problème de M. Erdős, *Elem. Math.*, **15**(1960) 84–85.

R. Stauduhar, Problem 7, *Proc. Number Theory Conf.*, Boulder, 1963, p. 90.

F12 How often are a number and its inverse of opposite parity?

For each x $(0 < x < p)$ where p is an odd prime, define $\bar{x}$ by $x\bar{x} \equiv 1 \bmod p$ and $0 < \bar{x} < p$. Let N_p be the number of cases in which x and $\bar{x}$ are of opposite parity. E.g., for $p = 13$, $(x, \bar{x}) = (1, 1)$, $\underline{(2, 7)}$, $(3, 9)$, $(4, 10)$, $\underline{(5, 8)}$,

$(\underline{6,11})$, $(12,12)$ so $N_{13} = 6$. D. H. Lehmer asks us to find N_p or at least to say something nontrivial about it. $N_p \equiv 2$ or $0 \bmod 4$ according as $p \equiv \pm 1 \bmod 4$.

p	3	5	7	11	13	17	19	23	29	31	37	41	43	47	53	59	61
N_p	0	2	0	4	6	10	4	12	18	4	14	18	20	16	30	32	30

Yi & Zhang generalize to an odd integer $q > 1$ in place of p and k an integer ≥ 1. If $N(k,q)$ is the number of $a \bmod q$ with $(a,q) = 1$ such that
$$q\left\{\frac{a^k}{q}\right\} \quad \text{and} \quad q\left\{\frac{\bar{a}^k}{q}\right\}$$
are of opposite parity, where $\{x\}$ means the fractional part of x and $\bar{a}$ is the inverse of $a \bmod q$, then they show that
$$N(k,q) = \tfrac{1}{2}\phi(q) + O_k\left(q^{3/4} d(q)^{1/2} (\log q)^2\right),$$
Chaładus has verified three conjectures of G. Terjanian, namely that if L_p is the set of x having the *same* parity as $\bar{x}$, then L_p are all residues just if $p = 5$, 13 or 29; all residues are in L_p just if $p = 3$, 5, 7 or 13; and all nonresidues are in L_p just if $p = 3$ or 7.

Ruzsa & Schinzel have shown that the cardinalities of the four sets, the intersections and the differences of L_p with the sets of residues and of nonresidues are each $\tfrac{1}{4}p + O(\sqrt{p}(\ln p)^2)$.

Zhang Wen-Peng considers the case with q odd and not necessarily prime in place of p. He shows that $N_q = \tfrac{1}{2}\phi(q) + O(q^{1/2+\epsilon})$ when q is a prime power or the product of two primes.

He also lets
$$L(p) = \#\{n : a\bar{a} = np + 1, 1 \leq a \leq p-1\}$$
and asks what can be said about the asymptotic properties of $L(p)$. In an unpublished paper Andrew Granville proved that

$$\frac{p}{\ln p} \ll L(p) \ll \frac{p}{(\ln p)^{c+o(1)}}$$

where $c = 1 - \tfrac{1}{2}e\ln 2 \approx 0.057915$

S. Chaładus, An application of Nagell's estimate of the least odd quadratic non-residue, *Demonstratio Math.*, **28**(1995) 651–655; *MR* **97b**:11003.

Imre Z. Ruzsa & Andrzej Schinzel, An application of Kloosterman sums, *Compositio Math.*, **96**(1995) 323–330; *MR* **96a**:11099.

Wang Hui, Gau Li & Hu Zhi-Xing, The distribution of the difference between the D. H. Lehmer number and its inverse in the primitive roots modulo q, *J. Northwest Univ.*, **26**(1996) 281–284; *MR* **98e**:11111.

Wang Hui, Hu Zhi-Xing & Gau Li, The distribution of the D. H. Lehmer numbers among the primitive roots modulo q, *Pure Appl. Math.*, **12**(1996) 84–88; *MR* **98i**:11082.

Yuan Yi & Zhang Wen-Peng, On the generalization of a problem of D. H. Lehmer, *Kyushu J. Math.*, **56**(2002) 235–241; *MR* **2003g**:11112.

Zhang Wen-Peng, On a problem of D. H. Lehmer, *Analytic Number Theory and Related Topics (Tokyo, 1991)*, 139–142, World Sci. Publishing, River Edge NJ, 1993; *MR* **96f**:11122.

Zhang Wen-Peng, On a problem of D. H. Lehmer and its generalization, *Compositio Math.*, **86**(1993) 307–316; *MR* **94f**:11104. II, **91**(1994) 47–56; *MR* **95f**:11079.

Zhang Wen-Peng, On a problem of D. H. Lehmer and a generalization of it, *J. Northwest Univ.*, **23**(1993) 103–108; *MR* **94f**:11099.

Zhang Wen-Peng, The distribution of D. H. Lehmer numbers in an arithmetic progression, *J. Northwest Univ.*, **24**(1994) 103–108; *MR* **96d**:11102.

Zhang Wen-Peng, On the difference between an integer and its inverse modulo n, *J. Number Theory*, **52**(1995) 1–6; *MR* **96f**:11123; II. *Sci. China Ser. A*, **46**(2003) 229–238; *MR* **2004a**:11100.

Zhang Wen-Peng, On the difference between a D. H. Lehmer number and its inverse modulo q, *Acta Arith.*, **68**(1994) 255–263; *MR* **96a**:11100.

Zhang Wen-Peng, A problem of D. H. Lehmer and its mean square value formula, *Japan J. Math. (N.S.)*, **29**(2003) 109–116.

Zhang Wen-Peng, Zongben Xu & Yuan Yi, A problem of D. H. Lehmer and its mean square value formula, *J. Number Theory*, **103**(2003) 197–213.

F13 Covering systems of congruences.

A system of congruences a_i mod n_i $(1 \leq i \leq k)$ is called a **covering system** if every integer y satisfies $y \equiv a_i$ mod n_i for at least one value of i. For example: 0 mod 2; 0 mod 3; 1 mod 4; 5 mod 6; 7 mod 12. If $c = n_1 < n_2 < \cdots < n_k$ then Erdős offers \$500.00 for a proof or disproof of the existence of covering systems with c arbitrarily large. Davenport & Erdős, and Fried, found systems with $c = 3$; Swift with $c = 6$; Selfridge with $c = 8$; Churchhouse with $c = 10$; Selfridge with $c = 14$; Krukenberg with $c = 18$; and Choi with $c = 20$. $c = 24$ is reputed to have been obtained by a Japanese. Erdős posthumously offers \$1000.00 for a proof or disproof that there are systems with arbitrarily large c. Simpson has also reformulated open problems and offers cash prizes.

Erdős offers \$25.00 for a proof of the nonexistence of covering systems with all moduli n_i odd, distinct, and greater than one; while Selfridge offers \$900.00 for an explicit example of such a system. Berger, Felzenbaum & Fraenkel showed that the l.c.m. of the moduli of such a system must contain at least six prime factors. More generally, "odd" could be replaced by "not divisible by the first r primes." Simpson & Zeilberger showed that if the moduli are odd and squarefree then at least 18 primes are required.

Jim Jordan offers comparable rewards to those mentioned above for solutions to the analogous problems for Gaussian integers (**A15**).

Erdős noted that you can have a covering system with all moduli n_i distinct, squarefree, and greater than one by using the proper divisors of 210:

a_i	0	0	0	1	0	1	1	2	2	23	4	5	59	104
n_i	2	3	5	6	7	10	14	15	21	30	35	42	70	105

Krukenberg used 2 and squarefree numbers greater than 3. Selfridge asks if you can have such a system with $c \geq 3$ in place of $c = 2$. He observes that the n_i cannot all be squarefree with at most two prime factors, but the above example shows that you do not need more than three.

It is easy, but not trivial, to prove that, for a covering system with distinct moduli, $\sum_{i=1}^{k} 1/n_i > 1$. The sum can be arbitrarily close to 1 if $n_1 = 3$ or 4. Sun Zhi-Wei showed that $\sum_{i=1}^{k-1} 1/n_i \geq 1$, where the omitted n_k is the largest of the distinct moduli. Selfridge & Erdős conjecture that $\sum 1/n_i > 1 + c_{n_1}$ where $c_{n_1} \to \infty$ with n_1.

Schinzel has asked for a covering system in which no modulus divides another. This would not exist if a covering with odd muduli does not exist.

Simpson calls a covering system **irredundant** if it ceases to cover the integers when one of the congruences is removed. He has shown that if the l.c.m. of the moduli of such a system is $\prod p_i^{\alpha_i}$ then the system contains at least $1 + \sum \alpha_i(p_i - 1)$ congruences.

Erdős conjectures that all sequences of the form $d \cdot 2^k + 1$ ($k = 1, 2, \ldots$), d fixed and odd, which contain no primes can be obtained from covering congruences (see **B21** for examples). Equivalently, the least prime factors of members of such sequences are bounded.

Sherman Stein asked if there exists an infinite set of congruences such that every integer satisfies just one congruence, the moduli are distinct with sum of reciprocals 1, and with at least one modulus not a power of 2. Krukenberg found such a set with all moduli of the for $2^i 3^j$. Ethan Lewis showed that no other prime divisors of the moduli are possible, unless their number is infinite.

Shiro Ando & Teluhiko Hilano, A disjoint covering of the set of natural numbers consisting of sequences defined by a recurrence whose characteristic equation has a Pisot number root, *Fibonacci Quart.*, **33**(1995) 363–367; *MR* **96f**:11022.

Marc Aron Berger, Alexander Gersh Felzenbaum & A. S. Fraenkel, Necessary condition for the existence of an incongruent covering system with odd moduli, *Acta Arith.*, **45**(1986) 375–379; *MR* **87h**:11005; II. *Acta Arith.*, **48**(1987) 73–79; *MR* **88j**:11002.

Chen Yong-Gao & Štefan Porubský, Remarks on systems of congruence classes, *Acta Arith.*, **71**(1995) 1–10; *MR* **96j**:11014.

S. L. G. Choi, Covering the set of integers by congruence classes of distinct moduli, *Math. Comput.*, **25**(1971) 885–895; *MR* **45** #6744.

R. F. Churchhouse, Covering sets and systems of congruences, in *Computers in Mathematical Research*, North-Holland, 1968, 20–36; *MR* **39** #1399.

Todd Cochrane & Gerry Myerson, Covering congruences in higher dimensions, *Rocky Mountain J. Math.*, **26**(1996) 77–81; *MR* **97c**:11003.

Fred Cohen & J. L. Selfridge, Not every number is the sum or difference of two prime powers, *Math. Comput.*, **29**(1975) 79–81; *MR* **51** #12758.

P. Erdős, Some problems in number theory, in *Computers in Number Theory*, Academic Press, 1971, 405–414; esp. pp. 408–409.

J. Fabrykowski, Multidimensional covering systems of congruences, *Acta Arith.*, **43** (1984) 191-298; *MR* **85d**:11005.

Michael Filaseta, Coverings of the integers associated with an irreducibility theorem of A. Schinzel, *Number Theory for the Millenium II (Urbana IL, 2000)* 1–24, AKPeters, Natick MA,2002; *MR* **2003k**:11015.

M. Filaseta, K. Ford & S. Konyagin, On an irreducibility theorem of A. Scinzel associated with coverings of the integers, *Illinois J. Math.*, **44**(2000) 663–643; *MR* **2001g**:11032.

J. Haight, Covering systems of congruences, a negative result, *Mathematika*, **26**(1979) 53–61; *MR* **81e**:10003.

Hu Zhong & Sun Zhi-Wei, On n-dimensional covering systems, *Nanjing Daxue Xuebao Ziran Kehue Ban*, **37**(2001) 486–492.

J. H. Jordan, Covering classes of residues, *Canad. J. Math.*, **19**(1967) 514–519; *MR* **35** #1538.

J. H. Jordan, A covering class of residues with odd moduli, *Acta Arith.*, **13**(1967-68) 335–338; *MR* **36** #3709.

C. E. Krukenberg, PhD thesis, Univ. of Illinois, 1971, 38–77.

Ethan Lewis, Infinite covering systems of congruences which don't exist, *Proc. Amer. Math. Soc.*, **124**(1996) 355–360; *MR* **96d**:11017.

Štefan Porubský, Identities involving covering systems, I, II, *Math. Slovaca*, **44**(1994) 153–162, 555–568; *MR* **95f**:11002, **96e**:11006.

A. Schinzel, Reducibility of polynomials and covering systems of congruences, *Acta Arith.*, **13**(1967) 91–101; *MR* **36** #2596.

A. Schinzel, On homogeneous covering congruences, *Rocky Mountain J. Math.* **27**(1997) 335–342; *MR* **98b**:11003.

R. J. Simpson, Regular coverings of the integers by arithmetic progressions, *Acta Arith.*, **45**(1985) 145–152; *MR* **86j**:11004.

R. J. Simpson, Covering systems of homogeneous congruences, *Rocky Mountain J. Math.*, **28**(1998) 1125–1133; *MR* **99k**:11007.

R. J. Simpson & D. Zeilberger, Necessary conditions for distinct covering systems with squarefree moduli, *Acta Arith.*, **59**(1991) 59–70; *MR* **92i:11014**.

Sun Zhi-Wei, Covering the integers by arithmetic sequences, *Acta Arith.*, **72**(1995) 109–129; *MR* **96k**:11013.

Sun Zhi-Wei, Covering the integers by arithmetic sequences II, *Trans. Amer. Math. Soc.*, **348**(1996) 4279–4320; *MR* **97c**:11011.

Sun Zhi-Wei, On covering multiplicity, *Proc. Amer. Math. Soc.*, **127**(1999) 1293–11300; *MR* **99h**:11012.

Yang Siman & Sun Zhi-Wei, Covers with less than 10 moduli and their applications, *J. Southeast Univ.*, **14**(1998) 106–114; *MR* **2000j**:11014.

Zhang Ming-Zhi, A note on covering systems of residue classes, *Sichuan Daxue Xuebao* **26**(1989) 185–188; *MR* **92c**:11003.

Stefan Znám, A survey of covering systems of congruences, *Acta Math. Univ. Comen.*, **40-41**(1982) 59–79; *MR* **84e**:10004.

F14 Exact covering systems.

If a system of congruences is both covering and disjoint (each integer covered by just one conguence) it is called an **exact covering system**. Necessary, but not sufficient, conditions for a system to be exact are $\sum_{i=1}^{k} 1/n_i = 1$ and $(n_i, n_j) > 1$ for all i, j, where the notation is as in the first sentence of **F13**. There is a theorem, variously attributed to subsets of {Davenport, Mirsky, Newman, Rado}, that if a set of distinct numbers > 1 are the moduli of congruences, then either there is a number which is not in any of them or there is a number which is in more than one of them. The ingenious proof used a generating function and roots of unity. Combinatorial proofs were later given by Berger, Felzenbaum & Fraenkel and by Simpson.

Znám notes that $(n_1, n_2, \ldots, n_k) > 1$ is *not*, as stated in the first edition, a necessary condition, as is evinced by the example 0 (mod 6), 1 (mod 10), 2 (mod 15) and 3, 4, 5, 7, 8, 9, 10, 13, 14, 15, 16, 19, 20, 22, 23, 25, 26, 27, 28, 29 (mod 30). He confirmed a conjecture of Mycielski by further proving that if p is the least prime divisor of n_k, then $n_k = n_{k-1} = \ldots = n_{k-p+1}$. He conjectured that if there exist only *pairs* of equal moduli, then the moduli are all of the form $2^\alpha 3^\beta$, but later he and Burshtein & Schönheim and Joel Spencer each gave counter-examples, such as

	0, 1	2, 7	3, 8	13, 28	4, 9	14, 34	19, 39	59, 119
mod	5	10	15	30	20	40	60	120

Stein proved that if there is a single pair of equal moduli, the rest being distinct, then $n_i = 2^i$ ($1 \le i \le k - 1$), $n_k = 2^{k-1}$. Znám proved analogously that if there is a triple of equal moduli, the rest being distinct, then $n_i = 2^i$ ($1 \le i \le k - 3$), $n_{k-2} = n_{k-1} = n_k = 3 \cdot 2^{k-3}$. Sun Zhi-Wei, and independently Beebee, have extended Stein's result by showing that a system is exact just if

$$\sin \pi z = -2^{k-1} \prod_{i=1}^{k} \sin \frac{\pi}{n_i}(a_i - z).$$

Simpson extended work by Burshtein & Schönheim and showed that if the primes $p_1 < p_2 < \cdots < p_t$ are those dividing the moduli of an exact covering system in which no modulus occurs more than N times, then

$$p_t \le N \prod_{i=1}^{t-1} \frac{p_i}{p_i - 1}.$$

The main outstanding problem is to characterize exact covering congruences.

Porubský asked if there is an "exactly m times covering system" which is *not* the union of m exact covering systems. More generally, call such a system S **reducible** if there is a partition $S = S_1 \cup S_2$ such that S_1 and S_2 are exactly l times and exactly $m - l$ times covering systems for some l, $0 < l < m$, and **irreducible** if there is no such partition. Zhang Ming-Zhi answers Porubský's question affirmatively by showing that for every $m > 1$ there is an irreducible exactly m times covering system. This had already been shown for $m = 2$ by S. L. G. Choi (Keszthely, 1973) and by Zeilberger, e.g.:

1(2); 0(3); 2(6); 0,4,6,8(10); 1,2,4,7,10,13(15); 5,11,12,22,23,29(30).

Infinite disjoint covering systems with all moduli distinct can exist. If the sum of the reciprocals of the moduli is 1, such systems exist with moduli $\{2, 2^2, 2^3, \ldots\}$ and with sets of moduli of shape $2^\alpha 3^\beta$. Fraenkel & Simpson conjecture that these are the only types. Lewis showed that the only possible exceptions had an infinite set of distinct primes dividing their moduli.

Questions can also be asked about covering systems of Beatty sequences (**E27**). Graham showed that if $\lfloor m\alpha_i + \beta_i \rfloor$; $m \in \mathbb{Z}$; $1 \le i \le k$ is such a system with $k > 2$ and at least one α_i irrational, then some two α_i must be equal. This is not so if all the α_i are rational, since

$$\left\lfloor m \frac{2^k - 1}{2^{k-i}} + 1 - 2^{i-1} \right\rfloor \qquad (1 \le i \le k)$$

are exactly covering systems. Fraenkel conjectures that the only such systems with distinct α_i are of this form.

There are connexions with pseudoperfect numbers (**B2**) and with Egyptian fractions (**D11**). For a recent survey, with 139 references, see the paper of Porubský & Schönheim.

John Beebee, Examples of infinite, incongruent exact covers, *Amer. Math. Monthly* **95**(1988) 121–123; errata **97**(1990) 412; *MR*, **89g**:11013, **91a**:11013.

John Beebee, Some trigonometric identities related to exact covers, *Proc. Amer. Math. Soc.*, **112**(1991) 329–338; *MR*, **91i**:11013.

John Beebee, Bernoulli numbers and exact covering systems, *Amer. Math. Monthly* **99**(1992) 946–948; *MR* **93i**:11025.

John Beebee, Exact covering systems and the Gauss-Legendre multiplication formula for the gamma function, *Proc. Amer. Math. Soc.*, **120**(1994) 1061–1065; *MR* **94f**:33001.

Marc Aron Berger, Alexander Gersh Felzenbaum & A. S. Fraenkel, A non-analytic proof of the Newman-Znám result for disjoint covering systems, *Combinatorica*, **6**(1986) 235–243; *MR* **88d**:11012.

Marc Aron Berger, Alexander Gersh Felzenbaum, A. S. Fraenkel & R. Holzman, On infinite and finite covering systems, *Amer. Math. Monthly*, **98**(1991) 739–742; *MR* **92g**:11009.

N. Burshtein, On natural exactly covering systems of congruences having moduli occurring at most N times, *Discrete Math.*, **14**(1976) 205–214; *MR* **53** #2886.

N. Burshtein & J. Schönheim, On exactly covering systems of congruences having moduli occurring at most twice, *Czechoslovak Math. J.*, **24**(99)(1974) 369–372; *MR* **50** #4521.

J. Dewar, On finite and infinite covering sets, in *Proc. Washington State Univ. Conf. Number Theory*, Pullman WA, 1971, 201–206; *MR* **47** #6597.

P. Erdős, On a problem concerning systems of congruences (Hungarian; English summary), *Mat. Lapok*, **3**(1952) 122–128.

A. S. Fraenkel, The bracket function and complementary sets of integers, *Canad. J. Math.*, **21**(1967) 6–27; *MR* **38** #3214.

A. S. Fraenkel, Complementing and exactly covering sequences, *J. Combin. Theory Ser. A*, **14**(1973) 8–20; *MR* **46** #8875.

A. S. Fraenkel, A characterization of exactly covering congruences, *Discrete Math.*, **4**(1973) 359–366; *MR* **47** #4906.

A. S. Fraenkel, Further characterizations and properties of exactly covering congruences, *Discrete Math.*, **12**(1975) 93–100; erratum 397; *MR* **51** #10276.

A. S. Fraenkel & R. Jamie Simpson, On infinite disjoint covering systems, *Proc. Amer. Math. Soc*, **119**(1993) 5–9; *MR* **93k**:11006.

R. L. Graham, Covering the positive integers by disjoint sets of the form $\{\lfloor n\alpha + \beta \rfloor : n = 1, 2, \ldots\}$, *J. Combin. Theory Ser. A*, **15**(1973) 354–358; *MR* **48** #3911.

R. L. Graham, Lin Shen & Lin Chio-Shih, Spectra of numbers, *Math. Mag.*, **51**(1978) 174–176; *MR* **58** #10808.

Hu Zhong & Sun Zhi-Wei[1], On n-dimensional covering systems, *Nanjing Daxue Xueban Ziran Kexue Ban*, **37**(2001) 486–492; *MR* **2002j**:11006.

H.Kellerer & G.Wirsching, *Discrete Math.*, **85**(1990) 191–206; *MR* **92d**:11007.

I. Korec, On a generalisation of Mycielski's and Znam's conjectures about coset decomposition of abelian groups, *Fundamenta Math.*, **85**(1974) 41–48; *MR* **50** #10025.

I. Korec, Multiples of an integer in the Pascal triangle, *Acta Math. Univ. Comenian. (N.S.)*, **46/47**(1985) 75–81; *MR* **88b**:11004.

I. Korec, On number of cosets in nonnatural disjoint covering systems, *Colloq. Math. Soc. János Bolyai* **51** (Number Theory, Vol. 1, Budapest, 1987), North-Holland, 1990, 265–278; *MR* **91k**:11016.

Ethan Lewis, A variation on the method of Cartier and Foata via covering systems for combinatorial identities, *European J. Combin.*, **15**(1994) 491–511; *MR* **95k**:05013.

Ethan Lewis, Infinite covering systems of congruences which don't exist, *Proc. Amer. Math. Soc.*, **124**(1996) 355–360; *MR* **96d**:11017.

Ryozo Morikawa, Some examples of covering sets; On a method to construct covering sets; On eventually covering families generated by the bracket function, *Bull. Fac. Liberal Arts Nagasaki Univ.*, **21**(1981) 1–4; **22**(1981) 1–11; **23**(1982/83) 17–22; **24**(1983) 1–9; **25**(1985) 1–8; **36**(1995) 1–17; *MR* **84c**:10051; **84i**:10057; **84j**:10064; **87m**:11010,11011ab,11012; **98b**:11006.

Ryozo Morikawa, Disjointness of sequences $[\alpha_i n + \beta_i]$, $i = 1, 2$, *Proc. Japan Acad. Ser. A Math. Sci.*, **58**(1982) 269–271; *MR* **83m**:10096.

Morris Newman, Roots of unity and covering sets, *Math. Ann.*, **191**(1971) 279–282; *MR* **44** #3972 & err. p. 1633.

Břetislav Novák & Štefan Znám, Disjoint covering systems, *Amer. Math. Monthly* **81** (1974) 42–45.

I. Polách, A new necessary condition for moduli of non-natural irreducible disjoint covering system, *Acta Math. Univ. Comenian.* (*N.S.*), **63**(1994) 133–140; *MR* **96f**:11021.

Štefan Porubský, On *m* times covering systems of congruences, *Acta Arith.*, **29**(1976) 159–169; *MR* **53** #2884.

Štefan Porubský, Results and problems on covering systems of residue classes, *Mitt. Math. Sem. Giessen*, **150**(1981) 1–85; *MR* **83j**:10008.

Štefan Porubský, Covering systems, Kubert identities and difference equations, *Math. Slovaca*, **50**(2000) 381–413; *MR* **2002h**:11013.

Štefan Porubský & J. Schönheim, Covering systems of Paul Erdős, Past, present and future, *Paul Erdős and his mathematics*, I (Budapest, 1999), 581–627, Bolyai Soc. Math. Stud., **11**, János Bolyai Math. Soc., Budapest, 2002.

R. J. Simpson, Disjoint covering systems of congruences, *Amer. Math. Monthly*, **94**(1987) 865–868; *MR* **89b**:11006.

R. J. Simpson, Exact coverings of the integers by arithmetic progressions, *Discrete Math.*, **59**(1986) 181–190; *MR* **87f**:11011.

R. J. Simpson, Disjoint covering systems of rational Beatty sequences, *Discrete Math.*, **92**(1991) 361–369; *MR* **93b**:11023.

Sherman K. Stein, Unions of arithmetic sequences, *Math. Ann.*, **134**(1958) 289–294; *MR* **20** #17.

Sun Zhi-Wei, Systems of congruences with multipliers, *Nanjing Univ. J. Math. Biquarterly*, **6**(1989) 124–133; *MR* **90m**:11006.

Sun Zhi-Wei, Several results on systems of residue classes, *Adv. Math.* (*China*), **18**(1989) 251–252;

Sun Zhi-Wei, Finite coverings of groups, *Fundamenta Math.*, **134**(1990) 37–53; *MR* **91g**:20031.

Sun Zhi-Wei, A theorem concerning systems of residue classes, *Acta Math. Univ. Comenianae*, **60**(1991) 123–131; *MR* **92f**:11007.

Sun Zhi-Wei On exactly *m* times covers, *Israel J. Math.*, **77**(1992) 345–348; *MR* **93k**:11007.

Sun Zhi-Wei, Exact *m*-covers and the linear form $\sum_{s=1}^{k} x_s/n_s$, *Acta Arith.*, **81**(1997) 175–198; *MR* **98h**:11019.

Sun Zhi-Wei, Exact *m*-covers of groups by cosets, *European J. Combin.*, **22**(2001) 415–429; *MR* **2002a**:20026.

Sun Zhi-Wei, Algebraic approaches to periodic arithmetical maps,*J. Algebra*, **240**(2001) 723–743; *MR* **2002f**:11009.

Sun Zhi-Wei, On covering equivalence, *Analytic number theory* (*Beijing/ Kyoto* 1999), *Dev. Math.*, **6**(2002) 277–302; *MR* **2003g**:11014.

R. Tijdeman, Fraenkel's conjecture for six sequences, *Discrete Math.*, **222** (2000) 223–234; *MR* **2001f**:11039.

R. Tijdeman, Exact covers of balanced sequences and Fraenkel's conjecture, *Algebraic number theory and Diophantine analysis* (*Graz*, 1998), 467–483, de Gruyter, Berlin, 2000; *MR* **2001h**:11011.

Charles Vanden Eynden, On a problem of Stein concerning infinite covers, *Amer. Math. Monthly*, **99**(1992) 355–358; *MR* **93b**:11004.

Doron Zeilberger, On a conjecture of R. J. Simpson about exact covering congruences, *Amer. Math. Monthly* **96**(1989) 243; *MR* **90a**:11008.

Zhang Ming-Zhi, Irreducible systems of residue classes that cover every integer exactly m times (Chinese, English summary), *Sichuan Daxue Xuebao*, **28**(1991) 403–408; *MR* **92j**:11001.

Štefan Znám, On Mycielski's problem on systems of arithmetical progressions, *Colloq. Math.*, **15**(1966) 201–204; *MR* **34** #134.

Štefan Znám, On exactly covering systems of arithmetic sequences, *Math. Ann.*, **180** (1969) 227–232; *MR* **39** #4087.

Štefan Znám, A simple characterization of disjoint covering systems, *Discrete Math.*, **12**(1975) 89–91; *MR* **51** #12772.

Joachim Zöllner, A disjoint system of linear recurring sequences generated by $u_{n+2} = u_{n+1} + u_n$ which contains every natural number, *Fibonacci Quart.*, **31**(1993) 162–165; *MR* **94e**:11011.

F15 A problem of R. L. Graham.

Szegedy won the prize that Graham offered for settling (affirmatively) the question: does $0 < a_1 < a_2 < \ldots < a_n$ imply that $\max_{i,j} a_i/(a_i, a_j) \geq n$? His proof, and that of Zaharescu, is for n sufficiently large. Cheng & Pomerance have given the specific bound 10^{4275} but there is still a fair amount of ground to be covered.

Andrew Granville wrote that Graham's conjecture was proved by Balasubramanian & K. Soundararajan when 'Sound' was an undergraduate.

Ramachandran Balasubramaniam & K. Soundararajan, On a conjecture of R. L. Graham, *Acta Arith.*, **75**(1996) 1–38; *MR* **97a**:11139.

Cai Tian-Xin, On a conjecture of Graham, *J. Hangzhou Univ. Natur. Sci. Ed.*, **19**(1992) 360–361; *MR* **94b**:11020.

Fred Cheng Yuanyou & Carl Pomerance, On a conjecture of R. L. Graham, *Rocky Mountain J. Math.*, **24**(1994) 961–975; *MR* **96b**:11009.

Paula A. Kemp, A conjecture of Graham concerning greatest common divisors, *Nieuw Arch. Wisk.*(4), **8**(1990) 61–62; *MR* **91e**:11003.

Rivka Klein, The proof of a conjecture of Graham for sequences containing primes, *Proc. Amer. Math. Soc.*, **95**(1985) 189–190; *MR* **86k**:11002.

J. W. Sander, On a conjecture of Graham, *Proc. Amer. Math. Soc.*, **102**(1988) 455–458; *MR* **89c**:11004.

R. J. Simpson, On a conjecture of R. L. Graham, *Acta Arith.*, **40**(1981/82) 209–211; *MR* **83j**:10062.

Sun Zhi-Wei, Covering the integers by arithmetic sequences, *Acta Arith.*, **72**(1995) 109–129; *MR* **96k**:11013.

Sun Zhi-Wei, Covering the integers by arithmetic sequences II, *Trans. Amer. Math. Soc.*, **348**(1996) 4279–4320; *MR* **97c**:11011.

M. Szegedy, The solution of Graham's greatest common divisor problem, *Combinatorica*, **6**(1986) 67–71; *MR* **87i**:11010.

Alexandru Zaharescu, On a conjecture of Graham, *J. Number Theory*, **27** (1987) 33–40; *MR* **88k**:11009.

F16 Products of small prime powers dividing n.

Erdős defines $A(n, k)$ as $\prod p^a$ where the product is taken over primes p less than k with $p^a \| n$ and asks: is $\max_n \min_{1 \leq i \leq k} A(n+i, k) = o(k)$? He remarks that it is easy to show that it is $O(k)$. Is

$$\min_n \max_{1 \leq i \leq k} A(n + i, k) > k^c$$

for every c and sufficiently large k? Is

$$\sum_{i=1}^{k} \frac{1}{A(n + i, k)} > c \ln k \quad ?$$

F17 Series associated with the ζ-function.

Alf van der Poorten had asked for a proof that

$$\zeta(4) \left[= \sum_{n=1}^{\infty} \frac{1}{n^4} = \frac{\pi^4}{90} \right] = \frac{36}{17} \sum_{n=1}^{\infty} \frac{1}{n^4 \binom{2n}{n}}$$

before he and others showed that

$$\frac{1}{2} \sum_{n=1}^{\infty} \frac{1}{n^4 \binom{2n}{n}} = \int_0^{\frac{\pi}{3}} \theta \left(\ln 2 \sin \frac{\theta}{2} \right)^2 d\theta = \frac{17\pi^4}{6480}.$$

It is also known that

$$\sum_{n=1}^{\infty} \frac{1}{\binom{2n}{n}} = \frac{2\pi\sqrt{3} + 9}{27}, \qquad \sum_{n=1}^{\infty} \frac{1}{n \binom{2n}{n}} = \frac{\pi\sqrt{3}}{9},$$

$$\zeta(2) \left[= \sum_{n=1}^{\infty} \frac{1}{n^2} = \frac{\pi^2}{6} \right] = 3 \sum_{n=1}^{\infty} \frac{1}{n^2 \binom{2n}{n}},$$

$$2(\sin^{-1} x)^2 = \sum_{n=1}^{\infty} \frac{(2x)^{2n}}{n^2 \binom{2n}{n}} \quad \text{and} \quad \zeta(3) \left[= \sum_{n=1}^{\infty} \frac{1}{n^3} \right] = \frac{5}{2} \sum_{n=1}^{\infty} \frac{(-1)^{n-1}}{n^3 \binom{2n}{n}}$$

Some remarkable identities discovered by Gosper include

$$\sum_{k \geq 1} \frac{30k - 11}{4(2k - 1)k^3 \binom{2k}{k}^2} = \zeta(3), \qquad \sum_{n \geq 1} \frac{2^{-n}}{1 + x^{2^{-n}}} = \frac{1}{\ln x} + \frac{1}{1 - x}$$

Yeung Kit-Ming proved a congruence for the Apéry numbers

$$\sum_{k=0}^{n} \binom{n}{k}^2 \binom{n+k}{k}$$

but was anticipated (and generalized) by (Coster &) Van Hamme.

Allouche shows that some of Radoux's statements are not correct.

Jean-Paul Allouche, A remark on Apéry's numbers, *J. Comput. Appl. Math.*, **83**(1997) 123–125; *MR* **98m**:11007.

Tewodros Amdeberhan & Doron Zeilberger, *q*-Apéry irrationality proofs by *q*-WZ pairs, *Adv. in Appl. Math.*, **20**(1998) 275–283; *MR* **99d**:11074.

David H. Bailey, Jonathan Borwein & Roland Girgensohn, Experimental evaluation of Euler sums, *Experimental Math.*, **3**(1994) 17–30; *MR* **96e**:11168.

David H. Bailey, Jonathan Borwein & Roland Girgensohn, Explicit evaluation of Euler sums, *Proc. Edinburgh Math. Soc.*, **38**(1995) 277–294; *MR* **96f**:11106.

David Borwein & Jonathan M. Borwein, On an intriguing integral and some series related to $\zeta(4)$, *Proc. Amer. Math. Soc.*, **123**(1995) 1191–1198; *MR* **95e**: 11137.

Jonathan Borwein & David Bradley, Empirically determined Apéry-like formulae for $\zeta(4n+3)$, *Experiment. Math.*, **6**(1997) 181–194; *MR* **98m**:11142.

Louis Comtet, *Advanced Combinatorics*, D. Reidel, Dordrecht, 1974, p. 89.

M. J. Coster & L. Van Hamme, *J. Number Theory*, **38**(1991) 265–286; *MR* **92h**:11049.

R. E. Crandall & J. P. Buhler, On the evaluation of Euler sums, *Experimental Math.*, **3**(1994) 275–285; *MR* **96e**:11113.

John A. Ewell, A new series representation for $\zeta(3)$, *Amer. Math. Monthly*, **97**(1990) 219–220; *MR* **91d**:11103.

Ira Gessel, *J. Number Theory*, **14**(1982) 362–368; *MR* **83i**:10012.

R. William Gosper, Strip mining in the abandoned orefields of nineteenth century mathematics, *Computers in Mathematics* (Stanford CA, 1986), *Lecture Notes in Pure and Appl. Math.*, Dekker, New York, **125**(1990) 261–284; *MR* **91h**:11154.

Nobishige Kurokawa & Masato Wakayama, On $\zeta(3)$, *J. Ramanujan Math. Soc.*, **16**(2001) 205–214; *MR* **2002i**:11083.

Leonard Levin, *Polylogarithms and Associated Functions*, North-Holland, 1981 [§7.62, and foreword by van der Poorten].

Christian Radoux, *C.R. Acad. Sci. Paris Sér. A-B*, **291**(1980) A567–A569; *MR* **82g**:10031.

Christian Radoux, Some arithmetical properties of Apéry numbers, *J. Comput. Appl. Math.*, **64**(1995) 11–19; *MR* **97c**:11005.

Tanguy Rivoal, Irrationalité d'au moins un des neuf nombres $\zeta(5)$, $\zeta(7)$, ..., $\zeta(21)$, *Acta Arith.*, **103**(2002) 157–167; *MR* **2003b**:11068.

Alfred van der Poorten, A proof that Euler missed ... Apéry's proof of the irrationality of $\zeta(3)$, *Math. Intelligencer*, **1**(1979) 195–203; *MR* **80i**:10054.

Alfred J. van der Poorten, Some wonderful formulas ... an introduction to polylogarithms, *Proc. Number Theory Conf.*, Queen's Univ., Kingston, 1979, 269–286; *MR* **80i**:10054.

L. Van Hamme, *Proc. Conf. p-adic analysis* (*Houthalen*, 1987), Vrije Univ. Brussel, Brussels, 1986, 189–195; *MR* **89b**:11007.

Yeung Kit-Ming, On congruence for Apéry numbers, *Southeast Asian Bull. Math.*, **20**(1996) 103–110; *MR* **97a**:11010.

Wadim Zudilin, One of the numbers $\zeta(5)$, $\zeta(7)$, $\zeta(9)$, $\zeta(11)$ is irrational, *Uspekhi Mat. Nauk*, **56**(2001) 149–150; *MR* **2002g**:11098.

V. V. Zudilin, One of the eight numbers $\zeta(5)$, $\zeta(7)$, ..., $\zeta(17)$, $\zeta(19)$ is irrational, *Mat. Zametki*, **70**(2001) 472–476; *MR* **2002m**:11067.

F18 Size of the set of sums and products of a set.

If a_1, a_2, ..., a_n are n numbers (not necessarily integers), how big is the set of their sums and products in pairs?

$$\text{¿} \quad |\{a_i + a_j\} \cup \{a_i a_j\}| > n^{2-\epsilon} \quad ?$$

Erdős & Szemerédi have proved that the cardinality is greater than n^{1+c_1} and less than $n^2 \exp(-c_2 \ln n / \ln \ln n)$. $A + A$ can be small, e.g., if $A = \{1, 2, \ldots, n\}$, and $A \cdot A$ can be small, e.g., if $A = \{1, 2, \ldots, 2^{n-1}\}$, but the conjecture is that they can't be simultaneously small.

Erdős offers \$100.00 for a proof or disproof that the cardinality is greater than $n^{2-\epsilon}$ and \$250.00 for a more exact bound.

Nathanson & Tenenbaum prove the conjecture in the special case where the number of sums is very small. Nathanson has shown that
$$|(A + A) \cup (A \cdot A)| \geq n^{\frac{32}{31}}.$$

Chang Mei-Chu has shown that if $|A \cdot A| < cn$, then the h-fold sum hA has cardinality at least $c_h n^h$. See also her paper with Bourgain, and his with Konyagin.

Solymosi has shown that $|A + A|$ or $|A \cdot A|$ is at least $n^{14/11}$.

Jean Bourgain & Chang Mei-Chu, On multiple sum and product sets of finite sets of integers, *C. R. Math. Acad. Sci. Paris*, **337**(2003) 499–503.

Jean Bourgain & S. V. Konyagin, Estimates for the number of sums and products and for exponential sums over subgroups in fields of prime order, *C. R. Math. Acad. Sci. Paris*, **337**(2003) 75–80.

Chang Mei-Chu, New results on the Erdős-Szemerdi sum-product problems, *C. R. Math. Acad. Sci. Paris*, **336**(2003) 201–205; *MR* **2004a**:11015.

Chang Mei-Chu, The Erdős-Szemerédi problem on sum set and product set, *Ann. of Math.*(2), **157**(2003) 939–957; *MR* **2004c**:11026.

P. Erdős, Some recent problems and results in graph theory, combinatorics and number theory, *Congress. Numer.*, **17** Proc. 7th S.E. Conf. Combin. Graph Theory, Comput., Boca Raton, 1976, 3–14 (esp. p. 11).

P. Erdős & E. Szemerédi, On sums and products of integers, *Studies in Pure Mathematics*, Birkhäuser, 1983, pp. 213–218; *MR* **86m**:11011.

Jia Xing-De & Melvyn B. Nathanson, Finite graphs and the number of sums and products, *Number Theory* (*New York*, 1991–1995) 211–219, Springer, New

York, 1996.

Melvyn B. Nathanson, On sums and products of integers, *Proc. Amer. Math. Soc.*, **125**(1997) 9–16; *MR* **97c**:11010.

Melvyn B. Nathanson & G. Tenenbaum, Inverse theorems and the number of sums and products, Structure theory of set addition, *Astérisque* No. **258**(1999) xiii, 195–204; *MR* **2000h**:11110.

J. Solymosi, On the number of sums and products, *Bull. London Math. Soc.*, (to appear)

F19 Partitions into distinct primes with maximum product.

In the first edition we asked: if n is large and written in the form $n = a+b+c$, $0 < a < b < c$ in every possible way, are all the products abc distinct, but Leech observed that **D16** answers this negatively. See also Kelly's paper. Bernardo Recamán suggests that the opposite may be true, i.e., that any sufficiently large n, when expressed as a sum $n = a+b+c$ in every possible way, *always* yields two equal products abc. The analogous problem of maximizing the l.c.m. in place of the product was studied algorithmically by Drago.

J. Riddell & H. Taylor asked if, among the partitions of n into *distinct primes*, the one having the maximum *product* of parts is necessarily one of those with the maximum *number* of parts, but Selfridge answered this negatively with the example

$$319 = 2 + 3 + 5 + 7 + 11 + 13 + 17 + 23 + 29 + 31 + 37 + 41 + 47 + 53$$
$$= 3 + 5 + 11 + 13 + 17 + 19 + 23 + 29 + 31 + 37 + 41 + 43 + 47$$

but the partition with the smaller number of parts gives the largest possible product. Is this the least counterexample? Jud McCranie confirmed that it is:

n		terms	log of product
319	13	3+5+11+13+17+19+23+29+31+37+41+43+47	38.321162101737697
	14	2+3+5+7+11+13+17+23+29+31+37+41+47+53	38.224872250045075
372	14	3+5+11+13+17+19+23+29+31+37+41+43+47+53	42.291454015289819
	15	2+3+5+7+11+13+19+23+29+31+37+41+43+47+61	42.237879951470051
492	16	3+5+11+13+17+19+23+29+31+37+41+43+47+53+59+61	50.479865323368850
	17	2+3+5+7+11+13+19+23+29+31+37+41+43+47+53+61+67	50.412864484413139
666		no others for $n \leq 666$.	

Can the cardinalities of the two sets differ by an arbitrarily large amount?

Antonino Drago, Rules to find the partition of n with maximum l.c.m., *Atti Sem. Mat. Fis. Univ. Modena*, **16**(1967) 286–298; *MR* **37** #180.

J. B. Kelly, Partitions with equal products, *Proc. Amer. Math. Soc.*, **15**(1964) 987–990; *MR* **29** #5803; II. **107**(1989) 887–893; *MR* **90e**:11148.

OEIS: A053020.

F20 Continued fractions.

A number x may be expressed as a **continued fraction**

$$x = a_0 + \cfrac{b_1}{a_1 + \cfrac{b_2}{a_2 + \cfrac{b_3}{a_3 + \cdots}}}$$

which, out of kindness to the typesetter, is often written

$$x = a_0 + \frac{b_1}{a_1+} \frac{b_2}{a_2+} \frac{b_3}{a_3+} \cdots$$

When the numerators b_i are all 1 the continued fraction is called **simple**, and may be written

$$x = [a_0; a_1, a_2, a_3, \cdots]$$

It may be finite or infinite, but if x is rational it is finite. In this case there are two possible forms, one of which has its last **partial quotient**, a_k, equal to 1:

$$\tfrac{7}{16} = [0; 2, 3, 2] = [0; 2, 3, 1, 1]$$

Zaremba conjectured that given any integer $m > 1$, there is an integer a, $0 < a < m$, $a \perp m$ such that the simple continued fraction $[0; a_1, \cdots, a_k]$ for a/m has $a_i \leq B$ for $1 \leq i \leq k$ where B is a small absolute constant (say $B = 5$). He was only able to prove $a_i \leq C \ln m$.

T. W. Cusick, Zaremba's conjecture and sums of the divisor function, *Math. Comput.*, **61**(1993) 171–176; *MR* **93k**:11063.

S. K. Zaremba, La méthode des "bos treillis" pour le calcul des intégrales multiples, in *Applications of Number Theory to Numerical Analysis* (*Proc. Symp. Univ. Montréal*, 1971) Academic Press, 1972, 93–119 esp. 69 & 76; *MR* **49** #8271.

F21 All partial quotients one or two.

Not every number n can be expressed as the sum of two positive integers $n = a + b$ so that the continued fraction for a/b has all its partial quotients equal to 1 or 2. For 11, 17 and 19 we can have

$$\tfrac{4}{7} = [0; 1, 1, 2, 1], \qquad \tfrac{5}{12} = [0; 2, 2, 2], \quad \text{and} \quad \tfrac{7}{12} = [0; 1, 1, 2, 2]$$

but 23 can't be so expressed. However Leo Moser conjectured that there is a constant c such that every n can be so expressed with the *sum* of the partial quotients, $\sum a_i < c \ln n$.

Bohuslav Divis asked for a proof that in any real quadratic field there is always an irrational number whose simple continued fraction expansion has all its partial quotients 1 or 2. He also asks the same question with 1 and 2 replaced by any pair of distinct positive integers.

F22 Algebraic numbers with unbounded partial quotients.

Is there an algebraic number of degree greater than two whose simple continued fraction has unbounded partial quotients? Does *every* such number have unbounded partial quotients? Ulam asked particularly about the number $\xi = 1/(\xi + y)$ where $y = 1/(1 + y)$.

Littlewood observed that if θ has a continued fraction with bounded partial quotients a_n, then $\liminf n|\sin n\theta| \le A(\theta)$, where $A(\theta)$ is not zero (though it is for almost all θ). He also asks if $\liminf n|\sin n\theta \sin n\phi| = 0$ for all real θ and ϕ ? It is for almost all θ and ϕ. Cassels & Swinnerton-Dyer treat a dual problem and show incidentally that $\theta = 2^{1/3}$, $\phi = 4^{1/3}$ does *not* provide a counterexample. Davenport suggested that a computer might help with proving that $|(x\theta - y)(x\phi - z)| < \epsilon$ has solutions for *every* θ, ϕ when, for example, $\epsilon = \frac{1}{10}$ or $\frac{1}{50}$.

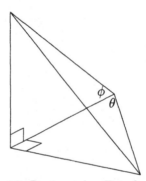

Figure 18. Rectangular Tetrahedron.

J. W. S. Cassels & H. P. F. Swinnerton-Dyer, On the product of three homogeneous linear forms and indefinite ternary quadratic forms, *Philos. Trans. Roy. Soc. London Ser. A*, **248**(1955) 73–96; *MR* **17**, 14.

Harold Davenport, Note on irregularities of distribution, *Mathematika*, **3** (1956) 131–135; *MR* **19**, 19.

John E. Littlewood, *Some Problems in Real and Complex Analysis*, Heath, Lexington MA, 1968, 19–20, Problems 5, 6.

F23 Small differences between powers of 2 and 3.

Problem 1 of Littlewood's book asks how small $3^n - 2^m$ can be in comparison with 2^m. He gives as an example

$$\frac{3^{12}}{2^{19}} = 1 + \frac{7153}{524288} \approx 1 + \frac{1}{73}$$

(the ratio of D$^\sharp$ to E$^\flat$).

The first few convergents to the continued fraction (see **F20**)

$$1 + \cfrac{1}{1+} \cfrac{1}{1+} \cfrac{1}{2+} \cfrac{1}{2+} \cfrac{1}{3+} \cfrac{1}{1+} \cdots$$

for $\log 3$ to the base 2 are

$$\frac{1}{1}, \ \frac{2}{1}, \ \frac{3}{2}, \ \frac{8}{5}, \ \frac{19}{12}, \ \frac{65}{41}, \ \frac{84}{53}, \ \cdots$$

so Victor Meally observed that the octave may conveniently be partitioned into 12, 41 or 53 intervals, and that the system of temperament with 53 degrees is due to Nicolaus Mercator (1620–1687; not Gerhardus, 1512–1594, of map projection fame).

Ellison used the Gel'fond-Baker method to show that

$$|2^x - 3^y| > 2^x e^{-x/10} \quad \text{for} \quad x > 27,$$

and Tijdeman used it to show that there is a $c \geq 1$ such that $|2^x - 3^y| > 2^x/x^c$.

Croft asks the corresponding question for $n! - 2^m$. The first few best approximations to $n!$ by powers of 2 are

5!	20!	22!	24!	61!	63!	90!
2^7	2^{61}	2^{70}	2^{79}	2^{278}	2^{290}	2^{459}
-1.34	$+0.13$	-0.10	$+0.046$	$+0.023$	-0.0017	-0.0007

where the third row is the percentage error in the exponent.

In the "Stellingen" that accompanied Benne de Weger's PhD thesis he observed that if the primes $p_1, \ldots, p_t$ are given, then there is an effectively computable constant C, depending only on the p_i, such that for all n, k_1, $\ldots$, k_t with $n! \neq p_1^{k_1} \cdots p_t^{k_t}$ it is true that

$$\left| n! - p_1^{k_1} \cdots p_t^{k_t} \right| > \exp(Cn/\ln n).$$

There is some experimental support for the conjecture that the right side could be replaced by $\exp(C'n\ln n)$. For a fixed m, the methods of his thesis will determine all solutions of $n! - p_1^{k_1} \cdots p_t^{k_t} = m$.

Erdős believes the conjecture. He also observes that $n! = 2^a \pm 2^b$ only when $n = 1, 2, 3, 4$ and 5.

F. Beukers, Fractional parts of powers of rationals, *Math. Proc. Cambridge Philos. Soc.*, **90**(1981) 13–20; *MR* **83g**:10028.

A. K. Dubitskas, A lower bound on the value of $\|(3/2)^k\|$ (Russian), *Uspekhi Mat. Nauk*, **45**(1990) 153–154; translated in *Russian Math. Surveys*, **45**(1990) 163–164; *MR* **91k**:11058.

W. J. Ellison, Recipes for solving diophantine problems by Baker's method, *Sém. Théorie Nombres*, 1970–71, Exp. No. 11, C.N.R.S. Talence, 1971; *MR* **52** #10591; and see *Publ. Math. Univ. Bordeaux Année I* (1972/73) no. 1, Exp. no. 3, 10 pp; *MR* **53** #5486

Maurice Mignotte, Sur les entiers qui s'écrivent simplement en différents bases, *Europ. J. Combin.*, **9**(1988) 307–316; *MR* **90g**:11015.

R. Tijdeman, On integers with many small prime factors, *Compositio Math.*, **26** (1973) 319–330; *MR* **48** #3896.

F24 Some decimal digital problems.

Sin Hitotumatu asks for a proof or disproof that, apart from 10^{2n}, $4 \cdot 10^{2n}$ and $9 \cdot 10^{2n}$, there are only finitely many squares with just two different decimal digits, such as $38^2 = 1444$, $88^2 = 7744$, $109^2 = 11881$, $173^2 = 29929$, $212^2 = 44944$, $235^2 = 55225$ and $3114^2 = 9696996$.

Pierre Barnouin, in a 96-08-30 letter, lists the only specimens with more than 2 digits, in the following range:

n	12	15	21	22	26	38	88	109	173	212	235	264	3114	81619
n^2	144	225	441	484	676	1444	7744	11881	29929	44944	55225	69696	9696996	6661661161

Another decimal digital problem was submitted to the MONTHLY by Wu Wei Chao: prove that the sum of the decimal digits of 2^n is less than or equal to the sum of the decimal digits of 5^n, with equality only if $n = 3$.

The decimal representation of 2^n contains no zero digit for $n = 0$, 1, 2, 3, 4, 5, 6, 7, 8, 9, 13, 14, 15, 16, 18, 19, 24, 25, 27, 28, 31, 32, 33, 34, 35, 36, 37, 39, 49, 51, 67, 72, 76, 77, 81 and 86. Shall we ever know if this sequence is complete? Dan Hoey has checked that there are no others with $n \leq 2500000000$.

Bill Gosper asked about zeroless powers of n. Dean Hickerson observed that if $n \equiv 2 \pmod 3$ and $n > 2$, then the cube of $N = \frac{1}{3}(2 \cdot 10^{5n} - 10^{4n} + 17 \cdot 10^{3n-1} + 10^{2n} + 10^n - 2)$ is zerofree. For example, $6666633333900003333366666^3 =$
296291851949629281489792513816596763211148776994823592461481829631851896296

If the digits of a number are monotonic, as in 24779, call it **sorted**. John Guilford & John Hamilton have proved the equivalence of "$b-1$ is divisible by a square" and "there are infinitely many integers n such that the base-b digits of both n and n^2 are sorted". So base 10 admits infinitely many solutions. In fact Hamilton can find infinitely many n such that n, n^2, n^3, ..., n^m are all sorted in base b provided $b-1$ is divisible by an m-th power.

See also **F31** and **F32**.

Don Coppersmith, Ratios of zerofree balanced ternary integers, Report RC 21131, 03/13/1998; IBM T.J. Watson Research Center, Yorktown Heights NY.

OEIS: A016069-016070, A018884-018885, A057659.

F25 The persistence of a number.

In the sequence 679, 378, 168, 48, 32, 6, each term is the product of the decimal digits of the previous one. Neil Sloane defines the **persistence** of a number as the number of steps (five in the example) before the number collapses to a single digit. The smallest numbers with persistence 1, 2, ..., 11 are 10, 25, 39, 77, 679, 6788, 68889, 2677889, 26888999, 3778888999,

277777788888899. There is no number less than 10^{50} with persistence
greater than 11. Sloane conjectures that there is a number d such that
no number has persistence greater than d.

In base 2 the maximum persistence is 1. In base 3 the second term is
zero or a power of 2. It is conjectured that all powers of 2 greater than 2^{15}
contain a zero when written in base 3. This is true up to 2^{500}. The truth
of this conjecture would imply that the maximum persistence in base 3 is
3.

Sloane's general conjecture is that there is a number $d(b)$ such that the
persistence in base b cannot exceed $d(b)$.

Erdős modifies the problem by letting $f(n)$ be the product of the
nonzero decimal digits of n, and asks how fast one reaches a one-digit
number, and for which numbers is the descent slowest. He says that it easy
to prove that $f(n) < n^{1-c}$, so that at most $c \ln \ln n$ steps are needed.

N. J. A. Sloane, The persistence of a number, *J. Recreational Math.*, **6**(1973)
97–98.

F26 Expressing numbers using just ones.

Let $f(n)$ be the least number of ones that can be used to represent n using
ones and any number of $+$ and $\times$ signs (and parentheses). For example,
$$80 = (1+1+1+1+1) \times (1+1+1+1) \times (1+1+1+1)$$
so $f(80) \leq 13$. It can be shown that $f(3^k) = 3k$ and $3\log_3 n \leq f(n) \leq$
$5 \log_3 n$ where the logs are to base 3. Does $f(n) \sim 3\log_3 n$?

Daniel Rawsthorne has shown that $f(n) = 2a + 3b$ when n is of the
form $2^a 3^b$ and not greater than 3^{10}. Is this true for larger such n ?

Is it always true that for a prime p, $f(p) = 1 + f(p-1)$? And that
$f(2p) = \min\{2 + f(p), 1 + f(2p-1)\}$?

Don Coppersmith notes that $f(n)$ has an upper bound of $3\log_2(n)$
which is slightly less than $5\log_3(n)$. The former comes from Horner's rule
on binary representation; the latter from Horner's rule on ternary representation. Jeff Lagarias reports that a problem of Shub & Smale is to
let $f_w(m)$ be the minimum number of steps needed to generate m using
1, addition and multiplication, and $f_s(m)$ the minimum number using 1,
-1, addition, subtraction and multiplication, then Conjecture A is that
$f_s(n!) \gg_k (\ln n)^k$ for each $k \geq 1$ and Conjecture B is the same for $f_w(n!)$
Conjecture A implies P$\neq$NP in the Blum-Shub-Smale real number model
of computation.

J. H. Conway & M. J. T. Guy, π in four 4's, *Eureka*, **25**(1962) 18–19.

Richard K. Guy, Some suspiciously simple sequences, *Amer. Math. Monthly*,
93(1986) 186–190; and see **94**(1987) 965 & **96**(1989) 905.

K. Mahler & J. Popken, On a maximum problem in arithmetic (Dutch),
Nieuw Arch. Wiskunde, (3) **1**(1953) 1–15; *MR* **14**, 852e.

Daniel A. Rawsthorne, How many 1's are needed? *Fibonacci Quart.*, **27**(1989) 14–17; *MR* **90b**:11008.

OEIS: A003037, A005245, A005520, A025280.

F27 Mahler's generalization of Farey series.

The **Farey series** of order n consists of all positive rational numbers in their lowest terms, with numerators and denominators not exceeding n, arranged in order of magnitude. For example, the Farey series of order 5 is

$$\frac{1}{5} \quad \frac{1}{4} \quad \frac{1}{3} \quad \frac{2}{5} \quad \frac{1}{2} \quad \frac{3}{5} \quad \frac{2}{3} \quad \frac{3}{4} \quad \frac{4}{5} \quad \frac{1}{1} \quad \frac{5}{4} \quad \frac{4}{3} \quad \frac{3}{2} \quad \frac{5}{3} \quad \frac{2}{1} \quad \frac{5}{2} \quad \frac{3}{1} \quad \frac{4}{1} \quad \frac{5}{1}$$

The determinant formed from the numerators and denominators of two adjacent fractions is -1. Mahler regards the members of the sequence as positive real roots of linear equations whose coefficients have g.c.d. 1 and do not exceed n, and obtained the following apparent generalization to quadratic equations.

Table 10. Segment of Generalized Farey Series of Order 3.

a	b	c	root	determinant
0	1	-1	1	
3	-1	-3	$(1+\sqrt{37})/6$	0
3	-2	-2	$(1+\sqrt{7})/3$	1
2	0	-3	$\sqrt{6}/2$	-1
3	-3	-1	$(3+\sqrt{21})/6$	1
2	-1	-2	$(1+\sqrt{17})/4$	0
1	1	-3	$(\sqrt{13}-1)/2$	-1
2	-2	-1	$(1+\sqrt{3})/2$	1
3	-2	-3	$(1+\sqrt{10})/3$	0
1	0	-2	$\sqrt{2}$	-1
3	-3	-2	$(3+\sqrt{33})/6$	1
0	2	-3	$3/2$	

List the coefficients (a, b, c) of the quadratic equations
$$ax^2 + bx + c = 0, a \geq 0, (a,b,c) = 1, b^2 \geq 4ac, \max\{a, |b|, |c|\} \leq n$$
which have positive real roots, in order of size of the roots. Then the third-order determinant (see **F28**) formed from any three consecutive rows of a, b, c appeared always to take the value 0 or ± 1. In the first edition, Table 10 illustrated this for $n = 2$, with initial and final entries $(0, 1, 0)$ and $(0, 0, 1)$, corresponding to roots 0 and ∞, just as the Farey series could include the terms $\frac{0}{1}$ and $\frac{1}{0}$. We also adopted a suggestion of Selfridge of duplicating rational roots to avoid trivial exceptions. The present Table 10 is an excerpt from the generalized series for $n = 3$. The entry in the last column is the value of the determinant formed from that row and its immediate neighbors.

The conjecture was verified for $n \leq 5$, but Lambertus Hesterman of Canberra discovered counterexamples when $n = 7$; for example

a	b	c	root	determinant
2	-7	-7	$(7 + \sqrt{105})/4$	
1	-3	-6	$(3 + \sqrt{33})/2$	-2
1	-6	7	$3 + \sqrt{2}$	

Can this be rescued, or is it yet another example of the Strong Law of Small Numbers? Lewis Low proved that the absolute value of the determinant cannot exceed n. Can this bound be substantially reduced?

What can be said about the fourth-order determinants associated with cubic equations?

H. Brown & K. Mahler, A generalization of Farey sequences: some exploration via the computer, *J. Number Theory*, **3**(1971) 364–370; *MR* **44** #3959.

Lewis Low, Some lattice point problems, PhD thesis, Univ. of Adelaide, 1979; *Bull. Austral. Math. Soc.*, **21**(1980) 303–305.

Kurt Mahler, Some suggestions for further research, Res. Report No. 20, 1983, Math. Sci. Res. Centre, Austral. Nat. Univ.

F28 A determinant of value one.

The third order determinant

$$\begin{vmatrix} a_1 & a_2 & a_3 \\ a_4 & a_5 & a_6 \\ a_7 & a_8 & a_9 \end{vmatrix}$$

may be defined as $a_1(a_5 a_9 - a_6 a_8) - a_2(a_4 a_9 - a_6 a_7) + a_3(a_4 a_8 - a_5 a_7)$.

Find whole numbers $a_1, a_2, \ldots a_9$, none of them 0 or ± 1, so that

$$\begin{vmatrix} a_1 & a_2 & a_3 \\ a_4 & a_5 & a_6 \\ a_7 & a_8 & a_9 \end{vmatrix} = 1 = \begin{vmatrix} a_1^2 & a_2^2 & a_3^2 \\ a_4^2 & a_5^2 & a_6^2 \\ a_7^2 & a_8^2 & a_9^2 \end{vmatrix}$$

In the first edition we attributed this to Basil Gordon. It was asked by Molnar, who only required that $a_i \neq \pm 1$, and did not restrict the order of the determinants to 3. A topological significance was given, with references to Hilton. Solutions by Morris Newman, Peter Montgomery, Harry Applegate, Francis Coghlan and Kenneth Lau appeared, some of orders greater than 3, including several parametric families, for example

$$\begin{vmatrix} -8n^2 - 8n & 2n + 1 & 4n \\ -4n^2 - 4n & n + 1 & 2n + 1 \\ -4n^2 - 4n - 1 & n & 2n - 1 \end{vmatrix}$$

Richard McIntosh gave examples with a high proportion of Fibonacci numbers:

$$\begin{vmatrix} 1167 & 2 & 5 \\ 1698 & 3 & 8 \\ 2866 & 5 & 13 \end{vmatrix} \qquad \begin{vmatrix} 610 & 5 & 13 \\ 1054 & 8 & 21 \\ 1665 & 13 & 34 \end{vmatrix}$$

Rudolf Wytek restricted himself to integers > 1 and in the closing days of 1987 used a computer to find

$$\begin{vmatrix} 2 & 3 & 2 \\ 4 & 2 & 3 \\ 9 & 6 & 7 \end{vmatrix} \begin{vmatrix} 2 & 3 & 5 \\ 3 & 2 & 3 \\ 9 & 5 & 7 \end{vmatrix} \begin{vmatrix} 2 & 3 & 6 \\ 3 & 2 & 3 \\ 17 & 11 & 16 \end{vmatrix} \begin{vmatrix} 5 & 7 & 6 \\ 6 & 4 & 7 \\ 17 & 16 & 20 \end{vmatrix} \begin{vmatrix} 8 & 7 & 8 \\ 12 & 11 & 7 \\ 17 & 15 & 16 \end{vmatrix} \begin{vmatrix} 10 & 7 & 12 \\ 4 & 2 & 7 \\ 17 & 12 & 20 \end{vmatrix}$$

the second of which had been found earlier by Kenneth Lau. The others were new and are not special cases of any of the parametric solutions. Solutions are evidently more numerous than might at first be thought.

Dănescu, Vâjâitu & Zaharescu solve the problem for determinants of any order in which all the elements are $\geq k$ for any given k.

Will the problem extend to cubes?

Alexandru Dănescu, Viorel Vâjâitu & Alexandru Zaharescu, Unimodular matrices whose entries are squares of those of a unimodular matrix, *Rev. Roumaine Math. Pures Appl.*, **46**(2001) 419–430; MR **2003d**:15020.

P. J. Hilton, On the Grothendieck group of compact polyhedra, *Fundamenta Math.*, **61**(1967) 199–214; MR **37** #3557.

P. J. Hilton, General Cohomology Theory & K-Theory, *L.M.S. Lecture Notes*, **1**, Cambridge University Press, 1971, p. 58.

E. A. Molnar, Relation between wedge cancellation and localization for complexes with two cells, *J. Pure Appl. Alg.*, **3**(1973) 141–158; MR **47** #5870.

E. A. Molnar, A matrix problem, *Amer. Math. Monthly*, **81**(1974) 383–384; and see **82**(1975) 999–1000; **84**(1977) 809 and **94**(1987) 962.

Sadao Saito, Third-order determinant: E. A. Molnar's problem, *Acta Math. Sci.*, **8**(1988) 29–34; MR **89j**:15031.

F29 Two congruences, one of which is always solvable.

Given a prime p, find pairs of functions $f(x)$, $g(x)$ such that one of the congruences $f(x) \equiv n$, $g(x) \equiv n \bmod p$ is solvable for all integers n. A trivial example is $f(x) = x^2$, $g(x) = ax^2$ where a is a quadratic nonresidue (**F5**) of the odd prime p. Mordell gives the further example $f(x) = 2x + dx^4$, $g(x) = x - 1/4dx^2$, where d is any integer prime to p and $1/z$ is defined as $\bar{z}$, where $z\bar{z} \equiv 1 \bmod p$.

F30 A polynomial whose sums of pairs of values are all distinct.

It was noted in **D1** that no nontrivial solution of $a^5 + b^5 = c^5 + d^5$ is known. In fact x^5 is a likely answer to the following unsolved problem of Erdős. Find a polynomial $P(x)$ such that all the sums $P(a) + P(b)$ $(0 \leq a < b)$ are distinct. Rusza has proved that there is a real number ξ such that the set
$$\{\lfloor n^5 + \xi n^4 \rfloor\} \; : \; n > n_0$$
is a Sidon set (see **C9**) for a suitable constant n_0. In fact he proves a more general theorem in which the exponents 5 and 4 are respectively replaced by α and β where $\alpha > 4$ and $\beta > 3 + 1/(\alpha - 1)$.

I. Z. Rusza, An almost polynomial Sidon sequence, *Studia Sci. Math. Hungar.*, bf38(2001) 367–375; *MR* **2002k**:11030.

F31 Miscellaneous digital problems.

Express the integers in base 4, using the digits 0, 1, 2 and $\bar{1}$ $(= -1)$. Let L be the set of integers which can be written in this way using the digits 0, 1 and $\bar{1}$, but not 2. Can every odd integer be written as the quotient of two elements of L? Loxton & van der Poorten show that, given an odd k, there is indeed a multiplier m such that m and km are both in L, but their analysis is ineffective in the sense that they do not know how to estimate the smallest such m. It may be that there is an absolute constant C such that there is always a multiplier less than $|k|^C$. Examples requiring large multipliers are $k = 133 = 2011_4$, $m = 333 = 111\bar{1}1_4$ and $k = 501 = 20\bar{1}11_4$, $m = 2739 = 1\bar{1}\bar{1}\bar{1}11\bar{1}_4$.

John Selfridge & Carole Lacampagne ask if every $k \equiv \pm 1 \mod 3$ can be written as the quotient of integers which can be represented in base 3 using the digits 1 and $\bar{1}$, but not 0. Experiments suggested that the answer is yes, but Don Coppersmith answers negatively with the example $k = 247 = 100011_3$. He claims that there is no m such that both m and km are expressible in base 3 using digits +1 and -1 but not 0. If we allow the digits 0 and 1, but not 2, then which integers can be written as such a quotient?

See also **F24** and the next section.

F. M. Dekking, M. Mendès France & A. J. van der Poorten, Folds! *Math. Intelligencer*, **4**(1982) 130–138, 173–181, 190–195; *MR* **84f**:10016abc.

D. H. Lehmer, K. Mahler & A. J. van der Poorten, Integers with digits 0 and 1, *Math. Comput.*, **46**(1986) 683–689; *MR* **87e**:11017.

J. H. Loxton & A. J. van der Poorten, An awful problem about integers in base four, *Acta Arith.*, **49**(1987) 193–203; *MR* **89m**:11004.

OEIS: A068196.

F32 Conway's RATS and palindromes.

It is not known if one repeatedly adds a number to its reversal, whether a palindrome is always produced. For example, $37 + 73 = 110$, $110 + 011 = 121$, a palindrome. But $196 + 691 = 887$, $887 + 788 = 1675$, $1675 + 5761 = 7436$, $7436 + 6347 = 13783$, $13783 + 38731 = 52514$, 94039, 187088, 1067869, 10755470, 18211171, ... ?

Conway's 'RATS' is 'Reverse, Add, Then Sort'. E.g., $16+61=77$, $77+77=154$, sort into 145, $145+541=686$, sort into 668, $668+866=1534$, $1345+5431=6776$, $6677+7766=14443$, and after ten more terms we have the divergent, period-two, patterns, 1334434432, 5677667765, 13344334432, 56776667765, ..., in which the numbers of threes and of sixes in the middles of alternate terms steadily increase. Conway's conjecture is that any initial number leads either to this pattern, or to a cycle, such as one of:

Initial number	3	29	69	3999	6999	27888
Cycle length	8	18	2	2	14	2
Least member	11	1223	78	11127	11144445	11667

Many other cycles have been discovered by Curt McMullen and by Cooper & Kennedy. Shattuck & Cooper found divergent RATS sequences in bases 50, 99, 148, 962, $18n + 1$, $18n + 10$ and $(2^t - 1)^2 + 1$, where t is a prime or pseudoprime, base 2. Conway's conjecture, that in base 10, all RATS sequences either cycle or are tributary to the sequence
$$1\,2\,3^m\,4^4, \quad 5^2\,6^m\,7^4, \quad 1\,2\,3^{m+1}\,4^4, \quad 5^2\,6^{m+1}\,7^4, \quad \ldots,$$
(where superscripts denote numbers of repetitions of the digit) remains an open question.

See also **F24, F31**.

Curtis Cooper & Robert E. Kennedy, Base 10 RATS cycles and arbitrarily long base 10 RATS cycles, *Applications of Fibonacci numbers*, **8** (Rochester NY, 1998) 83–93, Kluwer Acad. Publ., Dordrecht, 1999; *MR* **2000m**:11011.

Richard K. Guy, Conway's RATS and other reversals, *Amer. Math. Monthly*, **96**(1989) 425–428.

V. Prosper & S. Veigneau, On the palindromic reversal process, *Calcolo*, **38**(2001) 129–140; *MR* **20021**:11011.

Steven Shattuck & Curtis Cooper, Divergent RATS sequences, *Fibonacci Quart.*, **39**(2001) 101–106; *MR* **2002b**:11016.

Index of Authors Cited

The names appearing here are those of authors whose works referred to in this volume. References occur at the end of each problem (e.g. D11, pp.252–262); in the Introduction, I (pp. 1–2) and at the beginning of Sections A (pp. 3–7), D (p. 209), E (p. 311) and F (p. 365). Mentions unsupported by references are listed in the General Index.

General Index

Names appear here if their mention in this volume is unsupported by references. Single letter entries refer to the Introduction (pp. 1–2) and to the beginning of Sections A (pp. 3–7), E (p. 311) and F (p. 365).

$3x + 1$ problem, E16

A-sequence, E28
ABC conjecture, A3
abc-conjecture, B19, D10
Abe, Nobuhisa, D17
abundance, B2
abundant, B2
addition chain, C6
additive basis, C12
additive sequence, C4, E32
Adler, Andrew, C20
admissible partition, E11
Alanen, Jack, B10
algebraic number, F22
aliquot cycle, B7
aliquot parts, B, B4
aliquot sequence, B6
Allenby, Reg, E34
almost perfect, B2
amicable numbers, B4
amicable triples, B5
Andersen, Jens Kruse, A8, A9
Anderson, Claude, D13
Andrica, Dorin, A8
Ankeny-Artin-Chowla conjecture, B16
Apéry number, F17
Apéry numbers, F17

Applegate, Harry, F28
arithmetic number, B2
arithmetic progression, A, A5, A6, A19, E10, E23, E33
Artin's conjecture, F9
Ashbacher, Charles, D21
associates, A16
asymmetric primes, A20
asymptotic, A1
asymptotic density, A17

B_2-sequence, C9, E28
Bailey, David F., D11
Baillie, Robert, A3
Baker's method, D10
Baker, Alan, C4
balanced words, F14
ballot numbers, B33
Bang, Thøger, B46
Baragar, Arthur, D23
Barnouin, Pierre, F24
barrier, B8
basis, E32
Bateman-Horn conjecture, A
Beatty sequences, E27, F14
Benkoski, S. J., F10
Bergmann, Horst, D18
Bergum, Gerald E., D19

Problem Books in Mathematics *(continued)*

Algebraic Logic
by *S.G. Gindikin*

Unsolved Problems in Number Theory, (Third Edition)
by *Richard K. Guy*

An Outline of Set Theory
by *James M. Henle*

Demography Through Problems
by *Nathan Keyfitz and John A. Beekman*

Theorems and Problems in Functional Analysis
by *A.A. Kirillov and A.D. Gvishiani*

Exercises in Classical Ring Theory, (Second Edition)
by *T.Y. Lam*

Problem-Solving Through Problems
by *Loren C. Larson*

Winning Solutions
by *Edward Lozansky and Cecil Rosseau*

A Problem Seminar
by *Donald J. Newman*

Exercises in Number Theory
by *D.P. Parent*

Contests in Higher Mathematics:
Miklós Schweitzer Competitions 1962–1991
by *Gábor J. Székely (editor)*

CPSIA information can be obtained
at www.ICGtesting.com
Printed in the USA
BVOW06*0824280717

490494BV00003B/9/P